Handbook of Phytosanitary Risk Management

Theory and Practice

_____

Handbook of Phytosanitary Risk Management

Theory and Practice

Charles Yoe, Ph.D.

Robert Griffin

and

Stephanie Bloem

CABI is a trading name of CAB International

CABI
Nosworthy Way
Wallingford
Oxfordshire OX10 8DE
UK

CABI
WeWork
One Lincoln St
24th Floor
Boston, MA 02111
USA

Tel: +44 (0)1491 832111
Fax: +44 (0)1491 833508
E-mail: info@cabi.org
Website: www.cabi.org

Tel: +1 (617)682-9015
E-mail: cabi-nao@cabi.org

References to internet websites (URLs) were accurate at the time of writing.

A catalogue record for this book is available from the British Library, London, UK.

Library of Congress Cataloging-in-Publication Data

Names: Yoe, Charles E., author. | Griffin, Robert, author. | Bloem, Stephanie, author.
Title: Handbook of phytosanitary risk management : theory and practice / Charles Yoe, Robert Griffin, Stephanie Bloem.
Description: First edition. | Boston : CAB International, 2020. | Includes bibliographical references and index. | Summary: "Phytosanitary risk management is essential to the global economy. This book is about managing phytosanitary risks of trade and non-trade issues and the fundamentals of how to manage those risks in an effective and efficient manner. It covers risk management models, best practice, phytosanitary measures and current risk management issues"-- Provided by publisher.
Identifiers: LCCN 2020023957 (print) | LCCN 2020023958 (ebook) | ISBN 9781780648798 (hardback) | ISBN 9781780648804 (ebook) | ISBN 9781780648811 (epub)
Subjects: LCSH: Produce trade. | Phytosanitation. | Risk.
Classification: LCC HG6046 .Y64 2020 (print) | LCC HG6046 (ebook) | DDC 630.68/1--dc23
LC record available at https://lccn.loc.gov/2020023957
LC ebook record available at https://lccn.loc.gov/2020023958

ISBN-13: 9781780648798 (hardback)
       9781780648804 (ePDF)
       9781780648811 (ePub)

Commissioning Editor: Rachael Russell
Editorial Assistant: Lauren Davies
Production Editor: Tim Kapp

Typeset by SPi, Pondicherry, India
Printed and bound in the UK by Severn, Gloucester

Contents

Contributors

Chapters 19, 21, 24
Stephanie Bloem, PhD
Executive Director
North American Plant Protection Organization
1730 Varsity Drive, Suite 145
Raleigh, NC 27606
USA
stephanie.bloem@nappo.org

Chapters 20, 22, 23, 25, 26, 31, 32, 33
Robert L. Griffin (retired)
National Coordinator for Agriculture Quarantine Inspection (2014–2019)
Director, Plant Epidemiology and Risk Analysis Laboratory (2003–2014)
USDA, Animal and Plant Health Inspection Service, Plant Protection and Quarantine
Raleigh, North Carolina
USA
Coordinator, International Plant Protection Convention (1997–2003)
Food and Agriculture Organization of the United Nations, Rome, Italy
rlgriffin53@gmail.com

Chapters 1–18, 27–30
Charles Yoe, PhD
Professor of Economics
Department of Business and Economics
Notre Dame of Maryland University
505 Bathurst Road
Baltimore, MD 21228
USA
cyoe1@verizon.net

The authors would like to gratefully acknowledge the contributions of Christina Devorshak in developing the concept of this book and preparing written materials frequently sourced in this volume including but not limited to the companion volume to this one: Christina's *Plant Pest Risk Analysis: Concepts and Application*, published by CAB International, Wallingford, UK in 2012.

Preface

_____

The seeds of an idea that eventually led to the creation of this book were planted in 2015 at the Annual Meeting of the North American Plant Protection Organization in Memphis, Tennessee. The program for the meeting included a one-day symposium on "Innovations in Risk Management" that opened with a keynote presentation by Dr. Charles Yoe. Several of us who also contributed to the symposium and thought we knew something about risk management, were struck by how much we learned from Charlie's presentation, and how much valuable academic background was missing from the discussions of risk management in the phytosanitary community. As we started connecting new dots, we also realized that despite two decades of very dynamic evolution in pest risk analysis, there was a dearth of good information available to students, regulators, policy-makers, and the private sector to understand phytosanitary risk management and there was no guidance available as a contemporary reference for risk managers. The idea to create such a thing was an easy decision. The process was another thing.

Ultimately, we need and do risk assessment to determine whether risk management is appropriate, the possibilities for mitigating risk, and the best option to be applied. That is to say that risk management is the objective. And although the WTO-SPS Agreement says not one word about risk management, that's what the Agreement is all about. That's what phytosanitary measures are all about! So how is it that the phytosanitary community has created a large and ornate PRA cart with a very small risk management horse to pull it?

It turns out that mighty efforts to create and enhance pest risk analysis concepts don't necessarily translate into great risk management strategies, or if they do, there are ways to work around the uncomfortable pieces. For instance, the best possible PRA may argue for something entirely different than what two trading partners negotiate bilaterally. The science may provide a place to start and even support an important argument, but the risk management is ultimately determined by agreement, not by science. This is reality, but it results in a lack of seriousness around risk management that adversely affects its development and advancement.

So, while part of the impetus for this book was filling an obvious gap, there was also a strong desire to give risk management its due respect and a kick into the future. The simple idea from the WTO-SPS Agreement—now more than 20 years old—that phytosanitary measures must be risk-based, opens dozens of new paths for the evolution of risk management. We endeavored to highlight some of those in this book with special attention to the theoretical background which lends a high level of credibility to the technical discussions. We wanted to stick our heads above the rut of phytosanitary tradition to study the landscape of opportunities on the horizon.

Finally, the authors wanted a book that would stand proudly alongside its CABI companion, *Plant Pest Risk Analysis Concepts and Application*. Their shared DNA results in some overlap that is intentionally designed to make them complementary and at the same time individually valuable.

So dear reader, you have here an exhaustive reference on the theoretical underpinnings of risk management with detailed technical discussions, including past, current, and possible future applications for phytosanitary risk management. You have here a new stone for the foundation that supports safe trade.

Robert Griffin

List of Abbreviations

_____

Acronym	Details
ADI	acceptable daily intake
ALOP	appropriate level of protection
ACO	Accredited Certifying Official
ALPP	areas of low pest prevalence
APHIS	Animal and Plant Health Inspection Service
ALPP	area of low pest prevalence
ASTA	American Seed Trade Association
BLM	Bureau of Land Management
CPHST	Center for Plant Health Science and Technology
CNS	citrus nursery stock
COSO	Committee of Sponsoring Organizations of the Treadway Commission
CoP	community of practice
CT	concentration-time
CGIAR	Consultative Group for International Agricultural Research
CBD	Convention on Biological Diversity
CEA	cost-effectiveness analysis
CCP	critical control point
DFIU	diversion from intended use
EIA	economic impact analysis
ERS	Economic Research Service
ERM	enterprise risk management
EO	executive order
EDI	estimated daily intake
EPPO	European and Mediterranean Plant Protection Organization
EFSA	European Food Safety Authority
FAO	Food and Agricultural Organization (of the United Nations)
FDA	Food and Drug Administration
FSO	Food Safety Objective
FAVIR	Fruits and Vegetables Import Requirements database
GeNS	Generic ePhyto National System
GE	genetically engineered
GMOs	genetically modified organisms
GISP	Global Invasive Species Program
GSPP	Good Seed Production Practices Program

Acronym	Details
HACCP	hazard analysis critical control point
HAZOP	hazard operability study
ICA	incremental cost analysis
IPM	integrated pest management
IAEA	International Atomic Energy Agency
ICMSF	International Commission on Microbiological Specifications for Foods
IDCT	International Database on Commodity Tolerance
IDIDAS	International Database on Insect Disinfection And Sterilization
ILSI	International Life Sciences Institute
IPPC	International Plant Protection Convention
ISO	International Organization for Standardization
ISPMs	International Standards for Phytosanitary Measures
IPP	International Phytosanitary Portal
IUCN	International Union for the Conservation of Nature
LMOs	living modified organisms
MOE	margin of exposure
MRLs	maximum residue levels
NPPOs	national plant protection organizations
NGOs	non-governmental organizations
NTBs	non-tariff barriers
NTMs	nontariff measures
NAPPO	North American Plant Protection Organization
OSTP	Office of Science and Technology Policy
OMAF	Ontario Ministry of Agriculture and Food
PCIT	PC Issuance and Tracking System
PFPP	pest-free places of production
PFPs	pest-free production sites
PRA	pest risk assessment / pest risk analysis
PERAL	Plant Epidemiology and Risk Analysis Laboratory
PHOs	plant health objectives
PPQ	Plant Protection and Quarantine
PPOs	plant protection officers
PC	phytosanitary certificate
PexD	Phytosanitary Export Database
PI	phytosanitary irradiation
PEQ	post-entry quarantine
PHA	preliminary hazard analysis
QMS	quality management systems
RSIs	radiation-sensitive indicators
RPPO	regional plant protection organization
RSPM	Regional Standard for Phytosanitary Measures
SPS	sanitary and phytosanitary
SPS Agreement	Agreement on the Application of Sanitary and Phytosanitary Measures
XML	Extensible Markup Language

Part One:

Background Materials

The international trade and international plant protection communities are concluding the first quarter century during which the Agreement on the Application of Sanitary and Phytosanitary (SPS) Measures or SPS Agreement sought to link phytosanitary decisions to risks. This linkage marked a bold initiative in the advancement of the risk sciences, which have made deep inroads into the international communities of practice for food safety and animal health over the same time period. The so-called "three sisters" comprising the International Plant Protection Convention (IPPC), World Organization for Animal Health (OIE), and Codex Alimentarius are standard setting organizations, which have been responsible for the advancement of international standards based on risk assessment.

Much of the first quarter century was given over to interpretation of the SPS Agreement and development of the means by which it would be implemented by the various communities of practice. The primary emphasis in these early years was on risk assessment and considerable progress has been made in structuring phytosanitary risk assessment approaches. A significant contributor to this progress is *Plant pest risk analysis: concepts and application*, a predecessor of and companion to this handbook. With the task of phytosanitary risk assessment well underway, it is time to turn more seriously toward the consideration of the role of risk management in international trade and phytosanitary risks.

As applications of risk analysis continue to multiply, it has become evident that risk management, even phytosanitary risk management, is not just for regulatory agencies and national plant protection organizations. Successful management of phytosanitary risks involves a variety of stakeholders, as many people can own a part of the risk and each must do their part to manage it. This handbook seeks to provide these stakeholders with resources to support their risk management efforts. The first eight chapters that comprise Part 1 of this handbook address the essential language, concepts, models, and issues that stakeholders must understand to become more effective risk managers.

Chapter 1 begins with the importance of international trade, especially in plants and plant products as a segue to a discussion of the need and audience for this handbook, as well as its organization. Chapter 2 provides an important introduction to the language of risk. This chapter goes well beyond the phrases found in the SPS Agreement and integrates some of the phytosanitary risk management language with the language of the broader community of the three sisters, as well as the private sector.

Uncertainty, the topic of the third chapter, is critically important for risk managers to understand. Though a legitimate focal point for most risk analysis communities of practice, it remains somewhat subtle in its role in phytosanitary risk management. There is an opportunity for significant improvement and that opportunity begins with a clear understanding of the nature and nuances of the language of uncertainty.

Risk management is and must be a team effort from start to finish. Chapter 4 presents an argument for the need for a team. It describes a traditional interaction between phytosanitary risk managers and risk assessors and follows this up with a description of an ideal set of interactions. It concludes with the qualities of a good team and a discussion of the importance of vertical team alignment in pest risk management.

Chapter 5 identifies the risk managers in two broad categories of risk, production risk, which is borne and managed by the private sector and phytosanitary risk, which is borne by society at large and is managed by the public sector in concert with the private sector. The interdependent risk management roles of the private and public sectors are a recurring theme in this handbook. The essential duties and roles of risk

managers are discussed along with the consideration of the qualities of a good risk manager.

The SPS Agreement is the seminal document for the regulatory risk management function and the establishment of international standards. As such, it and the IPPC's efforts to aid the application of the principles of the SPS Agreement are the subject of Chapter 6.

Chapters 7 and 8 return the handbook's attention to international trade. The positive and negative consequences of international trade define the universe of risk effects that can result from phytosanitary issues. It is essential that all phytosanitary risk management stakeholders have a firm understanding of the range of consequences that can occur and of the stakeholders with interests in those consequences. Chapter 8 looks at the opportunities that are presented to a variety of stakeholders by efficient and effective phytosanitary risk management.

1

Introduction

Every man lives by exchanging.

Adam Smith

1.1. Introduction

Trade is an intrinsic part of human nature and international trade has transformed the world. Plants and plant products have always been and remain a significant part of international trade. The tangible benefits of plant trade over time have been remarkable, the intangible benefits may be even greater. An unintended consequence of this trade has been the introduction of non-native invasive pests and pathogens that have caused dramatic destruction and remarkable change to the environmental, economic, and other resources of natural systems, nations, industry, and individuals. Controlling the damage done by these pests needs to be a shared responsibility of the international community. Phytosanitary risk management has emerged as the most viable framework for addressing the damage done by exotic pests introduced via trade.

Phytosanitary risk management is a global concern because it facilitates trade among nations and all the benefits that stem from that trade. It is essential to the world's ability to feed itself because it helps control the damage alien pests and pathogens can do to food crops. It is an urgent need because of its potential to limit the damage done globally by invasive pests.

This handbook provides knowledge on how to manage phytosanitary risks of trade and non-trade issues. Its purpose is to provide the global phytosanitary community and its principal stakeholders with a practical guide to best phytosanitary risk management practices. This chapter proceeds by examining the global benefits of international trade in plants and plant products. The problem of exponentially increasing introduction of phytosanitary pests and pathogens via the plant trade pathway is then explored to point us toward phytosanitary risk management as the global community's responsibility and the solution to the problem caused by alien pests. The need for and purpose of this handbook are

---

International trade is the exchange of goods, services, and capital among various countries and regions

The World Trade Organization (WTO) Sanitary and Phytosanitary (SPS) Agreement establishes rights and obligations to adhere to the discipline of scientific risk assessment to ensure that SPS measures are applied only to the extent required to protect human, animal, and plant health, and do not constitute arbitrary or unjustifiable technical barriers to trade.

(Andersen *et al.*, 2004)

---

articulated, and its audiences are identified before the chapter concludes with a discussion of
the organization of this book.

1.2. The Importance of Trade

Many global supply chains for agricultural
products result from the colonial era of trade
when raw materials were produced in one part
of the world and converted into added-value
goods consumed in other parts of the world.
These traded agricultural commodities now
form the basis of multi-billion-dollar global

industries (Flood and Day, 2016). International
trade has flourished over the centuries as a
direct result of the many benefits it offers to
the nations of the world. In 2017, total world
merchandise exports were US$17.198 trillion
while total imports were US$17.572 trillion
(see Tables A4 and A5, WTO 2019, accessed
October 28, 2019). In 2017, agricultural trade
comprised 10% of world merchandise trade.
Global agricultural trade, a significant share of
total trade, is trending up, as Fig. 1.1 illustrates.
Fig. 1.2 shows 71% of all agricultural trade is of
processed, e.g., chocolate or processed coffee
or semi-processed goods such as oilseed cake or

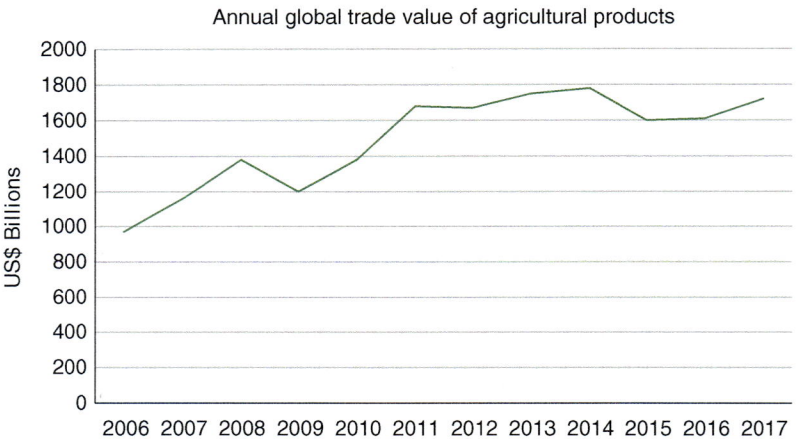

Fig. 1.1. Value of agricultural products in international trade from 2006–2017 (*World Trade Statistical Review
2017* https://www.wto.org/english/res_e/statis_e/wts2017_e/wts17_toc_e.htm accessed October 28, 2019).

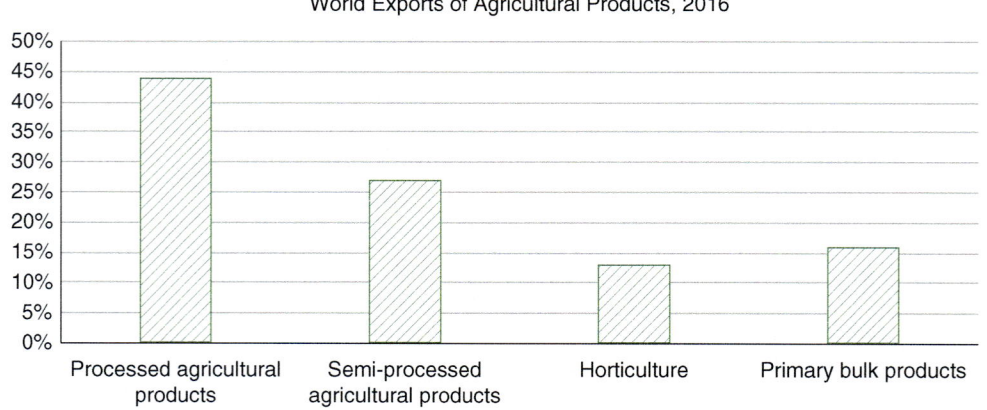

Fig. 1.2. Share of world exports by agricultural product type for 2016 (*World Trade Statistical Review
2018*, Box 4.1).

The WTO's 10 benefits of trade— Part I

---

1. The system helps promote peace—This sounds like an exaggerated claim, and it would be wrong to make too much of it. Nevertheless, the system does contribute to international peace, and if we understand why, we have a clearer picture of what the system actually does.
2. Disputes are handled constructively—As trade expands in volume, in the numbers of products traded, and in the numbers of countries and companies trading, there is a greater chance that disputes will arise. The WTO system helps resolve these disputes peacefully and constructively.
3. Rules make life easier for all—The WTO cannot claim to make all countries equal. It does reduce some inequalities, giving smaller countries more voice, and at the same time freeing the major powers from the complexity of having to negotiate trade agreements with each of their numerous trading partners.
4. Freer trade cuts the costs of living—We are all consumers. The prices we pay for our food and clothing, our necessities and luxuries, and everything else in between, are affected by trade policies.
5. It provides more choice of products and qualities—Think of all the things we can now have because we can import them: fruits and vegetables out of season, foods, clothing and other products that used to be considered exotic, cut flowers from any part of the world, all sorts of household goods, books, music, movies, and so on.

http://www.apeda.gov.in/apedawebsite/about_apeda/10%20benefits.pdf
(accessed October 28, 2019)

vegetable oils. Bulk products include such products as wheat and other grains; horticultural products include things like fruits, vegetables, and cut flowers. Plants and plant products are a major part of the world's agricultural trade and its ability to feed itself.

Modern production techniques, highly advanced transportation systems, transnational corporations, regional trade agreements, technological advances, outsourcing of manufacturing and services, and rapid industrialization have spurred the rapid growth and spread of international trade over the last several decades. This growth, summarized in Figs 1.3 and 1.4, brings with it an impressive array of benefits for the global community. International trade creates jobs

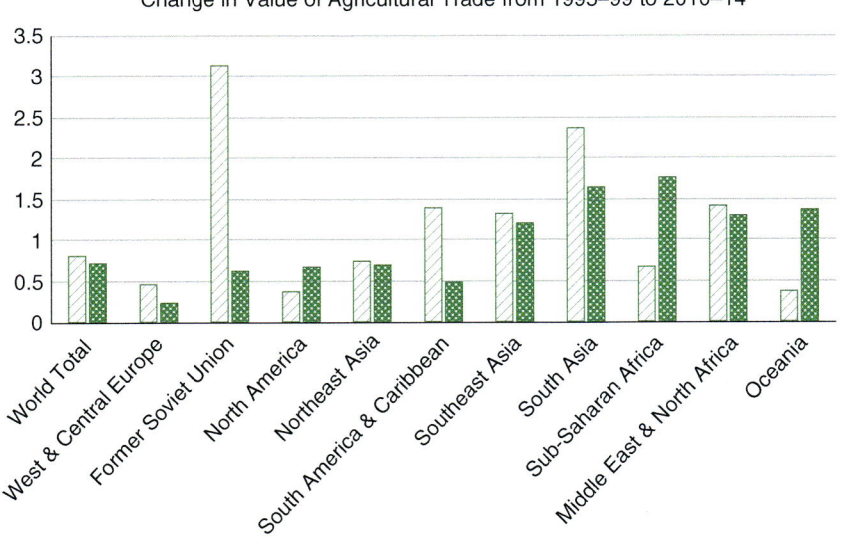

Change in Value of Agricultural Trade from 1995–99 to 2010–14

☐ Change in Export Value   ▦ Change in Import Value

Fig. 1.3. Percentage growth in value of agricultural trade by global region (Beckman *et al.*, 2017).

and income. It produces a greater variety of goods at lower prices and assures more year-long supplies of products. Trade is essential to economic growth and is a major component of the gross domestic product for developed and developing economies. It is a major source of revenue and foreign currency, especially in the developing world. It enables industry to specialize and extend its sales and profit potential, sometimes by providing essential inputs for manufacturing domestic goods and exports. Companies can diversify their risk by participating in more and larger markets. International competition improves the quality of products and trade requirements often provide incidental benefits for domestic populations.

Trade has proven essential to the growth of globalization; it has opened nations and reduced poverty in many nations around the world. The spread of goods, services, and the technologies used to produce them has helped to integrate cultures. Trade has also helped to limit economic dislocations, social readjustments, and unemployment around the world.

Trade helps promote peace if for no other reason than buyers and sellers are reluctant to fight each other. Political conflict is less likely when nations enjoy healthy commerce, commerce that helps make people all over the world better off. Prosperous and contented people are less likely to want to fight. The World Trade Organization (WTO) and other organizations provide opportunities to lessen tension through the establishment of common rules for trade and an impartial mechanism for the resolution of trade conflicts.

Trade is important and phytosanitary risk management is important to trade. To the extent that it succeeds, risk management facilitates trade in plants and plant products while protecting essential resources.

1.3. An Unintended Consequence of Plant Trade

Plants and plant products support pests and pathogens, which are a significant negative externality associated with trade. The trade of plants and plant products has been identified as the most important global pathway for alien pests and pathogens that cause damage to plant health (Hantula *et al.*, 2013). For example, recent findings suggest that over 70 of the exotic

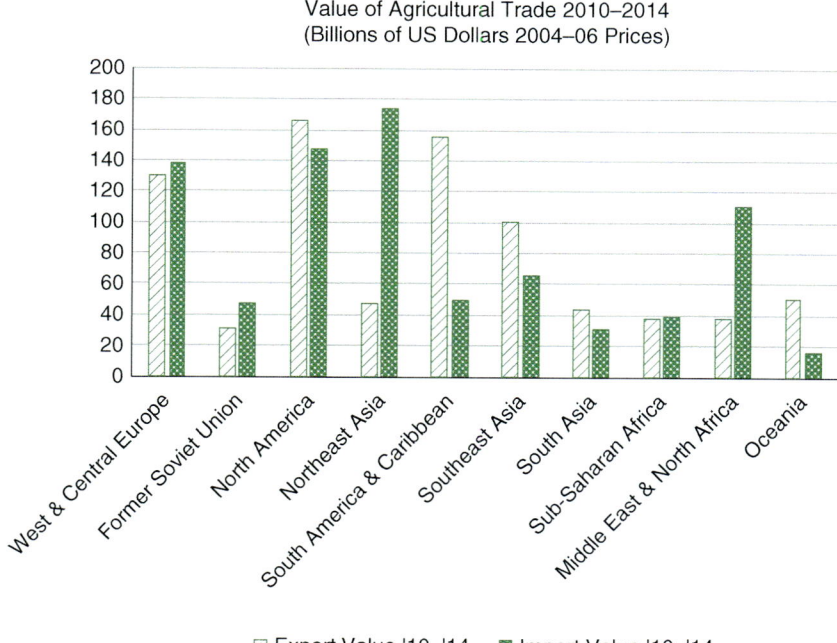

Fig. 1.4. Value of agricultural trade over a five-year period by global region (Beckman *et al.*, 2017).

plant pathogens identified in Europe gained entry via infected live plants since the 1980s (Santini *et al.*, 2013). Extreme quarantine efforts have been implemented to stem the incremental increases of introductions in Europe and Australia. Myrtle rust has arrived in Australia despite an advanced warning system and strict border control and is said to be causing an economic holocaust (FABI, n.d.). These pest introductions can be complicated by the fact that taxpayers and landowners often bear the financial burden of the introduction. The party responsible for the presence of the pest is rarely required to bear the costs of the introduction. Thus, private parties in international plant trade may, in the absence of effective phytosanitary risk management, lack a strong motive to reduce pest and pathogen risks.

Plant pests have been shown to cause extensive damage to environmental, economic, and other resources. The economic costs of introductions can be substantial. The Asian longhorn beetle epidemic in the large cities of the United States could potentially cause as much as $669 billion in tree damage (Nowak *et al.*, 2001). Sweden is estimated to have

total annual alien invasive species control costs of between approximately 1,620 and 5,080 million SEK.

At the same time, the number of introduced pests with extreme impacts is relatively small compared to the number of introduced pests with little or no effect. This makes phytosanitary risk management far too complex to follow the traditional "ounce of prevention" approach to managing pests, which can preclude beneficial trade. Risk managers need to balance and prioritize issues. They need to determine where, when, what, and how to invest in risk controls for maximum beneficial effects while also realizing that there may be some tolerance for the background "failure rate."

Plant exporters and importers responsible for the introductions of plant pests are not usually held economically responsible for the damages those introductions cause. It is typically government, in the form of taxpayers, landowners and/or affected industry that must bear the costs of these introduction risks. These costs represent a substantial burden to taxpayers and other private parties affected by the pest introduction. This break in the cause and effect chain of events presents special challenges to those who would manage the risk associated with the import and export of plants and plant products.

1.4. Risk Management Along the International Trade Continuum

Fig. 1.5 presents a big picture view of the risk management tasks involved in international trade of plants and plant products. These commodities have a complex supply chain that involves the interactions of many participants, including producers, buyers, traders, "middle men," processors, shippers, manufacturers, retailers, and consumers (Gilmour, 2004) as well as associated stakeholders such as agri-dealers, the financial service sector, legislators and other policymakers. Developed countries and their "high-consuming lifestyles" anchor one end of this supply network with producers who often struggle to make a livelihood, often found at the other end. In such a large and complex network risk and risk management will mean very different things to different people (Flood and Day, 2016).

Throughout this handbook, risk management along the international trade continuum for plants and plant products is considered the whole of the process displayed in Fig. 1.5. There are three distinct phases of risk management involved in the international trade of plants and plant products. First, exporters assume opportunity risks,

i.e., they initiate trade in anticipation of reaping uncertain potential benefits from trade, such as those indicated in the figure.

"Pest risk management" or "phytosanitary risk management" is the second phase of risk management. It represents the sanitary and phytosanitary (SPS) risk management responsibilities of the world's National Plant Protection Organizations (NPPOs). The most common view of phytosanitary risk management is the one promulgated by the World Trade Organization Agreement on the Application of Sanitary and Phytosanitary Measures (the SPS Agreement) and put forth by the International Plant Protection Convention (IPPC) as described in the Convention and several International Standards for Phytosanitary Measures (ISPMs). This is the collaborative work of the world's National plant protection organizations.

The need for pest risk management can also arise independently of international trade. Smuggling, natural spread, domestic disease outbreaks, and other domestic risk management issues can also give rise to the need for phytosanitary risk assessment. The circle identifies this view of risk management as the primary focus of this handbook.

A third phase of risk management is the responsibility of the importer, who also assumes the management of opportunity risks as suggested by the figure. Exporter and importer risks, together, will be referred to as "production risk management."

The figure identifies the holistic view of risk management associated with international trade

Fig. 1.5. Risk management along the international plant and plant product trade continuum comprises production risk management and phytosanitary risk management.

of plants and plant products. It encompasses growers, processors, exporters, importers, distributors, and sales as well as NPPOs. This big picture view of production and phytosanitary risk management offers the promise of facilitating international trade and therefore promoting the many significant economic and other benefits of that trade. Risk management can be an effective tool for deciding when to seek the uncertain potential gains—jobs, income, profits, product variety, and availability, consumer prices, and the like—that can directly result from the movement of plants and plant products across national boundaries. By supporting and facilitating the free, fair, and safe trade of goods, risk management also helps assure trade's contributions to the more intangible benefits like peace, constructive settlement of disputes, principles making life better for everyone and so on. This alone would constitute a compelling argument for the risk-informed management of the import and export of plants and plant products.

The most developed need for risk management arises, however, from the requirements of the SPS Agreement and its potential as an effective tool for minimizing the undue burden of economic and other costs associated with pest introductions on society at large, whether from trade or non-trade issues. Risk management is the most effective means devised to date for minimizing the harm that can be caused by the introduction of harmful alien species in new territories. It also offers the most promise for making the best use of the limited resources available to protect plants, while at the same time the need to do so is expanding in volume, diversity, and complexity. Risk management as required by the SPS Agreement is the primary focus of this handbook but it is not the only focus because there are many partners along the plant product supply chain who own some part of the risk associated with international trade in plants and plant products.

The international trade continuum has many levels of risk, pest risk being but one of them. Pest risk analysis (PRA) is very narrowly circumscribed by the SPS Agreement. Its practice has been prescribed by ISPM 2 in a manner that meets the needs of NPPOs, but which fails to satisfy the rationality test of those who would like to see the benefits of trade taken explicitly into account when phytosanitary risks are managed.

Risk management can be an effective strategy for maximizing the potential gains from trade while minimizing the potential losses associated with pest introductions. That role for risk management compels this handbook. The international phytosanitary community has a responsibility to lead the world in its desire to facilitate international trade while at the same time protecting plant health for current and future generations through risk management best practices. To do that, the phytosanitary community needs to know what that best practice is. Filling that gap is the purpose of this handbook.

Phytosanitary risk management is relevant for non-trade issues as well. Pest risk issues that are not for trade purposes do not have to adhere to the SPS Agreement. Thus, if a pest known to be present in the western part of a country threatens to spread to other parts of the country, it should be addressed through a pest risk management process. The model and processes introduced in Part 2 of this handbook are equally appropriate for trade and non-trade pest issues. It is worth bearing in mind that the SPS Agreement and the framework and guidance of the ISPMs need not apply for nontrade pest issues.

This handbook focuses on phytosanitary or pest risk management. It aspires, however, to keep the linkages between production risk and phytosanitary risk management, where risk management responsibilities are spread across the supply chain, in front of the reader. In an ideal world, risks along the international plant and plant product trade continuum might be managed in an integrated and collaborative manner. That ideal does not yet exist. Risks are compartmentalized by phase rather than integrated across all phases. Even so, cooperation among industry and government risk managers is in everyone's mutual best interests. Cooperation between government and industry is essential in a competitive, secretive world where the protection of confidential information is essential to business success. A systems approach to risk management requires the ability to apply effective independent risk management measures anywhere they arise in the risk continuum. Global allocation of scarce resources can be improved through the integration of production and phytosanitary risk management objectives. Consequently, an effective handbook cannot afford to dismiss or ignore risk management responsibilities at any part of the continuum.

1.5. The Need for and Purpose of this Handbook

The SPS agreement mentions "risk" 17 times, "risk assessment" four times, and "risk management" not at all. PRA guidance has focused primarily on how to conduct good risk assessments. The international community has been loath to develop risk management standards that bear directly on decision-making, which has remained in the domain of each sovereign nation. Consequently, notions of what constitutes risk management and what is good risk management practice have not been well developed. Meanwhile, risk management is one of the most rapidly developing aspects of risk analysis in many other communities of practice. This handbook is needed to start the conversation required to move pest risk management forward.

The WTO's Agreement on the Application of Sanitary and Phytosanitary Measures brought risk assessment to the forefront for the phytosanitary community. The SPS Agreement specified that phytosanitary measures must be based on either international standards or risk assessment. The phytosanitary community of practice has made great strides in the practice of risk assessment since that time. *Plant pest risk analysis, concepts and application* (Devorshak, 2012) marks a significant milestone in the evolution of thinking on that subject.

> Risk analysis comprises risk management, risk assessment, and risk communication.

As has been common in many communities of practice, the evolution of thinking about risk assessment has outstripped the thinking on risk management. There is a relatively broad assumption in the plant protection community that risk mitigation means identifying risk reduction methods and selecting the ones appropriate to include in a given PRA. Risk management, by contrast, is often considered to be the operational actions taken by field personnel to reduce risk, for example, fumigations or other treatments, pre-clearance programs, commodity inspection, and so on. These are myopic views; risk management is so much more than this. The understanding of risk management needs to be expanded to include:

- identifying problems
- setting priorities
- allocating resources to these priorities
- commissioning risk assessments
- evaluating the assessed risks
- identifying and evaluating measures to reduce risks
- overseeing the risk communication process
- negotiating and making decisions
- identifying outcomes to monitor
- implementing control measures
- monitoring and evaluating the outcomes of the risk management measures
- modifying the measures as needed
- directing and managing the entire process.

Risk management, as done by other communities of practice, such as food safety, finance, engineering, environmental services, and others has become a sophisticated and mature way to manage the risk analysis process. There are risk managers and there is a risk management process. If you do not have both you may need this handbook.

> A **negative externality** is the cost to a party that did not choose to incur that cost. Thus, a negative externality occurs when an individual does not pay the full cost of the decision and is able to externalize or impose that cost on a third party to the decision. When trade in plants or plant products results in the negative externality of a pest introduction, then the cost of that trade to society may be much greater than the cost to the importer. In fact, it is possible that the cost to society exceeds all the benefits associated with the trade. Negative externalities from trade result in market inefficiencies unless proper risk management action is taken. Phytosanitary risk management measures may, themselves, constitute a negative externality in the form of non-tariff barriers (NTBs) that may restrict trade.

Arguably, the SPS Agreement is all about risk management, despite the fact that it is never mentioned in the Agreement. That is one reason why it is time for a phytosanitary risk management handbook. Another major reason for this handbook is the ubiquity of uncertainty. There is never enough information to make a decision with complete confidence. Decision-making under uncertainty is a constant condition for the plant protection community. Risk management is honest about and pays intentional attention to the uncertainty that could affect a decision or its outcome. The time has come for the phytosanitary community to move the phytosanitary risk management process forward.

A purpose of this handbook is to articulate for the international plant protection community the best practice of risk management. The handbook is not intended to provide a point-by-point accounting of how nations currently manage phytosanitary risk. Instead, it describes what phytosanitary risk management could look like if it was newly revised to incorporate some of the best thinking about phytosanitary risk management as well as other government risk management practices and the practice of enterprise risk management by private industry. Thus, this handbook straddles the worlds of risk management practice and risk management promise.

No member of an NPPO will read this handbook and say "this describes how things are done in our office." Hopefully, any member of an NPPO will read this handbook and say "this describes many ways that we might improve our risk management process and practice." If industry risk managers have the same reaction, phytosanitary risk management cannot help but evolve and improve.

This handbook is intended to help its audiences to:

- understand the language and models of risk
- identify the key uncertainties in a PRA
- differentiate knowledge uncertainty from natural variability
- distinguish between risks of loss to plants and risks of potential gains from trade
- identify the major components of an idealized risk management framework
- identify opportunities for effective cooperation and collaboration between risk assessors and risk managers

- identify the most effective risk management control strategies available
- discuss the most significant issues associated with the most effective risk management strategies
- list the advantages of risk-based sampling
- identify several emerging issues of specific interest to risk managers
- use risk information to better manage plant protection and to better encourage international trade.

1.6. The Audience for this Handbook

This book is for anyone with an interest in promoting trade in plants and plant products or with an interest in protecting plant health from harmful non-indigenous pests that can be introduced via trade. The primary market for this handbook includes the professionals, practitioners and policymakers of the NPPOs of the 183 parties to the IPPC. It should be especially useful to those nations interested in establishing a modern and effective risk analysis approach to phytosanitary trade issues. It will be of interest to all nations, however, because it is the first book to carefully document what the phytosanitary risk management process ought to do.

There are two secondary audiences for this book. The first comprises industry's plant and plant product importers and exporters, non-governmental organizations (NGOs), and other stakeholders who work with the phytosanitary regulators the world over. Many of these organizations are co-owners of the pest risk management responsibility. It is in their self-interest to become and remain aware of evolving risk management practices so they can most effectively participate in the risk management process and assume their own responsibilities for promoting the many benefits of international trade as well as protecting plants. This handbook will empower them to do that.

The other secondary market for this handbook is academia. The many universities that support or participate in phytosanitary risk research and issues will have a keen interest in the content. That content is described in the next section.

1.7. The Organization of This Book

The book is organized into four parts, shown in Fig. 1.6. The first part provides all the essential background material, i.e., language, concepts and basic mental models necessary for understanding risk management. Part 2 presents a risk management model with a detailed description of the structure and processes necessary for good risk management practice in the global economy. The third part provides a detailed look at the phytosanitary measures that are available to risk managers to manage risks to plant health that arise. The final section of the book is devoted to an examination of the most compelling risk management issues of the day. The book will help governments and stakeholders establish a modern and effective risk analysis approach to all phytosanitary trade issues. It will add value to the experienced risk manager and to those getting started in the field.

The contents of the sections and short chapter summaries follow.

Part 1: Background materials

Chapter 1: Introduction—this introduces the book's purpose and its organization.

Chapter 2: Risk—this chapter provides a description of all the risk concepts and their relationships that risk managers need to understand to function in their jobs. These include definitions, language, and models as well as necessary skills such as risk identification. It also includes an orientation to the risk analysis decision-making framework and the manager's role as owner of that process.

Chapter 3: Uncertainty—this chapter introduces the all-important concept of uncertainty. It includes the types and causes of uncertainty and focuses on how it affects decision-making for risk management.

Chapter 4: Risk analysis team—this chapter introduces the need for a team approach to risk analysis in general and risk assessment in particular. It includes topics like the composition, processes, roles and workings of good teams. It introduces the need for risk managers and assessors to work closely and collaboratively with one another.

Chapter 5: Risk managers—this chapter introduces the role of the risk manager. It includes such fundamentals as who they are, what they do, where risk management is best located within an organizational structure, how to interact with risk assessors and the like.

Chapter 6: Risk management and the SPS Agreement—this chapter addresses the things that a risk manager must do to be compliant with the SPS Agreement. It develops the SPS principles associated with risk management. It considers non-trade issues as well as trade issues.

Chapter 7: Consequences and risk management—this chapter addresses the range of risk-related impacts risk managers are concerned with as well as the range of impacts associated with risk management options. Although the focus is heavily on trade, it is not exclusively on trade.

Chapter 8: Risk management opportunities—this chapter addresses the many exciting opportunities there are for a good risk management approach to improve the quality of phytosanitary decision-making. This includes such basic things as promoting safe trade, understanding impacts, managing uncertainty, and allocating resources, to more advanced topics like how to organize for risk management, harmonization, and reconciling pathways.

Part 2: Pest risk management model

Chapter 9: Thinking about risk management—this chapter defines the nature of the risk manager's job and explores a few models used to help define it.

Chapter 10: Establish a risk management framework—this chapter argues the need

Sequential structure of this handbook

Fig. 1.6. Sequential structure of this handbook.

for every NPPO to establish a risk management process. Then it illuminates the content of that process by focusing on the responsibilities of risk managers, risk assessors, and risk communicators.

Chapter 11: Risk manager and risk assessor interaction—this chapter discredits the notion that risk management and risk assessment should be separated from each other. It describes the points in the process where managers and assessors need to interact closely as well as describing the nature of that interaction.

Chapter 12: A phytosanitary risk management model—this chapter introduces a model prototype that can be adopted or adapted by NPPOs.

Chapter 13: Pest risk management through Stages 1 and 2—this chapter describes the risk manager's role in Stages 1 and 2 of the ISPM 2 pest risk analysis framework. It is the first of two chapters describing the phytosanitary risk management model introduced in Chapter 12.

Chapter 14: Pest risk management through Stage 3—this chapter details the structured process by which a recommended risk management option is chosen. This includes formulation, the identification of risk management options, evaluation of individual risk management options to establish their efficacy and the selection of the most appropriate risk management option. The model then shifts its focus to the risk manager's implementation responsibilities.

Chapter 15: Uncertainty and pest risk management—this chapter focuses on the role of uncertainty in risk management decision-making. It reviews some of the sources of uncertainty confronted in a pest risk assessment and addresses the need for a systematic method of characterizing and addressing uncertainty in the risk management process.

Chapter 16: Stakeholders and risk communication—this chapter describes the ways that stakeholders need to be involved in a risk management process. It identifies typical stakeholders as well as opportunities for input and feedback. There is a special section on risk communication.

Chapter 17: Enterprise risk management—this chapter describes a risk management model for the private industry side of international plant trade by providing an overview to an enterprise risk management model.

Part 3: Risk management controls

Chapter 18: Risk management strategies—this chapter examines the appropriate level of protection (ALOP) principle before turning to the risk-taker and risk-avoider roles of the risk manager through the lens of who owns what part of the risk. Generic strategies, including the systems approach, for reducing the likelihood and consequences of risks are reviewed before several specific strategies for formulating risk management options are offered.

Chapter 19: Certification—this chapter examines the role of certification in pest risk management. It provides practical advice on why certificates are needed and how to learn the requirements for a certificate. International guidance is reviewed, ePhyto is described along with certification procedures.

Chapter 20: Inspection and risk-based sampling—this chapter explores inspection as a deterrent, a procedure, and a data source. Inspection's reliance on sampling is introduced, then risk-based sampling is described and an example is provided.

Chapter 21: Treatments—this chapter begins by examining the IPPC/SPS framework for treatments, which requires that treatment shall be commensurate with the risk identified by PRA. There is an emphasis on assuring the treatments are in reasonable relationship to the pest risk. Relevant international standards are reviewed before the role of extrapolation and Probit 9 in treatment strategies are discussed.

Chapter 22: Pest-free concepts—this chapter examines the use of pest free windows, production sites, places of production, areas, and shipments as risk management measures. It also examines the potential use of low-pest prevalence and non-host status as potential risk management controls.

Chapter 23: Irradiation—the chapter begins with a careful examination of what makes irradiation different from other phytosanitary treatments. It progresses through a discussion of the role of international and national regulatory frameworks and then examines several policy issues including irradiation as a food additive, labeling, and dose ceilings for food. The chapter concludes with an examination of programmatic issues including oversight, inspection, mistakes, and fraud.

Chapter 24: Post-harvest processing and handling—this chapter defines the risk management measures used in this segment of the supply chain as well as their objectives. Definitions are offered and extensive review of international guidance is offered.

Chapter 25: Post-entry measures—this chapter describes a set of under-utilized measures that are under-appreciated for their effectiveness because risk managers are reluctant to allow potentially infested articles to enter their country even when risk analysis supports the use of such measures.

Chapter 26: Prohibition—this chapter focuses on the no trade scenarios that result from the use of prohibition. A declared prohibition is one way to forestall trade. The use of prohibition on items not authorized and during emergency actions is also examined.

Chapter 27: Systems approaches—this chapter details the growing reliance on the use of distinctly different pest mitigation measures and safeguards. Topics include mitigation systems, quality systems, control point systems, redundancy and the evaluation of systems as risk management options.

Part 4: Issues in pest risk management

Chapter 28: Hazard analysis vs. risk analysis—this chapter examines the problems that can arise when risk management focuses on entry as the risk rather than establishment and its consequences. It emphasizes potential trade issues and resource misallocation.

Chapter 29: Economic consequence assessment—this chapter examines the assessment of economic consequences. It introduces the different perspectives an analysis can take and reviews the economic guidance provided in the SPS agreement and by IPPC. Specific challenges to the SPS economic framework are examined as are types of economic impact and economic analyses used for consequence assessment.

Chapter 30: Knowledge management—this chapter develops the importance of collecting, organizing, and analyzing data to reduce uncertainty. Mining the extensive amounts of information generated by the pest risk management community to learn about pathways, pests, and risk management efficacy is one of the most promising opportunities for this community of practice.

Chapter 31: Commodities for consumption—this chapter describes how the full and accurate identification of a commodity is essential for regulatory processes and risk management. Processes, processing, and packaging associated with creating the commodity should be considered by risk managers. Risk management works best when trading partners collaborate honestly and freely about the risks and solutions associated with commodities they trade.

Chapter 32: Genetically modified organisms and invasive species—this chapter begins with careful attention to terminology and follows that with a look at the precautionary approach to managing GMO risks. It then turns to a more detailed discussion of GMO/LMO issues before concluding with an examination of invasive species.

Chapter 33: A new framework—this chapter looks at central concepts that break from the past and redefine the future of pest risk management. It takes a careful look at how the SPS agreement has and continues to affect phytosanitary risk management. It focuses on the need to consider strength of measures based on a rational relationship and the bold opportunities that using the continuum of risk management makes possible for the future.

1.8. Summary and Look Forward

Here are five things to remember from this chapter.

1. International trade has transformed the world and is essential to our way of life.

2. Trade in plants and plant products is a significant part of international trade.

3. The introduction of alien pests to new territories is a destructive and expensive negative externality of plant trade.

4. Risk management is an effective means of facilitating international plant trade.

5. Risk management is a necessary means for controlling the risks associated with the introduction of alien species.

In the next chapter, you will learn all you need to know about the language and mental models of risk to be able understand the language of this handbook.

1.9. References

Andersen, MC., Adams, H., Hope, B. and Powell, M. (2004) Risk assessment for invasive species. *Risk Analysis* 24, 4.

Beckman, J., Dyck, J. and Heerman, K. (2017) The global landscape of agricultural trade, 1995–2014. *Research in Agricultural and Applied Economics, November* 13, 2017. Available at: https://ageconsearch.umn.edu/record/265270/ (accessed April 16, 2020).

Devorshak, C. (ed.) (2012) *Plant Pest Risk Analysis, Concepts and Application.* CABI International, Wallingford, UK.

FABI Articles (n.d.) Disease darning—*Puccinia psidii,* the *Eucalyptus* rust pathogen: Australia in 2010; *South Africa next*? Available at: http://www.forestry.co.za/disease-warning-puccinia-psidii/ (accessed January 29, 2019).

Flood, J. and Day, R. (2016) Managing risks from pests in global commodity networks—policy perspectives. *Food Security* 8, 89–101, DOI 10.1007/s12571-015-0534-x.

Gilmour M. (2004) Towards sustainable cocoa production. In Flood, J. and Murphy, R. (eds) *Cocoa futures—a source book for important issues facing the cocoa and chocolate industry.* CABI Commodities Press, 150–161.

Hantula, J., Müller, M.M. and Uusivuori, J. (2013) International plant trade associated risks: laissez-faire or novel solutions. *Environmental Science & Policy* 37, 158–160.

Nowak, D.J., Pasek, J.E., Sequeira, R.A., Crane, D.E. and Mastro, V.C. (2001) Potential effect of *Anoplophora glabripennis* (Coleoptera: Cerambycidae) on urban trees in the United States. *Journal of Economic Entomology* 94, 116–122.

Santini, A., Ghelardini, L., De Pace, C. *et al.* (2013) Biogeographical patterns and determinants of invasion by forest pathogens in Europe. *New Phytologist* 197, 238–250.

WTO (World Trade Organization) (2019) *World Trade Statistical Review 2019.* WTO, Geneva. Available at: https://www.wto.org/english/res_e/statis_e/wts2019_e/wts2019_e.pdf (accessed April 16, 2020).

2

Risk

Creative risk taking is essential to success in any goal where the stakes are high. Thoughtless risks are destructive, of course, but perhaps even more wasteful is thoughtless caution which prompts inaction and promotes failure to seize opportunity.

Gary Ryan Blair

2.1. Introduction

Risk management requires a common terminology and a good understanding of the relevant risk concepts. That presents a challenge in a world where many well-established but different risk dialects already exist. The International Plant Protection Convention (IPPC) has defined some terms but their lexicon is not as complete as that of the International Organization for Standardization (ISO) and that vocabulary differs from that of CODEX and the World Organization for Animal Health (OIE), the two so-called sisters of the IPPC. All of this varies from the emerging language of enterprise risk management. Given all of these different dialects, how does one gain competence in understanding and speaking about risk?

This chapter provides a description of the risk concepts that risk managers need to understand in order to function in their jobs. It does not adhere strictly to the SPS and IPPC notions of risk management. Phytosanitary risk management does not exist in a vacuum and it is important for all involved in or touched by phytosanitary risk management decisions to understand not only the risk management language of the SPS Agreement and IPPC but also the broader risk management community of practice. For example, the conceptual risk assessment model used by national plant protection organizations (NPPOs) differs from the conceptual risk assessment model used by much of the rest of the world, including stakeholders of the NPPOs. Therefore, it is useful to be familiar with that broader model and some of its terminology. The principles behind the two models are remarkably similar. There are many such examples where a broader understanding of risk concepts would be beneficial to NPPOs.

That logic extends in both directions. It is useful for importers and exporters to be familiar with SPS and IPPC risk concepts. Above all else, this handbook aspires to be a practical guide to phytosanitary risk management and so it proceeds in this chapter by developing a broad and sound understanding of the language of risk and risk analysis concepts. It concludes by focusing on the risk language of the IPPC.

2.2. The Language of Risk

Risk managers must be conversant in the language of risk. This section introduces the meaning of the most frequently encountered terms a risk manager needs to know.

2.2.1. Risk

Risk is a measure of the consequence and probability of uncertain future events. It is the chance of an undesirable outcome. Phytosanitary risk managers face two broad categories of risk, risks of loss and risks of unrealized potential gains. A risk of loss is sometimes called a "pure risk," the obvious example being the risk of loss as a result of the establishment of a pest. The losses include such things as crop loss, reduced yields, impaired quality, ecological changes, market impacts, trade barriers and the like. The risk of an unrealized potential gain is sometimes called a "speculative" or "opportunity risk." Examples of potential gains that may not be realized that are of interest to consumers and industry include things like lower prices, greater variety, year-round availability, increased profits, jobs and income, as well as other benefits of international trade.

Consequences

Consequences can be good or bad. The consequence of a pure risk is something we do not want to happen. The consequence of a speculative risk is something we do want to happen. An undesirable consequence is either a loss that occurs or a gain that fails to occur.

It is a lack of information about events that have not yet occurred that gives rise to risks. This lack of information stems from two sources: there are facts we do not know and the universe is inherently variable. Let us call these two sources, together, uncertainty. Uncertainty is discussed at length in the next chapter.

It is sometimes convenient to differentiate the nature or status of a risk by considering the following kinds of risk:

- existing risk
- future risk
- historical risk
- risk reductions
- new risks
- residual risk

- transferred risk
- transformed risk.

Facts we do not know

How many kinds of pests are associated with an import, are they already present in the importing country, what damage are they capable of?

Inherently variable universe

How many pests will enter the country per shipment, how many will escape detection, how many will establish, what are the associated crop losses per year?

An existing risk is the risk that exists now. A future risk is a forecast of a risk at some point in the future. A historical risk is a hindcast of the risk at some point in the past. A risk reduction is the extent to which an existing, future or historical risk is or might be reduced by a risk management option. A new risk is a risk that did not heretofore exist. A residual risk is the amount of existing, future, or historical risk that remains or might remain after a risk management option has been implemented. When a risk management option reduces risk at one point in time or space for one kind of event or activity while increasing risk at another time or space for the same event or activity this is called a transferred risk. When a risk management option alters the nature of a hazard or opportunity or a population's exposure to that hazard/opportunity that is called a transformed risk.

Risk is often described by the following simple equation:

$$Risk = Consequence \times Probability \qquad (2.1)$$

This is not literally a formula for calculating risks. Most risk calculations are more complex. Think of it as a conceptual model that helps us think about risk. It tells us there are two essential elements to a risk, a consequence and its probability of occurring. The multiplicative form of this model tells us that if a consequence, i.e., a loss or an opportunity for gain, has no probability of occurring there is no risk. Likewise, no matter how probable an event is, if there is no consequence or undesirable outcome, the risk goes to zero. Risks can be estimated and described qualitatively or quantitatively.

For phytosanitary risks it may be convenient to modify equation 2.1 to:

$$Risk = Hazard \times Introduction \qquad (2.2)$$

This equation says if there is introduction, i.e. entry and establishment, but the pest represents no hazard, there is no risk. Likewise, if there is a hazard but no introduction, there is no risk.

In phytosanitary risk management it is common to speak of the economic, environmental, social, and other consequences of pest introduction. The probability of introduction embodies the sequence of events necessary for the entry and establishment of pests sufficient in number to cause the consequences. In equation 2.2, the hazard, which produces the consequences, is a quarantine pest. The probability of these consequences is embodied in the events that result in introduction of the pest.

2.2.2. Hazard

A "hazard" is anything that is a potential source of harm to a valued asset. Assets can be plant, human, animal, natural, economic, or social. Hazards include all natural/anthropogenic events capable of causing adverse effects on plants, people, property, economy, culture, social structure, or environment and the term is readily expanded to include biological, chemical, physical, and radiological agents. Phytosanitary risk managers are primarily engaged with natural biological agents, specifically organisms that can harm plants. A quarantine pest is a hazard.

> "A pest" (as defined by the IPPC) is "any species, strain or biotype of plant, animal or pathogenic agent injurious to plants or plant products."

2.2.3. Opportunity

An opportunity is any situation that causes, creates, or presents the potential for a positive consequence. It is any set of circumstances that presents a good chance for progress, advancement, or other desirable gain to a valued asset. The gain may be personal, communal, societal, local, regional, national, or global. Industry risk managers are primarily concerned with opportunities for gain through international trade.

> New products, cheaper products, new trade partners, year-round products, and better-quality products are some opportunities for phytosanitary risk managers to consider. For the importers and exporters, opportunities include the potential for increased income, employment, and profits. These potential gains would improve the quality of life for many people.

2.2.4. Uncertainty

Uncertainty, in a trade context, reflects a lack of awareness, knowledge, understanding, data, evidence or facts about circumstances germane to a trade-related decision problem. Most simply stated, when you are not sure you are uncertain. One can be unsure about the nature of the hazard or the opportunity, the consequences that can result, or the sequence of events necessary for the consequences to occur. The reasons for uncertainty may be several but they can all be categorized as either knowledge uncertainty or natural variability. All of these terms are carefully explained and developed in the next chapter.

2.2.5. Risk identification

It is essential that a phytosanitary risk manager be able to identify a risk. There are four essential steps to good risk identification and a fifth recommended step, either done by or for risk managers. The steps are:

1. Identify the trigger event.
2. Identify the hazard that can cause harm or the opportunity for uncertain gain.
3. Identify the specific harm or harms that could result from the hazard or the specific gain(s) from an opportunity that might not be realized.
4. Specify the sequence of events that is necessary for the hazard or opportunity to result in the identified harm(s) or potential gains.
5. Identify the most significant uncertainties in steps 2, 3, and 4.

Trigger. Something initiates the need to identify a risk. It could be a new trade agreement, a new trading partner, a new product, a significant change in an existing product, or expanding trade with existing partners. It could be a discrete event like information obtained from

---

Risk manager responsibilities

In subsequent chapters you will learn of the risk manager's responsibilities in the grander risk analysis process we have only begun to describe. For now, we speak only of the risk manager because of the risk manager's overall responsibility for the risk process. That should be interpreted to mean that the risk manager is responsible for seeing that the described tasks get done and not necessarily that it is the risk manager's responsibility to do them.

---

stakeholders, a trending risk in other countries, the accumulation of scientific knowledge, an intentional search for risks, and the like. It helps to document the event or circumstances that trigger a specific risk coming to light.

Hazard or opportunity? Once the trigger has been figuratively pulled to initiate risk identification it is critically important to identify the hazard(s) or opportunity(ies) relevant to the situation. Hazards may already be well recognized and identified, for example in lists of pests of quarantine concern like the European and Mediterranean Plant Protection Organization (EPPO) A1 and A2 lists of pests recommended for regulation as quarantine pests as approved by the EPPO Council in September 2015 (https://www.eppo.int/ACTIVITIES/plant_quarantine/A1_list and https://www.eppo.int/ACTIVITIES/plant_quarantine/A2_list, accessed October 28, 2019**)** or they may be latent or speculative pests of quarantine concern. Hazards may come to light in any number of ways, from an intentional search by risk assessors to any of the sources in the text box.

Hazards may be identified by the following.

- Regulators, plant protection officers (PPOs), farmers, property owners, importers, exporters, inspectors, the food industry, science providers, and interested consumer groups.
- Disease surveillance and monitoring information, risk-based sampling, entomological studies, laboratory studies, or changes in production practices.
- Hazards may be well recognized or unknown, new or latent. Hazards or pests may be insects, mites, mollusks, nematodes, plant diseases, or weeds.

An opportunity is any situation that causes, creates or presents the potential for an uncertain positive consequence. International trade is primarily concerned with opportunities for ecological, economic, and financial gain. Opportunities may include any subset of the benefits associated with international trade identified in Chapter 1. Some of the more common potential gains for consumers of an importing country are product quality, variety, availability, and price. Common opportunities for importers and exporters include profit, income, and employment.

Consequence. Determining the specific harm in a risky situation must precede an assessment of the probability of that harm. Thus, consequence comes before probability in the risk identification task. Risk managers must identify the specific harm or harms that can result from a hazard. This is normally done in risk assessment and it may be done for rather than by risk managers. Likewise, they should identify the disappointing and unwelcomed results that can occur when an opportunity fails to produce the hoped-for gains.

Loss risk

- Trigger—New product import from new trading partner.
- Hazard—Ferocies species (Fs is a generic hypothetical species).
- Harm— Reduced potato crop.
- Sequence—Pathway exists–> Fs arrives at pathway–> Fs survives passage through pathway–> Fs colonizes–> Fs spreads and destroys potato crop.
- Uncertainty—Survival through pathway, effectiveness of controls, probability of crop loss, value of crop loss.

Speculative risk

- Trigger—New product import from new trading partner.
- Opportunity—Reduced cost of fresh produce.
- Harm—Price reductions not realized.
- Sequence: Product denied entry to importing country-> Price reductions not realized.
- Uncertainty—Demand for new product, reliability of supply.

There may be more than one undesirable outcome. If so, identify all the relevant harms to be assessed. Pest introductions can damage plants, perhaps including cash crops, resulting in environmental and economic damage. New trade denied could result in unrealized profits, jobs and income, as well as lost opportunities for consumers. Risk managers operating within the SPS framework will define risks differently than those who are not. Risks can look different at different points in the international trade continuum. These differences will be detailed in subsequent chapters.

Sequence of events. For each relevant harm identified, risk managers should identify the specific sequence of events that is necessary for the identified hazard to result in the identified harm(s). The likelihood of that precise sequence of events occurring defines the probability of the risk. When there is more than one pathway from the hazard to the harm, each relevant pathway ought to be identified. In a similar fashion, the sequence of events from an opportunity to an undesirable outcome ought to be identified.

Uncertainty. The initial identification of a risk is likely to be uncertain. Some consequences, i.e., harms, may be uncertain and the sequence of events that leads to them may, likewise, be uncertain. It is the risk assessor's job to identify the most significant uncertainties that attend a risk.

Some of the factors to consider in the sequence of events that leads up to the introduction of a pest include entry and establishment. Entry depends upon a pest being on the pathway, the type of commodity and its intended use, and its survival during transit. Establishment depends upon:

- availability, quantity, and distribution of hosts in the pest risk assessment area
- environmental suitability
- other pest characteristics which enable the pest to reproduce and effectively adapt in a new environment
- pest mobility
- cultural production practices and control measures.

The probability of introduction is the most complete measure of pest risk probability because it entails the most complete sequence of events required to produce consequences.

2.2.6. Acceptable risk

A risk is acceptable when its probability of occurrence is so small, its consequences are so slight,

or its benefits (perceived or real) are so great, that individuals or groups in society are willing to take or be subjected to the risk that the event might occur. A risk that is not acceptable is unacceptable by definition. Risk managers will ultimately have to consider whether an assessed risk is acceptable or not. This is a subjective judgment, not a scientific determination. A risk that is judged acceptable requires no additional risk management. A risk that is unacceptable must be managed whenever possible.

It is conceptually possible for a risk control strategy to reduce an unacceptable level of risk to an acceptable level. Methyl bromide, heat treatments, and cold treatments are examples of risk mitigation measures that may reduce an unacceptable risk to an acceptable level. There may, however, be a tolerable level of risk between acceptable and unacceptable that risk managers must understand.

Acceptable and tolerable opportunity risks

An acceptable or tolerable level of opportunity/ speculative risk looks a little different. An acceptable speculative risk is one with a negligible probability of a negative outcome or with positive consequences so large they offset the chance of a negative outcome. Alternatively, the negative consequences may be so slight that individuals or groups in society are willing to take the risk. Permitting new trade that has zero chance of a negative outcome, e.g., introduction of a quarantine pest, is an example of an acceptable opportunity risk. Allowing trade that will produce widespread benefits despite some small chance of introduction is another example.

A tolerable opportunity risk is one that decision-makers or society are/is willing to take. Risk-taking is essentially different from risk avoidance. Risk-taking decisions are conscious decisions to expose oneself to a risk that could have otherwise been avoided.

2.2.7. Tolerable risk

A tolerable risk is not an acceptable risk. It is a non-negligible risk that has not yet been reduced to an acceptable level. Such a risk is tolerated for one of three general reasons: it may be impossible to reduce the risk further; the costs of further reduction are considered excessive; or the magnitude of the benefits associated with the remaining risky activity are too great to reduce it further.

A tolerable risk is an unacceptable risk whose severity has been reduced to a point where it is tolerated. Risk managers might decide to expose society to some tolerable risk of crop damage by allowing a new product with great potential benefits to enter the country.

2.2.8. Hazard-based decision-making

When a decision to take action is made based on the likely presence of a hazard with 'or without' an explicit evaluation of the consequences of that hazard, that is hazard-based decision making. Hazard-based decision-making is not risk-based decision-making because it does not take the probability of the consequences occurring explicitly into account. Hazard-based decisions are based solely on hazard information. That is not good risk management.

2.2.9. Risk-based decision-making

A risk-based decision is one that is based solely on risk criteria. In food additive safety analysis, the estimated daily intake (EDI) is divided by the acceptable daily intake (ADI) to create a risk metric. If this metric is greater than 1, risk management measures are taken to reduce the ratio below 1. Typically, if a risk metric exceeds a certain threshold, a specific response is triggered. Risk-based decisions consider only risk metrics. When an NPPO makes a decision based solely on the probability and consequence of one or more pests it is making risk-based decisions.

2.2.10. Risk-informed decision-making

In risk-informed decision-making, risk metrics are but one of the criteria used to determine if risk management is warranted. The other criteria may be economic, environmental, political, or other. Risk-informed decisions reflect a mix of risk metrics and other social values and the decisions based on them reflect subjective trade-offs among these criteria. When an NPPO makes a decision based on the probability and consequence of one or more pests, as well as other factors like the benefits of trade, restraint of trade, harmonization, equivalence, and so on required to achieve their appropriate level of protection, it is risk-informed decision making.

2.3. Risk Analysis

We need a term for the overarching risk process we have begun to discuss. Although the ISO favors risk management for this role we will use risk analysis, a term already in use by the international phytosanitary community. Risk analysis is a systematic way of identifying decision problems, then gathering and evaluating evidence that can lead to recommendations for a decision or action in response to an identified hazard or opportunity for gain. It is a process that has evolved specifically for decision-making under uncertainty. It consists of three tasks: risk management, risk assessment, and risk communication seen in Fig. 2.1. It examines the whole of a risk by assessing the risk and its related relevant uncertainties for the purpose of efficacious management of the risk, facilitated by effective communication about the risk.

Terje Aven, in 2018, reinforced by Yoe (2019) make an effective argument for risk analysis as a new emerging science. As a paradigm, it is capable of producing knowledge about risks and risky activities in the real world; as a science it also produces knowledge about concepts, theories, frameworks, methods, and the like to understand, assess, communicate, and manage risks. This latter knowledge set makes risk analysis as much a science as statistics is, for example.

Risk analysis, as used here, has evolved from a paradigm for decision-making under uncertainty into its own fledgling science. It recognizes that we may be uncertain about one or more aspects of the probability or the consequence of a risk of concern. Consequently, risk analysis is intentional in the way it directs analysts and decision-makers alike to base their decisions

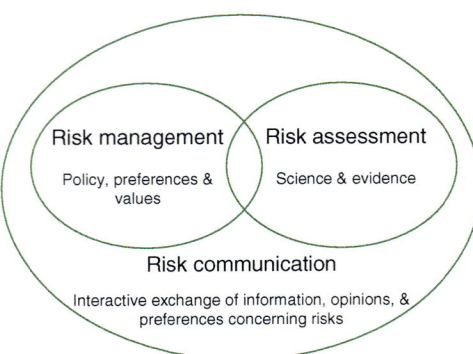

Fig. 2.1. Risk analysis comprises three tasks (based on and adapted from an FAO illustration http://www.fao.org/food/food-safety-quality/capacity-development/risk-analysis/en/, accessed October 28, 2019).

on the available science, while paying appropriate attention to the remaining uncertainty. Each of the three risk analysis tasks is introduced in the paragraphs that follow.

2.3.1. Risk management

Risk management consists of identifying a problem, defining the decision context, requesting the information needed to make that decision, evaluating assessed risks, and initiating action to identify, evaluate, select, implement, monitor, and modify actions taken to alter levels of unacceptable risk to acceptable or tolerable levels as opposed to taking no action. More informally, risk management is the work one has to do to ask and then answer the following kinds of questions (Yoe, 2019):

1. What's the problem?
2. What information do we need to solve it, i.e., what questions do we want risk assessment to answer?
3. What can be done to reduce the consequence of the risk described?
4. What can be done to reduce the probability of the risk described?
5. What are the trade-offs of the available options?
6. What is the best way to address the described risk?
7. Once implemented, is it working?

Risk management includes scientifically sound, cost-effective, integrated actions that reduce risks of loss or pursue opportunities for gain, while taking into account economic, environmental, social, cultural, ethical, political, and legal considerations.

The phytosanitary risk manager has a dual role. When managing risks of loss, the manager is a risk-reducer whose primary responsibility is to avoid, prevent or reduce losses. When managing opportunity risks, the manager is a risk-taker whose job is to prudently take those risks that can better society or advance the objectives of the organization for which the risk manager works. NPPO risk managers function primarily as risk-reducers while industry risk managers function primarily as risk-takers.

There are several distinct risk management strategies for both risk-taking and risk reduction. The risk reduction strategies and their risk-taking equivalents are shown in Table 2.1. These strategies are described in the sections that follow below. Risk management is taken up again at length in Part 2 of this handbook.

Risk reduction

Risk management strategies that address risks of loss are called risk reduction strategies in common usage. More precise definitions are provided below. Risks can be reduced by removing the source of the risk, i.e., the hazard; changing the nature and magnitude of its probability; changing the nature, magnitude, duration, or frequency of the consequences; transferring the risk to another party or parties; and retaining the risk by choice. Each of these strategies is described below.

Risk avoidance reduces or eliminates risk by removing the source of the risk (i.e., the hazard)

Table 2.1. Risk control strategies and their risk-taking equivalents.

Risk strategy	
Risk reduction	Risk-taking
Avoidance	Creation
Prevention	Enhancement
Mitigation	Exploitation
Transfer	Sharing
Retention	Ignoring

or executing the project or activity in a way that achieves the desired outcome while insulating valued assets from the effect of the risk. Risk avoidance either reduces the probability or the impact of the consequences to zero so that the risk no longer exists. Prohibiting the entry of a product into the importing country is an example of risk avoidance.

Risk prevention strategies reduce the probability of adverse consequences. Although the likelihood of the risk may not be reduced to zero, it may be reduced to a level considered acceptable or at least tolerable. This can be done through a number of risk control strategies including treatment, risk-based sampling, inspection, and examination.

Risk mitigation strategies reduce the magnitude of a risk by reducing the impact of the consequences. This may be done by changing the nature, magnitude, duration, or frequency of the negative consequences. When the consequences cannot be eliminated in their entirety it may be possible to reduce them to a level that is acceptable or at least tolerable. Containment and eradication measures are examples of mitigation strategies.

Risk transfer is a strategy for identifying stakeholders better able to manage the risk or finding a way to share a risk among many stakeholders. In the extreme this means passing the liability and responsibility for action to another stakeholder. In other applications insurance can be a means of transferring a risk.

Risk retention is necessary when no means exist for reducing a risk or when the residual risks cannot be reduced to a tolerable level. Thus, risk retention generally refers to the situation where stakeholders are forced to live with an unacceptable and intolerable level of risk. In such cases monitoring the status of the risk may be the only viable response to the risk.

In reality, these terms are not used quite so carefully. They are often used as synonyms.

Risk-taking

When faced with an opportunity for an uncertain gain, the risk manager must decide whether or not to assume the risk of taking an action that could result in less than the expected, desired, or even necessary outcome. This is the risk-taking role of the risk manager. The kind of risk-taking strategy the risk manager employs in decision-making can directly affect the likelihood and/or consequences of the opportunity's outcome. Hilson's (2001) four risk-taking strategies are modified and supplemented in Table 2.1. They are discussed below.

> **A natural tension**
>
> Industry representatives and exporting country government officials function as risk managers when they choose to pursue an opportunity for gain through plant trade with another country. They are risk-takers.
>
> In the importing country there is another set of risk managers. Representatives of industry and some government officials may be functioning as risk-takers as well, anxious to participate in the proposed trade. The importing nation's PPO, however, functions primarily as a risk-reducer charged with the primary responsibility for protecting plant life and health in the importing country.
>
> Ideally, the PPO, who occupies a critical position in the overall decision-making process, will be aware of the sometimes-conflicting nature of these risks. There is a natural tension between the risks of loss and the opportunity risk of potential gains. It is ultimately the phytosanitary risk manager's responsibility to weigh and tradeoff these conflicting interests in a transparent decision-making process.

Creating an opportunity with an uncertain outcome is, in a sense, creating a risk. It is the opportunity equivalent of avoiding a pure risk. This means creating circumstances for a known desirable consequence to occur. In the extreme, risk creation causes both the likelihood and positive consequence to come into being. Introducing a new product to a new market is an example of risk creation. The risk manager who does this has created an opportunity for potential gain as a result of the proposed trade.

Enhancing a risk is the opportunity equivalent of prevention strategies for a pure risk. It involves increasing the likelihood that an event or desired outcome will occur. Enhancement seeks to eliminate uncertainty in the likelihood and tries to make the desired event definitely happen. This is, effectively, seeking to increase the probability of the desired outcome to as close to 100% as possible. Under an enhancement strategy, aggressive measures are taken to ensure the necessary conditions arise as a result of the action taken. Pest-free concepts and post-harvest handling and processing are examples of this strategy.

Exploiting a risk is the opportunity equivalent of mitigating a risk. Exploiting a risk seeks to increase the impact of the opportunity in order to maximize the benefit to the project or activity undertaken. A risk exploitation strategy operates primarily on the consequence dimension of a risk to ensure that the benefits from the opportunity are as fully realized as possible. This strategy can be achieved by industry risk managers through product quality or availability enhancements along with price advantages for the importing country.

Risk sharing is the opportunity equivalent of a pure risk transfer. Risk sharing seeks a partner able to manage the opportunity, so that the likelihood of it happening or the potential benefits from it can be maximized. A successful risk sharing strategy results in mutual enjoyment of the benefits of trade.

NPPO risk managers

Three kinds of risk management functions can be identified in the NPPO risk management responsibility. There are risk managers with responsibility for the analytical process of identifying, evaluating, and recommending risk management options. A second responsibility covers the policy-based process of deciding and prescribing measures to mitigate pest risk. The third responsibility covers the operational aspects of implementing programs and activities to manage pest risk.

Ignoring a risk is choosing to take no action to realize an opportunity. It requires risk managers to take a reactive approach to risk management. Not initiating international trade would be an example of risk ignoring.

Who are the risk managers?

The prototypical phytosanitary risk manager works for the NPPO that makes decisions about what products will and will not enter their country and under what circumstances. These risk managers are usually employees of government organizations. They are the most visible and principal group of phytosanitary risk managers. However, they are not the only risk managers. Effective risk management is and must be a shared responsibility. The farm to table food chain is full of people with varying shares of the risk management responsibility. Likewise, the plant supply chain has many risk managers.

Fig. 2.2 shows a simple stylized supply chain. Even if plant protection regulators control and oversee this process there are many people along the way who must implement any mandated risk control measures and each of these people owns a piece of the risk management responsibility. The person applying a pesticide is a risk manager as are the packing house workers culling stems and leaves from the product, the shippers maintaining proper temperatures and the inspectors looking for evidence of pests. A risk management strategy will not work well unless growers, processors, exporters, transporters, importers, and others are implementing the strategy. This includes people in government, industry, and even consumers and end users in some cases. Risk management regulators may not be able to exercise control over the entire supply chain but they must be aware of the need for cooperation and support along that chain when making risk management decisions.

Risk managers who work for the NPPOs are inclined to focus on risk reducing activities as they strive to protect plants from invasive pests. Risk managers who work for private industry are more likely to focus on the risk-taking activities of seeking and promoting new international trade opportunities for plants and plant products. Ideally, both sets of risk managers would attend to both sets of roles. That is not the current design, however. The SPS Agreement does not include the benefits that accrue to exporters and importers of plants and plant products in its

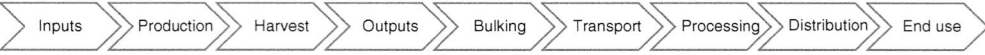

Fig. 2.2. Simple plant product supply chain.

consideration of the consequence of risk. In a perfect world, NPPO risk managers would consider both pest risks and potential gains from trade when they function in their decision-making roles. Industry risk managers would actively consider the pest risk potential of their trade proposals. That is not the current reality nor is it likely to be so anytime soon.

In many government organizations, "risk managers" are distinguished from "decision-makers." Risk managers deal with the technical issues and trade-offs associated with designing risk mitigations in these organizations. Decision-makers deal with policy and include other factors in their analysis. This handbook considers any phytosanitary decisions about matters of risk to be made by risk managers, thus breaking down the artificial distinction between risk managers and decision-makers.

2.3.2. Risk assessment

Risk assessment is a systematic process for describing the nature, likelihood and magnitude of risk associated with pests of quarantine concern as well as trade decisions, including consideration of the relevant uncertainties. It provides an understanding of risks, their causes, consequences, and likelihoods or probabilities. Risk assessment provides a basis for decisions about the most appropriate risk management option to be used to treat the risks. Risk assessment outputs are to be used as inputs to phytosanitary risk management decision-making processes. Risk assessment can be qualitative, quantitative, or a blend (semi-quantitative) of both. Risk assessment is informally described by asking and answering the following questions, that build on the Kaplan and Garrick triplet (1981):

1. What can go wrong?
2. What are the consequences?
3. How can it happen?
4. How likely is it to happen?

Risk assessment is the step that gathers the evidence, answers the risk manager's questions, and identifies and addresses the uncertainty that remains in the decision problem. It is the positive task of risk analysis, whereas risk management

is the normative task of risk analysis. Good risk assessment analyzes the causes of the risk to determine their contribution to the consequences and their likelihoods of occurrence. This analysis provides valuable insight into the most effective ways to further treat unacceptable risks.

Risk assessment—The evaluation of the likelihood of entry, establishment or spread of a pest or disease within the territory of an importing Member according to the sanitary or phytosanitary measures which might be applied, and of the associated potential biological and economic consequences; or the evaluation of the potential for adverse effects on human or animal health arising from the presence of additives, contaminants, toxins, or disease-causing organisms in food, beverages or feedstuffs.

(Annex A to the SPS Agreement)

Fig. 2.3 provides a generic four-step risk assessment process to help us think about both the risk reducing and risk-taking aspects of phytosanitary risk management. Some version of each of these four steps is found in most risk assessment models. The first step simply requires a clear identification of the source and nature of the risk. What is the hazard that threatens a loss or the opportunity that promises a gain? The next two steps require the assessor to identify the consequences of the specific hazard or potential gain and the probability of those consequences occurring. The final step, risk characterization, is where the analysis of the three preceding steps is pulled together to characterize the risk, qualitatively or quantitatively, for the purpose of supporting decision-making. Throughout this process the risk assessor is to carefully consider and address the uncertainty at each stage of the assessment, most importantly in the characterization of the risk itself.

This generic risk assessment model has been refined and specialized for pest risk assessment. Devorshak (2012) offers the pest risk analysis overview of Fig. 2.4 which includes a pest risk assessment model on the top part of the figure. Some sort of trigger causes the initiation of a pest risk assessment. Identifying the potential pests is the first step. It has a direct parallel with the hazard identification step. The pest risk assessment's next two steps provide for the likelihood and consequence assessments. The generic

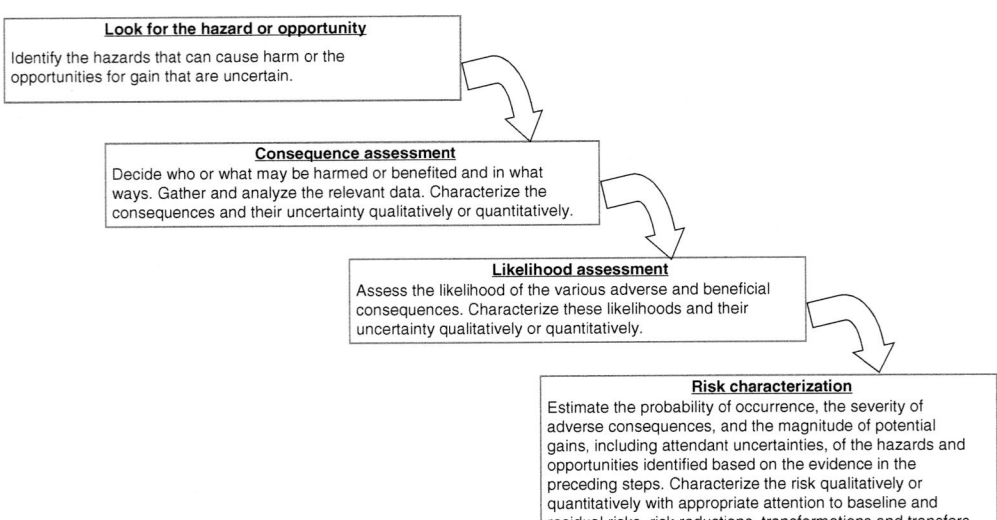

Fig. 2.3. Four-step risk assessment process.

risk characterization step of Fig. 2.3 takes place in the last step where conclusions are developed and uncertainty is described. The conduct of a pest risk assessment is described expertly and in great detail in the Devorshak text.

Qualitative risk assessment

Qualitative risk assessment is distinguished by its lack of reliance on numerical expressions of risk. Instead, qualitative risk assessment depends on risk descriptions, narratives, and relative values often obtained by ranking or separating likelihoods, consequences, and risks into descriptive categories like high, medium, low, and negligible or no risk. When the relative values are numeric but nominal or ordinal in character, such as when index numbers are used, the risk estimate is said to be semi-quantitative, but they remain more qualitative than quantitative in character.

Following are examples of qualitative and semi-quantitative risk assessment. The category ratings are qualitative. The numerical values reflect point scores for components of likelihood and consequences. The cumulative total expresses the risk on the following scale: Low = 11 to 18, Medium = 19 to 26 and High = 27 to 33.

Pest	Consequences of introduction (cumulative risk rating)	Likelihood of introduction (cumulative risk rating)	Pest risk potential (Total)
Arthropod			
Anastrepha fraterculus	High 14	High 16	High 30
Ceratitis capitata	High 14	High 16	High 30
Parlatoria cinereae	High 13	Medium 13	Medium 26
Parlatoria ziziphi	High 13	Medium 13	Medium 26
Bacteria			
Xanthomonas axonopodis pv *citri*	Medium 11	Low 9	Medium 20
Fungi			
Elsinoë australis	Low 6	Medium 13	Medium 19
Guignardia citricarpa	Medium 9	Medium 13	Medium 22

(Risk assessment for the importation of fresh lemon (Citrus limon (L.) Burm. F.) fruit from northwest Argentina into the continental United States, August, 2007 revised original.)

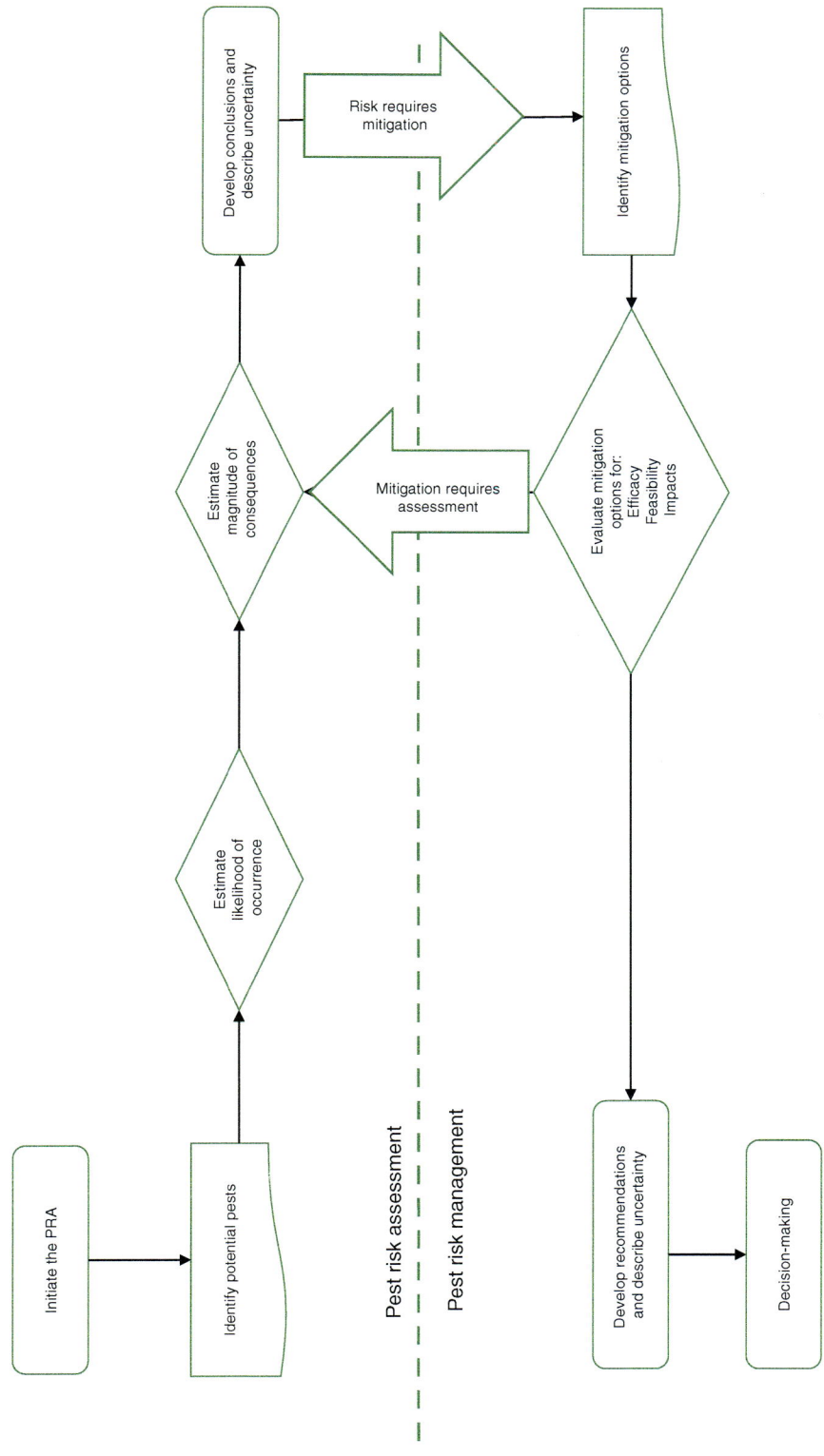

Fig. 2.4. Pest risk analysis process (Devorshak, 2012).

Quantitative risk assessment

Quantitative risk assessment relies on numerical expressions of risk. Both the probabilities and consequences of risk are expressed numerically in specific units defined by the needs of risk managers. A quantitative risk estimate can be a deterministic point estimate or a probabilistic estimate, such as an interval or probability distribution. Quantitative analysis is preferred when it is possible. Full quantitative analysis may not always be possible due to insufficient information about the decision problem, lack of data, unclear social preferences, and many other factors. In some instances, the resources and effort needed for quantitative assessment may not be warranted or required, even though the analysis is possible.

Risk estimate

A risk estimate is an estimate of the likelihood and severity of the adverse effects or opportunities, which addresses key attending uncertainties. A risk estimate combines the estimate of the consequences and their likelihood to describe the overall level of risk. Quantitative estimates are numerical in nature and are preferred over narrative qualitative estimates. Risk estimates should include all the relevant aspects of the risk, which may encompass existing, future, historical, reduced, residual, new, transformed, and transferred risks.

Risk description

A risk description is a narrative explanation and depiction of a risk that bounds and defines a risk for decision-making purposes. It is the story that accompanies the risk estimate that places it in a proper context for risk managers and others to understand.

Risk assessment for plant pests and invasive plants

In January of 1997, the Harvard Center for Risk Analysis (HCRA) held an invitation-only workshop in Washington, D.C. that brought experts in risk analysis and pest characterization together to develop general principles for pest risk analysis (Gray, *et al.*, 1998). The panel recognized there was no one best approach to be applied to pest problems but they considered the probabilistic approach, i.e., using distributions to characterize knowledge about pests, to be the ideal approach to pest risk analysis. Nonetheless, traditional risk assessment approaches to plant pests and invasive plant species have been primarily qualitative (Powell, 2004). The risk assessors' task has been to use the best available alternatives while working toward improvement in risk assessment data, models, and procedures.

Pest risk assessment guidelines tend to describe procedures to develop categorical ratings like low, medium, or high, of risk elements associated with the consequences and likelihood of introduction or spread of a plant pest or noxious weed (see box for examples). An overall risk measure is determined by combining the scores for the likelihood and consequences of introduction risk elements.

Examples of risk elements

- Pest prevalence on the harvested plant part(s).
- Likelihood of surviving post-harvest processing before shipment.
- Likelihood of surviving transport and storage conditions of the consignment.
- Likelihood of coming into contact with host material in the endangered area.
- Likelihood of arriving in the endangered area.
- Combined likelihood of establishment.
- Damage potential in the endangered area.
- Spread potential.
- Determining export markets at risk.
- Likelihood of trading partners imposing additional phytosanitary requirements.
(USDA/APHIS/PPQ, 2012)

2.3.3. Risk communication

Risk communication is the open, two-way exchange of information and opinion among risk analysts, their trading partners, stakeholders and various publics about risks, intended to lead to a better understanding of the risks and better risk management decisions. It provides a forum for the interchange of information with all concerned about the nature of the risks, the risk assessment, and how risks should be managed (Yoe, 2019). Risk communication may be informally characterized by asking and answering the following questions (Chess and Hance 1994):

- Why are we communicating?
- Who are our audiences?
- What do our audiences want to know?
- How will we communicate?
- How will we listen?
- How will we respond?
- Who will carry out the plans? When?
- What problems or barriers have we planned for?
- Have we succeeded?

Done well, risk communication helps partners and stakeholders understand the nature and magnitude of the risk. Risk communication is essential to developing credible and acceptable risk management responses. It can enhance trust and confidence in the decision-making process while promoting the participation and involvement of interested parties.

Risk communication is needed to explain actions required to avoid pest risks or to take trade risks. It is needed to explain the rationale for the risk management option chosen. Both the workings and the effectiveness of specific options needs to be communicated so partners and stakeholders understand their own risk management implementation responsibilities and know what actions they must take to reduce the risk or realize the gain. The benefits of a risk management option as well as the costs of managing the risk and who will bear them are additional information conveyed by good risk communication.

Phytosanitary risk communication needs to pay special attention to describing residual risks, i.e., risks that remain after the risk management option is implemented. There is sometimes a mistaken notion that efforts to manage a risk reduce the risk to zero and that is rarely the case. The uncertainty that could affect the magnitude of the risk or the efficacy of the risk management option must be carefully communicated to partners, stakeholders and the public. This should include the weaknesses, limitations of or inaccuracies in the available evidence. It should also include the important assumptions on which risk estimates are based so partners and stakeholders can understand the sensitivity of both risk estimates and the efficacy of an risk management option to changes in those assumptions and how those changes can affect risk management decisions (Yoe, 2019).

Risk communication does not require consensus or an agreement. It should, however, provide people with meaningful opportunities for

> **Goals of risk communication**
>
> **1.** Promote awareness and understanding of the specific issues under consideration during the risk analysis process, by all participants.
> **2.** Promote consistency and transparency in arriving at and implementing risk management decisions.
> **3.** Provide a sound basis for understanding the risk management decisions proposed or implemented.
> **4.** Improve the overall effectiveness and efficiency of the risk analysis process.
> **5.** Contribute to the development and delivery of effective information and education programs, when they are selected as risk management options.
> **6.** Foster public trust and confidence in the safety of the food supply.
> **7.** Strengthen the working relationships and mutual respect among all participants.
> **8.** Promote the appropriate involvement of all interested parties in the risk communication process.
> **9.** Exchange information on the knowledge, attitudes, values, practices and perceptions of interested parties concerning risks associated with food and related topics.
> (Source: *The Application of Risk Communication to Food Standards and Safety Matters*, a Joint FAO/WHO Expert Consultation, Rome, Italy, 2–6 February 1998)

input before decisions are made and for feedback as evidence is accumulated and uncertainty reduced. Risk communication requires listening to and understanding people's concerns about risks so they can be considered in decision-making. This is essential if the public is to respect the process even if they disagree with some of its decisions and outcomes.

Hazard and outrage

Experts and the public interpret risk in very different ways and this complicates risk communication. Risk involves scientific facts and people's feelings in response to the manner in which the risk is perceived. That perception may or may not align well with the facts of the risk. These competing dimensions of risk, the objective vs. the subjective, create some unique communication challenges.

Peter Sandman (1999) said the technical side of risk, usually the concern of phytosanitary risk managers, focuses on the magnitude and probability of undesirable outcomes. For pest risk management this includes such things as the life history of a pest, the pathways to introduction, the probability a pest is present on the pathway, the probability it survives the pathway, the probability the pest colonizes in the new location, the probability it spreads from that colony to do damage, and the nature of the damage that is done. Sandman categorizes such technical details and facts as "hazard."

The public by contrast focuses on the non-technical side of the risk. This is the social context of the risk that involves values and emotions. For example, the public cares more about how a pest risk decision affects them than they do about the facts of the risk themselves. Losing a citrus tree in one's yard or neighborhood to an invasive pest, or paying more for commodities because of protective trade restrictions matters more to people than the probability that a heat treatment will be effective. Concerns such as is the risk voluntary or coerced, familiar or exotic, dreaded or not dreaded and whether the NPPO is considered trustworthy or untrustworthy, responsive or unresponsive, are important to the public. Sandman calls all this "outrage."

The experts of NPPOs are more occupied by and concerned with the hazard aspects of a risk. As experts they think about these problems and know things that others do not. The public is less concerned with the science, numbers, and facts of the risk than they are with the outrage factors, i.e., the personal and social context of the risk. The public, when they are even aware of a pest risk issue, can have feelings about the risk. They believe things to be true or not, often without respect to the facts of a situation. The public is less concerned with the probabilities of introduction than they are with the relative importance of what might be lost, e.g., greater availability of product, lower prices, higher quality or more diversity of choices.

These two distinct dimensions of a risk, hazard and outrage, can lead to a disconnect between the phytosanitary risk management experts and the public. While experts tend to focus on what they *know and think*, the public focuses on what they *feel and believe*. Both dimensions of a risk are important but for different reasons.

The public may worry about things the experts do not even consider, like product diversity, quality, availability, and price. Other times they may not worry about things experts think they should worry about, such as the probability of colonization. Risk communication based on explaining the facts of the risk may well miss the greater concerns of the public, which tend to be the social and personal meaning of the risk. This means official risk communications are not always satisfying to the public.

Internal risk communication

The internal risk communication task ensures effective interaction between risk managers and risk assessors. Four rules of thumb for this task are:

- communicate continuously
- collaborate early
- coordinate often
- cooperate always.

More will be said about this relationship in Chapter 4 on the risk analysis team and again in Chapter 11 on the interactions between risk assessors and risk managers.

External risk communication

External risk communication requires phytosanitary risk managers, assessors, and communicators to interact with their partners, various publics, and external stakeholders. The extent of this interaction will depend on how "public" the risk management activity is. Not every risk management activity will require external risk communication. Public interest in many phytosanitary interests is minimal, and at times interest may be wholly contained within the NPPO. When external communication is warranted, four broad categories of tasks can be identified. These are:

- risk communication
- crisis communication
- public involvement
- conflict resolution.

These topics are covered at length in their own extensive professional literature. The external risk communication task is expanded on in Chapter 15.

2.4. Risk manager's role

The NPPO has the primary risk management role for pest risk issues but it is far from the only risk management role. There are benefits to be had from international trade and private industry as well as governmental trade-promoting organizations which have significant risk management responsibilities for taking those risks. Each party has a responsibility for supporting the risk management efforts of other risk managers involved in international plant trade and protection. Table 2.2 indicates the primary and secondary responsibilities of these parties in international trade.

Risk management is both an occupation and a role. There will be people with explicit risk management responsibilities and that is their occupation. Anyone who handles uncertainty while performing their job will assume the role of a risk manager from time to time. The risk manager's job and role are to make effective and practical decisions under conditions of uncertainty. As long as there is any uncertainty at all, a risk management decision is conditional. It is based on what is known and not known at the time a decision is made. As uncertainty is reduced during the decision-making process through risk assessment, in the future as more information is accumulated, or as the outcomes of the original risk management decision become known, it may be prudent to revise the decision to reflect the new knowledge. Hence, the risk management process is an iterative one. Every decision is subject to further revision in the future as long as uncertainty remains.

Although a formal risk management model is not introduced until Part 2 of this handbook, we can identify five basic roles for the manager in a risk management activity. These roles define the risk manager's job description. They are:

1. Initial risk management activities.
2. Estimating risk.
3. Evaluating risk.

4. Controlling risk.
5. Monitoring risk.

Fig. 2.5 shows the five responsibilities in a continuous loop to capture the iterative nature of the risk manager's job. As is true of any iterative process, the tasks, although presented in a sequential fashion, are not always executed sequentially. Activities in different tasks may occur simultaneously. Many activities and tasks can be repeated more than once. The entire sequence may be completed several times.

Risk management activities are triggered by some sort of event or are initiated in response to accumulated information inputs. A problem needs attention or an opportunity presents itself. The initial risk management activities include recognizing, accepting, and defining the risks on which to act. Risk managers are responsible for overseeing the NPPOs risk management process. They are its custodians. They assign personnel and provide the resources necessary to complete the risk estimation tasks, which are primarily the conduct of the risk assessment. They also guide the risk assessment process by requesting the specific information needed to solve the problems and realize the opportunities they identify.

Risk managers have an important, but limited, role in the risk estimation task. Estimating risk is the assessor's job; it is part of the evidence-based risk assessment process. Good risk assessment cannot be completed without direction and guidance from the risk manager.

SPS phytosanitary risk management is somewhat unique in that in some countries the risk assessor and risk manager are the same person. Even so it is useful to speak of the separate responsibilities. The text's narrative favors the language of separate people doing the separate jobs. This is not the case for all countries.

Once the risk assessment is completed, the risk manager must evaluate the assessed risk. The NPPO risk manager's first significant risk evaluation decision is to determine if the risk is

Table 2.2. Phytosanitary risk management responsibilities.

Risk management role	Primary responsibility	Secondary responsibility
Avoiding pest risks	NPPO	Industry and trade promoting organizations
Taking trade risks	Industry and trade promoting organizations	NPPO

Fig. 2.5. The risk manager's job in five tasks.

acceptable. Any unacceptable risk must be managed. If possible, unacceptable risks will be managed to an acceptable level. When this is not feasible the unacceptable risk should be managed to a tolerable level or simply be retained.

It may be misleading to suggest some risks can be controlled. Risk management options are identified or formulated, the options are evaluated and compared, the best risk management option is selected, conceptually measurable decision outcomes are identified and the best option is implemented as part of the manager's risk control responsibilities. In practice, parts of this task may be completed during risk estimation.

Decisions made under uncertainty can be more or less effective in reducing risks. The fifth responsibility of NPPO risk managers is to monitor decision outcomes, evaluate them, and then modify the decision as necessary. Once an option is implemented the NPPO can monitor decision information: inspection—detecting the pest of interest; decision implementation—has the option been properly implemented and are all stakeholders cooperating; and decision outcomes—are the desired risk reductions being realized? Monitored information is evaluated and judged and risk managers either hold the course or modify the risk management decision.

2.5. Risk and the IPPC

This chapter has provided a broad-based approach to risk language and concepts. Many of these terms are not defined by the IPPC. Where possible, for the remainder of this handbook, the terminology of the IPPC is used, specifically, the glossary of phytosanitary terms (ISPM 5).

The Agreement on the Application of Sanitary and Phytosanitary Measures (SPS Agreement) is an international treaty of the World Trade Organization (WTO). It first came into force in 1995 with the start of the WTO. The SPS Agreement identifies the IPPC as the organization that establishes internationally agreed upon standards for countries to protect themselves from phytosanitary (plant health and life) hazards that may be associated with international trade. The SPS Agreement specifies that phytosanitary measures that affect trade must be based on either international standards or risk assessment. Few international standards existed at the time so risk assessment became a primary means of establishing phytosanitary

Selected entries from ISPM 5 Part I

Pathway—Any means that allows the entry or spread of a pest.

Pest—Any species, strain or biotype of plant, animal or pathogenic agent injurious to plants or plant products.

Pest risk (for quarantine pests)—The probability of introduction and spread of a pest and the magnitude of the associated potential economic consequences (see glossary supplement No. 2) (ISPM 2, IPPC, 2019b).

Pest risk (for regulated nonquarantine pests)—The probability that a pest in plants for planting affects the intended use of those plants with an economically unacceptable impact (see glossary supplement No. 2) (ISPM 2, IPPC, 2019b).

Pest risk analysis (agreed interpretation) — The process of evaluating biological or other scientific and economic evidence to determine whether an organism is a pest, whether it should be regulated, and the strength of any phytosanitary measures to be taken against it (ISPM 2, IPPC 2019a).

Quarantine pest— A pest of potential economic importance to the area endangered thereby and not yet present there, or present but not widely distributed and being officially controlled.

measures that were no more trade restrictive than necessary. The SPS Agreement also meant that member countries of the WTO could challenge unjustified phytosanitary measures through the WTO.

The WTO identified three standard-setting organizations to protect human, animal, and plant life and health. They are the Codex Alimentarius for human health (food safety); the OIE for animal health, and the IPPC for plant health. None of these organizations had standards addressing risk analysis, so each embarked on its own journey to develop them.

Methods for assessing the uncertain potential gains from trade are well established within the global business community. Financial risk analysis is one of the more mature applications of risk analysis. Phytosanitary risk analysis or pest risk analysis is relatively early in its development and application. Like many other communities of practice, pest risk analysis has evolved in a rather unique way because of the needs of its practitioners and the path of its evolution.

The SPS, in Article 5.2 says: "In the assessment of risks, Members shall take into account available scientific evidence; relevant processes and production methods; relevant inspection, sampling and testing methods; prevalence of specific diseases or pests; existence of pest- or disease-free areas; relevant ecological and environmental conditions; and quarantine or other treatment" (WTO, 1994). Taking available scientific evidence into account means that risk assessment is scientific, i.e., it is based on science, but it is not pure science if for no other reason than that risk assessment is designed to address uncertainty, sometimes in ways that incorporate judgments about uncertain values that can change as uncertainty is gradually reduced.

Three essential references for pest risk assessment

The IPPC's ISPM 2 Framework for pest risk analysis (IPPC, 2019a), ISPM 11 Pest risk analysis for quarantine pests, including analysis of environmental risks and living modified organisms (IPPC, 2019c), and Devorshak (2012).

In the phytosanitary world, the hazard of concern is usually a harmful plant pest. It is common practice to estimate the likelihood of introduction next, followed by predicting potential consequences on a worst-case scenario basis (Devorshak, 2012). The likelihood is the "chance" that the pest of interest will traverse the pathway, become established and cause the identified consequences to occur. For this to happen a series of necessary events must occur, each of which has some probability of occurrence. If any of the necessary events does not occur, the consequences cannot occur. Consequences are usually economic, environmental, or other in nature. It is common practice to express all consequences, both market and non-market, in economic terms.

Most pest risk assessments are qualitative or semi-quantitative, if for no other reason than the sheer volume of risk assessments that are routinely conducted. Assessments of the likelihood and consequences of potential gains from trade are rarely completed explicitly. Instead, the gains from trade are usually implicitly assumed to be sufficient to justify the trade. Quantitative risk assessments are usually reserved for special cases of high visibility and importance to the

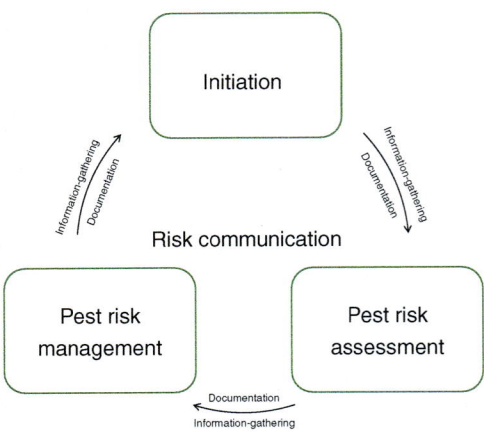

Fig. 2.6. Three stages of IPPC ISPM 2 pest risk analysis process.

trading partners. Qualitative or quantitative assessment of the uncertain gains from trade can be
considered in any risk assessment when deemed
appropriate.

Risk managers decide what is to be done
about the risk presented by a hazard. They consider the severity of the consequences and the
likelihood of their occurring, and weigh them
against the benefits of allowing the trade that
could introduce the pest. Thus, risk management is inexorably linked to the risk assessment.
Likewise, risk managers will rely on risk assessment to estimate the degree to which the hazard
can be mitigated by applying one or another risk
management measure. This means that risk assessment cannot be conducted in isolation of the
risk manager's needs for information.

2.6. Pest Risk Analysis

ISPM No. 2 (IPPC, 2019a) describes the pest risk analysis process as prescribed by the IPPC. Figure 2.6

illustrates the three parts of that process. The
PRA process consists of three stages: initiation,
pest risk assessment, and pest risk management.
PRA is an iterative process that does not necessarily proceed in a linear fashion. It is often necessary to go back and forth between the different
stages of a PRA.

A PRA process may be triggered by one or
more of the following:

- request made to consider a pathway that
  may require phytosanitary measures
- pest is identified that may justify phytosanitary
  measures
- decision made to review or revise phytosanitary measures or policies
- request made to determine whether an organism is a pest.

Once a PRA is triggered, the initiation stage, has
four steps as shown in Fig. 2.7.

Stage 2 of a pest risk analysis is the pest risk
assessment, it has the five primary steps shown
in Fig. 2.8 (IPPC, 2019c).

Notice that consequences and their probability are estimated in risk assessment. In general, the probability of introduction is the
product of the probability of entry and the probability of establishment. Some nations include
the probability of spread in this product, while
others do not. This handbook adopts the convention of using the terms as follows:

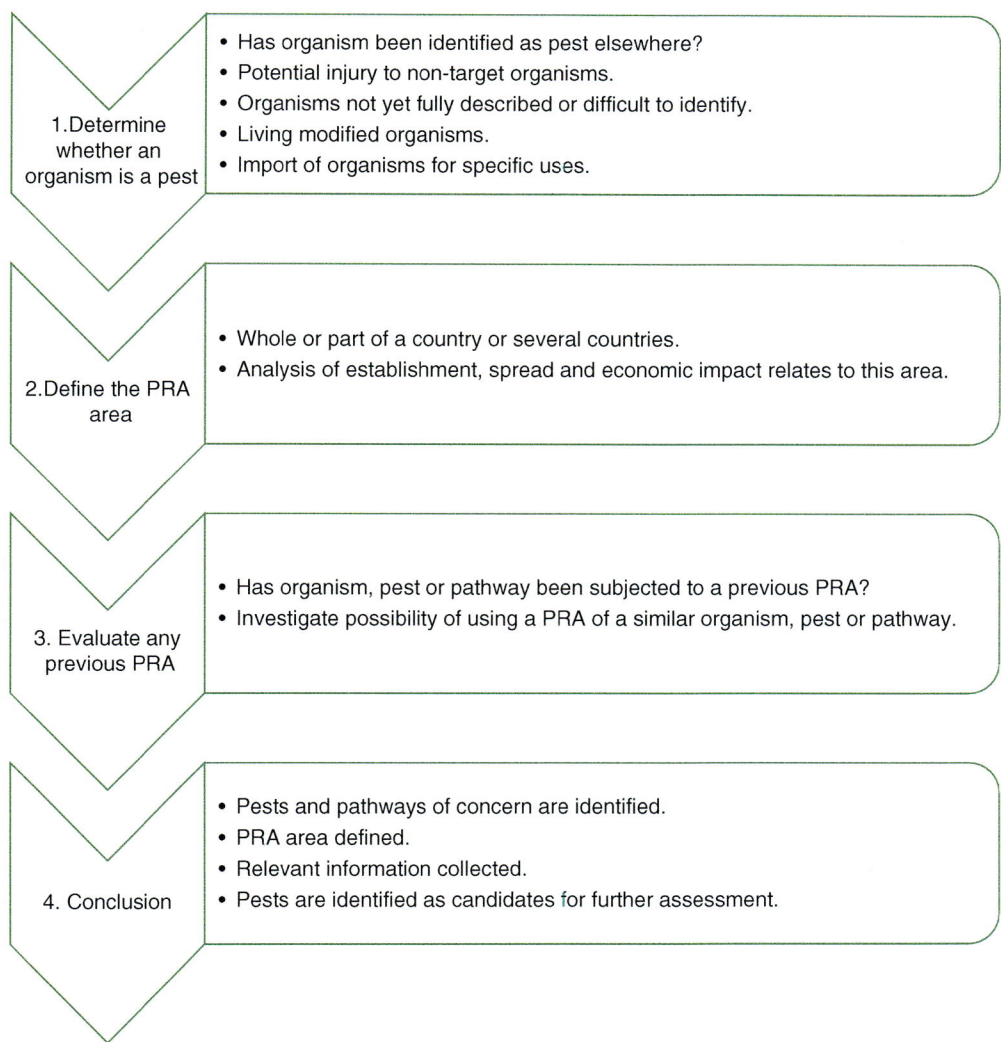

Fig. 2.7. Four steps of initiation stage of a pest risk analysis.

$$P(Introduction) = P(Entry) \times P(Establishment)$$
(2.3)

$$P(Introduction\ and\ Spread) = P(Entry) \times$$
$$P(Establishment) \times P(Spread)$$
(2.4)

Where, P(Introduction) means the probability of introduction and so on for like terms.

Pest risk management is the third PRA stage. Fig. 2.9 shows the seven steps that comprise this.

2.7. Summary and Look Forward

Here are five things to remember from this chapter.

1. There are many dialects in the language of risk; consequently, the language is messy and still evolving.

2. There are two kinds of risk: the risk of loss due to invasion by plant pests, and the risk of

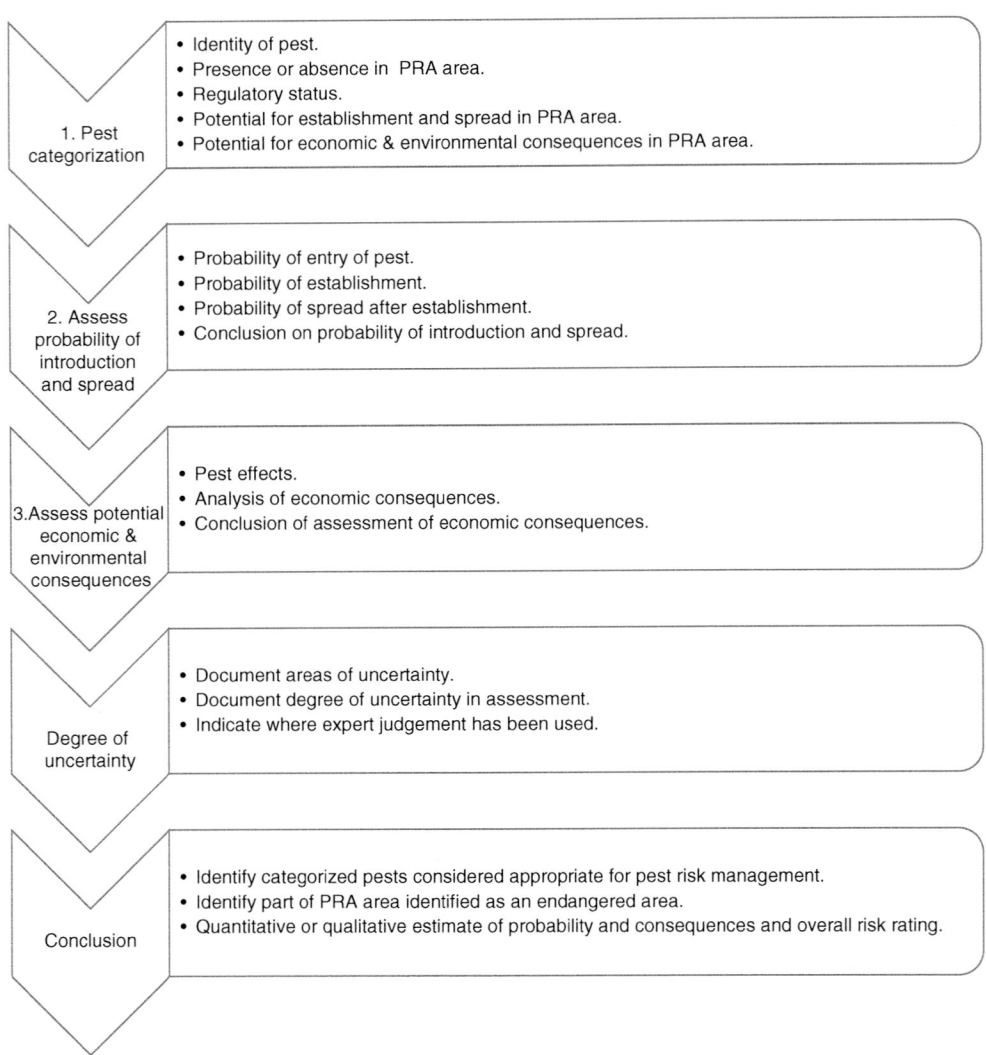

Fig. 2.8. Five steps of pest risk assessment stage of a pest risk analysis.

uncertain gains that result from trade in plants and plant products.

3. Risk analysis comprises risk management, risk assessment, and risk communication.

4. NPPOs are the primary risk managers for plant health, private industry, and trade promoting organizations are the primary risk managers for the uncertain gains from trade.

5. Risk managers have at least five roles to play in risk analysis: initiating risk management activities, risk estimation, risk evaluation, risk control, and risk monitoring.

Risk exists and risk analysis has arisen in response to risks for one primary reason, uncertainty. There is uncertainty in every trade decision risk managers make. The next chapter explains uncertainty in considerable detail because understanding it is essential to the risk management process.

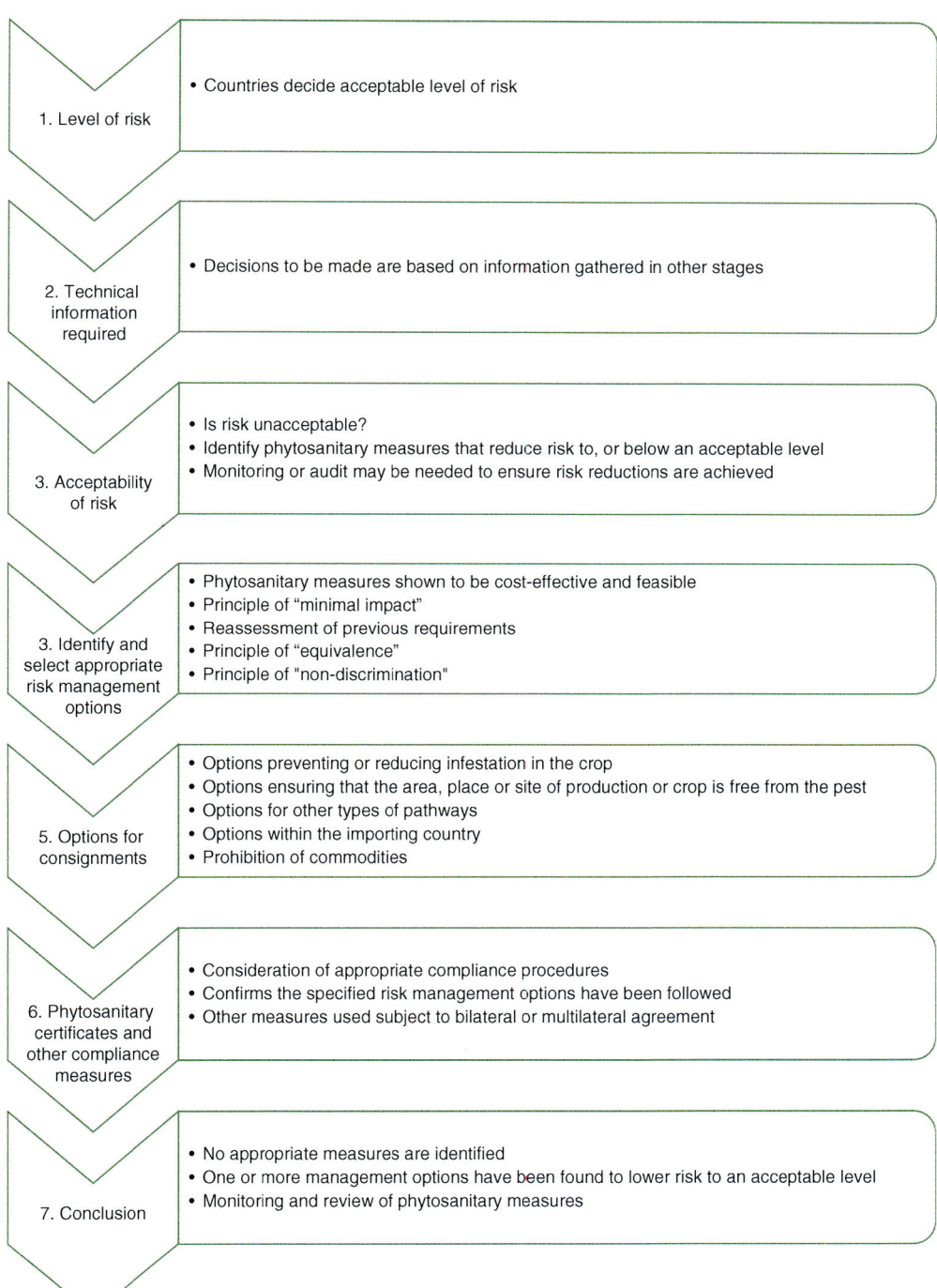

Fig. 2.9. Seven steps of pest risk management stage of a pest risk analysis.

2.8. References

Chess, C. and Hance, B.J. (1994) *Communicating with the Public: Ten Questions Environmental Managers Should Ask.* Center for Environmental Communication, New Brunswick, New Jersey, USA.

Devorshak, C. (ed.) (2012) *Plant Pest Risk Analysis, Concepts and Application.* CAB International, Wallingford, UK.

Gray, G.M., Allen, J.C., Burmaster, D.E. et al. (1998) Principles for conduct of pest risk analyses: report of an expert workshop. *Risk Analysis*, 18, 6.

Hilson, D. (2001) Extending the risk process to manage opportunities. *Proceedings of the Fourth European Project Management Conference*, PMI Europe 2001, 6–7 June 2001, London UK.

IPPC (2019a) International Standards for Phytosanitary Measures, Publication No. 2: *Framework for Pest Risk Analysis*. Secretariat of the International Plant Protection Convention (IPPC), Food and Agriculture Organization of the United Nations, Rome.

IPPC (2019b) International Standards for Phytosanitary Measures, Publication No. 5: *Glossary of Phytosanitary terms*. Secretariat of the International Plant Protection Convention (IPPC), Food and Agriculture Organization of the United Nations, Rome.

IPPC (2019c) International Standards for Phytosanitary Measures, Publication No. 11: *Pest Risk Analysis for Quarantine Pests Including Analysis of Environmental Risks and Living Modified Organisms*. Secretariat of the International Plant Protection Convention (IPPC), Food and Agriculture Organization of the United Nations, Rome.

Kaplan, S. and Garrick, B.J. (1981) On the quantitative definition of risk. *Risk Analysis* 1, 1, 11–27.

Powell, M R. (2004) Risk assessment for invasive plant species. *Weed Technology*, 18(sp1), 1305–1308.

Sandman, P.M. (1999) Risk = hazard + outrage: coping with controversy about utility risks. *Engineering News-Record*, October 4.

USDA/APHIS/PPQ (2012) *Guidelines for Plant Pest Risk Assessment of Imported Fruit and Vegetable Commodities*. Plant Epidemiology and Risk Analysis Laboratory, Center for Plant Health Science and Technology, Riverdale, MD, USA.

WTO (World Trade Organization) (1994) *The WTO Agreement on the Application of Sanitary and Phytosanitary Measures*. World Trade Organization, Geneva.

Yoe, C. (2019) *Principles of Risk Analysis Decision Making Under Uncertainty*, 2nd edn. CRC Press, Boca Raton, Florida, USA.

3

Uncertainty

I can live with doubt and uncertainty and not knowing. I think it is much more interesting to live not knowing than to have answers that might be wrong.

Richard P. Feyman

3.1. Introduction

Without uncertainty there would be no risk at all; risk assessments would provide complete and reliable answers and the effects of risk management measures would be perfectly predictable. The world is uncertain, however, and we must deal with that. One of the fundamental principles of risk analysis science is to base the assessment of risks on the best available science and evidence. A second, and distinguishing, principle of risk analysis science is to focus appropriate attention on those things that we do not know, which could affect decisions and decision-making outcomes. In best practice pest risk analysis, it is the risk assessor's responsibility to separate what they know from what they do not know in their assessment and to communicate the effects of uncertainty on decision criteria, which will include the risk of pest introduction and the effectiveness of risk management measures. It is the risk manager's job to take the effects of uncertainty into account when making a decision.

Risk management is decision-making under uncertainty. To understand risk management, it is necessary to understand uncertainty. In general, if we are not sure about any aspect of a pest risk, the potential gains from plant trade, or the efficacy of a proposed risk management option, we are uncertain. Uncertainty derives from one of two sources. There can be knowable facts that we, for any reason at all, may not know. This source of uncertainty we will call knowledge uncertainty. Other times the natural variability in the universe may prevent us from knowing a value even when we believe we have sufficient data and facts. This source of uncertainty we will call natural variability.

Uncertainty also occurs at two distinctly different levels of resolution. Macro-level uncertainty involves the social and physical environments in which we function, while micro-level uncertainty occurs at the level of the analyst's desktop. These levels present distinctly different challenges to pest risk management.

This chapter has six primary purposes. First, it distinguishes between the two levels of uncertainty. Second, it distinguishes between the two sources of uncertainty and explains why it may be helpful to do so. Third, it examines quantity uncertainty, the most common source of uncertainty encountered in pest risk analysis, in some detail. Fourth, it introduces seven causes of quantity uncertainty. Fifth, it explores the uncertainty encountered in a pest risk analysis or assessment in a general way. Sixth, it offers a strategy for being intentional about the instrumental uncertainty encountered in a risk management activity.

3.2. Two Levels of Uncertainty

Uncertainty occurs at two levels that affect phytosanitary risk management, the macro- and micro-levels. Macro-level uncertainty informs the environment and value systems in which risk managers operate. Micro-level uncertainty informs the specific problems risk managers work on. Each level of uncertainty is discussed, in turn, in the sections that follow.

3.2.1. Macro-level uncertainty

Pest risk managers operates in an uncertain environment. Decision-making takes place in an environment complicated by growing social complexity and an increasingly rapid pace of change. Society's size, its number of parts, the distinctiveness of those parts, and the variety of specialized social roles that are required to navigate in that society have grown immense over the last century (Tainter, 1996). There are millions of different roles and personalities. Add to that mix our cultural, political, economic, and other differences and our social systems become so complex they may defy understanding. The risk manager's problem-solving challenges are all the more demanding because of this complexity.

All of this complexity is further complicated by the increasingly rapid pace of change in almost every arena. Things, once impossible to conceive, are now made commonplace by scientific breakthroughs. National plant protection organization (NPPO) risk managers must now consider genetically modified organisms (GMOs), minichromosomes, biopesticides and transgenic organisms, for example. Rapid advances in technology that change social values and beliefs as well as the way we live and work, drive much of this macro-level change. Millions of people are open

to more exotic landscaping, they seek variety in fruits and vegetables and look for quality and year-round availability in once seasonal products. Change can sometimes be too rapid to be wholly understood or predicted by human beings.

Social, economic, and technological connectivity around the globe accelerates at a dizzying pace. This has forever changed the ways we communicate and the speed with which we can do that. That change continues in ways that are difficult to forecast. Social movements can be global in their pervasiveness. Global economic interdependence is a fact of life. A simple seed may have a complex global supply chain. Seed products may be developed in multiple research and development and screening sites to evaluate variety adaptation to different climates and ecologies. Once developed, they may be produced in several countries in the northern and southern hemispheres to improve quality and logistics. These seeds may then be processed, treated, enhanced, subjected to quality control, and packaged at a few central sites globally. The seeds may then be bulked, marketed, and distributed globally from a central logistic facility. The same seeds may be re-exported to many destination countries in multiple shipments over time.

Rising government deficits and debts produce relentless pressure on costs in all public decision-making. Patterns of competition are becoming unpredictable. International trade in plants and plant products has become an expectation for much of the global community. Patterns of trade continue to evolve, change, and grow more complex; lucrative markets are lost and replaced by risky new opportunities. Customers demand increasingly diversified plants and plant products. Rapid sequences of new tasks in business and government become more routine with the growing role for one-of-a-kind production. Transportation patterns shift, modes of transport change, priority projects are quickly displaced and budget commitments are unpredictable.

Gradually, we, as a civilization, have become aware of the existence of the unknown unknowns. Despite all we know, the unknown far outweighs the known in many of our most critical decision-making processes. There are some risks, e.g., climate change, that have no narrative closure. There is no ending by which the truth is recovered and the boundaries of the risk can be established for the foreseeable future. Phytosanitary risk managers are likely to continue to grapple with climate change, new pests, new plant products, and uncertain budgets for decades to come.

Public perception has become a palpable force, sometimes an irresistible one. Risks and uncertainty do not exist outside a social context. There would be fewer risks if there were no social and cultural judgments and these judgments are not always grounded in fact. The public, and even experts, sometimes equate the possibility of an undesirable outcome, like establishment of a phytosanitary pest, with the probability of such an outcome. This fuels our fears of uncertainty because it makes conceivable risks seem quite possible. As a consequence, zero tolerance may be the risk manager's recourse far more often than is warranted by the actual risk.

Discovery of a novel coronavirus (COVID-19) in China in 2020 reverberated around the world. Responsibility in this more connected world has become less clear. Who has to prove what and how do we prove anything under conditions of uncertainty? What are the norms of accountability and to whom are we accountable? Who is responsible morally, financially? These kinds of questions plague decision-makers nationally and transnationally.

As a result, phytosanitary risk managers must operate in a world where irreversible consequences, unlimited in time and space, are now possible. Some pest risks could have long latency periods. Global concerns like greenhouse gases, climate change, and sea level rise are clear examples of problems that took decades to emerge and be recognized. The implications for the solutions being formulated by phytosanitary risk managers may likewise take decades to be understood.

Phytosanitary risk management must navigate in this uncertain reality. Social values form, change, and re-form in this environment of macro-level uncertainty. With so much uncertainty, it is difficult to know what values a nation or stakeholder group hold dear at any one point in time. The phytosanitary risk manager functions in this changed and changeable environment. A "culture of uncertainty" is needed to survive in such an environment and risk analysis science provides the foundation for that culture. Decision-making under uncertainty is the NPPO risk manager's macro-level challenge. The NPPO's culture of uncertainty requires a collective will to act responsibly and to be accountable for our efforts to grapple with this fundamental uncertainty in furthering the causes of international trade and plant health. Inevitably, shortfalls will occur despite our best efforts to decide wisely in this environment. This is the environment the private sector and phytosanitary community face and it is this environment that bids phytosanitary risk management to the fore.

3.2.2. Micro-level uncertainty

Despite the macro-level uncertainty of geopolitics, the values of a trading partner, climate change, or IPPC policies these are not the uncertainties that command most of the attention in phytosanitary or production risk management. Neither is the uncertain environment in which NPPOs make decisions the most pragmatic challenge for NPPOs. The data gaps, the imperfect scenarios and models, the incomplete theory, and the inherent variability of the universe present the most immediate challenges to private sector managers, NPPO analysts and decision makers. It is the micro-uncertainty they deal with every day in their jobs that most challenges decision-making.

Rarely is a risk manager so well informed that a decision will yield a guaranteed outcome. There will always be a "pile of things" we know and a "pile of things" we do not know. Weather patterns, wind speeds and directions, origins of plants, prevalence and concentrations of pests, sizes of consignments, future commodity prices, transportation routes, soil conditions, the experience and conscientiousness of employees, they are but part of the myriad of routine details that land in that pile of things we do not know. Together they comprise micro-level uncertainty. Risk assessment enables assessors to sort through that pile of things to better understand the nature, causes and effects of the uncertainties the phytosanitary and production risk management communities face. The nature and cause of the micro-level instrumental uncertainties will dictate the best way to address those uncertainties in decision-making. So, we begin by understanding the nature of the uncertainty. The first distinction to make in the pile of things we do not know is which uncertainties do we need to be most concerned about. Then we proceed to consider the difference between knowledge uncertainty and natural variability.

3.2.3. Instrumental uncertainty

One purpose of risk assessment is to separate what we know from what we do not know. The image of assessors preparing "two piles" of information may be helpful. Into one pile goes all the science and evidence upon which the risk assessment, and subsequently, the risk management decision will be based. Into the other pile go the things that we do not know but would like to know. This is the uncertainty.

Uncertainty can contain many things that are relevant to one or another party to the decision. Not all of these uncertainties are equally important. So, we define a special category of uncertainties, instrumental uncertainty. Instrumental uncertainty can alter the decision that is made or the outcome of that decision. Non-instrumental uncertainty is uncertainty that would not alter a decision or its outcome if it was reduced (Yoe, 2019). The best risk assessors and managers are able to focus on the effects of instrumental uncertainty in decision problems.

3.3. The Nature of Uncertainty

Uncertainty, as used in this handbook, comprises knowledge uncertainty and natural variability. The two terms are defined below. Following the definitions is a discussion of the importance of distinguishing the two.

3.3.1. Knowledge uncertainty and natural variability defined

Knowledge uncertainty is attributed to a lack of knowledge on the part of the observer. Knowledge uncertainty can arise from incomplete theory,

incomplete understanding of a system, modeling limitations and/or limited data. It is reducible in principle, although it may be difficult or expensive to do so. Knowledge uncertainty has been called epistemic uncertainty in the literature (Yoe, 2019).

Examples of knowledge uncertainty abound in pest risk work. The pests associated with a particular product from a specific location may be unknown. The science may be unclear on whether a specific organism even has pest potential. The efficacy of a specific treatment may be uncertain. We may have a poor understanding of the linkages between inputs and outputs in an ecosystem. The costs of implementing a specific risk management measure may not be known. The presence of a pest, the point of origin, and the nature of post-harvest handling of a specific shipment may also be unknown.

Knowledge uncertainty has an important characteristic, there is a true and constant value for each source of knowledge uncertainty. Causes of knowledge uncertainty include outdated, conflicting, incorrect, vague, or missing information. Incorrect methods, faulty models, measurement errors, incorrect assumptions and the like are additional sources of knowledge uncertainty. When we do not know facts that are, conceptually, knowable, we are dealing with knowledge uncertainty. In theory, knowledge can be acquired and uncertainty can be reduced.

Natural variability is uncertainty that originates in the inherent variability of the universe. It includes differences in the attributes of a population element that are due to heterogeneity or diversity in the population. Random processes produce variability in a quantity over time or space or among members of a population. Unpredictable variations in the performance of a physical system can produce natural variability. More information cannot reduce or alter natural variability, but it may improve estimation of the natural variability that exists. A larger sample size can provide a more precise estimate of the standard deviation and, therefore, a better understanding of the natural variability, but the additional data does nothing to reduce the natural variability. Natural variability, also called variability, is known as aleatory uncertainty in the literature.

NPPOs work with many complex natural and manmade systems that are replete with examples of natural variability. The size of a consignment, the number of consignments per year, the number of pests per consignment, the price of a consignment, the time it takes a consignment to arrive, growing conditions and weather, the application coverage of a spray, the effectiveness of culling, the diligence of inspectors, and the randomness in a sample are a few such examples. They are uncertain because they vary naturally. Much of natural variability in the phytosanitary world may be attributed to the randomness of nature because we are dealing with living organisms whose DNA is adapting and mutating unpredictably.

3.3.2. Distinguishing knowledge uncertainty and natural variability

Distinguishing the nature of the uncertainty we face in a decision problem can be strategically important. Knowledge uncertainty, for example, can often be reduced by research, collecting data, consulting an expert, taking a course, and the like. Natural variability, on the other hand, cannot be reduced by gathering more information. This distinction can be important when choosing a strategy for addressing uncertainty in decision-making. Additional resources devoted to reducing knowledge uncertainty may provide more precise information. Additional resources devoted to natural variability cannot reduce natural variability, at best they provide better estimates of the natural variability of the quantity.

Fig. 3.1 sorts that hypothetical pile of things we do not know into natural variability and knowledge uncertainty, which are subsequently sorted into finer delineations of the nature of these "things." One reason for sorting the nature of the uncertainty is that there are specific tools and techniques appropriate for each endpoint in the figure. If you can identify the nature of your uncertainties you are much more likely to find an appropriate and effective way of addressing it than are those who cannot.

Let us illustrate the concepts with a simple example. Suppose we want to know the prevalence of pest E on imports of product D. We begin with an absence of information and that value, though it does exist, is unknown by us and uncertain. Imagine a consignment of 100 crates of

product D with a prevalence of pest E as shown in Table 3.1 where a zero indicates an absence of E and 1 its presence.

Ten of 100 samples have the pest, so we estimate the prevalence at 10%. The 100 crates have enabled us to greatly reduce our knowledge uncertainty. The true prevalence is a fact that is "out there." We did not know it but can now estimate it with some degree of confidence.

Now, think about the next crate of product D that arrives. Does knowing the prevalence rate help us know whether it will be contaminated by pest E or not? Knowing that 10% of all crates are contaminated provides some relative information but it does not really help us know the fate of a random box sitting before us. This example began with a significant knowledge uncertainty that was reduced by gathering data through a sample. In this simplistic example, it may be possible to reduce the knowledge uncertainty further by gathering data from more samples until the unknown prevalence of pest E can be calculated with certainty. Even so, uncertainty due to natural variability would remain—a complete census of prevalence data will not tell us with certainty if the next crate is contaminated or not.

Here is the key idea to understand when dealing with knowledge uncertainty. Ask yourself if a true value exists and if it is a constant. If the answers are yes, you are dealing with knowledge uncertainty. In the example, there is clearly a true prevalence for a given point in time. However, the contamination state of the next randomly selected crate is not a constant. For 10% of the time a crate will be contaminated, 90% of the time it will not be. To complicate the issue, rates of contamination in samples will naturally vary from the population rate.

Interestingly, once a crate is chosen, its contamination value becomes a constant. It is either not contaminated, 0, or contaminated, 1. Until it is chosen, natural variability is the source of the uncertainty, if we know the true prevalence. Although natural variability cannot be reduced by larger samples or more information, it is possible for risk managers to change the existing variability to a more desirable variability by altering the system that has produced the original variability. In other words, if risk managers require the imposition of measures that would, for example, reduce the prevalence to 2% or less,

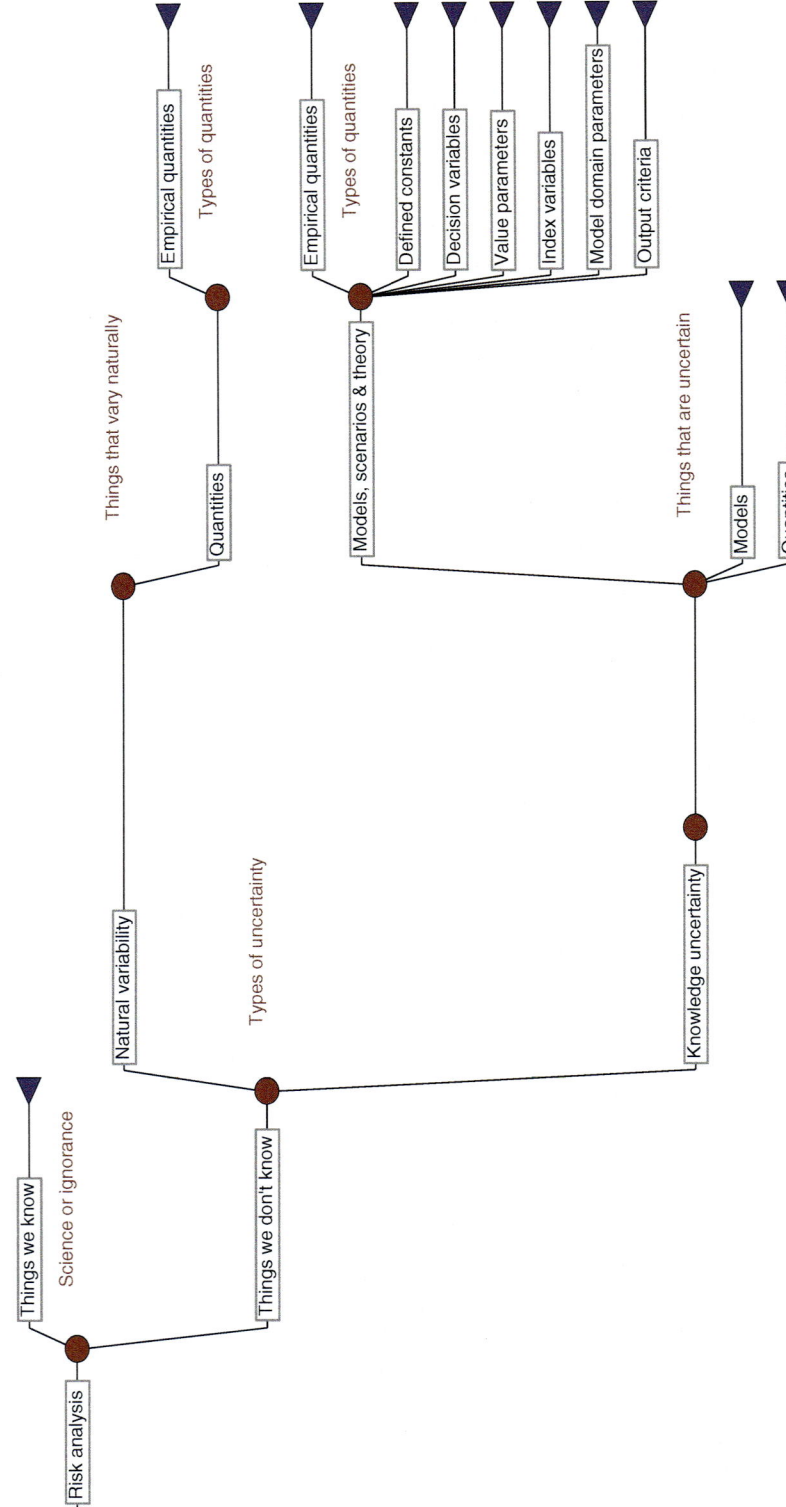

Fig. 3.1. Separating what we know from what we don't know then sorting what we don't know.

there would still be variability but there would be less of it and that would be an improvement.

We can have knowledge uncertainty about any facts, not just numerical values. Is organism X a pest, for example? What does it eat? What kind of environment does it require for reproduction? All of these things will be uncertain until someone gathers the facts.

3.3.3. Why is it important to distinguish the two?

There are at least two practical reasons for distinguishing the nature of the uncertainty you face in a decision problem. First, the risk assessor's choice of the most appropriate tool or technique for addressing uncertainty depends very directly on the nature of the uncertainty. Second, risk managers must make decisions under uncertainty that can, at times, affect decision outcomes. Thus, risk managers may want to know if the uncertainty can be reduced.

Consider Fig. 3.2. It shows two hypothetical outcomes that measure a single decision criterion. Imagine that it represents millions of currency units in damage that could be caused by introduction of a particular pest. The original estimate (solid line) shows considerable variation in potential damages. In more advanced probabilistic risk assessment, it is possible to attribute the variation in damage to quantities that are knowledge uncertainty and natural

Table 3.1. Hypothetical inspection results looking for pest E in product D.

0	0	0	0	0	0	0	0	0	0
0	0	0	0	0	0	0	0	0	1
0	1	0	1	0	0	0	0	0	0
0	0	0	0	0	0	0	0	0	0
0	0	0	0	0	0	1	0	0	0
0	1	0	0	0	0	0	0	1	0
0	0	0	0	0	0	0	0	0	0
0	0	0	0	0	0	0	0	1	0
0	0	1	0	0	0	0	0	0	0
0	0	0	0	0	1	0	0	0	1

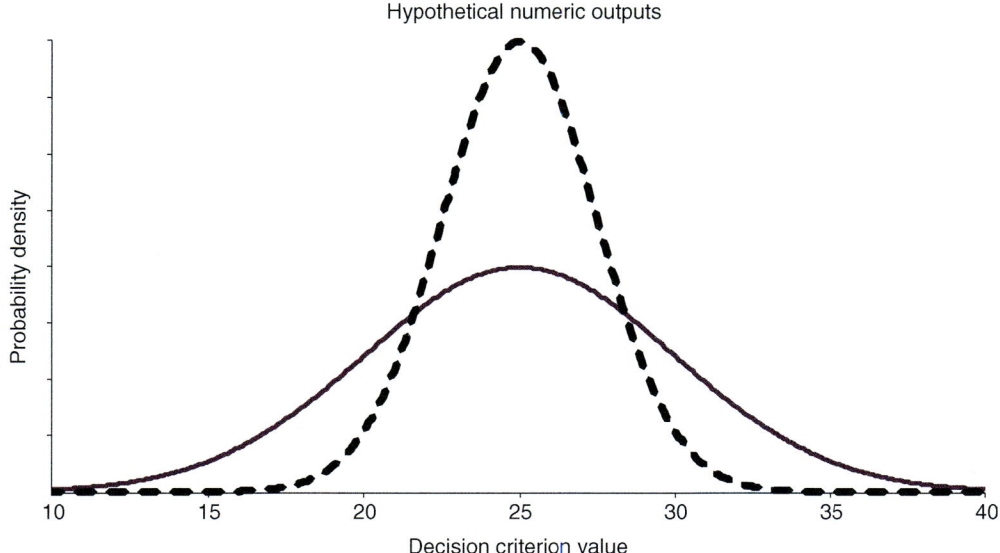

Hypothetical numeric outputs

Fig. 3.2. Two hypothetical distributions displaying uncertainty.

variability. If this information is communicated to risk managers, they can then decide whether or not to devote more resources to further reducing uncertainty.

Imagine the original variation is due to both knowledge uncertainty and natural variability, and decision-makers want the best characterization of damages possible. Ideally, that might be a known single point estimate or at least the tightest distribution possible. Risk assessors may be directed to gather the information needed to reduce the knowledge uncertainty as much as possible. The dashed curve shows the damages after they were re-estimated with the hypothetical addition of data gathered to reduce knowledge uncertainty. The tighter distribution provides a more precise and confident estimate of the damages.

Pleased with this improvement, the risk manager might seek additional reductions to the uncertainty. If the remaining uncertainty is due to natural variability alone there are no options for reducing the variation further. The dashed distribution, in this case, represents the true range of outcomes that is possible given the natural variability in the factors that lead to damages. There is no way to know if improvements in the estimate of the decision criterion are possible unless the assessor has distinguished between the two sources of uncertainty. Although the example offered here is quantitative, the same logic applies to qualitative estimates of decision criteria as well.

3.4. Three Kinds of Knowledge Uncertainty

Fig. 3.3 shows a hypothetical plant pest decision problem that illustrates three areas of knowledge uncertainty NPPOs are likely to encounter in their pest risk work. Knowledge uncertainty is encountered in theories, knowledge and scenarios; models; and quantities. It is rare to completely understand a system we are working with. The theories of our disciplines are still evolving, the scenarios we build to describe a decision problem are not often perfect. The way the components of our pest establishment scenarios relate to one another may be incomplete or even incorrect. Finally, our knowledge of the relevant quantities and facts is often incomplete (Yoe, 2019).

Model uncertainty is a common and persistent source of knowledge uncertainty. All NPPOs

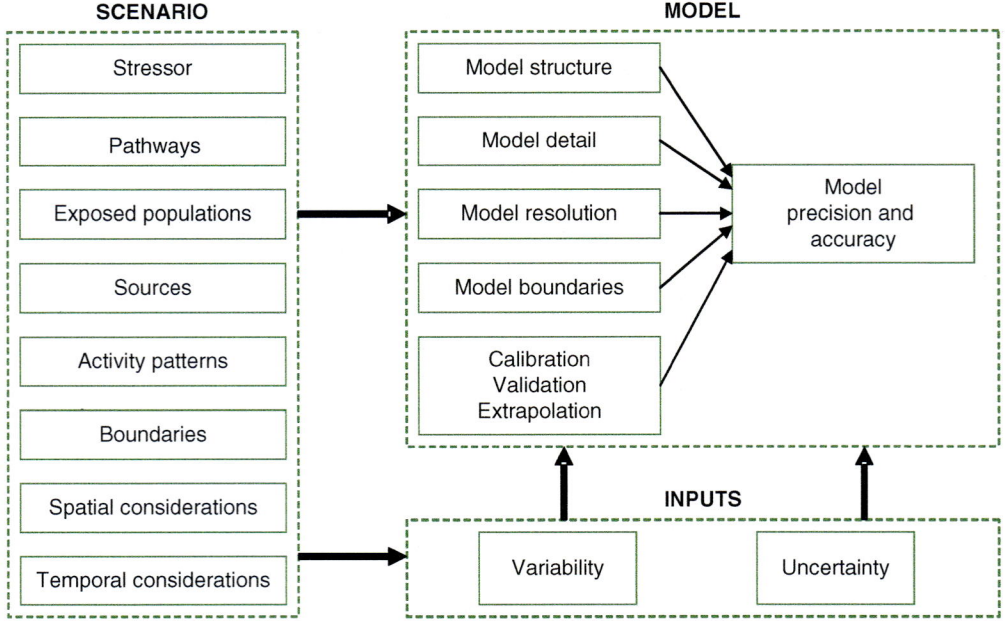

Fig. 3.3. Plant pest decision problem illustrating three major areas of uncertainty.

use more or less flawed models to assess pest risks. The flaws are owed to the simplifications of reality on which we are forced to rely. Every model builder has come face-to-face with knowledge uncertainty in models, but it is rarely addressed in a serious way in most risk analyses. In practice, it is often difficult to even get an imperfect model. It is exceedingly rare for a risk assessment to explore the significance of model uncertainty. The choice of a distribution, or probability model, to use to represent the uncertainty in a variable is another example of model uncertainty.

The most commonly encountered sources of knowledge uncertainty are found in the quantities with which risk assessors work. These include facts about the pests and the commodities they infest, risk elements, and inputs to models as well as the data used for day-to-day risk management.

3.5. Quantity Uncertainty

In pest risk management, the key areas of uncertainty include knowledge gaps about pests and commodities and uncertain quantities used in risk assessments as well as the efficiency of proposed risk management options. A quantity is a piece of information. It can include facts, parameters in a model, parameters of a population, statistics, variables, or data. Morgan and Henrion (1990) offer a taxonomy that is useful for considering the types of quantities that tend to be uncertain. Their classification of uncertain quantities includes:

- empirical quantities
- defined constants
- decision variables
- value parameters
- index variables
- model domain parameters
- outcome criteria.

Let us begin this discussion with an important distinction. Some uncertain quantities have a true or factually objective value while others do not. When a true value does not exist, assessors and decisions-makers should seek the best or most appropriate value. This will reflect some degree of subjective judgment. There is no reason to seek the true value for a quantity that does not

have one. There is an important distinction to make here—the search for a true value is an objective one while the search for a best value is subjective.

A true value can generally be measured, estimated or looked up. Best or most appropriate values cannot be. They are more appropriately varied systematically in some sort of sensitivity analysis. The effect of different values on model outputs, decision criteria and the like can then be examined and evaluated. The tools one uses for uncertainty analysis vary for quantities with true values and those with best values. Each Morgan and Henrion quantity is introduced below. Expanded discussion can be found in the original source as well as Yoe (1996, 1997, 2000, 2019).

True values

The prevalence of a pest on a product from a specific area, the number of pests associated with a product, the mortality rate for a specific product treatment, the number of days a fruit is susceptible to infection, spread potential, the size of a shipment, the sensitivity of a detection test, the number of contaminated fruit in a shipment, the average annual price of a product, the growing season for a product, these are all quantities that have a true value.

Best values

The best sample size, the temperature for shipping a product, the importance of year-round availability of a product, the money to be allocated for interception of a specific pest, an acceptable probability of introduction, a planning horizon, a mitigation goal, the ideal amount of international trade, not one of these quantities has a true value. An appropriate or best value must be determined by some degree of subjective judgment.

3.5.1. Empirical quantities

Empirical quantities are the most common type encountered in risk assessments. They include things that can be measured or counted and they always have true values. This includes parameters, e.g., ppm and pH of a chemical dip, rainfall, temperature, reproduction rates, number of native plant species, distances, times, sizes, statistics, and every kind of count imaginable.

An empirical quantity that has an exact value that is unknown but measurable in principle is an example of knowledge uncertainty. An empirical value that varies from place to place, time to time, and event to event is an example of natural variability. There is a full range of tools and techniques that can be used to characterize this quantity from narrative descriptions through probabilistic methods.

3.5.2. Defined constant

A defined constant has a true value that is fixed by definition. When a defined constant is unknown, the solution is to look them up. Examples include the number of square feet in an acre or square meters in a hectare.

3.5.3. Decision variables

Decision variables are quantities that a risk assessor or manager must choose or decide. They include such things as an acceptable level of protection, what constitutes a pest-free area, confidence intervals on a sampling plan, a mitigation goal, a tolerable level of risk, assumptions made about values in a PRA, and so on. Decision variables have no true value but decision-makers exercise direct control over them. They are subjectively determined. Uncertainty about a decision variable is most appropriately addressed through parametric variation and sensitivity analysis.

3.5.4. Value parameters

Value parameters express an important aspect of the social values that emerge from macro-levels of uncertainty in the phytosanitary risk manager's environment. These values like discount rates, the value of trade, and the weights used to tradeoff conflicting values represent aspects of the decision makers' preferences and judgments. These quantities do not have true values. They are subjective assessments of social values. Parametric variation and sensitivity analysis are the most common means of addressing uncertainty about these quantities.

3.5.5. Index variables

Index variables most often identify elements of a model, like a point in time or a location within a spatial domain. They do not usually have true values, so the choice of an index variable value is usually determined subjectively. Examples of index values include the number of years in the period of analysis, the location of a representative grid cell in a spatial model, the position of an object in a model where a sequence of events is initiated, e.g., the location of a product or pest on a pathway. Index variable uncertainty is most often addressed through parametric variation and sensitivity analysis.

3.5.6. Model domain parameters

Model domain parameters specify and define the scope of the systems considered in a decision problem. This could include definitions of origin or destination areas, PRA areas, impact areas, tributary areas to a port, pest-free regions, geographic extent of a quarantine pest, climate or growing condition boundaries, and so on. These parameters often describe the geographic, temporal, and conceptual boundaries, i.e., domains, of a PRA and define the resolution of its inputs, e.g., spatially or temporally, and outputs, and they may or may not have true values. They are usually based on a decision-maker's judgments about the level of resolution needed to assess risks adequately or to otherwise determine the model domain.

3.5.7. Outcome criteria

Outcome criteria are output values. These include things like the probability of pest introduction or the consequences of that introduction, benefit–cost ratios, cost-estimates and similar variables that are outputs of models and calculations. Their values are derived from the models used to estimate them and the quality of the model's input quantities. Characterizing the uncertainty about output criteria is the risk assessor's responsibility while addressing that uncertainty in decision-making is the risk manager's responsibility.

3.5.8. Qualitative assessment of uncertain quantities

Many of the types of quantities described above may be considered in qualitative risk assessment. In fact, persistent uncertainty is sometimes a reason for qualitative risk assessment; the data are not good enough to support quantitative risk assessment. In these instances, uncertainty may be characterized in different ways, but it should still be characterized. See Yoe (2019) for examples of how to characterize uncertainty quantitatively and qualitatively.

3.6. Causes of Uncertainty in Empirical Quantities

Quantities with true values that are subject to natural variability and knowledge uncertainty are likely to be the most common uncertain quantities for PRAs. The vast majority of these quantities are empirical quantities, the focus of this section. The discussion that follows would, however, be applicable to any quantity that has a true value. The work of Morgan and Henrion (1990) once again provides the structure for this discussion.

Empirical quantities are most often counted, measured, calculated, estimated, or described. There may be little or no knowledge uncertainty about the true value of an empirical quantity when good measurement data are available. Knowledge uncertainty may also be absolute and permanent because the data have been lost to history (we will never know the peak daily flow on the Tiber River for the month of August, 1476 in Rome, for example— nonetheless, that quantity has a true value). Even when knowledge uncertainty has been eliminated, natural variability may remain to be addressed. We need to know the cause(s) of uncertainty if we are to appropriately address it in a risk assessment.

3.6.1. Random error and statistical variation

Sample data are frequently used to estimate empirical quantities. Only valid probability samples can produce unbiased parameter estimates, but even these are subject to random error. Measurements of physical quantities are frequently inexact. A wide array of techniques and tools for quantifying this kind of uncertainty are found in classical statistics. They include estimators, standard deviations, confidence intervals, hypothesis testing, sampling theory, and probabilistic methods.

3.6.2. Systematic error and subjective judgment

When the measurement instrument, the experiment, or the observer are biased, systematic errors can arise. Both instruments and people can be imprecisely calibrated, resulting in this bias. If an inappropriate sample design is used for inspection or a consignment is improperly labeled, the solution is better calibration of the sample design or fixing the consignment paperwork. If an inspector tends to over- or under-estimate values the inspector needs to be calibrated or recalibrated or a more objective means of measurement is needed. The risk assessor's challenge is to reduce systematic error to a minimum. The preferred solution is to avoid, minimize, or correct the bias in instrumentation and personnel. Using risk-based sampling techniques to collect data can eliminate or at least minimize many forms of bias.

It is difficult to correct for unknown or merely suspected biases. Bias in subjective human estimates of unknown quantities is a topic covered extensively in the literature—see for example O'Hagan *et al.* (2006) or Yoe (2019).

3.6.3. Linguistic imprecision

Communication remains a critical challenge in risk work. People can use the same words to mean different things or they can use different words to mean the same things. This makes effective communication about complex risks especially challenging. What does it mean and what do people understand when we say infestation occurs frequently or the risk of infestation is unlikely?

Monte Carlo simulations are clearly useless for this source of uncertainty where the obvious solution is to carefully specify all terms and relationships and to clarify all language that is used. Using quantitative terms instead of qualitative terms whenever possible can also help.

3.6.4. Natural variability

Natural variability is a source of uncertainty that warrants repeating. Quantities may vary over space or time; they may vary from one individual or object in a population to another. A chemical dip kills some pests but not others. A given dose of radiation yields different mortality rates. Variability is inherent in the system of factors that produces the characteristic of a population we are interested in measuring. Frequency distributions, when available, can be used to characterize the values of interest. Other probabilistic methods may be used as well.

3.6.5. Randomness and unpredictability

Inherent randomness is irreducible in principle. This source of uncertainty gives rise to events that are, in practice, not predictable at the current time. Examples include where the next outbreak of citrus canker will occur, when inspectors will fail to inspect a consignment, or how the prevalence of a pest will change from consignment to consignment. Events like these can be treated as random processes. A full range of methods from narrative descriptions through probabilistic methods can be used to address this uncertainty.

It is worth noting that phenomena that appear random to one assessor may be well understood by a subject matter expert. Strong interdisciplinary risk assessment teams, centers of expertise, and peer involvement including peer review are processes that can provide a reasonable hedge against this sort of uncertainty.

3.6.6. Disagreement

Experts do not always agree on matters of uncertainty. Neither do organizations. There have been many spirited discussions over which, if

any, risk management measures to use during bilateral trade discussions. Widely disparate views of a problem are not uncommon to phytosanitary issues. Disagreements may arise from different analytical interpretations of the same data. There is also a real possibility of conscious or unconscious motivational bias.

Negotiation and other issue resolution techniques can be effective in resolving disagreements. Allowing disagreements to coexist is always an option in a process controlled by the importing country. Sensitivity analysis can be used to explore the effect of the different arguments on relevant decision criteria.

3.6.7. Approximation

In some circumstances, analysts may only be able to approximate a quantity because of its complexity or the extensive uncertainty that attends it. For example, what is the probability of pest introduction or establishment with a new risk management strategy in effect? Quantities like these can only be approximated because of model uncertainty and "spotty data." Methods for dealing with this source of uncertainty will depend on the specific limitations of the approximation.

3.7. Uncertainty in Pest Risk Analysis

It is not unusual for pest risk analysts to face the dilemma of having to make decisions about pests where very little information is published. It is quite common, in fact, it is the norm for risk analysts to lack information they would like to have. The uncertainty can take many forms. There may be data gaps in the information about a pest. The available evidence might indicate but not prove potentially high impacts by the pest. The provenance of the information may be hazy and its reliability questionable. Risk assessors must find a way to finesse these and like issues, while carefully communicating the significant implications of uncertainty to risk managers who will rely on this information for decision-making. Then the risk manager must consider all of this uncertainty when making a decision about how to handle a potential pest.

> Does the lack of information make the situation appear riskier than it is? Should the pest be considered high risk because there are so many unknowns? The SPS Agreement requires that a risk estimate be based on the available evidence, not on the uncertainty associated with the evidence.
>
> (Griffin, 2012)

Why does the phytosanitary community need to discuss uncertainty? The SPS Agreement requires measures to be based on scientific evidence. It also says SPS measures should be based on available information. The Agreement does not specify the minimum amount or quality of information needed to make a judgment about SPS measures. It provides no guidance for judging the quality, quantity, or relevance of evidence used in pest risk analysis. It is not possible to remain transparent about a risk without discussing what is known and what is not known. Risk analysis is decision-making under uncertainty and it is simply impossible to do risk analysis when ignorant of the significant uncertainties in a situation.

Fig. 2.6 (see p. 34) showed the IPPC pest risk analysis process comprising three tasks: initiation, risk assessment, and risk management. Uncertainty exists in each of these tasks, not just in risk assessment. In the initiation task there may be insufficient information to be able to identify all the relevant pests and a dearth of information on some of these pests may make it difficult to determine if an organism is or is not a pest. Likewise, there may be insufficient information to establish the PRA area, especially when the existence of a pest cannot be verified.

To begin to think holistically about the uncertainty in risk assessment, consider Fig. 3.4, which displays the conceptual nature of that uncertainty. The plant risk consists of consequences of introduction and the accompanying probability that those consequences will occur, i.e., the probability of introduction. There can be uncertainty about the consequences and different uncertainty about the probability of those consequences. There can be knowledge uncertainty about the consequences, i.e., the nature of the consequences may be unknown, and/or there can be natural variability in those consequences, e.g., their extent, duration and magnitude may all vary. The precise sequence of events necessary to result in establishment may be unknown (knowledge uncertainty in probability) or the circumstances of those events (i.e. the probability) may vary naturally from shipment to shipment, season to season, and so on (natural variability in probability).

It is the risk assessor's responsibility to address the uncertainty in the relevant evidence for a particular pest in the assessment. It is, then, the risk manager's responsibility to consider that uncertainty when making a risk management decision. New uncertainties may come to the fore in risk management. Risk managers often face substantial uncertainty about the efficacy of the risk management options they consider for adoption. Which measures will work best? How well will they work? Will there be any adverse impacts associated with the measures imposed?

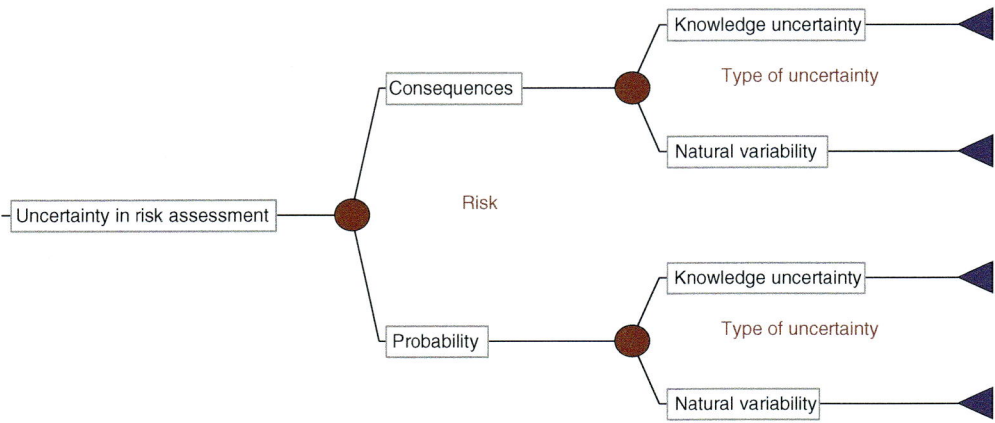

Fig. 3.4. Schematic of the different kinds of uncertainty that can be encountered in a risk assessment.

"Thousands of potential pests, hundreds of potential host commodities, and an unlimited number of possible trade scenarios contribute to uncertainty in PRA. Risk assessors need to recognize the source, type, and degree of uncertainty to completely understand the risk and what can be done to affect it. To be clear in an SPS context, risk analysis should separate the evidence from the uncertainty and characterize the degree to which uncertainty influences the judgment of risk."

(Zlotina, 2015)

3.8. Being Intentional About Instrumental Uncertainty

The reason to understand the nature of the uncertainty attending a risk issue is to be able to take effective steps to improve decision-making by intentionally addressing the uncertainty. Here are ten steps to take (Yoe, 2019) to become intentional about uncertainty in decision-making:

1. Recognize that uncertainty exists in the risk issues.
2. Specifically identify the relevant uncertainty and the sources of that uncertainty.
3. Identify those uncertainties that are instrumental to the risk issue(s). These are the uncertainties that have the potential to have a significant effect on the decision choice or the outcome of the decision.
4. Acknowledge the instrumental uncertainty and make stakeholders aware of its existence.
5. Choose appropriate tools and techniques to address each instrumental uncertainty.
6. Complete the analysis incorporating these tools and techniques.
7. Understand the results of the analysis.
8. Identify any options for further reducing the remaining instrumental uncertainty.
9. Convey the results, the significance of the instrumental uncertainty, and any options for reducing instrumental uncertainty to decision-makers.
10. Consider the effects of instrumental uncertainty on decision options and decision outcomes.

Uncertainty will always exist in a PRA, so being intentional in handling risk begins by recognizing it when it exists. Experienced professionals routinely underestimate the things they do not know and they routinely overestimate the quality of their data and the extent of their experience. The confidence of experts, like the rest of us, is often based as much or more on what they believe to be true as it is on what they actually know. Biases, mindsets and beliefs can obscure the existence of uncertainty. Fortunately, experts are often correct in their intuitive judgments but this strongly reinforces those biases. A sound starting point for all risk work is to recognize the existence of uncertainty. Risk assessors and managers, alike, are well served by frequently asking, "What is your evidence?" This can be an effective way to verify assertions, assumptions and beliefs. When evidence can be produced, it is reassuring. When it cannot be, it is enlightening.

Once the existence of uncertainty is recognized it is necessary to specifically identify what is known with certainty and what is not. It is time to repeat an important semantic distinction about things that are uncertain. Uncertainty is relevant if it could impact a decision in at least a subtle way. Relevant uncertainty may make one option less appealing than another or it might make a particular stakeholder uneasy. Relevant uncertainty can be categorized as instrumental or non-instrumental. Instrumental uncertainty can alter the decision that is made or it could change the outcome once a decision is made. Non-instrumental uncertainty refers to uncertainty that would not alter a decision if it was reduced. Non-instrumental holes in our data or gaps in information may be relevant in some way, but they would not affect the decision to be made or the outcome of the decision if they were filled. Good risk assessors and risk managers are able to focus on identifying, reducing or otherwise characterizing the instrumental uncertainty encountered in decision-making. The assessor's job is to identify all of the instrumental uncertainty. These might include scenarios, theories and knowledge, models, or quantities, which, if not certain, could influence the decision risk managers would make or the outcomes of their decision. In a PRA, any source of uncertainty that could affect the estimation of the probability of introduction, the consequences of introduction, or the efficacy of risk management options, is an instrumental uncertainty.

Some people may need to know about the instrumental uncertainties even before they are addressed in risk assessment and decision-making. If a trade partner lacks information about the harm a pest can cause, they need to be informed, for example.

Choosing a tool or technique that is appropriate for each instrumental uncertainty is an important analytical step. Some uncertainty can be addressed qualitatively, other uncertainty may warrant a probabilistic treatment. Risk assessors have many tools and techniques that lie between and beyond these two approaches. Pulling together the many and disparate approaches for addressing uncertainty to complete the uncertainty analyses in order to characterize the risks associated with the decision problem is the assessor's responsibility. Analysts should spend sufficient time with the results of their analyses to understand them and the ways that instrumental uncertainty could affect the decision or its outcome.

Risk assessors are responsible for addressing the instrumental uncertainties encountered in their PRA. Some simple options for doing this include narrative descriptions of the uncertainty, clarification of ambiguous language, negotiating differences of opinion, and confidence ratings for assessment results. Options for more quantitative analyses include parametric variation, bounding uncertain values, using sensitivity analysis, or probabilistic risk assessment.

In best practice, assessors will distinguish the effects of knowledge uncertainty from the effects of natural variability. This enables assessors to potentially identify options for further reducing the instrumental uncertainty.

Effectively conveying the results and the significance of the uncertainty, as well as any options for reducing instrumental uncertainty, to decision-makers is an area where there is a lot of progress to be made. An analysis of the instrumental uncertainty has failed if there is no success at this step. Risk assessors should also convey their confidence in the results they present.

The risk manager is responsible for addressing instrumental uncertainty in their decision-making. Adaptive management strategies may be implemented when the uncertainty is difficult or impossible to reduce and the consequence of making a wrong decision is a serious concern. Adaptive management strategies can reduce instrumental uncertainties, through research, experiments, test plots, trial and error, and so on, to produce information that better informs managers about the risks and the efficacy of the risk management options before they are irreversibly implemented.

The precautionary principle has been favored as an approach to decision-making under uncertainty by some, when the uncertainty is

---

Four tiers of confidence

High confidence decisions are those characterized by:

- high agreement and robust evidence
- medium agreement and robust evidence
- high agreement and medium evidence.

Medium confidence decisions are those characterized by:

- medium agreement and medium evidence
- low agreement and robust evidence
- high agreement and limited evidence.

Low confidence decisions are those characterized by:

- medium agreement and limited evidence
- low agreement and medium evidence.

No confidence decisions are those characterized by:

- low agreement and limited evidence.

Source: IPCC, 2010

great and could have significant consequences. A number of criteria have been developed for choosing from among alternative risk management measures under uncertainty (Yoe, 2019). They include the:

- maximax criterion—choosing the option with the best upside payoff
- maximin criterion—choosing the option with the best downside payoff
- Laplace criterion—choosing the option based on expected value payoff
- Hurwicz criterion—choosing an option based on a composite score derived from preference weights assigned to selected values, e.g., the maximum and minimum
- regret (minimax) criterion—choosing the option that minimizes the maximum regret associated with each choice.

Assessors, managers, and communicators must all be intentional in dealing explicitly with uncertainty, especially instrumental uncertainty, when carrying out their responsibilities. Good risk analysis demands no less.

3.9. Summary and Look Forward

Here are five things to remember from this chapter.

1. There are two levels of uncertainty: macro-level and micro-level uncertainty.
2. Uncertainty comprises knowledge uncertainty and natural variability.
3. Knowledge uncertainty can appear in scenarios, theories and knowledge, models, or quantities.
4. The thing that is uncertain and the cause of that uncertainty will largely determine which tools and techniques are best suited for addressing the uncertainty.
5. Risk assessors are duty bound to identify instrumental uncertainty and risk managers are duty bound to consider the instrumental uncertainty in a PRA when making decisions.

Now that we have covered the primary language and concepts that risk managers need to understand, it is time to begin to consider how risk managers work. Perhaps the single most important aspect of that work is the risk analysis team. The next chapter introduces the need for a team approach to risk analysis.

3.10. References

IPCC (2010) Guidance Note for Lead Authors of the IPCC Fifth Assessment Report on Consistent Treatment of Uncertainties. IPCC Cross-Working Group Meeting on Consistent Treatment of Uncertainties. Jasper Ridge, California, USA, 6-7 July. Available at: https://www.ipcc.ch/site/assets/uploads/2018/05/uncertainty-guidance-note.pdf (accessed August 11, 2020).

Griffin, R.L. in Devorshak, C. (2012) *Pest Risk Analysis: Concepts and Application*, CAB International, Wallingford, UK.

Morgan, M.G. and Henrion, M. (1990) *Uncertainty: A Guide to Dealing With Uncertainty in Quantitative Risk and Policy Analysis*. Cambridge University Press, Cambridge, UK.

O'Hagan, A., Buck, C.E., Daneshkhah, A., Eiser, R., Garthwaite, P.H. *et al.* (2006) *Uncertain Judgements: Eliciting Experts' Probabilities*. John Wiley & Sons, West Sussex, UK.

Tainter, J.A. (1996) *Getting Down to Earth: Practical Applications of Ecological Economics*. Island Press, Washington, District of Columbia, USA.

Yoe, C.E. (1996) *An Introduction to Risk and Uncertainty in the Evaluation of Environmental Investments*. Institute for Water Resources, Alexandria, Virginia, USA.

Yoe, C.E. (1997) *Risk and Uncertainty Analysis Procedures for the Evaluation of Environmental Outputs*. Institute for Water Resources, Alexandria, Virginia, USA.

Yoe, C.E. (2000) *Risk Analysis Framework for Cost Estimation*. Institute for Water Resources, Alexandria, Virginia, USA.

Yoe, C.E. (2019) *Principles of Risk Analysis Decision Making Under Uncertainty*, 2nd edn. CRC Press, Boca Raton, Florida, USA.

4

Risk Analysis Team

Alone we can do so little, together we can do so much.

Helen Keller

4.1. Introduction

Pest risk analysis is, ideally, a team effort. No one person knows enough to conduct a thorough pest risk assessment and to manage the risk efficaciously all on their own. Pest risk analysis requires managers and assessors to work together. They need to communicate, coordinate, cooperate, and collaborate early and often throughout a pest risk analysis. Ideally, that team would include representatives of the national plant protection organization (NPPO) as well as the producer and importer. The practice is a long way from this ideal. Consequently, this chapter focuses primarily on the NPPO.

In some countries, there may be one person to do the risk assessment and that same person may be the NPPO risk manager. Such situations are born of necessity due to limited resources. Even when a single person plays all the roles in a pest risk analysis there are multiple roles for that person to play. This chapter may be of limited value in situations where the option for teams does not exist because it presents the argument for a pest risk analysis team. It begins by considering the need for a team. It summarizes a traditional interaction between phytosanitary risk managers and risk assessors and follows this up with a description of an ideal set of interactions, an idea expanded upon in Chapter 16, where producers and importers re-enter the discussion. The chapter considers the qualities of a good team and concludes with a discussion of the importance of vertical team alignment in pest risk management.

4.2. Need for a Team

In many NPPOs risk assessors and risk managers function with near autonomy. Risk assessments are done in isolation from risk managers and risk management decisions are made independently of risk assessors. This is neither efficient nor efficacious. An uncoordinated group is not a team. A coordinated group is not a team. The first best step for correcting this situation is to form a risk analysis team. A team is a group of people working together to produce a specific result or outcome, like protecting plant life while promoting safe international trade.

A team of people with diverse expertise, talents, and skill sets can deliver better solutions than any individual could ever hope to do. The team's access to a wider range of skills and knowledge and a deeper well of energy enables them to solve problems faster than a collection of individuals can.

One of the main benefits of a team approach is the ability to share ideas among the group. Teams can develop more thoughtful ideas, because each person brings different experiences, knowledge, and information to the table. When there are several possible approaches for managing a pest risk, the varied perspectives of a team enable members to contribute a broader array of pros and cons of the different approaches, promising a higher quality outcome that is more thoughtful, efficient, and effective.

Another key advantage of teamwork is that things get done faster. Teamwork enables people to divide up the roles and tasks in ways that more thoroughly exploit the strengths of the team members. They can often complete tasks and activities in less time because they draw on the efforts of many contributors. The workload feels more manageable to the individuals on the team once it is shared than if they had to do everything themselves. This is especially helpful for PRA.

Teamwork between the risk assessors and risk managers is essential to effective risk management. Individual team members learn about the problem and its related issues in a team environment. Risk managers better understand the risk assessment and assessors better understand the information needs of the managers in that team environment. Well-functioning teams create camaraderie among assessors and between assessors and managers. People will often go to extreme lengths in effort when they know that they can rely on the support and encouragement of the team. The significance of this simple fact for achieving desired risk management outcomes should never be underestimated.

Ideally, a phytosanitary risk analysis team consists of three functions: risk assessment, risk management, and risk communication. *Risk assessment* requires one or more scientific experts, more complex risk assessments require more people. *Risk management* is one or more decision-makers—different nations have different risk management practices. *Risk communication* is one or more communication specialists. At a bare minimum, the team will be a risk assessor and a risk manager, with the risk communication covered by the two of them. More likely the risk assessment team will comprise several scientific experts and risk management consists of several decision-makers. Risk communication specialists are not always required for a PRA activity but it is always desirable to have access

to one. When a specialist is not available or not required, it is the risk manager's responsibility to see that risk communication is effectively accomplished.

Evidence-based pest risk analysis requires subject matter experts and decision-makers. A multidisciplinary team is a group of diverse experts who tackle a pest risk analysis together. The integration of their various disciplines is never a focus of the effort, and the work of such teams often has the flavor of a series of more or less well-connected analyses that hopefully add up to something meaningful to someone somewhere. Assembling the right disciplines is regarded as the hurdle to be cleared. Although a multidisciplinary team is a substantial improvement over an uncoordinated group's effort, little effort is expended to integrate the various disciplines.

Multidisciplinary teams are limited by the fact that disciplines have, over time, developed their own specific and occasionally peculiar way of looking at the world. Effective solutions to pest and trade issues require a better integrated view of the problems and their solutions. An interdisciplinary team also begins with a rich diversity of expertise. It differs, however, by its intentional effort to integrate the various disciplines in a way that better recognizes the big picture goals of a pest risk analysis. An interdisciplinary team crosses traditional boundaries between academic disciplines or schools of thought and weaves a more holistic viewpoint of risk issues among its members. Interdisciplinary solutions tend to be more responsive to social needs than multidisciplinary solutions.

The experts in an interdisciplinary team come to understand the language and basic concepts of the other disciplines and their perspectives. This enables entomologists to appreciate and consider the viewpoints of economists and environmentalists. This can lead entomologists to more economical and environment-friendly measures. Likewise, economists understand the importance of non-monetized damages that can be caused by pests. This may lead them to help risk managers to consider cost-effectiveness and incremental cost analysis of non-monetized pest effects.

A transdisciplinary team not only crosses traditional boundaries among disciplines, it erases those boundaries and integrates knowledge at the edges of disciplines. A transdisciplinary approach bridges many disciplines at once and develops a synergy among disciplines that can create new insights and ways of looking at problems and their solutions that transcend the abilities of traditionally-bounded disciplines. As new needs and new disciplines emerge, transdisciplinary approaches that bridge and accommodate these changes are increasingly valuable. The most exciting disciplines now are those that integrate the traditional ones. Transdisciplinary knowledge is greater than the sum of all the disciplines that comprise it.

The very best teams are transdisciplinary. A pest risk analysis team should strive to be at least interdisciplinary. What is absolutely certain is that no pest risk analysis team can afford to have separated and functionally independent risk assessors and risk managers. Managers and assessors must function as a team to be as efficient and efficacious as possible.

4.2.1. Staffing

Different disciplines are the key to interdisciplinary and multidisciplinary teams. NPPOs should be staffed with the disciplines they require. An informed consideration of the entire international trade risk continuum requires a great diversity of science and business disciplines. Traditionally strong in entomology and the plant and soil sciences, phytosanitary risk management organizations could use more statisticians, probability experts, model-builders and economists. These additional disciplines can provide NPPOs with the skills needed to conduct more rigorous analysis, build more quantitative models, conduct more economic analyses and thereby to make more rational decisions. These are disciplines that can also help develop the phytosanitary risk management community of practice's ability to deal effectively with uncertainty in both analysis and decision-making.

4.3. Typical Process

The nature of the interaction between risk assessors and risk managers will differ according to the way national or international organizations are structured. There has been no systematic survey of the methods used by phytosanitary risk managers, so the best evidence is anecdotal and, anecdotally, phytosanitary risk management appears to follow one of the early risk analysis models that argued for the complete separation of the risk assessment and risk management tasks. Early efforts to separate the science from the decision-making in risk analysis were undertaken out of an abundance of caution to insulate analysts from interference by decision-makers, who would later be called risk managers. *Risk assessment in the federal government: managing the process* (National Research Council, 1983), an early study, is well known for making a distinction between risk assessment and risk management. It says, "The importance of distinguishing between risk assessment and risk management does not imply that they should be isolated from each other; in practice they interact, and communication in both directions is desirable and should not be disrupted".

Early experience proved the separation of risk assessment and risk management to be ill-advised. By 1996 the National Research Council was saying, "We believe that acceptance of too strict a separation between risk assessment

---

Every individual brings a unique set of personal qualities to the team

Expertise: the technical background people began learning in school and now practice professionally.
Talent: skills and abilities not necessarily related to technical expertise, such as leadership, writing and speaking.
Affiliation: the groups people are formally associated with and represent including employers, political, religious, fraternal and other groups they belong to.
Personal values: what we each believe is right and wrong, good and bad, the answer or not the answer.
Personality: the essence of what makes each of us individuals. Are we introverts or extroverts, logical or intuitive, just how do we approach each day?

(Yoe, 2012)

Potentially useful team disciplines

Agriculture, agronomy, biology, biotechnology, botany, chemistry, computer scientists, consumer science, cytology, ecology, economics, engineering, entomology, ethology, genetics, immunology, legal, lepidoptery, meteorology, model-builders, mycology, probability, risk, statistics.

---

and risk management has contributed to an unworkably narrow view of risk characterization." The international community was concurring: "there is a need for frequent interaction between risk managers and risk assessors in order to arrive at effective risk management decisions. Active interaction is necessary to ensure that the assessment will meet the needs and answer the concerns of the risk manager ..." (United Nations/WHO 2000).

The modern view of these two separate functions is that risk assessment serves the decision-making needs of risk managers. The separation of the risk assessment and risk management tasks in phytosanitary risk analysis is, in many cases, too severe. This separation can be by organization as well as by function. The result in some cases is that separation has become *a barrier to* best practice risk analysis rather than *a guarantee of* best practice. Risk assessment and risk management have become so separated that there is insufficient communication, coordination, cooperation, and collaboration between the parties. This has, at times, led to an antagonistic relationship between assessors and managers within an NPPO. That is counterproductive. These two parties need substantial interaction and that interaction, the subject of Chapter 11, is best grounded in teamwork.

4.4. Where Are the Teams?

There are any number of conceptual models for organizing an NPPO and the risk assessment and risk management responsibilities, personnel, and tasks may be structured in any number of ways. Fig. 4.1 shows four conceptual structures for risk management and risk assessment to suggest the ways that organizational structures can enhance or challenge the use of a pest risk analysis team. Risk communication and other responsibilities of

an NPPO are omitted to sharpen the focus on management and assessment. The upper left of the figure suggests an organization that cleanly separates the assessment and management functions. When the separation is physical as well as conceptual, this can challenge the PRA team model. Opportunities for interaction must be provided for in this structure. The upper right shows the risk assessment task as subservient to the risk management task. There is built in communication in such a structure but power balances, roles and responsibilities have to be carefully defined.

The bottom of the figure shows NPPOs organized not around PRA responsibilities per se but around pest specialty or geographic region. For example, it is not difficult to imagine an NPPO organized around continents or other geographic regions. The lower left shows that each office has its own risk management and risk assessment personnel but they remain separated within those offices. The lower right suggests that risk managers and assessors are integrated in the same office. This sort of structure may best support and encourage the team model. In all of these situations the integrity of the risk assessment process must be safeguarded by the risk assessment policy. There is no one best way to organize the PRA resources of an NPPO.

4.5. An Ideal NPPO Process

Risk managers and risk assessors need to communicate, coordinate, cooperate, and collaborate from the start of a risk assessment through risk management. Although an explicit risk management process has not yet been introduced, we do have the generic five-step process of Chapter 2 reproduced as Fig. 2.5 (see p. 32). There are five distinct parts of the risk management process and there are ample opportunities for risk assessors and risk managers to function as a team in each of those steps.

4.5.1. Risk assessment policy

The functional separation of risk assessors and risk managers is assured and protected by a risk assessment policy. Risk managers are responsible

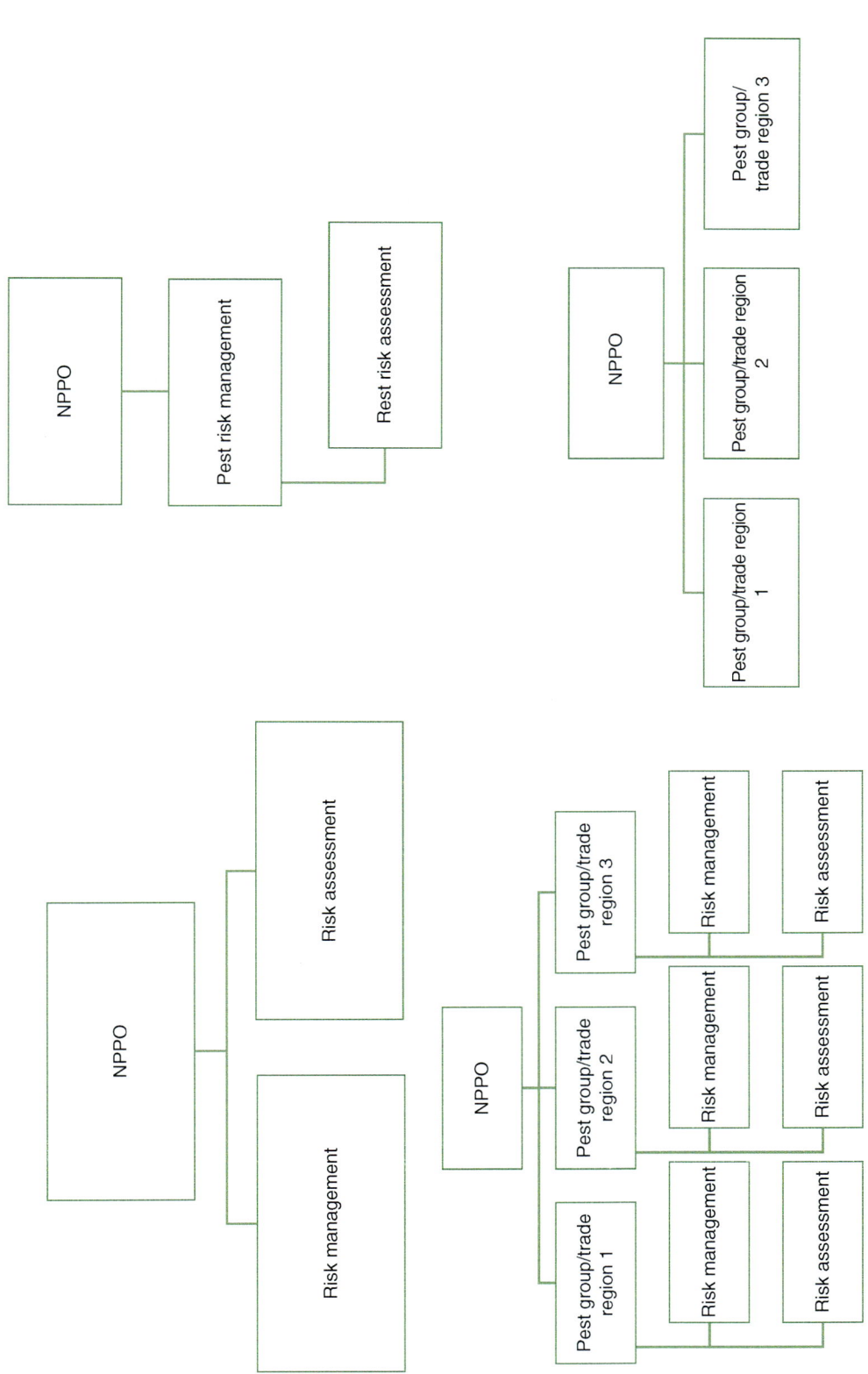

Fig. 4.1. Four conceptual risk analysis structures.

for establishing a risk assessment policy for the NPPO. This happens in two parts. The first part establishes the risk assessment policy or standard operating procedure (SOP) for the organization. This is a one-time task that applies to all risk assessments conducted by the office, which may be revised and updated from time to time. It includes such things as defining the roles and responsibilities of all members of the risk analysis team, defining the manner in which evidence is gathered and evaluated, identifying touchpoints between assessors and managers, establishing milestones to monitor and control the PRA process, and otherwise specifying the manner in which assessors and managers will interact. Risk assessment policy ought to also provide guidelines for addressing uncertainties encountered in the risk assessment process and for communicating their significance to risk managers. The policy will typically assign managers the responsibility to define the range and scope of the risk assessment. One of the primary goals of this policy must be to protect the scientific integrity of the assessment.

The party with authority to make value judgments or policy choices should be clearly articulated and the assessment review process ought to be addressed. The outcome of this process is a written risk assessment policy with a clear and commonly held understanding of the boundaries of the risk assessment and the manner in which it will be conducted. Risk assessors ought to have both input and feedback opportunities during the development of the organization's risk assessment policy.

The second part of the risk assessment policy deals with risk assessment specific issues. These might include unique information requests, identification or evaluation of risk management options, and the like. These issues ought to be resolved in coordination with risk assessor input and feedback as they arise.

4.5.2. Initial risk management activities

Risk managers should first meet with risk assessors at the outset of the initial risk management activities stage of risk management when managers are first identifying their information needs. Risk managers ought to identify the information they require for decision-making. Conceptually,

this could be accomplished by a series of questions risk assessors have been asked to answer for risk managers. The risk manager would be expected to identify the proposed trade applying for market access. This would identify the country of origin and the commodity as well as everything else that is known about the trade proposal. In return, NPPO risk managers would be predictably interested in some information that will usually include the following elements:

- identify pests of potential concern
- estimate the probability of introduction
- estimate the consequences of introduction
- identify alternative risk management options
- estimate the effectiveness of the risk management options considered
- identify any unique questions to be answered by the risk assessment.

Risk assessors would be expected to enter into a dialogue about this information. They would clarify the expectations for the requested information and negotiate whether a qualitative or quantitative assessment would be done.

4.5.3. Risk estimation

Managers and assessors need to meet during the conduct of the risk assessment, this happens during risk estimation. Risk managers must support the conduct of the risk assessment. Initially, this will be to assign personnel, allocate resources, and agree to a schedule for accomplishing the risk assessment. Once the risk assessment is underway managers and assessors should continue to meet to monitor and discuss progress toward the assessment's completion. These progress meetings may include continued clarification of the terms of the risk assessment. Changes in the risk assessment should be negotiated between the parties as they become necessary.

4.5.4. Risk evaluation

Once the risk assessment is completed the risk manager's risk evaluation responsibilities begin. They begin by deciding whether the risk assessment describes an acceptable or an unacceptable risk. Many NPPOs have defined risks rated medium or high in a qualitative risk assessment as unacceptable risks, rendering this decision automatic. Quantitative risks will, however, require an explicit evaluation. If the risk is judged to be unacceptable then measures to eliminate or reduce the risk must be considered. These measures may have been routinely identified and assessed in the course of the risk assessment or they may be collaboratively identified by risk managers and assessors after the initial assessment of the pest risk. In this latter event, the effects of the risk control measures must be assessed and subsequently evaluated. Do they avoid or eliminate the risk? If not, do they reduce the risk to a tolerable level? If there are any additional assessments, for example, economic analysis of measures or environmental assessments, they ought to be undertaken during the risk evaluation stage of the process. Risk managers must also evaluate the risk control options that have been identified. This is going to take close interactions between assessors and managers.

Risk assessors must communicate insights gained during their research, literature review, model-building, and analysis. Managers and assessors will need to meet during this evaluation to assure that risk managers understand the assessment, its uncertainty, its strengths and weaknesses and how decisions or decision outcomes might be affected by the remaining uncertainty. Risk managers should be made aware of key controversies concerning the data or assumptions used. It is also critical that risk managers understand the residual risk, i.e., the risks that will remain after risk controls are implemented. If there will be an independent review of the risk assessment, that takes place during the evaluation stage of the risk manager's job.

4.5.5. Risk control

Interaction between risk managers and risk assessors may be relatively limited during the risk control stage of the process. This is when risk control measures are decided and implemented. In other cases, risk assessors may possess the scientific expertise necessary to implement specific controls or they may know who must do what in order to implement controls. In these instances, close collaboration will be necessary. The definition of who is a risk manager is likely to be expanding at this point as field personnel who may be trained and instructed in the control, inspection and approval procedures when a product is allowed into the country subject to sanitary and phytosanitary (SPS) measures join the ranks of risk management.

Characteristics of a good team member

---

1. Works for consensus on decisions.
2. Shares openly and authentically with others regarding personal feelings, opinions, thoughts, and perceptions about problems and conditions.
3. Involves others in the decision-making process.
4. Trusts, supports, and has genuine concern for other team members.
5. "Owns" problems rather than blaming them on others.
6. When listening, attempts to hear and interpret communication from others' points of view.
7. Influences others by involving them in the issue(s).
8. Encourages the development of other team members.
9. Respects and is tolerant of individual differences.
10. Acknowledges and works through conflict openly.
11. Considers and uses new ideas and suggestions from others.
12. Encourages feedback on own behavior.
13. Understands and is committed to team objectives.
14. Does not engage in win/lose activities with other team members.
15. Has skills in understanding what's going on in the group.

(http://www.innovativeteambuilding.co.uk/characteristics-of-a-good-team-and-team-member/, accessed March 22, 2019)

4.5.6. Risk monitoring

The monitoring stage of the risk management process can range from an administrative function to science intensive sampling, inspection, and control measures. Risk assessors may be involved in the design of these measures. As the monitoring process accumulates data it is possible that risk assessors will play a role in transforming these raw data into risk characterization information. That information will subsequently be used by risk managers to evaluate the success of the proscribed measures in reducing risk to the desired level.

4.6. Good Teams

A risk analysis team will include risk managers and risk assessors; some may include risk communicators as well. What does it take for the individual members to interact well as a team, especially when some of the team members may be physically separated from one another by location? Good teams have common goals, well-defined roles, participation by all members, good communication, trust, mutual respect, and performance.

The goals of the team must be agreed upon and understood in common. Shared goals can help a group of people work together. When everyone uses their abilities to work toward a common goal, the result is greater than the efforts possible from any single person. In a strong team, the good of the common goal comes before the individual preferences and interests of the team members.

Each member of an effective team should have a clearly defined role in the risk analysis. Each person's skills should be used to achieve the risk analysis team's goals. It is important for each team member to participate in every stage of the risk analysis process. Team members need to feel important and understand their value to the group.

Every team member should be able to communicate with the group and individuals within the group. In effective teams, members are carefully listened to and receive thoughtful feedback. If there is anything they do not understand, they should feel free to ask questions and seek clarification. Each member should be able to contribute ideas and listen to others who have valuable

contributions. There should be communication between assessors and managers as often as needed in order to keep the team on task.

Because all of the team members have specific jobs, they need to trust that others are doing their part. Each person should follow through with what she or he has agreed to do in order to obtain trust from others. Even when team members disagree, they respect the contribution that a fellow team member is trying to make. A sense of commitment to the group and its goals helps build trust. Having mutual respect means team members keep their focus on the issue rather than the individual.

An effective team performs. It gets the job done and done well. Members take the initiative to think about the problems and their solutions. The team develops its own strategy of problem-solving and conflict management. When team members can work through problems together, the team will work smoothly even during rough times.

Teams will look different around the world. They will be influenced by organizational structures and cultures. There can and will be many different team structures. What is essential to the success of each of these is that risk managers and risk assessors share common goals in a trusting environment.

4.7. Vertical Team Alignment

Many NPPOs have separate risk management and risk assessment cultures that work at cross purposes. Where this is true, assessors and managers need to get out of their cultures and "stovepipes" in order to work together. Successful pest risk management requires the vertical engagement of the organization and so the risk analysis team should, in fact, be a vertical team. The exact makeup of the vertical team may vary from country to country or even from PRA to PRA, depending on the scope and complexity of the PRA. Among the nations that have different groups for risk assessment and risk management, it would not be unusual to find four levels represented on a vertical team as shown in Fig. 4.2.

The vertical team consists of risk assessors, risk communicators, risk managers and any other analysts who may play a role in the conduct of a PRA. Note that operational risk managers are

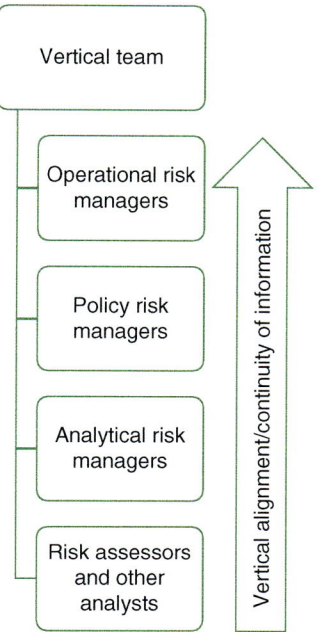

Fig. 4.2. Conceptualization of vertical team alignment for continuity of information.

not likely to become involved with a PRA until after it has been completed and implemented. At that time, their risk management role becomes of paramount importance, hence they are, conceptually, included in the vertical team.

The vertical team is jointly responsible for the PRA and any subsequent action undertaken by the NPPO. Vertical team alignment is not groupthink. It does not mean everyone will see things the same way or think the same things. The ultimate goal of aligning the vertical team is to identify and solve issues as early as possible in the PRA process and to maintain the continuity of information. When different individuals perform different functions possibly at different times

in a PRA, it is easy for information to get lost. Vertical team alignment means getting the entire vertical team engaged early in the PRA process and maintaining a consistent and effective level of engagement throughout that process.

The entire vertical team needs a common vision of how the PRA will proceed, how much uncertainty can be tolerated, and how resulting risks will be managed. Ideally, the vertical team jointly owns all the decisions made during the process. Having one common and shared NPPO decision at all four vertical levels of the team simultaneously can reduce the need to repeat steps and refine analyses. Vertical team alignment is essential to NPPO goals of saving time and money without sacrificing quality.

4.8. Summary and Look Forward

Here are five things to remember from this chapter.

1. Teams are more efficient and faster than individuals.
2. Pest risk analysis is a team activity.
3. Interdisciplinary teams are better than multi-disciplinary teams and transdisciplinary teams are better than interdisciplinary teams.
4. The traditional pest risk analysis process separates risk management and risk assessment too much.
5. Risk managers and risk assessors need to communicate, coordinate, cooperate, and collaborate throughout the risk management process.

Now that we have considered how risk managers and risk assessors ought to function as a team, it is time to consider the risk manager. The next chapter introduces the role of the risk manager and includes such fundamentals as who they are, what they do, and what makes a good risk manager.

4.9. References

National Research Council (1996) *Understanding Risk Informing Decisions in a Democratic Society* (C. Stern and H.V. Fineberg, eds). Committee on Risk Characterization Commission on Behavioral and Social Sciences and Education, National Academy Press Washington, D.C.

United Nations/WHO (2000) The Interaction between Assessors and Managers of Microbiological Hazards in Food. Report of a WHO Expert Consultation, Geneva, May 15, 2010. Available at: http://www.who.int/foodsafety/publications/micro/en/march2000.pdf (accessed August 11, 2020).

Yoe, C. (2012) *Introduction to Natural Resources Planning*. CRC Press, Boca Raton.

5

Risk Managers

The ones who make it are the ones who manage risk

Steve Burns

5.1. Introduction

Risk managers are found at many points in the international plant and plant product trade continuum. Chapter 1 identified two main groupings of risk managers, production risk management and phytosanitary risk management (see Fig. 1.5, p. 8). Exporters and importers bear and manage the production risks while national plant protection organizations (NPPOs) and the nations they represent bear and manage the phytosanitary risks. Although the responsibilities are often treated as if they are separate and independent, they are neither. No one group of risk managers can be successful without the success of the other groups. They are both interdependent and integrally related in the greater risk arena of international trade in plants and plant products.

Production risk management is a corporate responsibility, perhaps best expressed in enterprise risk management (ERM, the subject of Chapter 17), while phytosanitary risk management is a regulatory responsibility. To the extent that risk management marks the confluence of science and values, there should be close agreement on matters of science between the two groups and diverse views on matters of values.

One of the driving philosophical statements behind the rise of regulatory risk management is: prevention is better than cure. Rather than to rely on the reactive strategies of bygone days, regulatory risk management seeks to avoid threats and mitigate the effects of those threats which are essentially unavoidable. Directing this process is a new leadership role. No one was classically trained to be a risk manager. Industry lagged behind the regulatory sector in adopting risk management and came to this practice through ERM, which seeks to direct and control the effect of uncertainty on a firm's objectives. This means taking risks that present opportunities of uncertain gain, as well as avoiding risks of loss.

> Is there a single risk manager or is there a team of risk managers? The answer will vary from setting to setting. For the purposes of this handbook we adopt the convention of referring to a single risk manager, even though the tasks described may be accomplished in some cases by several risk managers.

Early definitions of risk management (see box below) tended to confine the risk manager's role to taking an appropriate regulatory action in consideration of risk assessment results and other considerations. Today risk management definitions and models are much richer. Decision-makers realized there was a risk management role in directing the risk assessment to produce the information necessary to address a

> Risk management is the process of weighing policy alternatives and selecting the most appropriate regulatory action, integrating the results of risk assessment with engineering data and with social, economic and political concerns to reach a decision.
>
> (National Research Council, 1983)

problem or to seize an opportunity. Risk managers had to get involved in direction-setting to make sure risk assessment produced the information needed to make a risk management decision. In a similar fashion, circumstances dictated the evolution of the risk manager's job in other directions, which will be described at length in subsequent chapters.

In the broader sense, production risk management involves businesses taking risks to meet their economic and other objectives. Phytosanitary risk management is about protecting plants and promoting international trade. That means we must understand both the NPPOs' and industry's objectives. No single risk manager currently takes that perspective. Phytosanitary risk managers concerned with safe trade tend to focus on plant health and safety. Industry risk managers concerned with trade tend to focus more on financial risks. If a risk manager sees his or her role as protecting plants it is clear that trade will suffer. If she or he sees the job as promoting trade, plants will suffer. In best practice, the risk manager who sees the job as all of these things will have the best opportunity of weighing the tradeoffs and making practical and successful decisions.

This chapter takes a high-level look at who the risk managers are, what they do, the roles they play, vertical and horizontal alignment of risk management and the qualities of a good risk manager. The concepts introduced here will be developed in detail in subsequent chapters.

5.2. Who Are the Risk Managers?

Risk managers comprise anyone who owns part of a risk. It takes a lot of people to facilitate the movement of plants and plant products around the globe in safe and beneficial ways. International plant trade risk managers comprise two

distinct types. First, there are the risk avoiders. These are the NPPOs and all those people they rely on to control pests by implementing required or recommended risk controls. Second, there are the risk takers. These include businesses establishing new markets, exporters seeking new markets, importers seeking new products and trade promoting organizations both in and out of government. These two groups have something important in common, i.e., seeing that no harm comes to plants as a direct or indirect result of their commerce while promoting the benefits of international trade.

Risk management is a shared responsibility with risk managers at several points in the supply chain. Fig. 2.2 in Chapter 2 (p. 24) shows nine components in a plant product supply chain. There are different risk managers at each point in the supply chain. Industry has a special role in supplying both strategic and tactical risk managers to the overall process. We focus on the pest risk management responsibilities of the NPPO risk managers, however, because they are the ones with the formal obligation for pest risk analysis. It may be helpful to differentiate the levels of pest risk management.

Fig. 5.1 shows the different tiers of responsibility commonly found among the pest risk management responsibilities of NPPOs. In more sophisticated organizations, the boxes of the figure can represent different groups of people. In simpler NPPO structures they may represent different roles for the same people or person. There is a positive role for risk managers in describing the world as it is. This comprises the risk manager's analytical role. It involves guiding and supporting the risk assessment as well as identifying, evaluating, and recommending risk management options. The normative and policy roles of risk managers are filled by those who function as policy risk managers deciding and prescribing measures to mitigate pest risk. Finally, there are operational risk managers who are responsible for implementing programs and actions to manage risks.

Parallels among the production risk managers can be drawn as seen later in Section 5.5 of this chapter. This handbook focuses primarily on the job and responsibilities of the NPPOs' risk managers. Theirs is the view most focused on risk and theirs has the greatest influence on the

An entity's risk management philosophy is the set of shared beliefs and attitudes characterizing how the entity considers risk in everything it does, from strategy development and implementation to its day-to-day activities … [It] is reflected in virtually everything management does in running the entity. It is captured in policy statements, oral and written communications, and decision making. Whether management emphasizes written policies, standards of behavior, performance indicators, and exception reports, or operates more informally largely through face-to-face contact with key managers, of critical importance is that management reinforces the philosophy not only with words but also with everyday actions.
(COSO's Enterprise Risk Management Framework, http://www.coso.org/erm-integratedframework.htm, accessed January 21, 2016)

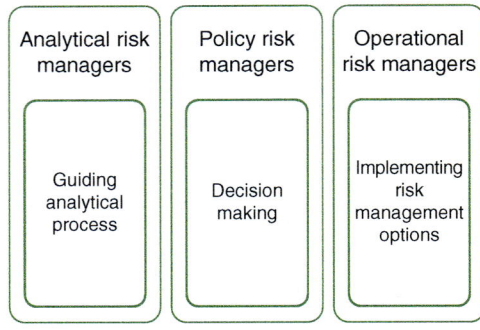

Fig. 5.1. Three distinct tiers of risk management can be identified in the pest risk management responsibilities of NPPOs.

decision-making process. The handbook does, however, attempt to keep the role and responsibilities of risk managers outside of the NPPOs in some focus, because they are of paramount importance in providing the context for pest risk management and they are responsible for much of the implementation of risk management measures. Although the International Plant Protection Convention (IPPC) constrains the focus and responsibilities of an NPPOs risk managers, it cannot constrain the greater reality of the interrelatedness of economic risks and their relationship to pest risks in the cause of international trade.

5.3. What Does the Risk Manager Do?

There are actually two distinct but inseparable risk managing processes here. One is the risk-taking activity of promoting international trade. The other is the risk-avoiding activity of plant protection. In an ideal world, both of these risks would be managed together. The world is far from ideal and responsibility for effecting international trade is largely left to the private sector while the public sector has responsibility for protecting plants. Consequently, no one has that idealized big picture view that would integrate and balance the management of these risks with the wisdom of Solomon.

Risks are currently managed in a myopic fashion, where risk managers respond to the self-interests of their organizations and often narrowly transcribed authorities. Trade interests seek to expand trade, often blissfully unaware of the negative externalities that trade might spawn. Plant protection is not, naturally, one of their decision criteria. Regulatory agencies with plant protection responsibilities seek to protect plants; although they ought to have a healthy awareness of the desire to promote international trade born of the Sanitary and Phytosanitary Agreement (SPS Agreement). There is no risk managing authority with a holistic vision that will unify the goals of these two interests across countries, nor is one on the horizon. It is nonetheless true that the interests of both groups are inexorably combined and would best be served by an integrated systems approach to risk management.

The risk manager's job requirements are, conceptually, the same for NPPOs and industry: they are responsible for the organization's overall risk management framework and they must manage each individual risk management activity undertaken by the organization. Risk managers are responsible for selecting and designing a process that is consistent with the organization's objectives and all applicable international agreements and standards as well as all domestic laws and policies. The risk manager owns the organization's risk analysis process and is expected to be its champion within the organization.

Risk managers determine how the organization will conduct risk assessments and implement risk management options on an ongoing basis. Pest risk assessment is generally conducted to inform two classes of phytosanitary risk management decisions: (1) those regarding the introduction of potentially invasive nonindigenous species, their vectors, or conveyances prior to establishment, which lead to decisions to authorize, prohibit, or permit activities under specified conditions; and (2) decisions regarding the allocation of scarce resources for the control of established invasive species, including rapid response to emerging threats (Andersen *et al.*, 2004). Private organizations conduct risk assessments to identify risks worth taking to further the objectives of the organization.

Risk managers promote the use of risk analysis throughout the organization by providing resources and support. They also ensure all personnel are aware of and trained in the use of the organization's risk analysis process. It is the risk manager's job to ensure that the organization is following its risk analysis process. Risk managers set the organization's risk priorities. They determine the organization's risk philosophy and establish and quantify the organization's "risk appetite," i.e. the level of risk they are prepared to accept (COSO, 2004). For NPPOs this means establishing a process for prioritizing risks to guide resource allocation. Given the priorities of the organization the risk manager, guided by these priorities, initiates a risk management activity. This process may or may not result in a recommendation of new risk management initiatives.

When it comes to individual risk management activities, risk managers have led responsibilities in each of the five tasks depicted in Fig. 2.5 (see p. 32). These are: initial risk management activities, risk estimation, risk evaluation, risk control, and risk monitoring. These responsibilities introduced in Chapter 2 will be developed in detail in Part 2 of this handbook. Risk communication, not shown in the figure, is also the risk manager's responsibility. Like risk assessment, the manager need not undertake it, but they must see that it is done well. Part of risk communication includes risk reporting in an appropriate way for different audiences. For phytosanitary risk management, this includes reporting to all NPPOs with a direct interest in the issue, business leaders affected by decisions, as well as individuals who need to understand their accountability for

controlling the risks in an appropriate and responsible manner.

The risk manager may or may not be the ultimate decision-maker. Some plant protection decisions may be made at a level above the NPPO. Likewise, industry decisions may be made at a corporate level above the risk manager.

5.4. The Risk Manager's Roles

The risk manager's job is making practical and useful decisions under uncertainty. Whether an NPPO employee charged with protecting a nation's plant life or an entrepreneur seeking new trade opportunities, the risk manager is working without all the information he or she would like to have. As a result, there are at least three distinct roles the risk manager must play as a decision-maker: these are positive, normative, and policy roles.

In the *positive* role, risk managers have a responsibility to see the world as it truly is. Whether stewards of a nation's plant life or protector of a company's bottom line, an evidence-based view of reality is essential. Thus, the risk manager has a role in the risk assessment and that role is to empower and support the assessment team in its pursuit of an objective science-based view of the reality of every decision problem. This is also where risk managers come face-to-face with the uncertainties that can plague their decision-making. In this role, NPPO and industry risk managers ought to have a similar view of a problem.

In their normative role, risk managers have some leeway to see the world as it ought to be. This is where social values appropriately enter the decision-making process along with other factors considered in combination with the objective facts of the risk assessment. In this role, NPPO and industry risk managers may begin to diverge in their views of the seriousness of the risk and the appropriate response to it. Some to much of this divergence may enter as different opinions about how the uncertainty in a decision problem ought to be resolved and regarded.

The policy role requires risk managers to serve the needs of the organization and its objectives. It is in this role that the NPPO and industry risk managers will most likely diverge. This is decision-making, where values are weighed and tradeoffs are made. A good risk assessment assures that the relevant science is considered in a decision process. Tradeoffs among time, resources, political pressure, and other social or organizational values inevitably will enter the decision-making calculus and the differences in values assure there will always be parties less than satisfied with every solution. The best risk managers are aware of these different roles and they serve each of them conscientiously.

5.5. Vertical and Horizontal Alignment of the Risk Management Function

Ideally, risk managers at all parts of the international trade continuum would achieve alignment of their actions to produce the benefits of trade and the protection of plants. Vertical and horizontal alignment applies to all the interlocking sets of activities production and phytosanitary risk managers must manage to successfully complete trade in plants and plant products (Fig. 5.2).

Horizontal alignment involves a unity of purpose in configuring goals, objectives, actions, and decisions of all production risk managers. Two kinds of risk management functions can be identified in the supply chain activities that are the responsibility of production risk managers. Some risk managers are responsible for the analytical process of identifying, evaluating, recommending, and deciding on risk management options. These will tend to be more corporate level personnel. Another important group of risk managers is those responsible for implementing programs and activities to manage risk. These will tend to be more production line and field level personnel with explicit risk management responsibilities. Horizontal alignment refers to the coordination of key activities across the plant and plant product supply chain.

At the phytosanitary risk management level, the personnel of the NPPOs for both the exporting and importing nation must be aligned. The SPS Agreement and the IPPC provide a framework to support that horizontal alignment. Horizontal alignment at this level is primarily relevant to cross-functional and intra-functional integration.

Vertical alignment is a bit more elusive at the current time. Vertical alignment is the alignment of different parts of the plant and plant product international trade continuum. It requires continual interaction through effective collaboration and coordination between production and phytosanitary risk managers. Vertical alignment, extending to World Trade Organization (WTO) intervention if needed, ensures improvements to performance.

Vertical alignment can occur at broad or specific levels of a trade continuum. Specific examples include businesses and NPPOs collaborating to export goods from one country, or NPPOs and businesses collaborating to bring new

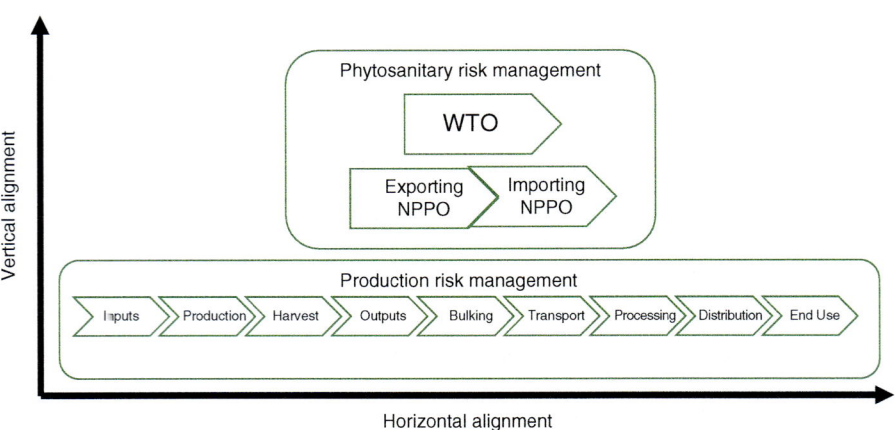

Fig. 5.2. Vertical and horizontal alignment of production and phytosanitary risk management.

goods into a country. Regulatory risk management establishes standards for successfully moving plants and plant products. Once a standard is established, vertical alignment is accomplished at a broad level when everyone accepts their responsibility and performs in accordance with the standard.

Both the standards and the assessments they are based upon must be vertically aligned with one another so that they reflect the science and logic that underpins successful international trade.

5.6. Qualities of a Good Risk Manager

What makes a good risk manager? The easy answer is a good risk manager gets the desired outcomes from the decisions she or he makes; but the job is more complex than that. A good risk manager must be multi-faceted. One of the authors has conducted an informal survey of food safety professionals, including NPPO personnel, in training events for a number of years. Following is a summary of what those surveys reveal about what the people who work most closely with regulatory risk managers consider the qualities of a good risk manager.

The many comments collected have been grouped under six topical headings. They are:

- perspective
- science
- analytical
- uncertainty
- communication
- leadership.

Each set of qualities is examined below in turn.

Government is focused primarily on plant health; industry is focused primarily on business interests. Both are essential and they can lead to different perspectives on a great many issues. Risk managers must have a broad perspective of the job. That means understanding the different perspectives of the many stakeholders to a decision. Open-mindedness and flexibility are among the most frequently identified qualities of a good risk manager.

Risk analysis is the interface of science and values and the risk manager must be able to see both of these perspectives clearly. Politics, of the political world or of the internal office, are often a powerful influence on the risk manager's world. They cannot preclude the science, but neither can the science preclude the politics or other important values. A risk manager cannot be narrowly focused and an interdisciplinary background contributes to this ability to see many perspectives. The ability to see the big picture is a "must have" quality for a risk manager.

> **Here are some frequently repeated phrases used to describe a good risk manager**
>
> Accountable, analytical, basic knowledge in the field, calm under pressure, capable, clear-sighted, comfortable in interdisciplinary setting, communicator,* competent, decisive, detailed, direction-setting, effective, efficient, experienced, facilitator, flexible, holistic vision, impartial, knowledgeable, leader, listener, mentor, methodical, multidimensional perspectives, objective, open-minded,* organized, persuasive, prompt, reasonable, resourceful, responsible, thought provoking, strategic.
> *Most common responses.

The ability to make connections between facts and people's concerns is important, so is the ability to see and make tradeoffs among conflicting values. It is important to recognize that plant safety is not the only priority, but it is sometimes difficult to commit to that perspective when plant safety is the main concern. Likewise, the promotion of trade is not the only priority in a risky situation. Above all else, a good risk manager sees all sides of an issue and understands the broad view of the issue.

Science is important. Risk managers must have experience and knowledge of the field in which they manage. A primary goal of risk management is to get the appropriate science and evidence focused into the decision-making process. It is essential to get the facts and the objective truths on the table well before making any subjective tradeoffs or decisions. It often helps if a risk manager has a science background, but a risk manager cannot afford to "be" an entomologist, a plant scientist, or any one discipline anymore. If she does, she will fail in that perspective piece. She must "be" a risk manager.

It is not necessary that a risk manager be a scientist. As noted below in the analytical discussion, she must know what kind of information is going to be needed to solve a problem. She need not understand each scientific fact in detail but she must be able to understand the whole of the risk issue. There is a growing number of risk managers who come from fields other than the natural sciences. In those instances, it is essential that a risk manager learn how to speak and understand all of the relevant "science." Risk assessment experience is frequently cited as a desirable quality in a risk manager.

An analytical mind is another important quality. This includes being strategic, detailed, methodical, and thought provoking. Everyone makes decisions and we all have adopted and adapted ways of doing so that make sense for us individually. We can be habitual, impulsive or even irrational in our personal decision making. However, NPPO risk managers are stewards of plant health, a great public trust, and as such they are no longer making personal decisions. Society deserves that their decisions be analytical. Risk analysis is intentionally designed to overcome our innate tendencies to overlook analytical thinking. Risk managers should have analytical minds, practitioners want them to understand and consider the science, articulate tradeoffs and make reasoned decisions.

Another skill a risk manager needs is the ability to deal with uncertainty. The person who figuratively says, "Why can't you just give me the number?" is not well suited to be a risk manager. There is no such thing as "the number" in most of the problems risk managers encounter. Uncertainty is everywhere and what makes risk analysis different from other decision-making frameworks is the forthright manner with which it addresses and pays attention to the things we do not know. If risk assessors are accounting for it in their work, they will describe the range of potential risk effects and the uncertainty about the efficacy of the available risk management options. Then the risk manager must make decisions taking the potential effects of this uncertainty into account. This is fundamentally the risk

manager's job: making good decisions without all the information.

Communication skills are probably the single most commonly mentioned attribute of a good risk manager. The abilities to listen and to communicate effectively with team members are considered critically important. A good risk manager is expected to ask a lot of helpful questions. This is the internal risk communication task. As far as external risk communication goes, risk managers do not have to do all the risk communication but they are ultimately responsible for seeing that risk communication gets done well.

Plant safety and trade issues can be complex, controversial, and trying. More than a few are pressure packed as well. Effective leadership is an important quality for navigating such tricky waters. The most prized leadership skills include openness, accountability, calmness under pressure, confidence, compassion, and being a mentor. Managers' work habits include being organized, resourceful, committed, enthusiastic, tireless, and patient.

5.7. Summary and Look Forward

Here are five points to take away from this chapter.

1. Risk management seeks to avoid threats and mitigate the effects of those threats which are essentially unavoidable. It generally prefers prevention to cure.
2. Risk managers comprise anyone who owns part of a risk and there are many different owners along a plant product supply chain.
3. Phytosanitary risk managers facilitate the movement of plants and plant products around the globe in safe and beneficial ways.
4. The NPPO risk manager must play positive, normative, and policy roles in the pest risk analysis process.
5. A good risk manager has perspective and a scientific background, is analytical, able to cope with uncertainty, and has good communication and leadership skills, along with effective work habits.

Now that we have begun to understand who the risk managers are and what their jobs entail, it is time to consider how the WTO *Agreement on the Application of Sanitary and Phytosanitary Measures* affects and informs the risk manager's job. That is done in the next chapter.

5.7. References

Andersen, MC., Adams, H., Hope, B. and Powell, M. (2004) Risk assessment for invasive species. *Risk Analysis*, 24, 4.

COSO (Committee of Sponsoring Organizations of the Treadway Commission) (2004) *Enterprise Risk Management—Integrated Framework, Executive Summary*. Available at: http://www.coso.org/documents/COSO_ERM_ExecutiveSummary.pdf (accessed January 31, 2017).

National Research Council (Committee on the Institutional Means for Assessment of Risks to Public Health Commission on Life Sciences) (1983) *Risk Assessment in the Federal Government: Managing the Process.* National Academy Press: Washington, District of Columbia, USA.

6

Risk Management and the SPS Agreement

The WTO has one of the most impressive records in global economic governance, by promoting trade liberalization and economic development.

Anna Lindh

6.1. Introduction

The World Trade Organization (WTO) was formed in 1995. Its primary purpose is to promote trade by limiting tariff and nontariff barriers. The WTO is self-described as a negotiating forum, where member governments go to try to sort out the trade problems they face with each other (https://www.wto.org/english/thewto_e/whatis_e/tif_e/fact1_e.htm, accessed March 26, 2019). The WTO Agreement on the Application of Sanitary and Phytosanitary Measures (the

SPS Agreement) was developed to strike a balance between the right of nations to protect their food supplies as well as their plant and animal health, and the desire to avoid the use of sanitary and phytosanitary measures to restrict trade unnecessarily (http://www.tradeforum.org/The-SPS-Agreement-WTO-Agreement-on-the-Application-of-Sanitary-and-Phytosanitary-Measures/, accessed March 26, 2019). Other authorities bear on the question of nonindigenous pests (see box), but none has a greater influence on the practice of phytosanitary risk management than the SPS Agreement. The Agreement went into effect on January 1, 1995, when the WTO began. It replaced the General Agreement on Tariffs and Trade—GATT (Gruszczynski, 2006). The SPS Agreement protects international trade (e.g., in plants), from artificial trade barriers.

The SPS Agreement recognizes three organizations as international standard-setting bodies. They are:

1. Codex Alimentarius: international food safety community.
2. World Organization for Animal Health (Office International des Epizooties, or OIE): international animal health community.
3. International Plant Protection Convention (IPPC): international phytosanitary community.

The Convention on Biological Diversity urges countries to control or eradicate nonnative species that threaten ecosystems, habitats, or species.

The Cartegena Protocol, Risk Assessment for Invasive Species 789 Biosafety Protocol, requires the use of risk assessment to support decisions regarding the international movement of living modified organisms.

Article 196 of the United Nations Convention on the Law of the Sea (UNCLOS) calls for measures to prevent, reduce, and control the intentional or accidental introduction of species, alien or new, that could cause significant and harmful changes to the marine environment.

(Andersen *et al.*, 2004)

The SPS Agreement allows member states to take scientifically based measures to protect human, animal, plant life, or health. The Agreement commits members to base these measures on internationally established standards and risk assessment. Thus, the Agreement permits countries to take measures to protect public health within their borders so long as those measures restrict trade as little as possible.

The purpose of this chapter is to link the SPS Agreement to the phytosanitary risk management process. This begins with an overview of the SPS Agreement that is followed by a reassessment of the Agreement in a risk management context. It concludes with the IPPC's role in implementing the SPS agreement.

6.2. The SPS Agreement

The structure of the SPS Agreement is shown in Fig. 6.1. The agreement is briefly summarized in the paragraphs that follow.

The *Note to Members* reaffirms the members' right to adopt or enforce measures necessary to protect human, animal, plant life, or health in order to improve the same in all countries. It notes the desire to establish a multilateral framework of rules and disciplines to guide sanitary and phytosanitary measures in order to minimize their negative effects on trade and to do so using international standards, guidelines, and recommendations developed by the relevant international organizations (in this instance the IPPC). The Agreement notes that trade with developing countries may present challenges.

Article 1 says this Agreement applies to all sanitary and phytosanitary measures which may, directly or indirectly, affect international trade. Article 2 articulates members' rights to take the sanitary and phytosanitary measures necessary for the protection of human, animal, plant life, or health. It also requires them to ensure that any sanitary or phytosanitary measure:

- is applied only to the extent necessary to protect human, animal, plant life, or health
- is based on scientific principles, and

- is not maintained without sufficient scientific evidence.

It also notes that no sanitary or phytosanitary measures shall be a disguised restriction on international trade.

Harmonization, the subject of Article 3, requires that members base their sanitary or phytosanitary measures on international standards, guidelines, or recommendations, where they exist. Members may seek a higher level of sanitary or phytosanitary protection than would be achieved by measures based on the relevant international standards, guidelines, or recommendations, if there is a scientific justification for doing so.

Equivalence (Article 4) requires that members accept the sanitary or phytosanitary measures of other members as equivalent if the exporting member can objectively demonstrate to the importing member that its measures achieve the importing member's appropriate level of sanitary or phytosanitary protection. This allows different members to use different measures if they have equivalent outcomes. Equivalence may result in different members using different measures for trade in the same product.

> The SPS agreement effectively adopts the economic perspective of domestic producers. This, no doubt, has had a profound impact on the nature and substance of pest risk analysis as it neglects to mention foregone opportunities for domestic consumers and potential gains for importers.

Article 5 addresses the assessment of risk and it is of some interest to note that neither risk analysis nor risk management are mentioned in this article. Members are to base their sanitary or phytosanitary measures on an assessment of the risks to human, animal, plant life, or health. This assessment is to be based on the available scientific evidence and relevant economic factors.

Minimizing negative trade effects should be considered when determining the appropriate level of sanitary or phytosanitary protection. This requires members to avoid arbitrary or unjustifiable distinctions in the levels they consider to be appropriate in different situations. Enacted

Agreement on the application of sanitary and phytosanitary measures	Note to Members
	Article 1: General provisions
	Article 2: Basic rights and obligations
	Article 3: Harmonization
	Article 4: Equivalence
	Article 5: Assessment of risk and determination of the appropriate level of sanitary or phytosanitary protection
	Article 6: Adaptation to regional conditions, including pest—or disease—free areas and areas of low pest or disease drevalence
	Article 7: Transparency
	Article 8: Control, inspection, and approval procedures
	Article 9: Technical assistance
	Article 10: Special and dfferential treatment
	Article 11: Consultations and dispute settlement
	Article 12: Administration
	Article 13: Implementation
	Article 14: Final provisions
	Annex A: Definitions
	Annex B: Transparency of sanitary and phytosanitary regulations
	Annex C: Control, inspection, and approval procedures

Fig. 6.1. Structure of the WTO SPS Agreement.

measures should be no more trade restrictive than necessary to achieve their appropriate level of sanitary or phytosanitary protection. Members are advised to take into account risk assessment techniques developed by the relevant international organizations.

Article 6 requires members to adapt their measures to regional conditions. Article 7 requires members to inform others of changes in measures and to explain measures as part of their commitment to transparency. Article 8 requires members to follow the control, inspection, and approval procedures consistent with more detailed guidance in one of the annexes. Article 9 provides guidance on technical assistance, especially to developing countries. In situations that require substantial investments by a developing country exporter to comply with the sanitary or phytosanitary requirements of an importing country, the importing member should consider providing technical assistance to the developing country to help it maintain and expand its market access opportunities for the product involved. Article 10 advocates members to take account of the special needs

of developing country members. Article 11 suggests that members follow the GATT Agreement for dispute settlement. Article 12 creates a Committee on Sanitary and Phytosanitary Measures, which among other things encourages the use of international standards. Article 13 indicates that all member nations are responsible for implementing the terms of the Agreement. The final article provides for some transition in implementing these measures.

The definitions in Annex A include only one risk term, risk assessment. The appropriate level of sanitary or phytosanitary protection is defined as, "The level of protection deemed appropriate by the Member establishing a sanitary or phytosanitary measure to protect human, animal, plant life, or health within its territory." This is an essential risk management concept that is discussed and developed in detail in Chapter 18.

Annex B concerns itself with transparency and addresses the need to promptly publish all regulations. Each member country is to have a single enquiry point responsible for answering all reasonable questions from other member states. It also includes notification procedures to follow when measures differ substantially from international standards, when an international standard does not exist, or when trade may be affected.

The final section, Annex C, addresses control, inspection, and approval procedures. This section ensures that the procedures for admitting a product into a country are expeditiously undertaken and no more disruptive to trade than necessary. Among other things, it says procedures to check and ensure the fulfilment of sanitary or phytosanitary measures must be undertaken and completed without undue delay and they should be limited to what is reasonable and necessary. Imported products should not be treated less favorably than similar domestic products.

6.3. SPS Agreement in a Risk Management Context

The term "risk management" is never mentioned in the SPS Agreement. The notion of risk had not evolved enough at that point in time for this community of practice for it to appear. It has only been since the turn of the 21st century that the emphasis on risk management has taken off. The original emphasis on using risk assessment to provide a scientific basis for decision-making has been so successful that organizations the world over have had to scramble to develop and refine decision-making frameworks capable of taking the results of risk assessment into proper account. Risk management is the confluence of science and values. The SPS Agreement articulates some common values for risk management in the plant health protection community even as it introduces risk assessment as the standard for presenting the scientific evidence required for decision-making.

With over a quarter of a century of progress on the risk analysis front since the SPS Agreement was implemented, it is now clear that the Agreement was and continues to be a significant proclamation of risk management principles. The *Note to Members* establishes the role of member states as risk managers. It articulates the primary global value to be shared by risk managers as the right to adopt or enforce measures necessary to protect human, animal,

plant life, or health. The IPPC is identified as a standard-setting body with substantial influence over the risk management process.

Article 1 articulates the member's sovereignty and the second article rearticulates the shared value of the protection of human, animal, plant life, or health. It goes on to emphasize the importance of scientific principles and evidence and the common value of making measures no more stringent than necessary. Restraint of trade is identified as an impact to avoid in enforcing measures.

Harmonization and equivalence, the topics of Articles 3 and 4, can be considered two evaluation criteria to consider when making risk management decisions. Article 5 establishes the role of scientific and economic evidence in making such decisions. Members are cautioned to minimize negative trade effects.

Articles 6 through 13 shift the focus to risk control aspects of risk management. Article 6 cautions members to adapt risk management measures to regional conditions, although it does not use such blatant risk management language. Some risk communication considerations are the subject of Article 7, and Article 8 explicitly considers several implementation procedures. The remaining articles largely address the manner in which the member states interact with one another.

Despite the lack of risk management terminology, the SPS Agreement lays a clear foundation for risk management decision-making. That foundation and the best experiences of the decades since the Agreement was enacted form the basis for the content of this handbook.

6.3.1. Article 5

Article 5, "Assessment of risk and determination of the appropriate level of sanitary or phytosanitary protection," may have the most direct impact on phytosanitary risk management practice of all the SPS Agreement articles. It begins in Section 5.1 by saying, "Members shall ensure that their sanitary or phytosanitary measures are based on an assessment." This has been interpreted rather narrowly by many in the phytosanitary community to suggest that risk management decisions must be based on the contents of the risk assessment and should not consider other factors.

Sections 5.2 and 5.3 identify what shall be taken into account (requirements) in the assessment of risks. Section 5.2 identifies "available scientific evidence; relevant processes and production methods; relevant inspection, sampling and testing methods; prevalence of specific diseases or pests; existence of pest- or disease-free areas; relevant ecological and environmental conditions; and quarantine or other treatment" as relevant components of the risk assessment. It is evident that this listing is not considered preemptive of other considerations as pest risk assessment routinely considers likelihood of establishment and a great many other things not enumerated by the Agreement.

Section 5.3 addresses economic factors to be considered in assessing the risk to plant life or health and determining the measure(s) to be applied to achieve the appropriate level of phytosanitary protection from the risk. The economic factors enumerated project the perspective of domestic producers of plants and plant products that would be affected by the proposed market access. These include:

1. Potential damage in terms of loss of production or sales in the event of the entry, establishment or spread of a pest or disease.
2. Costs of control or eradication in the territory of the importing member.
3. The relative cost-effectiveness of alternative approaches to limiting risks.

Section 5.4 identifies what should (recommended not required) be taken into account when determining the appropriate level of phytosanitary protection. Minimizing negative trade effects, presumably such as those identified in Chapter 7 of this handbook, is a recommended objective for risk managers. Prohibition or limitations of trade can affect consumer and producer well-being as measured by total surplus. Economic impacts on jobs and income can also be associated with trade effects. Thus, it would seem the Agreement permits consideration of the loss of potential gains from trade, as these are primary negative effects to be minimized.

Phytosanitary risk management decisions should be consistent. Section 5.5 deals with the objective of achieving consistency in the application of the concept of appropriate level of phytosanitary protection. It is recommended (shall) that risk managers (members) avoid arbitrary or unjustifiable distinctions in the levels considered to be appropriate in different situations. The concept of appropriate level of protection (ALOP) has to do with different levels of protection. It is helpful to conceptualize ALOP by imagining that levels of protection can be precisely and uniformly measured on a 1 to 10 scale, where 10 represents maximum protection. This section seems to suggest that if a member decides that a level of protection of 8 is appropriate for a commodity or pest in one case, then a level of protection of 8 would be used for the same commodity or pest in all other cases. The conceptual ideal is presumed to be a constant level of protection in all similar cases. This, of course, is not possible because there is no precise and uniform measure of risk that can be used in the manner of this example. As a result, the level of protection may be left to a matter of subjective argument rather than objective determination.

Section 5.6 requires members to ensure that their phytosanitary risk management measures are not more trade restrictive than required to achieve their ALOP, taking into account technical and economic feasibility. This means if there is another measure that achieves the ALOP that is technically and economically feasible and it is also less restrictive to trade, then that measure ought to be used. For example, imagine a system of five independent measures achieves the ALOP for an importing country. If irradiation is available and achieves the same or better level of protection at a lower cost, then the five independent measures may be considered more trade restrictive than required.

Uncertainty is addressed, somewhat indirectly, in Section 5.7. When the relevant scientific evidence is insufficient, risk managers may employ phytosanitary measures on a provisional basis using the pertinent available information. That information includes information from relevant international organizations and phytosanitary measures applied by other members in similar situations. When provisional measures are taken, the member is obligated to obtain the additional information necessary for a more objective assessment of risk. Members are also obligated to review the phytosanitary measure within a reasonable period of time.

The final section, 5.8, of Article 5 provides the right for a member to request an explanation of the reasons the chosen phytosanitary measure is required. This explanation can be requested if a phytosanitary measure constrains or has the potential to constrain a member's exports and it is not based on relevant international standards, guidelines, or recommendations.

6.4. IPPC and Pest Risk Management

The SPS Agreement recognizes the IPPC as the only acceptable international standard-setting body with regards to plant health. The IPPC is a legally binding international agreement that develops standards for addressing global phytosanitary concerns (Schrader and Unger, 2003). It aims to ensure common and effective action for preventing the spread and introduction of pests of plants and plant products and promoting measures for their control (IPPC, 2005). The IPPC has been addressing invasive species since 1951. A supplement to ISPM (International Standards for Phytosanitary Measure) 11 (Annex 1) in 2003 extended the scope of the IPPC's interests beyond "pests of plants" to include "plants as pests." This includes weeds or invasive plants and their impacts on the environment, biological diversity, ecosystems, and wild flora (Hedley, 2004).

> ISPM 2: Guidelines for Pest Risk Analysis, endorsed in 1995
>
> ISPM 11: Pest Risk Analysis for Quarantine Pests including Analysis of Environmental Risks and Living Modified Organisms, endorsed in 2001.

The IPPC has developed and adopted two ISPMs to provide guidance on pest risk analysis. These two standards are important because the WTO has adopted pest risk analysis as the basis for countries restricting trade for plant protection and biosecurity reasons. These ISPMs influence national and regional biosecurity policy and regulations. In any instance where the IPPC might differ from the SPS Agreement, the Agreement and the WTO shall prevail.

ISPM 5 defines a "pest" as any species, strain, or biotype of plant, animal, or pathogenic agent injurious to plants or plant products. Thus, a pest is an organism capable of harming plants

(Devorshak, 2012). It defines a "quarantine pest" as "a pest of potential economic importance to the area endangered thereby and not yet present there, or present but not widely distributed and being officially controlled."

The IPPC says that risk analysis shall be used to determine which pests are quarantine pests and that risk analysis should follow the IPPC framework (Lindgren, 2012). That pest risk analysis framework comprises initiation, risk assessment, and risk management. Once a pest or an invasive plant is identified as a potential threat to a country's plant life or natural environment, the scientific justification for prohibiting or allowing the import is to be based on pest risk analysis.

The IPPC view of risk management might best be summarized as protecting the world's plant resources from pests. This is done primarily by providing harmonized guidelines like ISPMs and facilitating implementation of these guidelines through capacity development. These contributions to risk management have been more in the arena of risk control measures than in the development of an all-encompassing risk management framework. The list below shows the ISPMs pertaining to risk management. The IPPC role in phytosanitary risk management has been integral to the progress of pest risk analysis and its spread around the world.

- ISPM 3: Guidelines for the export, shipment, import and release of biological control agents and other beneficial organisms.
- ISPM 4: Requirements for the establishment of pest-free areas.
- ISPM 7: Phytosanitary certification systems.
- ISPM 9: Guidelines for pest eradication programs.
- ISPM 10: Requirements for the establishment of pest-free places of production and pest-free production sites.
- ISPM 12: Phytosanitary certificates.
- ISPM 13: Guidelines for the notification of non-compliance and emergency action.
- ISPM 14: The use of integrated measures in a systems approach for pest risk management.
- ISPM 15: Regulation of wood packaging material in international trade.
- ISPM 17: Guidelines for the use of irradiation as a phytosanitary measure.
- ISPM 20: Guidelines for a phytosanitary import regulatory system.

- ISPM 22: Requirements for the establishment of areas of low pest prevalence.
- ISPM 23: Guidelines for inspection.
- ISPM 24: Guidelines for the determination and recognition of equivalence of phytosanitary measures.
- ISPM 26: Establishment of pest-free areas for fruit flies.
- ISPM 28: Phytosanitary treatments for regulated pests.
- ISPM 29: Recognition of pest-free areas and areas of low pest prevalence.
- ISPM 30: Establishment of areas of low pest prevalence for fruit flies.
- ISPM 32: Categorization of commodities according to their pest risk.
- ISPM 33: Pest-free potato micro-propagative material and mini-tubers for international trade.
- ISPM 35: Systems approach for pest risk management of fruit flies.
- ISPM 36: Integrated measures for plants for planting.

6.5. Non-tariff Measures

SPS regulations play a key role in protecting agricultural resources by protecting plant health. Such regulations can also be used as non-tariff measures (NTMs) to favor domestic producers or to discriminate against certain importers.

Under the SPS and Technical Barriers to Trade (TBT) Agreements, WTO members agreed to establish regulations and standards that achieve domestic goals in the least trade-distorting manner. Domestic pressures in some countries and differences in the interpretation of WTO rules have generated areas of conflict among trading partners, including:

- the appropriate level of tolerance for risks to health and safety
- the use of the precautionary principle, in which temporary regulations are issued to prohibit a new product or technique because the available scientific evidence on its possible risks is viewed as uncertain or incomplete

- the appropriate response to consumer demands for information beyond that necessary to ensure health and safety (Zahniser *et al.*, 2018).

The Agreement has had significant impacts on international trade. The signing of the Uruguay Round Agreement on Agriculture (URAA) marked a significant shift in the focus of agricultural trade policy concerns. Non-tariff obstacles replaced border-related costs such as tariffs, quotas, and export subsidies as a more obscure and potentially trade distorting set of behind-the-border regulatory policies. NTMs is a term used to describe the universe of standards and regulatory policies adopted by governments to meet public policy objectives such as food safety and the protection of plant, animal, and human health. NTMs are policy measures that can potentially cause significant economic consequences for international trade in plants and plant products (Grant and Arita, 2017).

SPS measures occupy a special place in terms of consequences for international trade of plants and plant products among the NTMs that affect agricultural trade. SPS measures are widespread in agri-food trade because of the sensitive nature of issues such as food safety and the protection of plant and animal health from pest and disease risks (Grant and Arita, 2017). The SPS Agreement allows countries to adopt their own standards provided they are based on a risk assessment, not discriminatory between countries with similar conditions, and are minimally trade distorting. These conditions are intended to prevent the disingenuous use of SPS measures as instruments of protectionism (Josling *et al.* 2004). SPS measures are the most frequently encountered NTMs in agri-food trade according to the United Nations Conference on Trade and Development (UNCTAD, 2013). A small sample of NTM business surveys conducted by the World Bank and International Trade Center confirms SPS measures as among the most relevant impediments to exports (World Bank, 2008; ITC, 2011).

If a country believes that a policy violates the SPS or TBT Agreement, it may initiate a dispute settlement case at the WTO. The WTO Committee on Sanitary and Phytosanitary Measures (WTO SPS Committee) is one place where WTO members try to resolve contentious issues regarding SPS regulations. Formally raised trade issues are known as Specific Trade Concerns (STCs). Most STCs are resolved before they escalate to the level of a formal dispute.

6.6. Summary and Look Forward

Here are five things to remember from this chapter:

1. The WTO enacted the SPS Agreement to protect international trade, e.g., in plants, from artificial trade barriers.
2. The roles of scientific principles, scientific evidence, and economic concerns are well established by the SPS Agreement as means for making risk management decisions.
3. Although the SPS Agreement does not mention risk analysis in general or risk management in particular, it provides a clear and sound foundation for the conduct of phytosanitary risk management.

4. The IPPC is established by the SPS Agreement as the only acceptable international standard-setting body with regards to plant health; it has greatly advanced the cause of risk control through its series of ISPMs which are devoted to some aspect of risk management.
5. The SPS Agreement has given rise to a set of NTMs that have altered the focus of agricultural trade policy concerns.

The SPS Agreement sets a powerful stage for considering risks associated with plant trade. The consequences of that trade are essential to managing phytosanitary risks. Too often parties to plant trade and pest risk analysis emphasize the probability of introduction of a pest over the consequences of that introduction. The next chapter explores the consequences associated with plant trade issues by focusing on the various positive and negative consequences of trade as well as of risk management measures.

6.7. References

Andersen, MC., Adams, H., Hope, B. and Powell, M. (2004) Risk assessment for invasive species. *Risk Analysis*, 24, 4.

Devorshak, C. (ed.) (2012) *Plant Pest risk Analysis: Concepts and Application*. CABI International, Wallingford, United Kingdom.

Grant, J. and Arita, S. (2017) *Non-tariff Measures: Assessment, Measurement and Impact*. IATRC Commissioned Paper Series 21.

Gruszczynski, L. (2006) The role of science in risk regulation under the SPS Agreement. *EUI Working Papers Law 2006/03 European Institute University Department of Law*.

Hedley, J. (2004) The International Plant Protection Convention and invasive species, Miller, M.L. and Fabian, R.N. (eds) *Harmful Invasive Species: Legal Response*. Environmental Law Institute, Washington, District of Columbia, USA, 185–201.

IPPC (2005) Identification of risks and management of invasive alien species using the IPPC framework. *Proceedings of the Workshop in Braunschweig, Germany 22–26 September 2003*. International Plant Protection Convention Secretariat, FAO, Rome, Italy.

IPPC (2019) International Standards for Phytosanitary Measures, Publication No. 11: *Pest Risk Analysis for Quarantine Pests Including Analysis of Environmental Risks and Living Modified Organisms*. Secretariat of the International Plant Protection Convention (IPPC), Food and Agriculture Organization of the United Nations, Rome.

ITC (International Trade Center) (ITC) (2011) "Sri Lanka: Company Perspectives." An ITC Series on NTMs, Geneva, ITC, http://www.intracen.org/uploadedFiles/intracenorg/Content/Publications/NTM%20Report_Sri%20Lanka.pdf.

Josling, T., Roberts, D. and Orden, D. (2004) *Food Regulation and Trade: Toward a Safe and Open Global System*. Institute for International Economics, Washington D.C.

Lindgren, C.J. (2012) Biosecurity policy and the use of geospatial predictive tools to address invasive plants: updating the risk analysis toolbox. *Risk Analysis*, 32, 1.

Schrader, G. and Unger, J. (2003) Plant quarantine as a measure against invasive alien species: the framework of the International Plant Protection Convention and the plant health regulation in the European Union. *Biological Invasions*, 5, 357–364.

UNCTAD (2013) *Non-tariff Measures to Trade: Economic and Policy Issues for Developing Countries.* United Nations Publication, ISSN: 1817-1214, http://unctad.org/en/PublicationsLibrary/ditctab20121_en.pdf.

World Bank (2008) *A Survey of Non-tariff Measures in East Asia and Pacific Region: Policy Research Report*, Washington, D.C., 'http://documents.worldbank.org/curated/en/2008/01/9065965/survey-non-tariff-measures-eastasia-pacific-region-policy-research-report.

Zahniser, S., Beckman, J. and Heerman, K.E.R. (2018) *World agricultural trade experiences sizable growth but still faces barriers*. *Amber Waves Magazine,* January/February, https://www.ers.usda.gov/amber-waves/2018/januaryfebruary/world-agricultural-trade-experiences-sizable-growth-but-still-faces-barriers/.

7

Consequences and Risk Management

A mountain is composed of tiny grains of earth. The ocean is made up of tiny drops of water. Even so, life is but an endless series of little details, actions, speeches, and thoughts. And the consequences whether good or bad of even the least of them are far-reaching.

Swami Sivanda

7.1. Introduction

There is no risk without an undesirable consequence. That consequence can be a loss of plant life and quality or it may be expected benefits from trade that are not realized. Absent a consequence, there is no risk; absent a risk, there is no need for risk management, ergo consequences are essential to risk management. As a matter of practice, the pest risk analysis (PRA) process does not currently take the economic benefits of trade explicitly into account in its decision-making. The consequences of interest in a PRA tend to be restricted to damage to plants in the importing country. This narrow focus has the ability to constrain the rationality of phytosanitary risk management decisions, resulting in economically inefficient decisions, as demonstrated in Chapter 29. If risk management is to ever become a truly shared responsibility of importers, exporters, end users, and national plant protection organizations (NPPOs) it must take a more holistic view of the full array of consequences

associated with international plant trade. The current chapter presents information about promising directions in which phytosanitary risk management can grow by describing the range of consequences of risks associated with international plant trade. These consequences are absorbed by individuals, companies, and countries.

This chapter proceeds by revisiting the notion of risk in a plant trade context. There are consequences for individuals, industry, regions, countries, and sometimes ecosystems associated with every risk management decision. Ideally, the risks associated with trade and with pests would be managed jointly by industry and government. Reality is not yet close to this ideal, therefore the consequences of the risk–risk tradeoffs inherent in plant trade risk management are noted. The remainder of the chapter is devoted to describing the range of economic consequences, positive and negative, associated with trade and pest risk management. This includes considering the positive and negative consequences of international trade as well as the consequences of risk management, which include potential losses attributable to pests through the costs of managing pest risks.

7.2. Risk Revisited

Pests are hazards but they do not develop into risks unless there is some probability that they

---

will cause adverse consequences. The mere entry of a pest into a new region is insufficient for a risk. That pest must have the ability to establish and cause harm to result in a risk. That does not always happen. Pests may arrive in insufficient numbers or be unable to find the life requisites necessary for a sustainable colony capable of causing harm.

There is a tendency on the part of some people to equate the presence of a pest with phytosanitary risk. Some would regard the mere entry or establishment of a nonindigenous pest as sufficient evidence of an adverse consequence. The global community demands more of a risk. There must be evidence of potential harm that results from the introduction of a pest. There is no phytosanitary risk if there is no probability of harmful consequences. Introduction alone is not sufficient evidence of risk. Risk management measures taken on the basis of entry or establishment alone cannot be supported as sound risk management decisions in the absence of evidence of resulting harm. Thus, risk requires more than the presence of a hazard, it requires evidence of harmful consequences.

International plant trade comes entwined in two significant categories of risk. There is the risk of potential but uncertain gains from trade and there is the risk of damage to plants and plant health from pests introduced by trade. The risk management responsibilities for these risks are cleanly and clearly divided. NPPOs are responsible for managing the latter risk and private industry is, generally, responsible for managing the former risk. Although a single unified approach to risk management makes a great deal of theoretical sense for any kind of optimizing behaviors, e.g., social well-being, profit, or cost, the reality is that these risks are currently managed by different entities with different objectives. At a minimum, this insures the existence of a risk–risk tradeoff with many phytosanitary risks. There is a potential for uncertain gains from trade that should be weighed against the potential losses from harmful pests when considering the net consequences of new international plant trade proposals. These trade offs exist whether they are considered or not.

An industry that is proposing new trade must be aware of the potential negative externalities the trade could impose on third parties. NPPOs that consider regulation of trade must be

aware of the potential reductions of beneficial impacts on domestic consumers as well as others who would benefit from that trade. The costs of risk management will, of course, always be a concern. Such a holistic approach to trade and all associated phytosanitary risks does not yet exist and these goals remain aspirational at this point in time. Nonetheless, a sound understanding of the potential consequences of the risks of loss, the risks of potential gains, and the costs of each is essential to good phytosanitary risk management. The best risk managers would clearly identify the potential losses and the potential gains associated with each of their decision options. Best practice phytosanitary risk management decisions should consider these trade offs explicitly to make effective risk management decisions. Such a broad view of phytosanitary risk management is not the current reality.

7.3. Consequences of Trade

Trade is often described as the exchange of goods between countries, but *countries* do not engage in international trade with the relatively rare exception of country-level commodity trade boards. Instead, it is the individual consumers and businesses within countries that engage in trade. When you buy a mangosteen from Malaysia in a supermarket you are engaging in international trade. The company that buys wheat from Australia for bread baked in New Zealand is engaging in international trade. International trade comes about as the result of the power of consumer preferences and the competitive pressure faced by retailers to satisfy them. The benefits of international trade are more obvious if we understand why individual firms and consumers choose foreign goods over domestic ones.

7.3.1. Competition and choice

Individual choice and competition among producers are two factors that compel international trade. All consumers face a budget that limits the choices they can make in satisfying their wants and needs through the goods and services they buy. Consumers obtain the products they most want at the best possible prices available to them. When purchasing strawberries in the local off-season, consumers are interested in availability, taste, and price more so than where the strawberries were grown. Consumers engage in international trade in plant products precisely so they can enjoy diversity, availability, quality, and price benefits.

Producers are faced with the competitive realities of the market, which include consumer demand for year-round availability of high-quality products at the lowest prices possible. Global competition in these markets pressures businesses to pursue cost-minimizing strategies. These strategies affect everything from the way production is engineered through the way the organization is managed. They also include obtaining factors of production at the lowest possible cost. When foreign firms can provide plant products at a lower cost, domestic firms are compelled to purchase them through international trade. Global competition from rival firms makes such decisions necessary if businesses are going to meet the demands of the marketplace and survive (Taylor *et al.*, 1996).

The discussion of trade consequences that follows considers consequences to individuals, firms, and countries. These consequences occur in a limited number of ways. A market access proposal may be granted, which will allow the plant product into the importing country. This will result in maximum trade benefits. At the other extreme, market access may be denied. This would deny all of the benefits that would have resulted from trade. In between these two extremes would be the identification of known or suspected quarantine pests requiring implementation of risk management controls before the product can enter the importing country. This would presumably result in benefits somewhere between the maximum benefits of unrestricted trade and the absence of benefits from trade denied.

7.3.2. Positive economic consequences

Domestic consumers tend to gain from international trade imports. Domestic producers tend to suffer from imports in direct competition with their own products. Firms engaged in the purchase, transportation, processing, distribution, and sale of imports would also gain from this trade.

Theory and experience both show that the gains from international trade usually exceed the losses, assuring that countries, as a whole, benefit from trade.

Consumers

When trade in plants and plant products is promoted, individuals have greater choice and quality of fruits and vegetables and ornamental plants. They also enjoy extended availability of products that are out of season domestically. Perhaps more importantly, global markets often result in lower prices for products that are imported. These lower prices enable consumers to purchase desired quantities of products for less income than would otherwise be possible, freeing income for the purchase of additional goods and services. The lower prices also make the purchase of more fruits and vegetables possible, bringing potential health benefits along with tastier meals. These advantages improve both the standard of living and the quality of life for people around the world. Lower prices for plant products result in an increase in consumer surplus, which can be informally defined as value that consumers receive for which they do not have to pay.

Consumer surplus

For one individual, consumer surplus is the amount she or he is willing to pay for something minus the amount actually paid. It is value received over and above the price of the good. For many individuals together, consumer surplus is the area under the demand curve above the price line. Fig. 7.1 illustrates this definition.

Let Q be the quantity of a plant product sold in a domestic market measured in thousands of kilograms and P the price per kilogram of the product, measured in currency units, C.[1] When the price is 40C the market demands 10,000 kilograms. The solid area above the line at 40C and below the demand curve is consumer surplus of 100,000C.[2] Thus, consumers of this product enjoy 100,000C in value for which they do not have to pay.

Imagine that trade results in an increase in the supply of the plant product that reduces the price from 40C to 30C. The downward sloping cross-hatched area represents real savings of

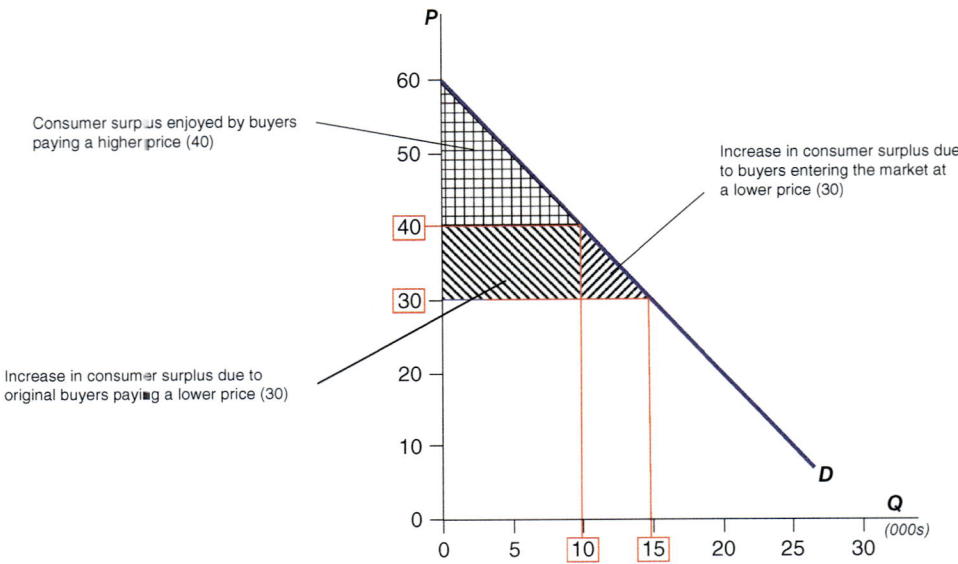

Fig. 7.1. Plant product trade consequences showing consumer surplus changes with a decrease in price.

100,000C[3] that current consumers of the product will realize because of the lower price. The positively sloped cross-hatched area shows consumer surplus accruing to new consumers of the product enticed to buy by the lower price. This consumer surplus amounts to 25,000C.[4] The total benefit to consumers as a result of trade in this example is the two cross-hatched areas, equal to 125,000C.

In Fig. 7.2 this effect is shown in a different way, where it is easier to see how trade enters the picture. Price is again in "C" currency units and quantity is measured in thousands of kg. The intersection of the supply (S) and demand (D) curves at a price of 40C and a quantity of 10,000 kg is the domestic equilibrium before trade. The world price for this product is 30C, lower than the domestic price. Global competition spurs firms to seek supply at the lowest cost and trade lowers the domestic price from 40C to the world price of 30C.

When the price was 40C consumer surplus was equal to area A. Producer surplus, not yet discussed, is the area below the price line and above the supply curve. In this instance, that would be an amount equal to area B+C. The total surplus, i.e. the benefits to consumers and producers for participating in the domestic market equals A+B−C.

When this country engages in international trade the price drops to 30C and consumption increases from 10,000 to 15,000 kg. But the real

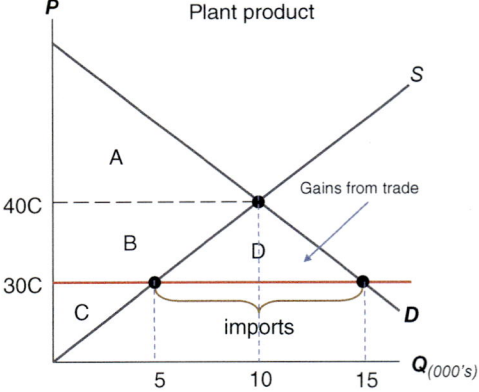

Fig. 7.2. Effect of imports on consumer and producer welfare.

change is more subtle. When the price drops to 30C some domestic producers will be unable to meet that price and will drop out of the market by ceasing to offer the good for sale. The figure shows domestic production falls to 5,000 kg from 10,000. Domestic demand and supply conditions result in a shortage of 10,000 kg. The additional product comes from imports. With trade, consumer surplus, which used to equal area A now equals area A+B+D. That is a substantial increase in benefits to consumers of B+D, above estimated to be 125,000C.

Producers who once enjoyed producer surplus of B+C have had their surplus reduced to C.

Thus, domestic producers could be expected to oppose this trade because they suffer negative consequences. With imports, domestic consumers gain and domestic producers lose. From an economic efficiency perspective, however, the economy as a whole is better off by the amount D and on this basis economists consider trade to be desirable. Imports denied or reduced by a risk management decision would result in the loss of a potential welfare gain by the amount of D.

Firms

Although the preceding discussion pointed out how domestic producers can be harmed by international trade, there are other firms who will gain. Importing firms receive the exports of other countries and sell them at wholesale or retail, realizing profits and income. Firms that transport the goods from nation to nation and those that distribute them within the nation realize gains as do the firms that sell the goods to consumers. To the extent that producers use imports as inputs in their own production process they would share in the consumer benefits as described above.

Countries

At a national level, the gains from international trade exceed the losses. In general, economic growth is enhanced by international trade. The balance of trade and exchange rates may be favorably affected by imports. Likewise, employment might increase or decrease, depending on how direct the competition is between the imported product and domestic markets. It is at the national level that many of the more intangible gains from trade are realized. (See the WTO's ten benefits from trade in Chapter 1, this volume.)

7.3.3. Negative economic consequences

Consumers can suffer from international trade when the world price of goods exceeds the domestic price of goods, spurring domestic producers to export products. Producers, however, benefit from this situation. Groups of people within a nation can suffer adverse consequences as their economy adjusts to changes in international trade patterns.

Consumers

Consumers benefit substantially from imports. A loss of imports, as well as exporting a plant product, however, can have adverse effects on consumers. Figure 7.3 shows these effects by assuming the initial price is 30C.

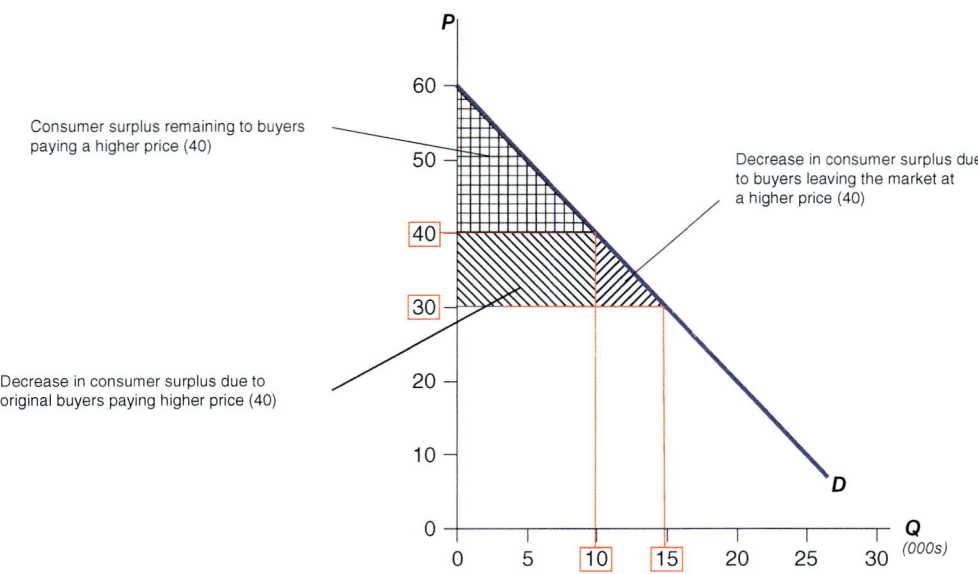

Fig. 7.3. Plant product trade consequences: consumer surplus changes with an increase in price.

Before trade, consumers pay a price of 30C and consume 15,000 kg of product. Exports of the product reduce domestic supplies and cause the price to rise from 30C to 40C. Consumption is reduced to 10,000 kg and consumer surplus falls.

Before trade consumer surplus was 225,000C.[5] After trade, consumer surplus is reduced by 25,000C because some consumers find the new price too high and no longer consume the product. Those who remain in the market now pay 10C more per kg and they lose 100,000C in consumer surplus. This leaves consumers with 100,000C of consumer surplus, 125,000C less well off than before trade. Exports have the opposite effect on consumers than imports do.

Consumers face higher prices for the product in the domestic market and this reduces income available for other purchases. Exports tend to reduce consumers' standard of living. The mix of international trade with imports increasing consumer surplus and the standard of living and exports having the opposite effect will affect different individuals differently, depending on the overall balance of effects.

Fig. 7.4 shows how exports affect consumer and producer surpluses. The intersection of the supply (S) and demand (D) curves at a price of 30C and a quantity of 10,000 kg is the domestic equilibrium before trade. The world price (40C) for this product is higher than the domestic price (30C). Global competition spurs firms to sell their product at the highest price and that means selling in the world market rather than domestically.

When the price was 30C consumer surplus was equal to area A+B. Producer surplus was given by C. The total surplus of domestic consumers and producers was equal to A+B+C. When this country engages in international trade the price rises to 40C and domestic consumption drops from 10,000 to 5,000 kg. Domestic producers, however, are willing to produce 15,000 kg. This would produce a domestic surplus of 10,000 kg of unwanted product at the world price of 40C. This surplus is sold on the world market as exports.

With trade, consumer surplus, which used to equal area A+B now equals area A. That is a substantial decrease in benefits to consumers of area B, calculated above as 125,000C. Producers who enjoyed producer surplus of C are substantially better off now with producer surplus of B+C+D. Domestic consumers might be expected to oppose this trade because they suffer negative consequences. This rarely occurs, however. Whereas losses to producers due to imports are concentrated among a relatively few firms, losses to consumers are widely dispersed over the entire domestic population. Those individuals are less inclined to be aware of how they are affected by exports and even less inclined to protest noticeably.

With exports, domestic consumers lose and domestic producers gain. From an economic efficiency perspective, however, the domestic economy as a whole is better off by the amount D and on this basis, economists consider trade to be desirable. Exports denied or reduced by a risk management decision would function in much the same manner as a price drop in Fig. 7.4.

Firms

Domestic producers actually benefit by the denial or reduction of imports as a result of risk management decisions in a manner similar to a price hike. Countries that export plant products contribute to industry growth, which has proven to be an engine for economic growth in every country that has engaged in exports. The additional markets spur additional production which generates increases in employment and profits. The additional wages earned by the new workers increase spending in the domestic economy. As workers spend their wages on food, clothing, entertainment, and the like they generate new income for these industries, which in turn hire new employees. These multiplier effects spur subsequent rounds of spending and growth, which gradually diminish in size until the impacts are extinguished. Firms that enjoy increased producer surplus are often able to expand and grow.

Prohibition of or reductions in plant product exports results in a failure to attain these benefits. The potential for such prohibitions and reductions comprises much of the uncertainty that renders the benefits of international trade as potential benefits.

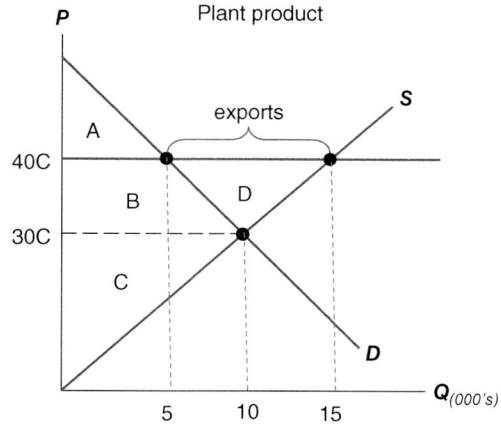

Fig. 7.4. Effect of exports on consumer and producer welfare.

Invasive pests have a very real impact on trade. When a contaminated container is found in port, the cargo owner, importer, or shipper can expect:

* delayed cargo release
* demurrage charges due to cargo holds, and
* unexpected costs associated with having the container quarantined, tarped and treated, cleaned, or re-exported back to origin at the cargo owner's expense.

By taking reasonable steps to keep containers and their cargo clean, shippers may experience:

* reduced port-of-entry inspections
* faster cargo release and
* fewer unexpected expenses, such as demurrage charges due to cargo hold or costs associated with having your container quarantined, tarped and treated, cleaned, or re-exported.

(NAPPO North American Sea Container Initiative, http://www.nappo.org/english/nasci/)

Crop losses

Growers have had to compete with harmful organisms since the beginnings of agriculture. These include animal pests like insects, mites, nematodes, rodents, slugs, snails, and birds; and plant pathogens like viruses, bacteria, fungi, chromistans, and competitive plants or weeds. Boote *et al.* (1983) classify the ways pests can reduce crop productivity by their impacts. There are:

* stand reducers (damping-off pathogens)
* photosynthetic rate reducers (fungi, bacteria, viruses)
* leaf senescence accelerators (pathogens)
* light stealers (weeds, some pathogens)
* assimilation sappers (nematodes, pathogens, sucking arthropods)
* tissue consumers (chewing animals, necrotrophic pathogens)
* inorganic nutrient competitors (weeds).

'For *T. indica* (Karnal bunt) ... Direct costs include yield losses and downgrading affected milling wheat to feed. Reaction costs include downgrading unaffected milling wheat to feed, price and export effects, seed industry costs and quality assurance costs. Costs associated with control include: surveillance and testing; administration and compliance procedures; income loss from cropping restrictions; yield reduction from growing tolerant cultivars (if permitted); additional fungicide inputs; value of standing crops destroyed; costs of destroying growing crops; value of affected grain destroyed; costs of destroying affected grain; treatment of mill by-products; grain processing costs (heat treatment); livestock industry costs; machinery cleaning costs and facility cleaning costs.'

(Sansford *et al.*, 2008)

Oerke (2006) describes how both the quantity and quality of crops can be affected by pests. Quantitative losses result from reduced productivity, which lead to reduced yield for a given area. Qualitative losses can result from:

- reduced content of valuable ingredients
- reduced market quality, e.g. due to aesthetic features like pigmentation
- reduced storage characteristics
- contamination of the harvested product with pests, parts of pests or toxic products of the pests like mycotoxins.

Crop losses are commonly expressed in absolute terms (kg/ha, financial loss/ha) or in relative terms (% loss). Oerke identifies two important loss rates. One is the potential loss from pests, which consists of expected losses without physical, biological, or chemical crop protection compared with yields from a no-loss scenario. The other is actual losses, which comprise the crop losses sustained despite the crop protection practices employed. Oerke *et al.* (1994) describe the calculation of these loss rates. The efficacy of crop protection practices may be calculated as the percentage of potential losses prevented. These losses are, most helpfully, quantified as financial (monetary) losses due to pest infestation. These losses reflect the lost market value of quantity and quality reductions.

The economic damages resulting from individual invasive species can be significant. In the United States the weed, leafy spurge (*Euphorbia esula*), has spread to more than 5 million acres of rangeland in the northern Great Plains, causing estimated production losses, control expenses, and other economic damages in excess of $100 million per year. The eradication program for citrus canker (*Xanthomonas campestris* pv. citri) is expected to exceed $200 million (Andersen *et al.*, 2004).

Non-native, wood-boring insects such as the emerald ash borer and the Asian long-horned beetle are costing an estimated $1.7 billion in local government expenditures and approximately $830 million in lost residential property values every year, according to a study by a research team that included scientists with the U.S. Forest Service, Northern Research Station. (Northern Research Service, 2011).

The U.S. Department of Agriculture budgets for invasive species activities for fiscal years 2013–2016 averaged $1.2 billion (NISC, 2015).

Countries

As was the case with imports, exports benefit the nation as a whole. Exports tend to improve a nation's balance of trade. They increase jobs and income and spur economic growth as noted. At a national level the gains from international trade exceed the losses. Denial of these exports denies the stream of benefits associated with them. Growth in the export industries may or may not exceed losses in other import industries.

Invasive species associated with imports can cause tangible economic damage and may also diminish the provision of nonmarket ecosystem services. These invasive species can cause major crop damage, damage to ornamental plants and subsequent property value loss, damage to commercial, public and private forests, loss of natural habitat, and all the cascading complications that result from these events.

7.3.4. Environmental consequences

Trade in plant products has the potential to cause environmental consequences. Positive consequences can result from trade in plants that are used to restore natural habitats and to enhance the human living environment. Of greater concern with international trade are the negative consequences associated with the introduction of invasive pests to new areas. Over half a century ago, biological invasions were described as ecological explosions (Elton, 1958). These biological invasions are now becoming ecological catastrophes with high costs to our ecology, environment, and biological diversity. In 2000 these costs were estimated to be US$137 billion per year in the US (Pimentel *et al.*, 2000) With increased globalization, the economic impact of biological invasions is only going to increase nationally as well as internationally unless prevention and control strategies are developed through comprehensive risk analysis frameworks (NISC, 2007).

Nonindigenous invasive species can impact native ecological communities directly through infestation, predation, grazing, parasitism, infection, competition, or hybridization. They can also impact communities indirectly by modifying ecosystem functions, e.g., by altering nutrient cycles, hydrologic cycles, and energy flows.

The introduction of nonindigenous infectious agents or exotic disease vectors presents risks to public health and to the health of domestic and wild animal and plant populations (Andersen *et al.*, 2004).

A biological invasion can be broken into four phases: entry, establishment, spread, and impact. Entry consists of one or more arrivals of a nonindigenous species at one or more points of entry into the new environment. During establishment, one or more of the arriving populations reproduces *in situ*, escaping immediate danger of local extinction. The species disperses from its initial site(s) of establishment and occupies available habitat or infects susceptible hosts within its new environment during spread. Impacts occur when an established species persists and competes in its new geographic range (Andersen *et al.* 2004).

Invasive species can cause irreversible changes to ecological communities by altering the composition and abundance of native species. This can have an impact on the flow of ecosystem services from those communities. Ecological services differ in their vulnerability to invasion and the values society attaches to them. Parker *et al.* (1999) suggest the impact of an invader can be measured at five levels: (1) effects on individuals including demographic rates such as mortality and growth; (2) genetic effects including hybridization; (3) population dynamic effects such as abundance, population growth, and so on; (4) community effects like species richness, diversity, trophic structure; and (5) effects on ecosystem processes including nutrient availability, primary productivity, and the like.

Effects on individuals include reduced growth or reproduction, in the face of predators or competitors (Cowie, 1992; Fraser and Gilliam, 1992) and change in the morphology of individual organisms in response to an invader (Crowder, 1984; Busch and Smith, 1995). The genetic impacts of invading species on native species can either be indirect, as a result of altered patterns of natural selection or gene flow within native populations, or direct, through hybridization and introgression. Species are granted special status in ecology and conservation, so it is natural to focus on the impact an invader has on the abundance and dynamics of particular native species, the most extreme population dynamic impact being complete extinction. Community-level impacts are often framed in terms of species number. Some invaders may cause overall reductions in biodiversity (Parker *et al.*, 1999).

Mimosa pigra, a noxious weed in northern Australia, converted thousands of hectares of open sedge wetland to shrubland, accompanied by a loss of particular native flora and fauna (Braithwaite and Lonsdale, 1987; Lonsdale and Braithwaite, 1988).

7.4. Consequences of Risk Management

There are costs of action and costs of inaction. This section develops the range of consequences that result directly from the actions taken or not taken by phytosanitary risk managers.

7.4.1. The big picture

Presented with a new market access proposal, the NPPO risk manager has three basic choices. The trade could be approved with no new risk control requirements, the trade could be denied, or the trade could be allowed subject to some risk control requirements. Each of these three decisions may have positive or negative consequences for stakeholders as shown in Fig. 7.5.

If trade is allowed with no additional risk control requirements, the benefits of trade may be realized by consumers, firms and countries.

SPS Measures: trade impact

A study by the International Agricultural Trade Research Consortium (Grant and Arita, 2017) performed an analysis on Specific Trade Concerns (STCs) recorded by the WTO over the period 1995–2014. STCs are voluntarily raised by members in an effort to avoid a formal resolution process. The study shows that plant health concerns have the highest resolution success rate at 53% based on a total of 99 concerns of which 47 were fully resolved and 5 were partially resolved.

The effects of these concerns on trade are striking and cast a rather dim light on SPS measures maintained by importers that have been raised as concerns by exporters. The average percentage trade difference that compares trade flows with and without an active STC for a given country-pair by product is –34%. SPS measures can have significant impacts on trade.

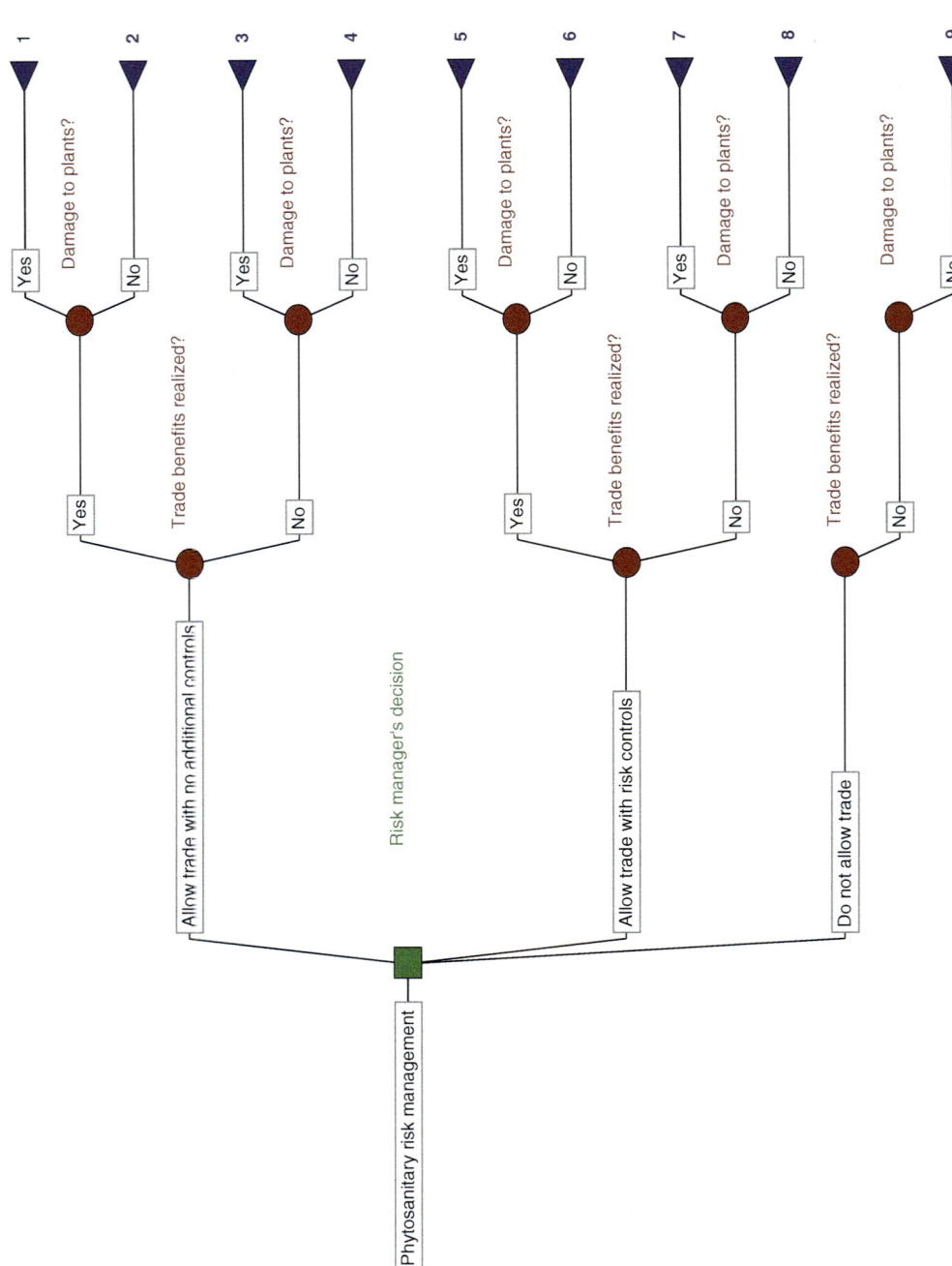

Fig. 7.5. Potential consequences of a pest risk manager's decision on trade.

If the benefits of trade are not realized this will result in negative consequences for all those parties that committed resources to promoting that trade. Trade benefits are the concern of industry's risk managers. Potential damage to plants is the second risk that is always present, which must be considered by the pest risk manager.

Pathways 1–4 show the range of potential outcomes from a decision to allow trade without additional risk controls. Pathway 1 shows benefits from trade being realized but damage to plants also results. In that case, risk managers must weigh the benefits from trade against the damage to plants. If the benefits to trade are tremendous and the damage to plants is minor this may still be an acceptable outcome. However, if the benefits to trade are minor and the damage to plants is extensive this outcome would be unacceptable. Pathway 2 shows the most desirable outcome with trade benefits and no plant damage. If there is an ideal outcome this might be it. Pathway 3 shows no benefits from trade, a negative consequence for all who committed resources to promoting that trade, along with damage to plants. This outcome produces no winners and is undesirable. Pathway 4 shows promoters of trade losing and no effect on plant community stakeholders. If trade is not allowed then it is clear that any potential benefits from trade will not accrue, but it is equally true that no plants will be damaged as a result of trade. This outcome is shown at pathway 9. Pathways 5–8 show the possible outcomes that would result from allowing trade with new risk controls in place.

A closer look at the figure shows the risk–risk tradeoffs a conceptually holistic risk manager faces. Pathways 1 and 5 show that allowing trade without or with additional risk controls could result in damage to plants along with realizing the benefits of trade. Ideally, risk managers would consider the magnitude of the trade gains relative to the plant losses before making such a decision. Pathways 2 and 6 represent best case scenarios where the benefits from trade are realized with no damage to plants. Pathways 3 and 7 represent worst case scenarios where trade benefits are not realized but plants are damaged. Pathways 4, 8, and 9 indicate scenarios where no trade benefits are realized, a negative consequence for those promoting trade, and no plants

are damaged. Not shown in the figure is the smuggling that can result when trade is denied.

What is, hopefully, evident is the fundamental consequence tradeoff that risk managers face in every trade decision they make. If the outcomes could be known before the decision is made, outcomes 3, 4, 7, 8, and 9 would clearly be avoided. Outcomes 2 and 6 would clearly be pursued. The best decision for outcomes 1 and 5 is not immediately obvious. In this case, the risk manager must weigh the gains of one group of stakeholders against the losses of another; clearly economic analysis would be useful for this decision scenario.

The outcomes cannot, of course, be known in advance because of the uncertainty that attends every trade proposal and decision context. There are four potential outcomes for a decision to allow trade without risk controls. Each of these outcomes will have a different likelihood of being realized and these likelihoods are not going to be known with certainty. Likewise, the magnitudes of the positive and negative consequences are not likely to be known with certainty. This reality complicates the risk assessor's job. The same arguments are true for a decision to allow trade with risk control requirements.

It is useful to consider one more aspect of a decision: its implementation cost. There is an extra implementation cost for Pathways 5–8. Industry, consumers, and taxpayers must bear the costs of implementing the required risk controls.

7.4.2. Pest damage and risk management costs

A risk management decision to deny trade can cost society the potential economic benefits of trade. These are described in general terms in preceding sections. This section covers the consequences of the risk manager's decision that have not yet been considered. These would include damage to plants and costs of implementation.

Damage caused by pests

The damage associated with invading species may be environmental, economic, or cost to human health. Environmental damages include potential costs to native species, populations, and genotypes, as well as costs to ecosystems' components and processes that can include rapid mortality loss of abundance, and loss of viable populations. Invasive species can have indirect environmental effects by degrading habitat quality for native species, affecting nutrient cycling, promoting disturbances like wildfire, and exacerbating floods (Mack *et al.*, 2000). When these environmental impacts affect the flow of ecosystem services, the economic and social impacts can be substantial. Economic damages can be incurred by consumers, taxpayers, firms, and governments. Crop losses and the costs of risk controls comprise the major forms of economic damage. The annual costs in lost agricultural production, expenditures for control, human health costs, and other losses in the United States alone was estimated to be about $120 billion/year (Pimentel, 2005). Globally, the cumulative costs would be staggering. Quantifying reductions in native species, loss of native genetic diversity, extinctions, and the like, requires nonmarket valuations. Costs to human health result when crop losses are severe or when human pests invade. Direct costs to health include injury, illness, and deaths due to such pests as West Nile virus and killer bees. Indirectly, secondary effects of pesticides, herbicides, and allergic reactions can occur (Stohlgren and Schnase, 2006).

Invasive pests can have direct and indirect effects. Direct effects include reductions in crop yield and degradation of crop quality. Crop losses can be devastating to industry. Indirect effects include the loss of export markets, changes in producer costs or input demands including control costs, changes in consumer demand, feasibility and cost of eradication and containment, and resources needed for additional research and advice as well as social and other effects (Sansford *et al.*, 2008).

Costs to implement

The economic costs to implement a risk management strategy include the costs of evaluating the containment potential, direct and indirect costs of required risk controls, and opportunity costs related to invasive species containment. There may also be legal mandates and social considerations for controlling various species to consider (Stohlgren and Schnase, 2006).

The risk assessment stage of the risk analysis process includes selecting priority species to control and that, in turn, depends on the potential effectiveness of the available risk controls and restoration efforts relative to their costs. Executing this analysis, alone, can be time consuming and costly, especially when it involves a complex quantitative risk assessment. Implementing the required risk controls may include the costs of manual, chemical, or biological control measures, which will have direct costs. Surveillance, sampling, inspection and monitoring costs can be substantial when required. These costs can include labor, materials, equipment, transportation, processing, and other risk management infrastructure costs. Implementing these controls can entail indirect costs of training personnel in the proper execution and deployment of risk control measures, as well as their administration and management. The effects of control agents on non-target species may entail indirect costs that must also be considered.

Direct effect of pests on plants:

- reduce abundance and size of keystone plant species or plant species that are major components of ecosystems
- reduce endangered native plant species
- significant reduction, displacement or elimination of other plant species.

Indirect effects of pests on plants:

- significant effects on plant communities or on designated environmentally sensitive or protected areas
- significant change in ecological processes and structure, stability or processes of an ecosystem
- effects on human use, e.g., water quality, recreational uses, tourism, animal grazing, hunting, fishing
- costs of environmental restoration.

(IPPC, 2004)

In 2007, the United States Department of Agriculture (USDA) allocated $1.2 billion USD to manage invasive species. Of this amount, about 22% was directed towards early detection, rapid response, and preemptive measures (NISC, 2007).

Imposing a quarantine can be a costly action, so minimizing the costs is always a concern. Quarantines disrupt economic activities in the affected areas. They impose additional costs of monitoring, shipping restrictions, and preventative treatments at the local level (USDA APHIS/PPQ, 2010) and can entail additional costs at the central government level.

Risk managers should consider the opportunity costs of their decisions. Imagine that the resources required to contain species A will be extensive. Suppose the diversion of resources as a result of that choice will enable species B, C, and D to expand unchecked. The expansion of these species is now a cost of the risk manager's decision.

Other social considerations might include harmful human pests like fire ants, killer bees, and West Nile virus. Additional examples of social considerations include threats to species listed as threatened or endangered, impacts on habitats or private property rights, or unfairly distributed economic costs of risk control.

Sometimes priorities for controlling invasive species will be dictated by legal mandates, local regulations, or a public sense of "urgency" based on social considerations that may be unrelated to the invasive species' threats to ecological endpoints. Some jurisdictions, for example, are legally required to address weeds classified as "noxious" regardless of their abundance and spread potential.

Risk analysis for invasive species places demanding requirements on information, science and technology. Gathering and assembling the biology, ecology, natural history, and associated habitat characteristics of a species is a nontrivial information management problem. The volume of biological information and data is massive and humans still play a critical role in processing this information and creating higher-order understanding from dispersed data sets (Stohlgren and Schnase, 2006).

Some risk control measures have higher costs than others. Importing biological controls can be a cost-effective tactic for long-term pest management that reduces insecticide use. While this strategy has great potential benefits, it also can pose a threat to native species, ecosystem function, crop production, forests, and other natural resources (McCoy and Frank, 2010).

7.5. Summary and Look Forward

Here are five things to remember from this chapter:

1. There is no risk without one or more undesirable consequences.
2. Every phytosanitary risk management decision has costs of action and costs of no action.
3. There are positive and negative consequences associated with trade decisions.
4. There are positive and negative consequences associated with risk control requirements.
5. Phytosanitary risk managers face risk–risk tradeoffs in every pest risk management decision they make.

Chapter 8 concludes the background section of this handbook with a look at the opportunities that risk management presents. Risk management is a fluid, living and learning thing. It will not always be practiced as it is practiced now. The potential of risk management is explored in the next chapter.

7.6. Notes

[1] C is a fictional unit used to avoid dependence on any particular currency.
[2] Calculated as the area of the triangle, $0.5 \times (60 - 40)C \times 10,000$.
[3] This area is $(40 - 30)C \times 10,000 = 10,000C$.
[4] This area is $0.5 \times (40 - 30)C \times 5,000 = 25,000C$.
[5] This is obtained by $0.5 \times (60 - 30)C \times 15,000$.

7.7. References

Andersen, M.C., Adams, H., Hope, B. and Powell, M. (2004) Risk assessment for invasive species. *Risk Analysis* 24, 4.

Boote, K.J., Jones, J.W., Mishoe, J.W. and Berger, R.D. (1983) Coupling pests to crop growth simulators to predict yield reductions. *Phytopathology* 73, 1581–1587.

Braithwaite, R.W. and Lonsdale, W.M. (1987) The rarity of *Sminthopsis virginiae* in relation to natural and unnatural habitats. *Conservation Biology* 1, 341–343.

Busch, D.E. and Smith, S.D. (1995) Mechanisms associated with decline of woody species in riparian ecosystems of the southwestern US. *Ecological Monographs* 65, 347–370.

Cowie, R.H. (1992) Evolution and extinction of Partulidae, endemic Pacific island land snails. *Philosophical Transactions of the Royal Society of London B* 335, 167–191.

Crowder LB (1984) Character displacement and habitat shift in a native cisco in southeastern Lake Michigan: evidence for competition? *Copeia* 1984, 878–883.

Elton, C.S. (1958) *The Ecology of Invasions by Animals and Plants*. Methuen, London, UK.

Fraser, D.F. and Gilliam, J.F. (1992) Nonlethal impacts of predator invasion: Facultative suppression of growth and reproduction. *Ecology* 73, 959–970.

Grant, J. and Arta, S. (2017) *Sanitary and phyto-sanitary measures: assessment, measurement and impact. International Agricultural Trade Research Consortium*. St. Paul, IATRC.

IPPC (2004) International Standards for Phytosanitary Measures, Publication No. 11: *Pest Risk Analysis for Quarantine Pests Including Analysis of Environmental Risks*. Secretariat of the International Plant Protection Convention (IPPC), Food and Agriculture Organization of the United Nations, Rome.

Lonsdale, W.M. and Braithwaite, R.W. (1988) The shrub that conquered the bush. *New Scientist* 120, 52–55.

Mack, R.N., Simberloff, D., Lonsdale, W.M., Evans, H., Clout, M. and Bazzaz, F. (2000) Biotic invasions: Causes, epidemiology, global consequences and control. *Issues in Ecology* 5, 1– 19.

McCoy, E.D., and Frank, J.H. (2010) How should the risk associated with the introduction of biological control agents be estimated? *Agricultural and Forest Entomology* 12, 1–8.

NISC (National Invasive Species Council) (2007) *Fact sheet: National Invasive Species Council Fiscal Year 2007 Interagency Invasive Species Performance Budget*. National Invasive Species Council. Available at: http://www.invasivespecies.gov/global/org_collab_budget/organizational_budget_ performance_based_budget.html (accessed January 31, 2017).

NISC (National Invasive Species Council) (2015) *Invasive Species Interagency Crosscut Budget*. Available at: https://www.doi.gov/sites/doi.gov/files/uploads/nisc_2015_crosscut_budget_summary.pdf (accessed 17 April 2020).

Northern Research Station (NRS) USDA Forest Service (2011) Local government, homeowners paying price for non-native forest insects, U.S. study finds. *ScienceDaily*, 9 September 2011, <www.sciencedaily. com/releases/2011/09/110909195132.htm>.

Oerke, E.C. (2006) Crop losses to pests. *The Journal of Agricultural Science* 144, 31–43.

Oerke, E.C., Denne, H.W., Schonbeck, F. and Weber, A. (1994) *Crop Production and Crop Protection: Estimated Losses in Major Food and Cash Crops*. Elsevier, Amsterdam, Netherlands.

Parker, I.M.D., Simberloff, W.M., Lonsdale, K., Goodell, M. and Wonham, P.M. (1999) Impact: toward a framework for understanding the ecological effects of invaders, *Biological Invasions* 1: 3–19.

Pimentel, D. (2005) Environmental and economic costs of the application of pesticides primarily in the United States. *Environment, Development and Sustainability* 7, 229–252, doi: 10.1007/s10668-005-7314-2.

Pimentel, D., Lach, L., Zuniga, R. and Morrison, D. (2000) Environmental and economic costs of nonindiginous species in the United States. *BioScience* 50(1), 53–64.

Sansford. C.E., Baker, R.H.A. Brennan, J.P.,Ewert, F. Gioli, A. *et al.* (2008) The new pest risk analysis for *Tilletia indica*, the cause of Karnal bunt of wheat, continues to support the quarantine status of the pathogen in Europe. *Plant Pathology* 57, 603–611.

Stohlgren, T.J. and Schnase, J.L. (2006) Risk analysis for biological hazards: what we need to know about invasive species. *Risk Analysis* 26, 1.

Taylor, T.G., Fairchild, Gary F. and Harris, H.M. Jr. (1996) *Economic Impacts of Trade:Perceptions and Perspectives*. AGRIS. Available at: https://www.ces.ncsu.edu/depts/agecon/trade/two.html (accessed Feburary 1, 2017).

USDA/APHIS/PPQ (2010) *Integrated Plant Health Inspection System (IPHIS)* Available at: www. aphis.usda.gov/aphis/ourfocus/planthealth/pest-detection/sa_iphis/ct_integrated_plant_health_ information_system (accessed March 30, 2020).

8

Risk Management Opportunities

_____

A wise man will make more opportunities than he finds.

Francis Bacon

8.1. Introduction

Phytosanitary risk management has been practiced for centuries but it is relatively young as a discipline. This chapter speculates on and anticipates its future, because phytosanitary risk management provides a framework that offers substantial future opportunities to industry, national plant protection organizations (NPPOs), and national interests. As the practice of risk management becomes more mature and sophisticated, we can expect better decisions, from setting corporate strategy at the industry, NPPO, and national interest level, to driving major trade initiatives, and operational decision-making. With reliable, timely, and current information on pure and speculative risks, people can make better quality decisions. Risk management also offers a substantial number of other opportunities for all plant trade stakeholders. For example, the systematic and continued practice of risk management can produce:

- opportunities to encourage and promote trade for the benefit of consumers, industry, and national economies
- more efficient resource allocation

- more useful methods for managing uncertainty
- new organizational opportunities to break down barriers between risk assessors and risk managers
- more harmonization of measures, their implementation, and monitoring
- reconciled pathways
- more and better risk management models
- more skilled risk management staff
- new partnerships between government and industry and across countries
- more effective and efficient use of technology
- a systems approach to pest risk management
- more integrated risk management across the plant product supply chain.

Phytosanitary risk management of international plant and plant product trade is currently a bifurcated responsibility, split between the public and private sectors. As a discipline, it is young and it brings with it many opportunities for continued growth and evolution. This chapter explores some of these opportunities in discussions about:

- systems theory
- partnerships
- resource allocation
- technology
- risk management models and skills.

A systems theory approach, made possible by an integrated risk management process, is one of the more promising innovations on the horizon. Partnerships among phytosanitary risk management stakeholders, especially private–public partnerships, might represent one of the more obvious and more challenging opportunities presented by a risk management approach to decision-making. Better alignment of the risk management goals of the public and private sectors could have substantial benefits for both international trade and plant protection. The resource allocation benefits that could result from a vertically aligned systems approach to risk management are reason enough to explore these partnerships more carefully. Technology has the potential to transform the face of risk management in the decades ahead by improving risk management measures as well as the risk management process. Finally, there is reason to hope that a few decades of experience with risk management will produce a risk management community of practice that is far more sophisticated than is now possible.

8.2. Systems Theory

A systems theory approach:

> focuses on the whole, not the parts, of a complex system. It concentrates on the interfaces and boundaries of components, on their connections and arrangement, on the potential for holistic systems to achieve results that are greater than the sum of the parts. Mastering systems thinking means overcoming the major obstacles to building the process-managed enterprise.
> (Jack Welch, Letter to Share Owners in General Electric 1990 *Annual Report*)

Pest risk management has begun a move toward a systems approach, an adaptation of systems theory. ISPM 14 (IPPC, 2019) says in part,

> Systems approaches, which integrate measures for pest risk management in a defined manner, could provide an alternative to single measures to meet the appropriate level of phytosanitary protection of an importing country. They can also be developed in situations where no single measure is available. A systems approach requires the integration of different measures, at least two of which act independently, with a cumulative effect ... A systems approach requires two or more measures that are independent of each other, and may include any number of measures that are dependent on each

other. An advantage of the systems approach is the ability to address variability and uncertainty by modifying the number and strength of measures to meet phytosanitary import requirements.

The transition from treatment or prohibition to a systems approach to pest risk management requires a change in perspective that can lead to different ways to organize institutions.

The Center for Ecoliteracy (2012) offers useful ideas for shifting one's perception to a systems theory approach. The shift in perception begins by reimagining the task, not as preventing pests from introduction into a country or finding new markets for plants and plant products but as an interlocking system of trade and plant health. By thinking about the whole of plant trade and health rather than its parts, it is possible to begin to understand the connections between the different elements. Pest management is no longer a singular task, it becomes part of a greater system of moving plants and plant products around the world for the benefit of the human and natural environments.

In this new systems-view of plant health and trade, the relationships between individual parts may be more important than the parts. The "objects" of interest in this system are not pests and quarantine pests, instead they are networks of relationships. There are pests, plants, people, and organizations interacting with each other and their nonliving environment. Relationship-based processes such as cooperation and consensus become possible means for managing risks.

When we shift our focus from the parts to the whole, this implies shifting our thinking from analytical thinking to contextual thinking. Risk management focuses on knowledge of the trade and plant health whole rather than on any one or two narrow aspects of the issue. Importers learn from exporters; government learns from industry and vice versa.

Western science focuses on things that can be measured and quantified. This leads many to believe that phenomena that can be measured and quantified are more important than those things which cannot be. Some aspects of this plant trade system, like some of the relationships in the food web/supply chain, cannot be measured. They must be mapped to be considered. Mapping these systems could lead to more comprehensive decision-making and more flexible forms of risk management as they identify more intervention points.

The living systems involved in trade and plant health develop and evolve. To shift our perception to a systems theory approach requires a shift in focus from structural to natural and human processes such as evolution, renewal, and change. Sometimes, this can mean that how a problem is solved is more important than the actual solution. Sometimes, the ways we make decisions are as important as the decisions.

Within a system, configurations of certain relationships appear over and over in cycles and feedback loops. The antagonism between government regulators and industry, between exporters and importers, or the implied cat and mouse antagonism between inspectors and pests are simple examples. Understanding how antagonism affects a natural or social system may help us understand how antagonism affects other systems. That understanding empowers us to reduce the antagonism in ways we have not yet begun to imagine.

Thus, instead of measures that focus on killing the pests, a systems theory approach reperceives the issue as a complex process where one group of people in one culture desire to send their goods to another set of people in another culture who, in turn, will sell or share those goods to consumers, potentially exposing different plant species to risk along the way. Such a system involves a great many relationships and practices, each of which represents a potential intervention point. Consequently, a systems theory approach to risk management can involve designated harvest or shipping periods, restrictions on the maturity, color, hardness, or other condition of the commodity, the use of resistant hosts, limited distribution or restricted use at the destination, changes in cultural practices, crop treatment, harvest methods, post-harvest disinfestation, inspection, maintaining the integrity of lots, requiring pest-proof packaging, screening packing areas, pest surveillance, trapping, sampling, and other procedures as well as an openness to accepting the same. Systems theory and the systems approach it gives rise to may be the most promising phytosanitary risk management opportunity for gain.

8.3. Partnerships

A partnership is a voluntary collaborative agreement between two or more parties in which all participants agree to work together to achieve a common purpose or undertake a specific task and to share risks, responsibilities, resources, competencies, and benefits (Kotelnikov, n.d.). The drivers of successful organizations in the knowledge economy reside in connectivity, the intersections of disciplines, and other intangibles that enable an organization to develop and manage complex ecologies around themselves. One of the most promising opportunities that comes with risk management is for new partnerships to emerge, grow, and flourish.

Regional plant protection organizations

- Asia & Pacific Plant Protection Commission.
- Comunidad Andina.
- Comité Regional de Sanidad Vegetal del Cono Sur.
- Inter-African Phytosanitary Council/Conseil phytosanitaire interafricain.
- North American Plant Protection Organization.
- European and Mediterranean Plant Protection Organization/Organisation Européenne et Méditerranéenne pour la Protection des Plantes.
- Organismo Internacional Regional de Sanidad Agropecuaria.
- Pacific Plant Protection Organization.
- Near East Plant Protection Organization.

Strategic partnerships for private firms are essential to survival in an increasingly competitive global business climate. Private–private partnerships between importing and exporting firms have long been a mainstay of international trade. Public–public partnerships are often subject to the vagaries of one nation's political interests vis-à-vis another's. Even so, plant protection organizations around the world have succeeded in establishing important partnerships for government agencies involved in international plant trade. Public–private partnerships are generally less numerous and are often impeded by obstacles, real and imagined. Nonetheless, there are precedents: governmental land management agencies like the United States Bureau of Land Management (BLM) and the United States Forest Service have substantial histories with public–private partnerships.

The BLM (2003) describes partnering as working together to create a common vision and implementation strategy to collectively solve problems. Partners pool certain resources, develop collaborative relationships among stakeholders, and use a consensual decision-making process. Interestingly for phytosanitary risk management, the BLM notes that partnerships have been driven by a field-level response to a pressing management problem. They go on to proclaim that partnerships are the way they do business today.

Building bridges between government, industry, and other stakeholders and across countries is both a significant challenge and an exciting opportunity. Public–private partnerships offer new organizational opportunities to break down barriers between regulatory agencies and those they regulate. The advantages of such partnerships would be similar to those of private partnerships.

Phytosanitary risk managers face wicked problems. A wicked problem is a problem that is ill-defined and complex. It may defy concise and complete description. Wicked problems do not have clear solutions. Their complexity often guarantees multiple views of the problem and its solutions. Its solutions are better or worse rather than right or wrong. Solving wicked problems most successfully requires partnerships.

Partnerships will echo several of the benefits of teamwork. Partners can help with the creative process. New opportunities for collaboration can inspire great ideas. Colleagues from a different field or interest group might complement your own strengths and weaknesses, adding a new dimension to your problem-solving skills. Partners may have experience with processes that are unfamiliar to you. They may also have expertise in areas of particular relevance to a problem or in areas where you lack expertise.

Partners increase your capacity to communicate. Partners in industry, for example, enable you to reach more of the relevant industrial sector, while partners in government may provide you with new channels of communication.

Risk management partnerships provide the opportunity for better alignment of the risk management objectives of government and industry. The objectives of private firms engaging in international trade are not automatically at odds with the risk management objectives of NPPOs. Tearing down some of the walls and the mistrust between government and industry would have long-run benefits for international trade.

Partnerships between government and industry offer opportunities to promote international trade without compromising their stewardship over plant life and health. New risk management approaches like some of those being considered in the pest risk management systems approach or, perhaps, the application of new technologies, might enable trade opportunities to grow while minimizing the risk to plants.

Businesses have a new set of needs in a changing world. Cavusgil (1990) found that a key weakness in the export assistance programs of the U.S. was the lack of coordination between private sector organizations and government. Schuster and Lundstrom (2003) found different levels of intervention by governments in stimulating export activity and concluded the best level of interventionist programs must be tailored to the needs of business. They also found that the success or failure of public–private partnerships depended on the user's knowledge of the partnership and the use of the services. Public–private partnerships are still in the formative stages in many countries and risk management can provide the opportunity to explore new forms of partnership.

8.4. Resource Allocation

One of the risk manager's responsibilities in the risk management model presented in Part 2 of this handbook is establishing risk management priorities and then allocating resources based on those priorities. NPPO authorities deal with numerous plant protection issues at the same time. Resources (budget, personnel, equipment, and time) are rarely sufficient to manage all the issues competing for the risk manager's attention at any given time. That means ranking the risk issues in priority order for risk assessment and risk management are important activities for plant protection regulators.

Integrated pest management (IPM) is a science-based, common-sense approach for reducing populations of disease vectors and public health pests. IPM uses a variety of pest management techniques that focus on pest prevention, pest reduction, and the elimination of conditions that lead to pest infestations. IPM simply means (1) don't attract pests, (2) keep them out, and (3) get rid of them, if you are sure you have them, with the safest, most effective methods.

(Centers for Disease Control, http://www.cdc.gov/nceh/ehs/Docs/Factsheets/What_Is_Integrated_Pest_Management.pdf, accessed November 3, 2019)

Usually, the perceived level of risk each trade issue presents to plants would be the primary criteria for ranking the risks. Then risk management resources can be optimally applied to reduce the overall plant trade risks. Issues could also be prioritized based on other factors. For example, impacts to industry and consumers, serious restrictions in international trade resulting from different measures, the relative ease or difficulty of resolving the issues, and pressing public or political pressure could be the basis for establishing priorities. Ideally, in a more holistic view of the relative risks to plants, impacts on industry and consumers would be one of the primary risk-ranking criteria.

Anecdotally, relationships between NPPOs and industry are more or less cooperative or adversarial. An adversarial risk management process requires more resources than an effective cooperative risk management process. A vertically integrated risk management process would produce better aligned risk management goals of the public and private sectors enabling them to take a more holistic view of the risk associated with plant trade. To the extent that collaboration displaces antagonism, risk management resources could be reduced. A horizontally integrated risk management process could reduce the time and resources required for pest risk assessment domestically, by providing for more efficient interactions between risk assessors and risk managers.

NPPOs that may be tempted to turn to more severe measures of control or prohibition may increase the overall level of resources required to provide a given level of plant protection. By contrast, a systems approach to pest risk management that considers alternative means of providing a given level of plant protection also offers an opportunity to reduce the overall level of risk management resources. Risk managers can make minimizing the cost of plant safety an objective. That means finding cost-effective ways to manage risks rather than relying on go to strategies. New systems approaches and new technologies may be expected to become even more effective strategies for doing this in the future.

8.5. Technology

Risk managers who embrace a more unified systems approach can be expected to use technology to improve risk management outcomes. Using novel technologies to manage pest risk is likely to complicate the risk management chain as these technologies may involve regulatory authorities from other government ministries.

Nanotechnology opens up an array of opportunities in pest management. The potential uses include the formulations of nano-materials-based pesticides and insecticides, producing pheromone gel to trap fruit flies, enhancement of agricultural productivity using bioconjugated nanoparticles (encapsulation) for slow release of nutrients and water, nanoparticle-mediated gene or DNA transfer in plants for the development of insect pest-resistant varieties, and use of nanomaterials for preparation of different kind of biosensors et al., which would be useful in remote sensing devices required for precision farming and monitoring (Bhattacharyya et al., 2011; Herlekar et al., 2014). Pesticides containing nanotechnology may allow for more effective targeting of pests, use of smaller quantities of a pesticide, and minimizing the frequency of spray-applied surface disinfection. These could contribute to improved human and environmental safety and could lower pest control costs (Shatkin, 2009). Bhattacharyya et al. predict that nanotechnology will revolutionize agriculture including pest management over the next two decades.

Currently, the most promising technology for protecting host plants from insect pests is nanoencapsulation. This is a process through which a chemical such as an insecticide is slowly and efficiently released to a particular host plant for insect pest control … The pesticide nanoparticles can be more properly absorbed than larger particles (Scrinis and Lyons, 2007). Nanoencapsulation can also deliver DNA and other desired chemicals into plant tissues to protect host plants against pests.

(Torney, 2009)

Genomic mapping is another technology that offers substantial promise for risk managers in the near future. Genomic resources will accelerate the ability of plant breeders to enhance productivity, pest resistance, and nutritional quality. Because most plants are not yet mapped, many secrets of their genome have yet to be revealed (Casler *et al.*, 2011). Genome mapping offers at least as much promise with the mapping of pest genomes, which can be expected to reveal new management strategies in the decades to come.

Insect and microbial biotechnology are promising environmentally compatible pest management solutions. Genetic engineering techniques have been refined and enhanced in several insect species that cause significant economic loss, making it clear that insect biotechnology will move forward as one of the key tools of pest management in agriculture (Wozniak, 2007). Genetically engineered (GE) insects require relatively novel technologies with inherent biological differences in the use of the two insect biotechnologies, i.e. direct genetic modification and paratransgenesis. These technologies are still in an early stage of development with respect to regulatory guidance in most parts of the world (Wozniak, 2007).

The Food and Agricultural Organization (FAO) of the United Nations defines biosecurity as a strategic and integrated approach that encompasses the policy and regulatory frameworks that analyze and manage risks in the sectors of food safety, animal life and health, and plant life and health, including associated environmental risk. In this respect, biosecurity, as an initiative, echoes the sentiments expressed in the discussion of vertical and horizontal alignment of risk management objectives along with a systems theory approach and prudent use of new technology. Biosecurity is included among the technology opportunities because it includes the introduction of plant pests, animal pests and diseases, and zoonoses, the introduction and release of genetically modified organisms (GMOs) and their products, and the introduction and management of invasive alien species and genotypes. FAO considers biosecurity a holistic concept of direct relevance to the sustainability of agriculture, food safety, and the protection of the environment, including biodiversity (FAO, 2007).

Digital technology will continue to affect risk management in ways already underway as well as in ways that we have not yet anticipated or imagined. As data and information bases grow globally, technology continues to provide more people with greater access to more pest risk data than ever before. Information-sharing has had a tremendous impact on the practice of risk assessment.

Database examples

CAB Abstracts http://www.cabi.org/publishing-products/online-information-resources/cab-abstracts/ (accessed November 3, 2019).
Global Population Dynamics Database https://www.imperial.ac.uk/cpb/gpdd2/secure/login.aspx (accessed November 3, 2019).
PLANTS Database https://plants.usda.gov/dl_all.html (accessed November 3, 2019).
Systematic Mycology and Microbiology Laboratory Fungus-Host Database https://nt.ars-grin.gov/fungaldatabases/fungushost/fungushost.cfm (accessed November 3, 2019).

Knowledge management is the process of capturing, distributing, and effectively using knowledge (Davenport, 1994). This definition has the virtue of being simple, stark, and to the point. Duhon (1998) expanded on this definition to call knowledge management a discipline that promotes an integrated approach to identifying, capturing, evaluating, retrieving, and sharing all of an enterprise's information assets. These assets include databases, documents, policies, procedures, and previously uncaptured expertise and experience in individual workers. Knowledge management, if pursued, offers the promise of helping to harness the vast experience held by the world's pest risk assessors and risk managers. Knowledge management could add substantially to the world's ability to manage its plant pest risks. Chapter 30 revisits knowledge management in considerable detail.

Unmanned drones have been used for pest surveillance and for the delivery of sterile insects. Drones can drastically decrease the cost of surveillance. They are more cost-effective than walking the fields or airplane fly-over filming. Dogs trained to detect wood-boring pests are being deployed in citrus groves.

In what has been pitched as the most ambitious civilian drone aircraft trial ever conducted in Australia, technology company Ninox Robotics used military-grade drones in surveillance roles in rural regions in the battle against invasive pests during July of 2015.

(http://www.beefcentral.com/production/technology-high-tech-drones-used-to-combat-invasive-pest-problem-photos/, accessed November 3, 2019)

As risk science matures into a discipline it can be expected to continue to offer new insights, tools and techniques to assist all those who do PRA. Risk-based sampling of consignments entering a country, the topic of Chapter 29, provides a good example of how risk science can improve the quality of pest risk management.

Before leaving the topic of technological risk management opportunities behind we must consider the potential of social media and communication technology. Social media enables risk managers to communicate directly with their stakeholders. This includes public–private, private–public, and private–private opportunities for communication including supplementing the notices and communications required by the SPS Agreement and risk communication. Social media's potential advantages (adapted from Finch, 2009) include the opportunity to:

- listen to what others are saying about your company/ministry online
- engage in one-to-one communication with your stakeholders
- gather feedback about new decisions and measures
- produce and distribute exclusive content that can be shared
- extend risk communication strategies
- provide platforms to build communities of practitioners

- support knowledge management initiatives
- engage stakeholders in risk management surveillance and monitoring efforts.

8.6. Risk Management Models and Skills

The formal use of risk management in PRA is still young. In many countries risk managers focus on the risk control decisions that arise from pest risk assessments. There is scant evidence of the widespread use of formal risk management models in the plant trade system. This creates the opportunity for more holistic risk management approaches across the plant product supply chain. The integration can be viewed both horizontally, as organizations adopt or adapt more of an enterprise approach to risk management, and vertically as the risk management efforts of the various stakeholders are more closely aligned.

One of the more promising opportunities is for organizations to formally adopt risk management frameworks. This handbook provides the community of practice with a significant reference resource for accomplishing that goal. One of the more immediate benefits of adopting a formal risk management framework would be to break down barriers between risk assessors and risk managers and promote better integrated risk analysis within the individual NPPO.

A second, perhaps more intermediate range benefit would be the production of more skilled risk management staff. As the risk manager position and its responsibilities become better defined, it stands to reason that risk managers will become more effective and efficient. The growing stock of risk management expertise, both organizationally and globally, offers the potential for more successful risk management solutions. More and better risk management models can also be expected as the phytosanitary risk management community of practice gains experience.

8.7. Summary and Look Forward

Here are five things to remember from this chapter:

1. A systems theory approach focuses on the whole, not the parts, of a complex system.

2. Partnering is working together to create a common vision and implementation strategy to collectively solve problems.

3. Risk management provides the opportunity for better alignment of the plant protection risk management objectives of government and the international trade objectives of private firms.

4. A vertically integrated risk management process would produce better aligned risk management goals of the public and private sectors, enabling them to take a more holistic view of the risk associated with plant

trade with a minimum expenditure of resources.

5. Technologies like nanotechnology, genomic mapping, biotechnology, biosecurity, digital technologies, and social media promise to reshape pest risk management in the decades ahead.

This chapter concludes Part 1 of the handbook. The reader is now sufficiently equipped with a background understanding of risk management in a plant protection and trade promoting context to consider that risk management process in detail. That process is described in Part 2.

8.9. References

Bhattacharyya, A., Datta, P.S., Chaudhuri, P. and Barik, B.R. (2011) Nanotechnology: a new frontier for food security in socio economic development. In: *Proceeding of disaster, risk and vulnerability conference 2011* held at School of Environmental Sciences, Mahatma Gandhi University, India in association with the Applied Geoinformatics for Society and Environment, Germany.

BLM (Bureau of Land Management, U.S. Department of The Interior) (2003) *Partnership Questions and Answers*. Available at: http://www.blm.gov/wo/st/en/prog/more/partnerships_home/tools/Frequent_Questions/frequent_questions.html#A2 (accessed February 1, 2017).

Casler, M.D., Tobias, C.M. Kaeppler, Shawn, M., Buell, C. *et al.* (2011) The switchgrass genome: tools and strategies. *The Plant Genome* 4, 3.

Cavusgil, S.T. (1990) Export development efforts in the United States: experiences and lessons learned. In S.T. Cavusgil and M. R. Czinktoa (eds) *International Perspectives on Trade Promotion and Assistance,* 173–183. Quorom Books, Westport, Connecticut, USA.

Center for Ecoliteracy (2012) *Systems Thinking*. Available at: http://www.ecoliteracy.org/article/systems-thinking (accessed February 1, 2017).

Davenport, T.H. (1994) Saving IT's soul: human centered information management. *Harvard Business Review*, March-April, 72 (2), 119–131.

Duhon, B. (1998) It's All in our Heads. *Inform*, September, 12 (8).

FAO (Food and Agricultural Organization of the United Nations) (2007) *Biosecurity Tool Kit*. Rome, Italy. Available at: ftp://ftp.fao.org/docrep/fao/010/a1140e/a1140e00.pdf (accessed February 1, 2017).

Finch, D. (2009) Can social media help your business? Available at: https://www.socialmediaexplorer.com/social-media-marketing/can-social-media-help-your-business/ (accessed August 17, 2016).

Herlekar, M., Barve, S. and Kumar, R. (2014) Plant-mediated green synthesis of iron nanoparticles. *Journal of Nanoparticles*, www.hindawi.com/journals/jnp/2014/140614/.

IPPC (2019) ISPM 14: *The use of Integrated Measures in a Systems Approach for Pest Risk Management*. Secretariat of the International Plant Protection Convention (IPPC), Food and Agriculture Organization of the United Nations, Rome.

Kotelnikov, V. (n.d.) Building partnerships. Available at: http://www.1000ventures.com/business_guide/partnerships_main.html (accessed February 1, 2017).

Schuster, C. and Lundstrom, W. (2002) Public-private partnerships in international trade: a lobbying effort from passive to aggressive in the USA? *Public Affairs*, 2, 3, 125–135.

Scrinis, G. and Lyons, K. (2007) The emerging nano-corporate paradigm: nanotechnology and the transformation of nature, food and agri-food systems. *International Journal for the Sociology of Food and Agriculture* 15 (2), 22–44.

Shatkin, J.A. (2009) *Risk analysis for Nanotechnology: State of the Science and Implications*. CLF Ventures, Inc. Boston, Massachusetts, USA.

Torney, F. (2009) Nanoparticle mediated plant transformation. Emerging technologies in plant science research. Interdepartmental plant physiology major fall seminar series. *Physics* 696.

Wozniak, C. (2007) Regulatory impact on insect biotechnology and pest management. *Entomological Research* 37, 4, 221–230.

Part Two:

Pest Risk Management Model

The challenge for this handbook is to describe how the phytosanitary risk management world currently functions while also presenting an aspirational view of how it could function more effectively. It is naïve to suggest that there is a broad consensus on best risk management practices for phytosanitary risk managers. It would be even more naïve to suggest that little progress has been made in identifying some good risk management practices. We now know many useful practices to aspire to and Part 2 of this handbook details these concepts in the next nine chapters.

Chapter 9 considers risk management as part of a larger process. It presents a number of models that are useful to aid one's big-picture thinking about risk management and it reviews some of the seminal phytosanitary risk management guidance. Chapter 10 offers an impassioned argument for using a formal risk management process. Few, if any, of the risk managers with some share of phytosanitary risk management responsibility have been *trained* as risk managers. Even fewer stakeholders with an interest in the process have formal training in risk management. Despite these facts, international trade and the need to protect plants from phytosanitary risks calls many people to function as phytosanitary risk managers. Risk managers must spend time doing things that manage risks. There is a critical need for risk managers with either stewardship or fiduciary responsibilities to have a process and to use that process.

Despite some common misconceptions, risk managers and risk assessors must communicate continuously, coordinate as needed, cooperate always, and collaborate regularly if risk management is going to reinforce the links between decisions and risk sciences. The nature of those interactions is the subject of Chapter 11.

A comprehensive pest risk management model for regulators is developed and presented in Chapters 12 through 14. Chapter 12 provides a holistic overview of the conceptual pest risk management model for regulators developed in this handbook. The model builds on the pest risk management framework developed by the International Plant Protection Convention (IPPC). Chapter 13 develops the risk manager's tasks in support of and while planning for the risk assessment in considerable detail. This covers Stages 1 and 2 of the IPPC Pest Risk Analysis Model. Stage 3 risk management responsibilities of this model are developed in detail in Chapter 14, which develops the concepts of formulation, evaluation, selection, and implementation of measures in considerable detail.

The role of uncertainty in risk management is the topic of Chapter 15, beginning with a consideration of the sources of uncertainty followed by the need to convey the relevant uncertainties to risk managers. The chapter concludes with some strategies for managing risks under conditions of uncertainty.

Stakeholders and risk communication are the focus of Chapter 16. The chapter begins by considering stakeholders of all types, a broader group than regulatory risk managers are accustomed to considering. The discussion of risk communication begins by considering what little the SPS Agreement has to say on the topic. It leans heavily on the risk communication lessons learned by the so-called "three sisters" since the advent of the SPS Agreement. The remainder of the chapter considers some PRA goals for risk communication before looking at the shortcomings of the current state of risk communication. The conversation is advanced by considering opportunities for improving PRA risk communication and then examining elements of, types of, barriers to, and strategies for risk communication

developed over the years of recent experience in food safety.

The handbook's treatment of conceptual risk management models concludes in Chapter 17 where an enterprise risk management model is presented. PRA stakeholders from the private industry side are less inclined to follow the more demanding risk management model presented for regulators. The model developed by ISO and the Committee of Sponsoring Organizations of the Treadway Commission (COSO) is a joint initiative of five primarily financial private sector organizations. Their enterprise risk management models are presented as a representative example of the type of risk management model likely to be employed by private industry.

9

Thinking About Risk Management

_____

The first step in the risk management process is to acknowledge the reality of risk. Denial is a common tactic that substitutes deliberate ignorance for thoughtful planning.

Charles Tremper

9.1. Introduction

Risk management is never mentioned in the SPS Agreement. This chapter begins to define the nature of the phytosanitary risk manager's job and explores a few models used to help define it. Fig. 2.5 (see p. 32) shows the five clusters of risk management activity that define the responsibilities of a modern risk manager, whether in private industry or an NPPO. There are a number of tasks that an entity's risk managers must attend to before a risk assessment is initiated. These are the initial risk management activities. They determine the decision context for a risk management activity and may give birth to and set the direction for the risk assessment. There are some risk management responsibilities that are necessary to support the conduct of the risk assessment and they are accomplished in the risk estimation activities. Once the risk assessment is completed, risk managers have a responsibility to understand and respond to its results in their risk evaluation activities. If a risk is judged to be unacceptable, risk control activities begin. Risk assessors may be tasked to assess risk

management controls in a new iteration of the risk assessment at this time. These risk control activities end with the implementation of a risk management decision. At that point, risk monitoring activities are initiated to assure that the desired risk reductions are obtained.

The chapter proceeds by presenting a conceptual holistic approach to risk management, a topic that is revisited repeatedly in this handbook. Each of the five groups of risk manager responsibilities is addressed briefly, ISPM 2's phytosanitary risk management model is presented, then the chapter concludes by presenting several variations of how this model is being implemented around the world. A single recommended pest risk management framework is developed and presented in Chapters 13 and 14.

9.2. A Holistic Risk Management Approach

Moving plants and plant products from one country to another requires a complex system as the international trade risk management continuum of Fig. 9.1 suggests. That system involves a fiscal pathway as well as a physical pathway that includes government and industry. Risks arise along both pathways. Quarantine pests are the hazards along the physical pathway or continuum, designated by blue arrows in the figure; profits, income,

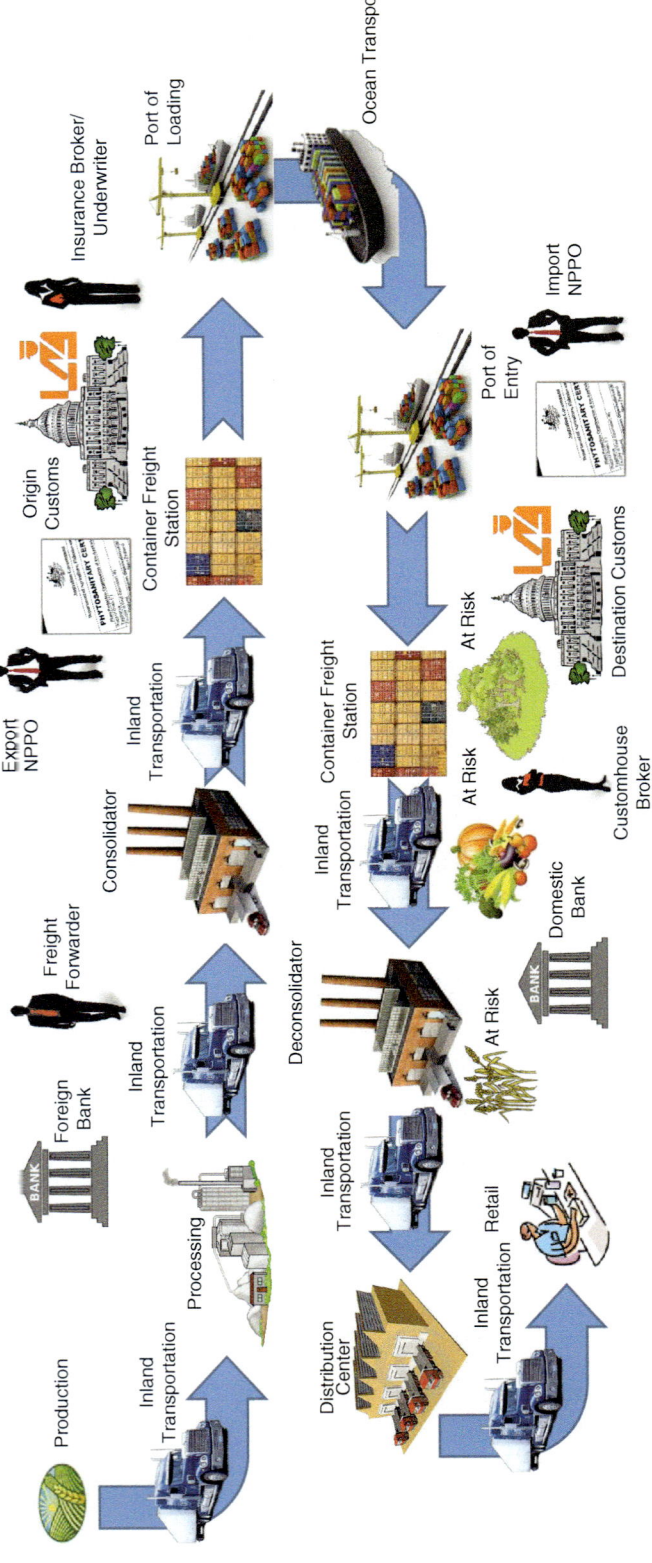

Fig. 9.1. Stylized representation of plant export–import system with physical and fiscal pathways.

jobs, consumer surplus, and satisfaction are the opportunities along the fiscal pathway, shown above the top half of the physical pathway and below the bottom half of the physical pathway. This system involves multiple risks, many potential risk managers, and a variety of possible control points.

The long run challenge for international plant trade risk management is to recognize this

risk management or enterprise risk management tasks lends itself better to adopting a holistic risk management approach.

It also better aligns the organization with the notions of a systems approach as described in ISPM 14. This international standard describes a systems approach as one that integrates measures to meet phytosanitary import

A system is any set of distinct parts that interact to form a complex whole. A closed system is not affected by its environment, think of the universe as a closed system. An open system is affected by its environment. International trade is an open system, so is a quarantine pest risk. From a risk management perspective, an organization, e.g., a company or an NPPO, is also an open system. Problems can arise when an organization attempts to function like a closed system.

An open system has inputs, throughputs, and outputs. For an organization, the inputs include equipment, natural resources, information, stakeholder views, and the work of employees. These inputs are transformed by processes, like risk assessment, risk management, international guidance, and global politics into throughputs. The throughputs yield decisions, products or service outputs, which are released back into the environment to affect other open systems sharing that environment. For example, NPPO decisions affect industry and other stakeholders.

Feedback loops connect the outputs to the inputs. A negative feedback loop indicates a problem. The failure of a risk management decision to control a pest or a risk management decision that unnecessarily restricts trade is an example of a negative feedback loop. Positive feedback loops, by contrast, identify outputs that have worked well. Examples, include risk management measures that prevent the establishment of quarantine pests and risk management decisions that are least trade restrictive.

greater system and the variety of risks it entails. More realistically, the near and midterm goals are to make all the risk managers in this system more aware of the complete array of risks that attach themselves to plant trade issues and to promote the spread of a more disciplined and systematic use of risk management to guide the PRA process.

The stakeholders to international plant trade risks tend to see themselves and their organizations as closed systems. Given the lack of a unified view of all the risks at the present time, it is possible to define three risk "systems" as shown in Fig. 9.2. Exporters face risks, importers face risks and the combined import–export interests give rise to plant health risks.

Risk managers can take a holistic approach to risk within each of these subsystems of risk. That is, even though risk managers may not be ready to see the grand system at the level of international plant trade, it is still preferable for risk managers in the subsystems to see their risk management responsibilities as part of a system rather than as a series of independent actions and projects. This sort of view of the phytosanitary

requirements. In other words, instead of relying on a small set of "go to" risk controls, risk managers are encouraged to see the risk as a system, with inputs, throughputs, and outputs, any one of which might be altered to achieve an acceptable or, at least, a tolerable level of risk.

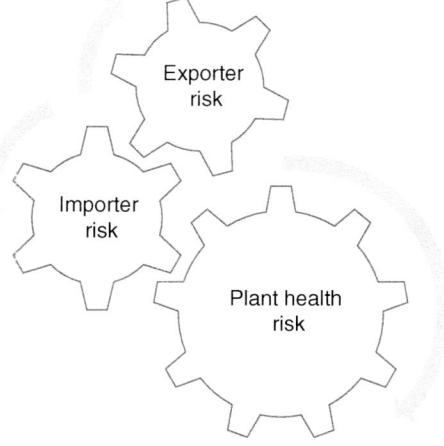

Fig. 9.2. Three systems of risk involved in international plant trade.

Fig. 9.3 (USDA/APHIS/PPQ, 2012) represents general harvest and post-harvest processes that comprise a stylized closed system view of a quarantine pest. It depicts multiple measures controlling a pest risk. The rounded rectangles in the center represent stages in a commodity's path to export. The ovals on the left represent ways a pest gets on a commodity. Preventing one or more of the left oval events is a potential risk control. The ovals on the right represent risk control measures. Implementing one or more of the control measures is also a risk control option. Finding combinations of measures, i.e., preventing the left ovals and implementing the right ovals, that provide the desired degree of risk reduction, is an illustration of the system approach that could be formulated on the export side.

Fig. 9.4 (USDA/APHIS/PPQ, 2012) illustrates a more comprehensive framework for a system approach. Risk control measures are identified beneath the triangle for stages of the

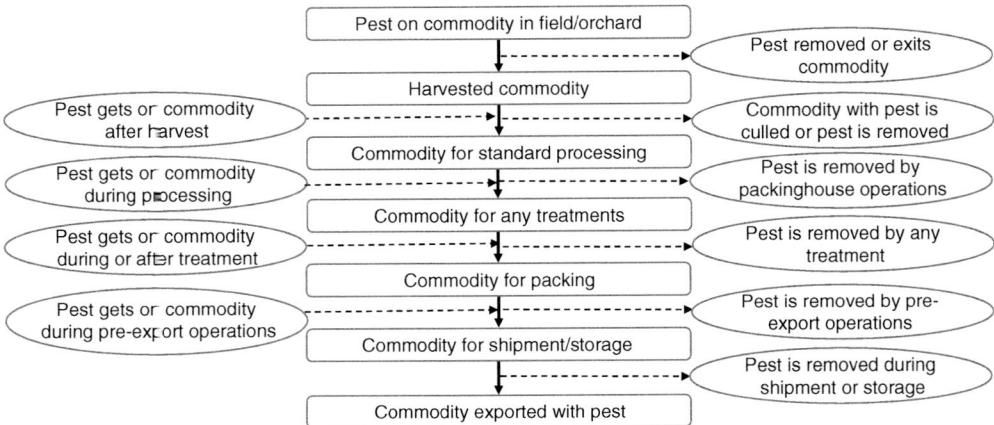

Fig. 9.3. System approach risk control opportunities for export side of the market (adapted from Fig. 1.2, USDA/APHIS/PPQ, 2012).

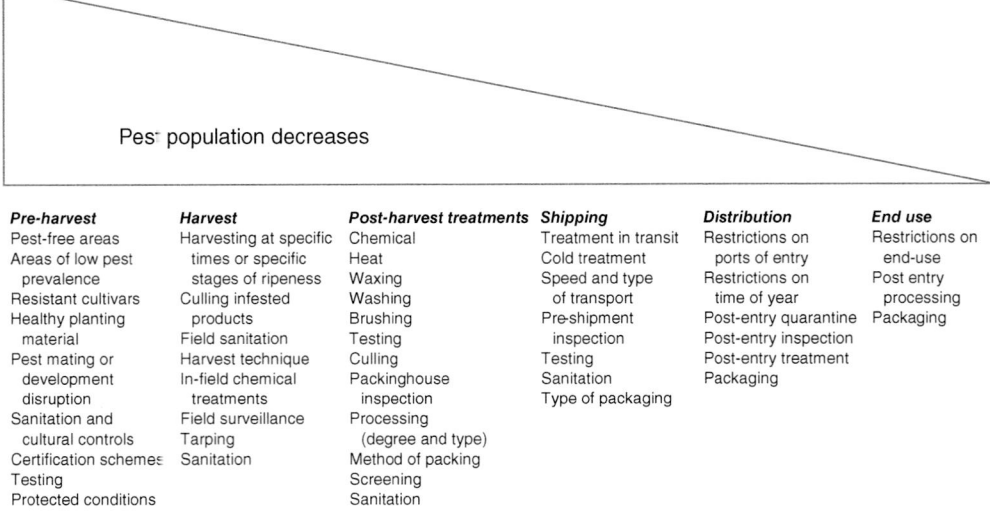

Pest population decreases

Pre-harvest	Harvest	Post-harvest treatments	Shipping	Distribution	End use
Pest-free areas	Harvesting at specific	Chemical	Treatment in transit	Restrictions on	Restrictions on
Areas of low pest	times or specific	Heat	Cold treatment	ports of entry	end-use
prevalence	stages of ripeness	Waxing	Speed and type	Restrictions on	Post entry
Resistant cultivars	Culling infested	Washing	of transport	time of year	processing
Healthy planting	products	Brushing	Pre-shipment	Post-entry quarantine	Packaging
material	Field sanitation	Testing	inspection	Post-entry inspection	
Pest mating or	Harvest technique	Culling	Testing	Post-entry treatment	
development	In-field chemical	Packinghouse	Sanitation	Packaging	
disruption	treatments	inspection	Type of packaging		
Sanitation and	Field surveillance	Processing			
cultural controls	Tarping	(degree and type)			
Certification schemes	Sanitation	Method of packing			
Testing		Screening			
Protected conditions		Sanitation			

Fig. 9.4. System approach risk control measures for entire commodity pathway (adapted from Figure 1.3 and Tables 2–3, USDA/APHIS/PPQ, 2012).

pre-harvest through end-use pathway of a commodity. In some instances, one measure may be sufficient to reduce a risk to an acceptable or tolerable level. In other cases, multiple measures may be required to reduce the risk to an acceptable or tolerable level. These measures may be chosen from the same or different commodity stage to obtain the desired degree of robustness and reliability.

With a systems approach to risk management in mind, let us now examine the general responsibilities of a risk manager.

9.3. Risk Manager's Responsibilities

A specific phytosanitary risk management framework is presented in Chapters 13 and 14. We approach that model in an iterative fashion, beginning with the generic model of Fig. 2.5 (see p. 32). It shows five groups of risk management responsibilities in a continuous loop to capture the iterative nature of the risk manager's job. In an iterative process, the tasks, although presented in a sequential fashion, are not always executed sequentially. Activities in a later task may precede activities in an earlier task. Activities in different tasks may occur simultaneously. Many activities and tasks can be repeated more than once. The entire sequence may be completed several times. That is the nature of an iterative process, and risk management is an iterative process. Iterations are often triggered by reductions in uncertainty, which occur sporadically rather than in a smooth and predictable manner. Although their interests and perspectives on risks may vary for exporters, importers, and regulators, the risk manager's job is conceptually the same, whether we speak of an NPPO risk manager or a private industry risk manager. Each will have responsibility for some version of every task shown in Fig. 2.5.

Initial risk management activities. A risk management activity begins when the risk manager becomes aware of and identifies one or more risks. That risk could be the opportunity for gain associated with the trade of a new item with existing trade partners, the addition of new trade partners for an existing commodity,

or new trade commodities and new trade partners. The potential for gain from these ventures is uncertain and to a great extent it may depend on how the importing country's NPPO regards plant risks associated with the market access proposal. This leads us directly to the other primary risk of concern, the risk of loss, such as damage to plant life that could result from pests associated with a traded plant or plant product.

In an ideal situation, all of these risks might be considered simultaneously by an integrated group of risk managers, so that risk–risk tradeoffs can be considered in decision-making. Such an ideal view is not yet realistic and risks are currently handled separately and by separate entities. Nonetheless, conscientious and effective risk management behooves all parties to consider the risk identification task as comprehensively as possible, even if they will ultimately only address a subset of these risks. Once the relevant risks are identified, the successful risk manager will establish the decision context for the risk management activity that has begun. That context will vary for public and private sector risk managers. Decision context is presented in detail in Chapter 13.

Risk estimation. Ideally, the NPPO risk manager has a significant role in guiding and supporting the pest risk assessment process. They establish the risk assessment policy, provide resources and support for the risk assessment, and identify information needs to be filled by the risk assessment. A similar role is played by the enterprise risk manager in private industry. These tasks in support of risk assessment are part of the risk manager's risk estimation responsibilities and are also described in Chapter 13.

Private industry may be more inclined to manage opportunity risks than government interests would be. The risk manager's role is largely the same for opportunity risks as for risks of loss. An unacceptable opportunity risk is one with potential gains so small or a likelihood of attaining desirable gains so low that the risk is not worth taking.

The risk control task is devoted to identifying measures that will either increase the potential gains or increase the probability of the hoped-for gains occurring. Risk monitoring is often easier for opportunity risks.

Risk evaluation. Once the risk has been estimated, qualitatively or quantitatively, it is the risk manager's responsibility to evaluate the risk. This is a subjective deliberation that considers the results of the risk assessment in concert with other decision criteria in order to determine whether the assessed risk(s) is (are) acceptable or not. Unacceptable risks must be reduced to an acceptable or tolerable level of risk whenever it is feasible and appropriate to do so. Risk assessment requires the risk assessor to account for the uncertainty encountered in the risk assessment process. Risk management requires the risk manager to take that uncertainty into appropriate consideration during decision-making. This first happens as risk managers consider the instrumental uncertainty in determining whether a significant risk of pests of quarantine concern exists.

Risk control. When a risk is judged to be unacceptable the risk manager's responsibility shifts once again, this time to risk control. The primary purpose of the risk control task is to identify a range of risk control measures that could reduce the unacceptable risk to a tolerable or acceptable level. This is where the ISPM 14 notion of the systems approach is most applicable. If the risk control measures were anticipated when the risk assessment was commissioned, their efficacy in reducing the inherent risk should be estimated. If risk control measures are not identified until after the initial risk assessment, additional risk estimation and evaluation work to determine if the expected efficacy of a risk management option achieves a sufficient level of risk reduction and an acceptable/tolerable level of residual risk may be required. The relevant risk control measures may be identified by the risk assessors or the risk managers. The risk manager will often, if not usually, be required to consider the uncertainty associated with the efficacy of the risk control measures considered.

Risk control includes the risk manager's decision-making responsibility. This is where risk managers recommend or decide the actual risk management measures that will be used to reduce the risk to at least a tolerable level. Implementation of the risk control measures is also part of the risk control task. It may take years to implement phased-in risk management measures.

Risk control measures are the focal point of Part 3 of this handbook.

Risk monitoring. Monitoring is rather unique to the risk management framework. Due to the existence of uncertainty throughout the PRA process, any risk management decision is conditional. It is conditioned on the evidence and the uncertainty that existed at the time the decision was made. This usually means there is some uncertainty about how well the risk control measures taken will reduce and manage the risk. Because of this uncertainty about the efficacy of the solution it is necessary to identify a desirable outcome that can be measured, monitored, and evaluated from time to time. When the evaluation shows the desired risk reduction outcomes are not being achieved, the risk management process is reiterated to consider new information and alternative responses to the persistent risk. When the desired outcomes are being obtained, it is usual to continue the successful measures.

9.4. ISPM 2

ISPM 2 (IPPC, 2019a) describes the *Framework for Pest Risk Analysis* that is adhered to by all contracting parties of the IPPC. That process is summarized in Fig. 9.5, adapted from the Appendix to ISPM 2. The figure as presented here suggests a rather circumscribed role for risk management in the PRA process.

The ISPM is prescriptive in its description of the PRA process but it is rather silent on the matter of roles for the risk manager in implementing this process. For example, ISPM 2 in Section 1 says, in part, "When the PRA process has been triggered by a request to consider a pathway ..." but it does not specify who makes that request. Subsequent subsections speak of "The need for a new or revised PRA may arise ...". The need for a PRA is a determination to be made by risk managers. This handbook provides ideas and suggestions about what the risk manager's role should be without altering, in any way, the ISPM guidance on risk assessment and risk management.

Section 1.5 of ISPM 2 describes the conclusion of the initiation stage. It begins with, "At the

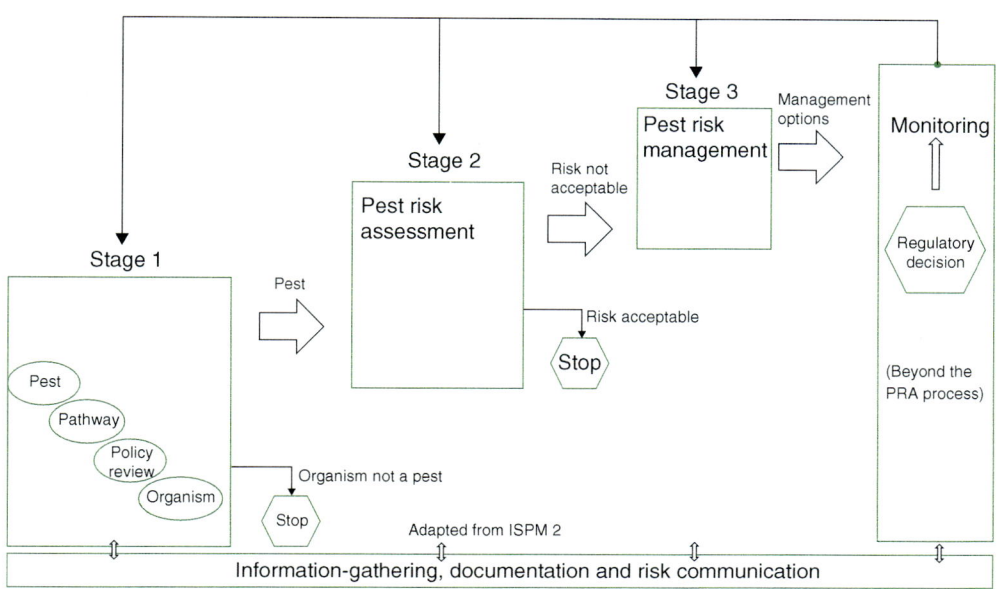

Fig. 9.5. ISPM 2 pest risk analysis flow chart.

end of PRA Stage 1, pests and pathways of concern will have been identified and the PRA area defined." This aligns rather nicely with what other risk assessment processes define as the "scoping or problem identification process," one of the initial risk management activities. Scoping is one of the specific tasks identified for risk managers in Chapter 13, which describes the risk manager's role in Stages 1 and 2 of the pest risk analysis process.

Section 2.3 of ISPM 2 summarizes Stage 3 of the PRA process, pest risk management. The lead sentence says, "Pest risk management Stage 3 involves the identification of phytosanitary measures that (alone or in combination) reduce the risk to an acceptable level," thus seeming to narrowly prescribe the risk manager's role. Chapter 13 will establish a significant role for the risk manager in Stages 1 and 2. Chapter 14 will establish the risk manager's role in Stage 3 of the PRA process.

ISPMs 3, 11, and 21 are linked to ISPM 2 but they follow the same basic model. Like ISPM 2 they do not specify roles for risk managers or assessors. Instead they describe more or less subtle differences in the framework for quarantine pests (ISPM 11), non-quarantine pests (ISPM 21), and biological agents (ISPM 3).

9.5. NPPOs' Refinements of the ISPM 2 Pest Risk Analysis Framework

The ISPM 2 framework is interpreted differently by different NPPOs. This section presents a few representative illustrations of how different entities have chosen to apply the framework. These are presented to illustrate the range of viewpoints on how to implement and apply the framework. The examples chosen illustrate the range of roles and responsibilities that can be assigned to risk managers, in order to legitimize the content of this part of the handbook.

The first representative model, shown in Figure 9.6, shows the PRA model of the United States Department of Agriculture's Animal and Plant Health Inspection Services (APHIS), Plant Protection and Quarantine Plant Epidemiology and Risk Analysis Laboratory. This model illustrates the three stages of the ISPM 2 framework and depicts the separation of the risk management and risk assessment tasks. The model suggests risk managers engage when the risk requires mitigation and that managers and assessors interact during the evaluation of mitigation options.

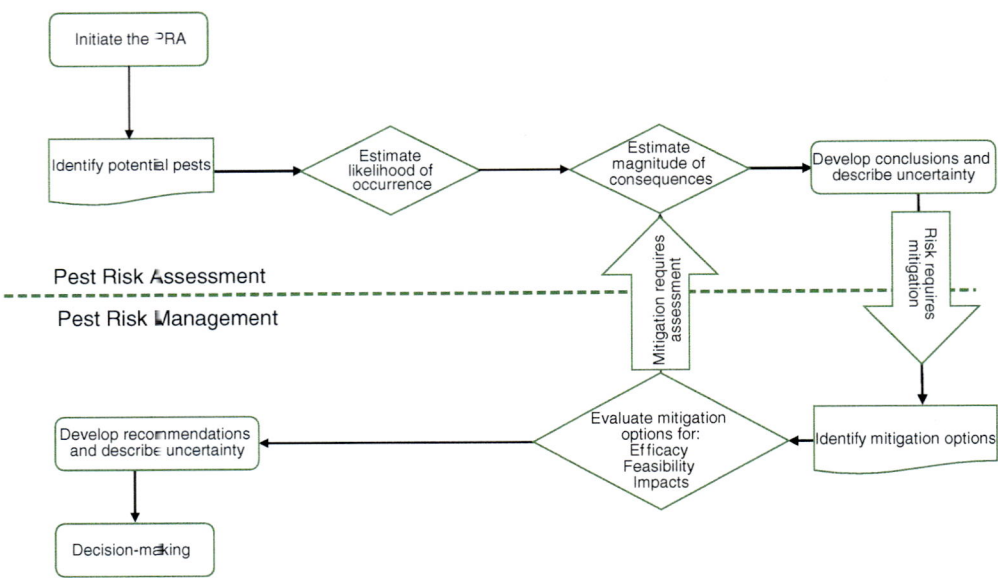

Fig. 9.6. USDA, APHIS Plant Protection and Quarantine PERAL pest risk analysis model.

A more updated look at the risk assessment process is provided in Fig. 9.7. This shows the US NPPO risk assessment model comprises ten processes. Each of these provides potential opportunities for the interaction of managers and assessors, but these interactions are not addressed in the source document. Process 10, risk mitigation is the only of the ten processes that is obvious y a risk management responsibility.

The Council of the European Union's model (2008) is summarized in Fig. 9.8. The basic model is shown on the left while the right displays a communication strategy for the PRA process. Each column on the right of the figure represents a risk analysis stakeholder to be included in communications at various points throughout the process—their identity is not essential to appreciate the model. This framework shows risk mitigation measures as an explicit part of the pest risk assessment document. Some other NPPOs handle pest risk mitigation measures in a document that is separate and distinct from the risk assessment.

The take-away points from this brief presentation of selected models are simple but important.

- Different NPPOs apply the ISPM 2 framework in a variety of ways.
- There is substantial leeway for parsing the ISPM 2 framework into varying levels of detail.
- The role for risk managers can vary substantially among the applications.

Given these simple points, Part 2 of this handbook presents a detailed description of what the phytosanitary risk manager's involvement in the PRA could be if some of the best practices of other communities of practice are adopted or adapted.

9.6. Summary and Look Forward

Here are five things to remember from this chapter.

1. Phytosanitary risks are managed within a complex system that is not always readily recognized by its participants.
2. The NPPO component of the phytosanitary risk community of practice follows the

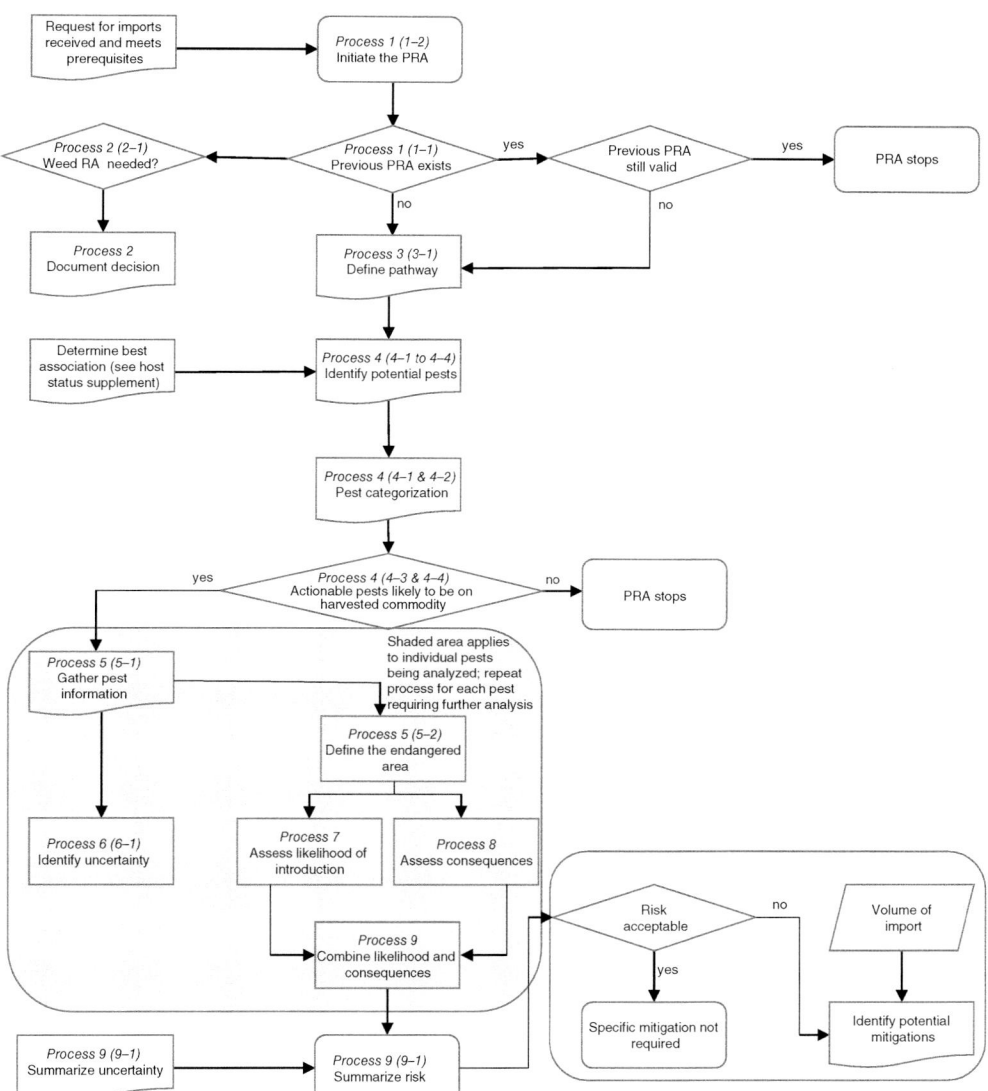

Fig. 9.7. Overview of the PPQ risk assessment model for commodity imports for consumption.

framework described in the IPPC's ISPM 2, which includes an initiation stage, a pest risk assessment stage, and a pest risk management stage.

3. There is no one single best way to implement this ISPM 2 framework.

4. Different NPPOs implement the framework in varying ways, with ample examples of striking differences in the roles of the risk manager in applying the framework.

5. This handbook presents a comprehensive discussion and risk management model that can improve the efficacy and efficiency of phytosanitary risk management.

The next chapter begins the development of a general risk management model by arguing for the need for NPPOs and other risk management entities to establish a risk management framework with clearly defined responsibilities for managers and assessors.

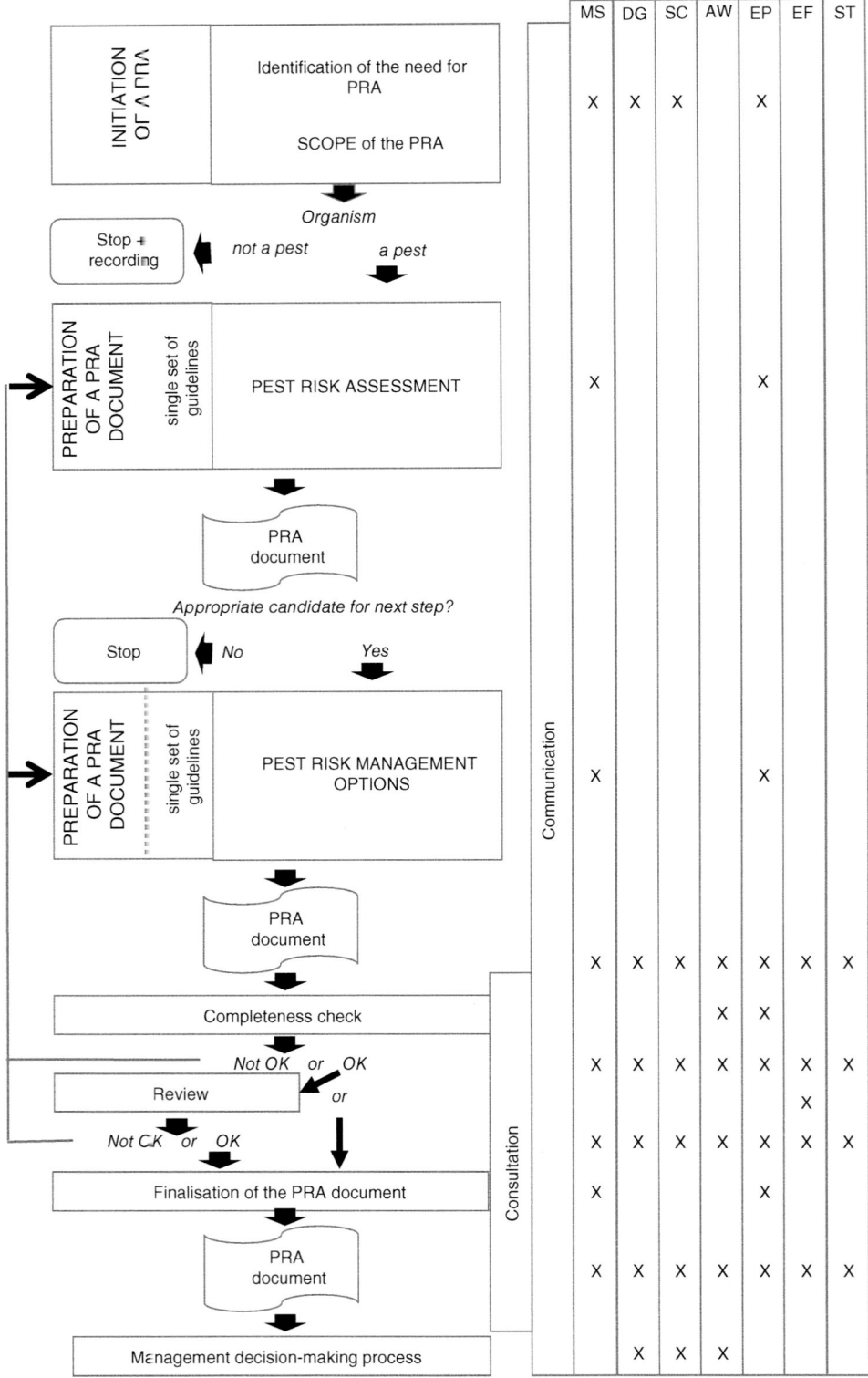

	MS	DG	SC	AW	EP	EF	ST
INITIATION OF A PRA — Identification of the need for PRA / SCOPE of the PRA	X	X	X		X		
Organism — not a pest → Stop + recording / a pest ↓							
PREPARATION OF A PRA DOCUMENT (single set of guidelines) — PEST RISK ASSESSMENT	X				X		
PRA document							
Appropriate candidate for next step? — No → Stop / Yes ↓							
PREPARATION OF A PRA DOCUMENT (single set of guidelines) — PEST RISK MANAGEMENT OPTIONS	X				X		
PRA document	X	X	X	X	X	X	X
Completeness check				X	X		
Not OK or OK — Review / or	X	X	X	X	X	X	X
						X	
Not OK or OK	X	X	X	X	X	X	X
Finalisation of the PRA document	X				X		
PRA document	X	X	X	X	X	X	X
Management decision-making process		X	X	X			

Communication / Consultation

Fig. 9.8. Flowchart of the Council of the European Union procedures of phytosanitary measures setting in regular process.

9.7. References

Council of the European Union (2008) *Pest risk analysis (PRA) process in the European Community*. Addendum to outcome of proceedings from: Working Party of Chief Plant Health Officers (COPHS), Brussels.

IPPC (2017) International Standards for Phytosanitary Measures, Publication No. 3: *Guidelines for the Export, Shipment, Import and Release of Biological Control Agents and Other Beneficial Organisms*. Secretariat of the International Plant Protection Convention (IPPC), Food and Agriculture Organization of the United Nations, Rome.

IPPC (2019a) International Standards for Phytosanitary Measures, Publication No. 2: *Framework for Pest Risk Analysis*. Secretariat of the International Plant Protection Convention (IPPC), Food and Agriculture Organization of the United Nations, Rome.

IPPC (2019b) International Standards for Phytosanitary Measures, Publication No. 11: *Pest Risk Analysis for Quarantine Pests Including Analysis of Environmental Risks and Living Modified Organisms*. Secretariat of the International Plant Protection Convention (IPPC), Food and Agriculture Organization of the United Nations, Rome.

IPPC (2019c) International Standards for Phytosanitary Measures, Publication No. 21: *Pest Risk Analysis for Regulated Non Quarantine Pests*. Secretariat of the International Plant Protection Convention (IPPC), Food and Agriculture Organization of the United Nations, Rome.

USDA/APHIS/PPQ (2012) *Guidelines for Plant Pest Risk Assessment of Imported Fruit and Vegetable Commodities*. Plant Epidemiology and Risk Analysis Laboratory, Center for Plant Health Science and Technology.

10

Establish a Risk Management Framework

The more important the subject and the closer it cuts to the bone of our hopes and needs, the more we are likely to err in establishing a framework for analysis.

Stephen Jay Gould

10.1. Introduction

Risk analysis is a powerful tool that should be used to enhance the scientific basis of regulatory decisions. It is conducted through the efforts of risk management, risk assessment, and risk communication teams. Risk management should be a thoughtful, formal and systematic process of decision-making under uncertainty that integrates the best available science with the relevant social values to manage risks. Risk assessment is a science-based analytical effort to provide risk managers with the evidence they need to make decisions. It should be conducted in an iterative manner that allows refinement of the risk assessment question(s), key assumptions, and data used in the assessment. Risk communication should include the exchange of information within and between the risk analysis teams, with other agencies, and stakeholders, including industry, consumer groups, and other interested parties.

There is no one best way to do risk analysis. Each NPPO will be influenced by its organizational structures and decision-making cultures

> Each private entity involved in international plant trade should adopt an enterprise risk management (ERM) framework to manage risks. An example of an ERM framework is provided in Chapter 17. Although private entities must commit to a risk management framework too, if they will be a risk managing organization, this chapter focuses on the national plant protection organization (NPPO).

necessitating that each NPPOs' risk management framework will look different from every other framework. Nonetheless, each NPPO should adopt a risk management policy or framework to guide their risk analysis process. Such a policy defines, in a general way, how risk analysis, and phytosanitary risk management in particular, are to be conducted in each nation.

Fig. 10.1 shows historical detail of the current risk analysis model used to comply with the SPS Agreement. This figure is constructed from figures one through three of ISPM 2 (IPPC 1996), which show the three stages of the risk analysis process. Risk management comprises Stage 3, which consists of the last three ovals shown at the bottom of Figure 10.1. This is neither adequate to meet the needs of risk managers nor sufficient to represent the evolving role of risk management among the "three sister" standard-setting organizations named in the SPS Agreement, CODEX Alimentarius for food safety, the World Organization for Animal Health

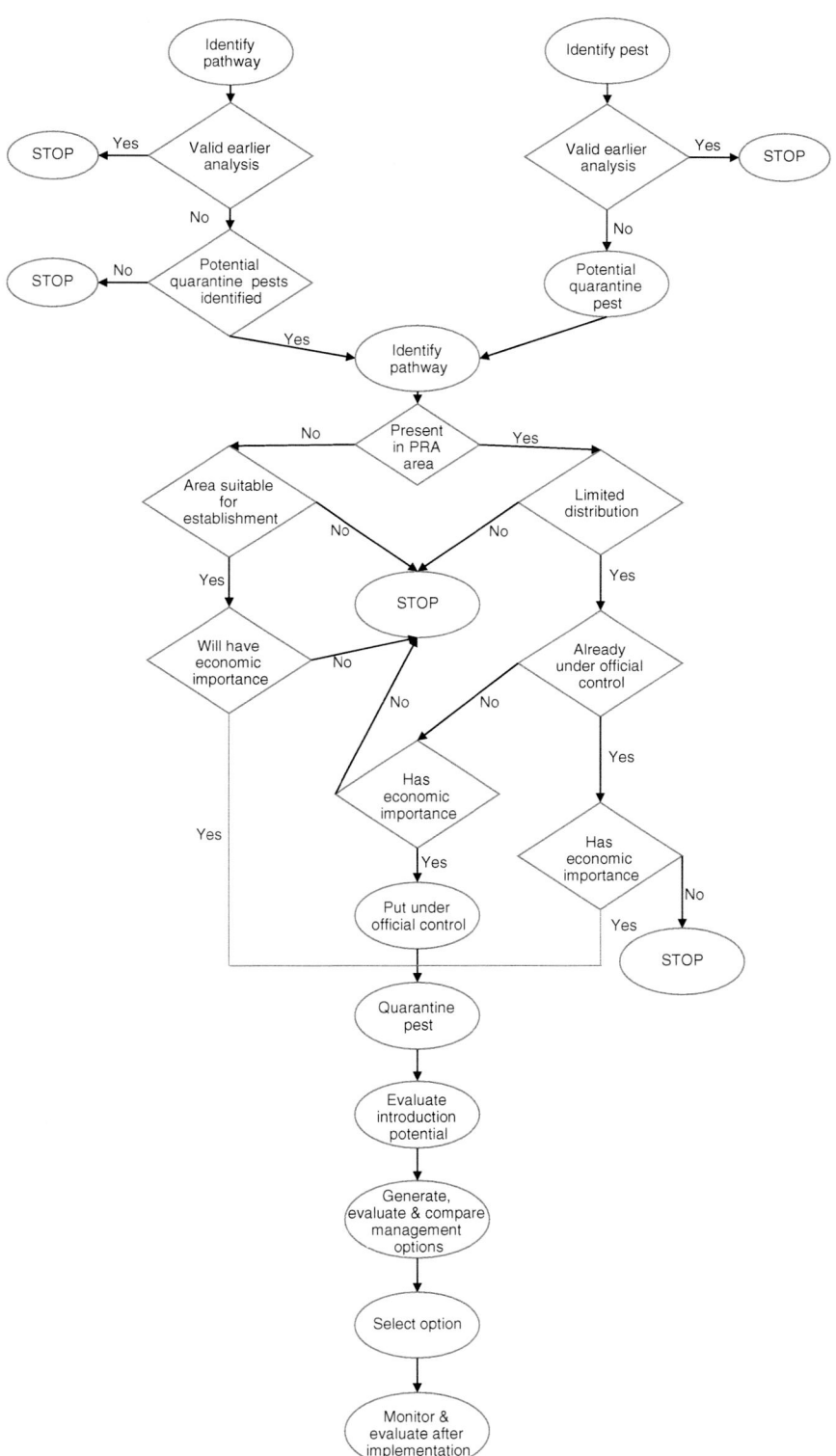

Fig. 10.1. IPPC pest risk analysis process stages 1, 2, and 3 from the since revised ISPM 2 of 1996.

(OIE) for animal health, and the International Plant Protection Convention (IPPC) for plant health. A more thorough and complete risk management process is needed by NPPOs.

This chapter proceeds to argue that every NPPO should establish a risk management process. Then it illuminates the content of that process by focusing, one by one, on the responsibilities of risk managers, risk assessors, and risk communicators in such a process.

10.2 Establish a Risk Management Process

Any entity currently involved in pest risk assessment has some form of risk management activities in place. These activities are, however, often ad hoc, informal, and uncoordinated. As a result, most organizations fall short of a complete, robust risk management process. In addition, many pest risk assessment activities lack the transparency that has become increasingly important to the international community (COSO, 2011). Establishing a formal and systematic risk management process for your organization is a critical and essential first step in moving toward a robust risk management process. It is important that every risk managing entity has a document that essentially says, "This is how we do risk management here."

Risk analysis or risk management?

For simplicity and to sharpen the focus on risk management, we assume that the process by which risk analysis proceeds is guided by the risk management process. Therefore, the risk management process can be considered equivalent in meaning to the risk analysis process.

There is no one best way to do risk analysis and there is no one best risk management process. The most common-sense rule seems to be to use what works best for you. Choosing a risk management process is a conscious and intentional choice that each organization needs to make. It establishes the risk management model the organization will use so there is an agreed upon framework for addressing risk problems and opportunities. The adopted risk management process establishes the roles and responsibilities

of everyone involved in the entire risk analysis process including risk assessment, risk management, and risk communication.

The risk management process, should, at a minimum, include both a specific risk management policy and a risk assessment policy. These should be carefully documented, publicized to and known by all with a legitimate interest in the organization's risk analysis activities. It may also include a risk communication policy. To successfully manage risk, the risk management process should be enterprise-wide and viewed as an important and strategic effort (COSO, 2011). Top management must actively demonstrate support for the new risk management process.

Implementing a risk management framework

Once an organization identifies and commits to a risk management process it must carefully introduce the process to the organization and its stakeholders. That entails an extensive training effort and providing the support needed to successfully use the process.

It is not sufficient to have a successful roll out of the new framework, it is absolutely essential that top management insist on the use of the framework and that risk assessors, risk managers, and others adhere to its use. A new process will need to be tweaked and revised as necessary.

10.2.1. Responsibilities of risk managers

A risk management policy serves to protect the scientific independence and integrity of the risk assessment process. Identifying and selecting a risk management process that includes a comprehensive model is the principle task associated with establishing a risk management process. An example of such a model is presented in Chapters 13 and 14.

Some elements of risk management policy may be broad and overarching; they are addressed before any actual risk management activities are initiated. They include choosing the risk management framework the organization will use and establishing the roles and responsibilities of managers, assessors, and communicators through the development of appropriate policies and guidance. Other elements of the policy may come into play once a

Adopting a risk management framework will represent a change for many organizations. Many theories about how to "do" change originate with John Kotter (1996) and his book, *Leading change*. He offers eight steps for leading change:

1) Create urgency—It helps if the whole organization really wants it.
2) Form a powerful coalition—Convince people that change is necessary.
3) Create a vision for change—A clear vision helps everyone understand why you're asking them to change.
4) Communicate the vision—Communicate the vision frequently and powerfully, embed it within everything that you do.
5) Remove obstacles—Removing obstacles empowers people to execute the vision and it helps the change move forward.
6) Create short-term wins—Quick wins help staff see the value of the change.
7) Build on the change—Quick wins are only the beginning, build on your momentum.
8) Anchor the changes in corporate culture—To make the change stick it must become part of the core of your organization.

specific risk management process is initiated. This might include resolving policy questions that arise as a result of unique aspects of a risk management activity or recurring tasks, such as establishing criteria for selecting options to manage the assessed risks. Risk management policy might establish guidelines for some or all of the following:

- Priorities for which risks to manage.
- Developing risk assessment and risk communication policies.
- Interaction between risk managers and risk assessors.
- Identifying and selecting options to manage the assessed risk.
- Determining which factors to use to evaluate and select risk management options, e.g., risk reductions, residual risk, economic feasibility, or technical feasibility.
- Allocating resources to complete risk assessments and other risk management activities.
- Adjudicating disputes with risk assessors.
- Peer review process and other quality control measures.
- Building capacity to conduct risk assessments and to manage pest risks.
- Risk communication.

The key point is that in order to do risk management you must have and use a specific risk management process. Committing your organization to identifying and adopting a process is an essential risk management decision. If you

do not have a structured risk management process for your organization, then you are not yet doing your best risk management.

Chapters 13 and 14 present and develop a risk management framework suitable for adoption, adaptation, and use by any NPPO. The framework presented is entirely compatible with the IPPC guidance that informs the practice of phytosanitary risk management. Fig. 10.2 presents a summary of this handbook's model in graphic form. The top row identifies four groups of tasks to be performed during stages 1 and 2 of the ISPM 2 framework risk analysis process. The bottom row identifies five categories of tasks for stage 3. The tasks are described in detail in subsequent chapters that, together, provide a clear and systematic framework for the phytosanitary risk management process. An enterprise risk management model suitable for use by any private industry involved in international plant trade is presented in Chapter 17. The desired output of this preliminary risk management responsibility is a well-defined, written and vetted risk management policy that will guide the organization through repeated risk management activities.

10.2.2. Risk assessor responsibilities

Establishing a risk assessment policy is the risk manager's responsibility. A risk assessment policy should serve to protect the scientific independence and integrity of the pest

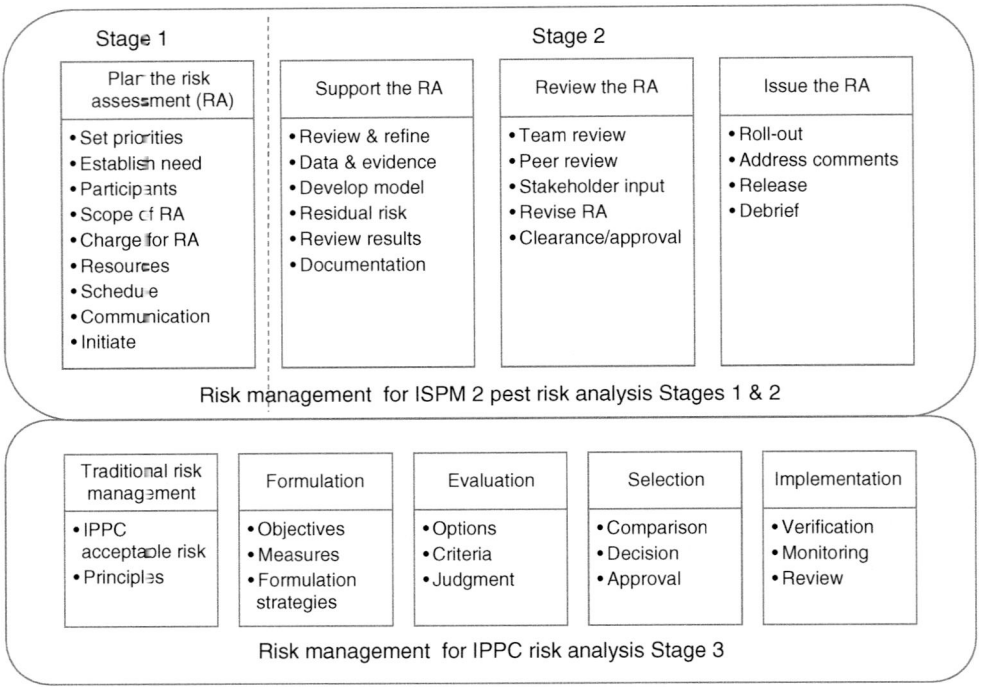

Fig. 10.2. Phytosanitary risk management model showing categories of tasks consistent with Stages 1, 2, and 3 of the ISPM 2 framework.

risk assessment that is so essential to pest risk management. Developing this policy needs to be a collaborative process for any organization that is engaged in making public policy or in the stewardship of public resources like plant health, public health, public safety, economic growth, trade development, natural resources, the environment, and so on. This collaboration should, at a minimum, include risk managers, risk assessors, and risk communicators. It should provide appropriate opportunities for input and feedback from relevant stakeholders. The risk assessment policy should be documented and made publicly available to ensure consistency, clarity, and transparency. For most private organizations establishing a risk assessment policy is wholly an internal affair.

A typical risk assessment policy will include some or all of the following:

- definition of key terms found in the policy
- roles and responsibilities of all pest risk assessment participants
- guidelines for establishing the scope, range, and policy of a specific risk assessment

- conduct of a risk assessment
- guidelines for dealing with uncertainties
- communication requirements.

One of the hallmarks of best practice risk analysis is insulating the science from the policy. In the early days of risk analysis many thought the risk management and risk assessment tasks should be completely separated from one another. This is not true; these two groups need to work closely together. A more appropriate relationship is discussed at length in Chapter 11. It is best when these functions are separate and handled by different people with the appropriate skill sets required by their jobs. However, it is absolutely essential that managers and assessors communicate, coordinate, cooperate and even, at times, collaborate throughout the entire risk management process. Risk managers begin and end the risk assessment process. They may collaborate with risk assessors in identifying a problem, in formulating risk management options, or on other tasks throughout the risk management activity. They will cooperate in the conduct of the risk assessment and must communicate continuously throughout the iterative risk analysis process.

The risk assessment policy ought to resolve any language issues for the organization by defining all potentially ambiguous terms employed in the policy. The roles and responsibilities of everyone involved in the PRA process should be defined in a manner that addresses transparency and ensures the assessment is free from bias and interference throughout the risk management process.

and managers must, together, ensure that the boundaries established for a risk assessment compromise neither the risk management objectives nor the risk assessment science.

The risk assessment process should focus on science-based evidence and analysis. For plant protection assessments associated with trade, the process will include compliance with the SPS Agreement and all relevant IPPC guidance.

Scope and range

Scope—Limits of what will be considered
Range—Extent to which elements within that range will be addressed
   Think of looking through a set of binoculars. Scope is analogous to the breadth of your field of vision; it defines what you do and do not see. Range is analogous to the depth of your vision; it defines the foreground and background.
   Scoping a risk assessment consists of bringing all essential stakeholder groups into the decision-making process to determine the extent of issues to be addressed, and to identify the major issues related to the risk. In other words, scoping determines what is and is not going to be considered in the risk assessment.
   The range of a risk assessment determines the depth to which a topic will be examined. It requires a determination of whether the assessment will be qualitative or quantitative. It includes what will be addressed in the risk characterization. The range of a risk assessment implicitly influences the uncertainty.

The scope, range and policy of a specific risk assessment will usually be established through the interactions of risk managers and risk assessors. That interaction may occasionally include stakeholder groups. Adjustments to the scope, range and policy may continue as the risk assessment develops.

The scope of the risk assessment depends on the identified risk management objectives and the reason for doing the assessment. The range of a risk assessment depends on the risk manager's questions and the degree of detail or depth required in their answers. It is wise to devise a means of resolving scope and range issues before they are encountered in practice. Assessors

The content and impact of these documents are well described in Devorshak (2012). It is important that risk assessors understand that the purpose of their work is to help the risk manager make an informed decision about the presenting problem rather than, say, to do research or pursue interesting avenues of investigation.

Generally, a risk assessment has been expected to provide an estimate of the relevant risk. It can, however, also estimate the risk reductions attributable to different risk management measures as well as the residual risks that remain once a measure is implemented. The scope of the risk assessment will define the parts of the plant product supply chain that need to be considered. It also provides the necessary focus for the scientific work involved in risk assessment.

Risk assessment guiding principles example

1. Risk analysis is a valuable tool to use to enhance the scientific basis of regulatory decisions. All risk assessments should be performed within the risk analysis framework, which includes risk assessment, risk management, and risk communication teams.
2. Risk assessment, especially of complex topics, should be conducted in an iterative manner. This allows refinement of the risk assessment question(s), key assumptions, and data used in the model. The iterations require active collaboration between risk managers and risk assessors.

Risk assessment guiding principles example *(Continued)*

3. An open exchange of information and ideas (communication) within and among the risk assessment, risk management, and risk communication teams is critical for successful risk assessments. Ideally, stakeholders including industry, other agencies, and consumer groups should be identified early in the process for frequent communication throughout the iterations of the risk assessment.
4. The resources needed to effectively and efficiently conduct risk assessments should be identified and allocated before an assessment is commissioned.
5. Realistic timeframes with intermediate milestones should be established and understood by all participants when an assessment is commissioned.
6. Risk assessments should provide risk managers with the information needed to make decisions, otherwise, they ought to be as simple as possible.
7. All assumptions, data, and decisions that impact risk assessment conclusions and risk management actions should be clearly documented and shared with interested parties to make the assessment transparent.
8. Draft risk assessment documents and models should be peer reviewed by government and non-government experts.
9. Risk managers should continue to build the capacity to conduct risk assessments through research, training, mentoring, hiring, promotion, contractors, and other means as necessary.

(Adapted from CFSAN, 2002)

A risk assessment will be constrained by the entity's available time, resources, expertise, and data. These constraints should be communicated to all stakeholder groups before risk assessment begins, especially if they will influence the outcome of risk characterization.

The risk assessment policy ought to establish guidelines for dealing with uncertainty in a risk assessment. This would include prescriptions for risk assessors to address and communicate the uncertainty encountered in the risk assessment as well as responsibilities for risk managers to address that uncertainty in their decision-making. A good policy will establish expectations for all internal and external communications concerning the risk assessment throughout the entire risk management process. The desired output of this preliminary risk management responsibility is a well-defined written and vetted risk assessment policy that will guide the organization through repeated risk assessment processes. For the NPPO, this means a written description of how risk assessments will be conducted in conformance with the ISPM 2 PRA process.

10.2.3. Risk communication responsibilities

It should be the responsibility of the risk managers to establish an entity's risk communication policy.

Risk communication in pest risk management should promote awareness and understanding of the specific issues under consideration in each risk management activity; promote consistency and transparency in formulating risk management options/recommendations; and, provide a sound basis for understanding the risk management decisions proposed (UN/FAO/WHO, 1998).

A scalable process

A good risk management process is perfectly scalable. You can use it when you have 30 minutes and no budget as easily as you can over years with millions of dollars.

One of its greatest values is it provides a systematic, science-based approach to solving problems. The risk management elements identified in a conceptual model should be able to be completed in any time frame with any budget, albeit at possibly vastly different levels of detail and confidence.

Building on the work of a WHO/FAO expert consultation (UN/WHO, 2000), such a policy might hold its central goal to be to provide meaningful, relevant, and accurate information, in clear and understandable terms targeted to specific audiences as required to realize specific risk management objectives. A risk communication policy might include some or all of the following goals articulated by the expert consultation:

- Promote awareness and understanding of the specific issues under consideration during the risk analysis process, by all participants.
- Promote consistency and transparency in arriving at and implementing risk management decisions.
- Provide a sound basis for understanding the risk management decisions proposed or implemented.
- Improve the overall effectiveness and efficiency of the risk analysis process.
- Contribute to the development and delivery of effective information and education programs, when they are selected as risk management options.
- Foster public trust and confidence in the safety of the plant product supply.
- Strengthen the working relationships and mutual respect among all participants.
- Promote the appropriate involvement of all interested parties in the risk communication process.
- Exchange information on the knowledge, attitudes, values, practices, and perceptions of interested parties concerning risks associated with plants, plant products, and related topics.

The major function of risk communication in SPS Agreement matters should be to ensure that all information and opinions required for effective risk management are incorporated into the decision-making process. That requires a transparent explanation of the results of any risk assessment, including the uncertainty and how it was dealt with. The output of this preliminary risk management activity may or may not include a written and vetted risk communication policy.

10.3. Summary and Look Forward

Here are five things to remember from this chapter.

1. Every NPPO needs a risk management process.
2. Every NPPO must follow their risk management process.
3. Risk managers should establish risk management and risk assessment policies.
4. Risk managers are responsible for seeing that risk communication gets done.
5. If your organization does not have its own structured and specific way to do risk management then you need to develop or adopt one.

There are many risk management communities of practice that have misinterpreted early guidance to keep risk management and risk assessment functionally separated, incorrectly interpreting it to mean managers and assessors should have limited contact with one another. That is a formula for disastrous risk management. In the next chapter we will examine the desired level of interactions between risk managers and risk assessors.

10.4. References

CFSAN (Center for Food Safety and Applied Nutrition) (2002) *Initiation and Conduct of all "Major" Risk Assessments Within a Risk Analysis Framework*. CFSAN Risk Analysis Working Group. Available at: http://www.fda.gov/Food/FoodScienceResearch/RiskSafetyAssessment/ucm475127.htm (accessed February 1, 2017).

COSO (Committee of Sponsoring Organizations of the Treadway Commission) (2011) *Embracing Enterprise Risk Management: Practical Approaches for Getting Started*. COSO. Available at: coso.org, (accessed February 3, 2017).

Devorshak, C. (ed.) (2012) *Plant Pest Risk Analysis, Concepts and Application.* CABI International, Wallingford, United Kingdom.

IPPC (1996) International Standards for Phytosanitary Measures, Publication No. 2: *Framework for Pest Risk Analysis*. Secretariat of the International Plant Protection Convention (IPPC), Food and Agriculture Organization of the United Nations, Rome.

Kotter, J.P. (1996) *Leading Change*. Harvard Business School Press, Boston, MA.

UN, FAO/WHO (1998) *The Applications of Risk Communication to Food Standards and Safety Matters*. A Joint FAO/WHO Expert Consultation, Rome, Italy.

UN/WHO (2000) The interaction between assessors and managers of microbiological hazards in food. *Report of a WHO Expert Consultation in collaboration with The Institute for Hygiene and Food Safety of the Federal Dairy Research Center the Food and Agriculture Organization of the United Nations*. Kiel, Germany.

11

Risk Manager and Risk Assessor Interaction

You cannot buy engagement, and you will pay for disengagement.

Adele du Rand

11.1. Introduction

Risk assessment n the federal government: managing the process (NRC, 1983) was a seminal document on government use of risk assessment. This so-called "Red Book" recommended a clear cut "conceptual separation" between the, then, two steps of risk analysis, i.e., assessment and management. This recommendation has, in turn, played a role in the "functional separation" of the two groups, which has, in the extreme, grown into an undesirable "institutional separation." The reason behind the recommended conceptual separation has always been the desire for the highest possible scientific integrity of the risk assessment process with minimum political pressure to undermine the objectivity and credibility of its conclusions. The misinterpretation of this recommended separation has unnecessarily complicated the pest risk assessment process in particular.

Little international guidance has been developed to describe the pest risk manager's job and responsibilities. In many nations there is a deep-seated notion that the job, whatever it is, ought to include little if any direct contact with risk assessors and has no connection whatsoever

with the conduct of the risk assessment. These are both mistaken notions. Although notions of pest risk management may not always be well developed, risk management in other communities of practice has been developing rapidly. The spread of enterprise risk management throughout the public and private sectors gives evidence of this development. One of the enduring lessons of the evolution of risk management is that risk managers and risk assessors must communicate consistently, coordinate as necessary, cooperate always, and collaborate when needed. This simple development is likely to be at odds with how many in the PRA community of practice view the ideal relationship between risk managers and assessors. Hence, the need for this chapter to address the degree of interaction between managers and assessors arises, before describing the proposed risk management model.

Phytosanitary risk analysis often practices a very severe form of separation of the risk management tasks from the risk assessment tasks, often to the detriment of the risk analysis process. One of the most effective, least-cost ways to improve the quality of phytosanitary risk analysis is to increase and improve the interactions between risk managers and risk assessors. The risk management model presented in Chapters 13 and 14 is predicated on increased effectiveness of manager/assessor interactions.

This chapter proceeds with a review of the "Red Book" legacy to set the historical record straight. Recent efforts to reconnect the risk management and assessment tasks are then considered before the chapter concludes by looking at opportunities for increased interactions with appropriate boundaries in place.

11.2. The "Red Book" Legacy

The notion that risk assessors and managers must not interact comes from the earliest thoughts on risk assessment, thoughts that have subsequently proven to be misinterpreted if not just wrong. The "Red Book" (NRC, 1983), one of the world's first policy documents on risk assessment, said in its opening sentence, "This report explores the intricate relations between science and policy in a field that is the subject of much debate—the assessment of the risk of cancer and other adverse health effects associated with exposure of humans to toxic substances." The lines were drawn early between risk assessment, the science, and risk management, the policy. The first of three objectives identified in the book was, "To assess the merits of separating the analytic functions of developing risk assessments from the regulatory functions of making policy decisions."

The notion of separating risk management and risk assessment has deep roots in the practice of risk analysis. The rapid development of the use of risk assessment in the 1970s and 1980s by US government agencies led to much criticism of risk assessments. That criticism stemmed from dissatisfaction with regulatory outcomes. Many proposals for change at the time were based largely on the naïve assumption that altering the administrative arrangements for risk assessment would lead to regulatory outcomes that critics would find more agreeable (NRC, 1983). It is from this context that the notion of separating management and assessment arose. This notion has led to a harsh division of risk assessment and risk management tasks in the phytosanitary world that has caused no end of problems for nations and trade, not to mention the risk managers and assessors themselves. This division needs to be reversed.

> **Recommendation 1 from the "Red Book"**
>
> Regulatory agencies should take steps to establish and maintain a clear conceptual distinction between assessment of risks and the consideration of risk management alternatives; that is, the scientific findings and policy judgments embodied in risk assessments should be explicitly distinguished from the political, economic, and technical considerations that influence the design and choice of regulatory strategies.

The committee that conducted the "Red Book" study concluded that risk assessment cannot be completely free of policy considerations, but that policy should not influence risk assessment unduly. They found that if risk management considerations, like the economic or political effects of a particular risk management action, are perceived to affect either the scientific interpretations or the choice of inference options in a risk assessment, the credibility of the assessment inside and outside the agency can be compromised. This could result in the risk management decision itself losing legitimacy. One suggestion for safeguarding the risk assessment was to restructure the formal organization, separating an agency's or program's risk assessment staff from its policymaking staff. This was but one of several suggestions offered.

The committee went on to say that their survey of agency structures failed to show that organizational structure was a factor in any undesired influence on risk assessment. In fact, the committee noted that formal separation like this has disadvantages. Although risk assessment and risk management functions are analytically distinct, in practice they do, and must, interact. One recognized danger was that completely isolating risk assessors from regulatory policymakers may inhibit important communication in both directions. The committee found communication essential for risk assessors to be able to evaluate risk management options and "to ensure that the regulatory decision-maker understands the relative quality of the available scientific evidence, the degree of uncertainty implicit in the final risk assessment, and the sensitivity of the results to the assumptions that have been necessary to produce the assessment." The committee concluded that the benefits of increased separation are uncertain and the disruption and confusion that could be caused by separating the management and assessment staff could be considerable. Alternatives to structural separation included the practice of preparing written risk assessments, independent peer review, and adhering to uniform guidelines for risk assessment. Thus, from the outset, the separation of the assessment and management tasks was recognized as problematic.

Anecdotally, early on in the sanitary and phytosanitary (SPS) risk analysis world, as people were trying to understand risk analysis, there was a lot of discussion about how risk assessment is a scientific or science-based process, whereas risk management includes elements of policy, i.e., aspects not based on science. This debate led to the literal separation of risk assessment and risk management. The thinking was that risk assessment, untouched by risk management, would remain pure and untouched by political and other values. Consequently, the desire to protect the integrity of the science that was to be used to support decision-making has hardened into an institutional separation of risk management and risk assessment (see box) that was neither intended nor as efficient as it could be with effective ongoing interaction.

Literal separation of assessment and management

The United States Department of Agriculture's (USDA) Animal Plant Health Inspection Service (APHIS) has a Plant Protection and Quarantine (PPQ) division which has a Center for Plant Health Science and Technology (CPHST) in its Science and Technology Office. CPHST houses the Plant Epidemiology and Risk Analysis Laboratory (PERAL), which is USDA's primary unit producing PRAs. PERAL, the APHIS PRA function, is located about 300 miles from PPQ's risk management function. As a leader in the field of PRA, many NPPOs around the world followed their lead in separating these two elements of risk analysis in a manner most risk experts would find unadvisable.

The IPPC ISPM 2 risk analysis framework consists of three stages. The first two are primarily considered risk assessment and the third is considered risk management. There still remains a strong belief in many NPPOs that risk assessment and risk management must be separated,

as if this is an immutable law of the universe. One direct result of this separation is that many opportunities to improve the risk analysis process of management, assessment, and communication go unmet. In too many instances, risk assessments act only as a trigger for risk management. They do not actually *inform* risk management, because they are not guided by risk managers to provide all of the scientific information that is needed to support decision-making. Risk management does not go back to the risk assessment often enough to see how effective the risk management actions under consideration are likely to be. There is not sufficient interaction between risk managers and risk assessors. Valuable feedback loops are missing in what often devolves into two parallel, independent and unidirectional processes of assessment and management that too often fail to arrive at a common destination.

11.3. Recent Efforts to Reconnect Management and Assessment

The 1983 "Red Book" stressed the importance of a "conceptual distinction" between risk assessment and risk management but rejected the concept of "institutional separation" between the processes. The authoring committee was concerned with protecting risk assessments from the inappropriate intrusions of policymakers and other stakeholders. From that concern came recommendations for the conceptual separation of assessment and management. This conceptual separation has been widely misinterpreted if not misunderstood (NRC, 2009).

The NRC (1996) pointed out how careful studies of the risk decision process have increasingly acknowledged the limitations of a strict separation. These limitations were first recognized in the "Red Book" itself, which pointed to the need to iterate between risk assessment and risk management so that assessment could incorporate analytical assumptions that may need to be different for functions such as initial screening and the evaluation of regulatory options. The "Book" noted that "a single risk assessment method may not be sufficient" and that the choice of appropriate assumptions required interaction between the assessment and management functions. Stern and Fineberg (NRC, 1996) also pointed out that separation of the risk assessment function from an agency's regulatory activities is likely to inhibit the interaction between assessors and regulators that is necessary for the proper interpretation of risk estimates and the evaluation of risk management options. They concluded that separation can lead to disjunction between assessment and regulatory agendas and cause delays in regulatory proceedings.

Room for improvement

Typically, the processes of risk assessment and risk management for commodity imports are conducted separately and by two different groups. Risk assessors are not usually involved in developing the risk management document for commodity imports even though the assessors may be in the best position to provide analytical information to inform the risk management decisions.

The NRC (1996) describes an analytic-deliberative process that leads to risk characterization, a final step in risk assessment. They believe the analytic-deliberative process should be mutual and recursive. Deliberation (risk management) frames analysis (risk assessment), analysis informs deliberation, and the process benefits from feedback between the two. As long as "analysis and deliberation" does not involve efforts by risk managers to shape risk assessment outcomes to match their policy preferences, but rather involves efforts to ensure that assessments (whatever their outcomes) will be adequate for decision-making, interactive processes involving "the spectrum of interested and affected parties" are seen as imperative (NRC, 2009).

The extreme institutional separation that plagues many NPPOs needs to change. Robinson and Levy (2011) succinctly state that one often-lamented problem in risk analysis is the need for better coordination between risk assessors and risk managers. Calow and Forbes (2013) identify four challenges that arise from the sharp separation of management and assessment. Their view is that risk assessments do not inform management decisions as effectively as they should and the outputs of risk assessment need to be more policy and management relevant. This can be facilitated by more dialogue. Their four challenges argue that risk assessment endpoints involve value judgments; risks need to be

expressed in metrics that matter to decision-making; consequences need to be better calibrated than simple effect/no effect determinations; and, regulatory decision-making faces a challenge in deciding how much science is "enough." They conclude that when scientists make decisions about what to measure in assessing risks and debate how much more work is needed to address uncertainties, they often exercise value judgments that should be "subjected to policy steer."

Gabbi (2007) points out that with scarce interaction between the parties, even the simplest matters can cause substantial inconsistencies in the final output of the risk analysis process. This has serious implications for the quality of decision-making. Following a few of these incidents, the European Commission and the European Food Safety Authority (EFSA) have decided to improve their cooperation and overall interaction. As a consequence, they have set up some structures and procedures on both the management and assessment sides of the process to improve interactions. Gabbi concludes that it is now clear active interaction is necessary to ensure the assessment will meet the needs and will answer the concerns of the risk managers. This point is well summarized when Gabbi says, "it is rather obvious that the importance of the interaction between the two institutions is now well recognized both in EFSA and in the EC ... We believe that only an enhanced cooperation between the risk assessors and the risk managers in the food safety domain can substantiate the choice of institutional separation of the two groups."

Lest the reader think the closer relationship between managers and assessors is universal, consider the case of the EFSA. Following the European food scares of the 1990s, the European Union sought to regain the confidence of consumers by opting for an institutional separation of risk assessment from risk management, when it created EFSA for the former while the latter was maintained by the Council, the European Commission and the Member States.

The NRC (2009), upon revisiting the "Red Book" found the original authors emphasized that the distinction between risk assessment and risk management is a conceptual one. It concerns the fact that the content and goals of the two activities are distinguishable on a conceptual level.

The NRC concluded the "Red Book" nowhere calls for any other type of "separation" of the two activities than this conceptual one, a point that clearly was misunderstood in the past.

A 2013 European Commission study by three independent non-food scientific committees (SCHER, 2013) took as its major motivation a review to ensure that risk assessments carried out by Commission scientific committees were meeting the needs of risk managers. The review found that risk assessments made little sense unless they informed risk management. This potential mismatch is recognized as one possible unintended consequence of the separation of assessment and management functions. The review also found the differences in perceptions and responsibilities of the two groups argue for a better dialogue between assessors and managers, especially for framing of the questions raised by the initial mandate and then iteratively through subsequent refinements. The challenge is to develop a dialogue in which assessment is informed but not influenced by management, in terms of how the assessments are carried out and what conclusions are drawn from them.

It is clear that the intentions of the "Red Book" regarding the functional separation of risk management and risk assessment were misinterpreted and misunderstood from the start. Furthermore, experience in many spheres of endeavor have proven the need for more and better interactions between managers and assessors. The challenge facing several communities of practice, including phytosanitary risk analysis, is how to ensure the integrity of the science in risk assessment while assuring that science plays an appropriate role in what is fundamentally a values-based decision-making process, where risk is but one of the values considered. The functional separation of risk assessment and risk management, i.e., the individuals who prepare the risk assessment should not normally be the same individuals who are responsible for the management of the risk, is one cornerstone of this process. It is essential to protect the scientific integrity of the risk assessment but that does *not* mean there can be no contact between assessors and managers.

Communication, coordination, cooperation, and collaboration are the hallmarks of a good relationship. Risk managers and assessors need to communicate at many points during the risk analysis process, before, during, and after the risk assessment. There are many occasions for them to coordinate their activities and needs. They need to cooperate at all points of contact throughout the risk analysis process. Finally, there are some points in the process where they must collaborate. All of this can be accomplished without impugning the integrity of the science or sacrificing any decision-making authority.

The risk management model described in the chapters that follow is posited on two assertions. First, risk managers need to interact with risk assessors before, during and after the risk assessment. Second, risk assessors can play a more active role in identifying and assessing risk management options. Risk assessment must produce the information that risk managers need to make decisions. That cannot happen with the requisite frequency unless risk managers communicate their information needs to risk assessors. That means risk managers do have some involvement in planning the risk assessment. They have an important support role in the conduct of the risk assessment that does not extend to interfering in any way with the scientific integrity of the assessment. Risk managers play an important role in reviewing and issuing the risk assessment. Risk managers should make use of the expertise of risk assessors when formulating risk management options. Ideally, risk managers will consider a number of alternative means of addressing an unacceptable phytosanitary risk. Risk assessors can be most helpful in providing a scientific assessment of the technical efficacy and economic efficiency of these different measures, especially with regard to estimating residual risks associated with the available risk management options.

11.4. Closer Interactions with Boundaries

There need to be clear boundaries between the duties of risk managers and assessors that recognize and respect the interdependence of their functions, where the functions are performed by different individuals. Clearly defined roles and responsibilities are essential to well maintained boundaries. The goal of a risk assessment is to provide the best available science to support the risk manager's decision-making. The assessment may, however, be only one of many sources of information risk managers use to make decisions, others may include cost–benefit analysis, legal opinions, environmental assessments, and the like.

Risk managers should identify the questions to be answered by a risk assessment, they establish the objectives of the risk management activity and provide the overall scope of the assessment. Risk assessors ensure the assessment is of high scientific quality, is conducted consistent with best practice, and is responsive to the needs of risk management. Assessors need to help risk managers to understand the assessment, especially the impact and significance of uncertainty on assessment findings. Risk managers must understand how the limitations of available data affect the outcome of the risk assessment. Assessors must carefully explain and describe how assumptions they made during the assessment affect the results of the assessment and the level of uncertainty associated with the risk estimate. All of this requires interaction.

Managers and assessors must be willing to exchange information and ideas and to collaborate on PRAs, while respecting the boundaries between their two sets of responsibilities. This collaboration is essential to the iterative refinements of objectives, assumptions, data, and decisions that result as more information is gathered and exchanged. Continual dialog is necessary to ensure that the risk manager's expectations of the risk assessment results are met.

Given those boundaries, active interaction is necessary to ensure that the risk assessment will meet the needs and answer the concerns of risk managers (Gabbi, 2007). Such interaction is not yet natural in phytosanitary risk analysis where managers and assessors often have different backgrounds, organizational hierarchies, needs, beliefs, and even locations. Nonetheless, such interaction is needed in the inception phase, when risk managers scoping the problem could collaborate with risk assessors and set up an ongoing process of communication, cooperation, coordination, and collaboration throughout the entire risk analysis process. Likewise, collaboration is needed to identify risk management options, especially those based on a systems approach to managing risk. Allocating these essential functions separately to managers and assessors creates more problems than solutions.

The fundamental issue of functional separation is clear in principle. Risk assessors use the best available science to establish the connections between introduction of a pest to the likely resulting consequences. The science excludes value judgments to the maximum extent possible. Risk managers, on the other hand, consider the facts of the matter, as determined by risk assessment, and other values in making decisions about interventions to alleviate consequences by managing likely causes of the risk.

There are ample opportunities for improved interactions that respect the functional separation of responsibilities. They occur in three distinct levels of interaction. Cooperation is the lowest level of interaction. It is the process of working together to the same end. It includes actions like ready

Table 11.1. Suggested opportunities for interaction among risk managers and risk assessors during a risk analysis process.

Task	Cooperation	Coordination	Collaboration
Scoping a risk assessment			X
Charge for risk assessment		X	
Risk assessment resources	X		
Risk assessment schedule		X	
Review risk assessment results		X	
Document draft risk assessment	X		
Team review of risk assessment			X
Peer review of risk assessment	X		
Clearing the risk assessment	X		
Address comments, revise risk assessment		X	
Identify risk management options			X
Assess efficacy of options	X		
Identify risk management option to implement		X	

compliance with requests and effective regular communications. It is followed by coordination, the second level of interaction. Coordination means to work together successfully in an effective relationship. It includes actions like providing and seeking input and feedback throughout the risk analysis process. Collaboration is working together to produce or create something in common. It includes such things as partnering on tasks and working together at the same time and place.

Table 11.1 suggests the level of interaction that is possible, if not desirable, for selected tasks in the risk analysis process. More details on the interaction opportunities are found in the next two chapters.

> Once a tentative risk management solution has been identified, risk managers and risk assessors need to meet to ensure that the solution is unambiguous, well founded scientifically, and answers the needs of the risk manager.
>
> (Adapted from SCHER, 2013)

Risk managers and risk assessors should collaborate on the scoping task, in the team review of the risk assessment and in identifying potential risk management options. They should coordinate closely on the charge for the risk assessment and in establishing a realistic schedule. Risk assessors will play an instrumental role in helping risk managers to understand the results of the risk assessment and the significance of the remaining uncertainty. Close coordination will also be required in addressing comments on the draft risk assessment and revising the assessment as well as when identifying the best risk management option to implement.

Assessors and managers must cooperate in establishing the risk assessment team and in allocating its resources. They will also need to cooperate in documenting the draft assessment and participating in its review and clearance. Risk assessors can be expected to play a significant role in the risk management decision by providing assessments of the risk management options under consideration.

Communication between the two groups needs to be continuous. Risk assessment and risk management are both iterative processes. Data needs may evolve as the team's understanding of a problem solidifies, new questions may emerge, or old ones may fade away. Changes in team composition, resource availability, or schedule can be expected and communication will be required throughout the risk analysis process.

11.5. Summary and Look Forward

Here are five things to remember from this chapter.

1. The conceptual separation of risk management and risk assessment can be traced to the 1983 "Red Book."
2. The conceptual separation recommended by the "Red Book" has been widely misinterpreted and misunderstood.
3. The severe institutional separation of the two tasks found in phytosanitary risk analysis has caused serious problems for risk analysis.
4. Reinterpretation of the "Red Book" along with wisdom gained from several decades of experience for several communities of practice indicate the need for close interactions among managers and assessors.
5. Phytosanitary risk analysis needs to take advantage of the numerous opportunities for managers and assessors to communicate, cooperate, coordinate, and collaborate over the course of a risk analysis.

The proposed phytosanitary risk management model is formally introduced in the next chapter. It is, then, developed and explained in subsequent Chapters 13 and 14.

11.6. References

Calow, P. and Forbes, V.E. (2013) Making the relationship between risk assessment and risk management more intimate. *Environmental Science Technology*, 47, 8095–8096.

Gabbi, S. (2007) The interaction between risk assessors and risk managers, the case of the European Commission and of the European Food Safety Authority. *European Food and Feed Law Review (EFFL)*, 3/2007, 126–135.

NRC (National Research Council) Committee on the Institutional Means for Assessment of Risks to Public Health Commission on Life Sciences (1983) *Risk Assessment in the Federal Government: Managing the Process.* National Academy Press: Washington, District of Columbia, USA.

NRC (National Research Council) (1996) Stern, P.C. and Fineberg, H.V. (eds) *Understanding Risk Informing Decisions in a Democratic Society.* Committee on Risk Characterization Commission on Behavioral and Social Sciences and Education, National Academy Press Washington, D.C.

NRC (National Research Council) (2009) *Science and Decisions: Advancing Risk Assessment. Committee on Improving Risk Analysis.* National Academy of Sciences, Washington, District of Columbia, USA.

Robinson, L. and Levy, J. (2011) The revolving relationship between risk assessment and risk management. *Risk Analysis*, 31(9), 1334–1344.

SCHER (Scientific Committee on Health and Environmental Risks), SCENIHR (Scientific Committee on Emerging and Newly Identified Health Risks), SCCS (Scientific Committee on Consumer Safety) (2013) *Making Risk Assessment More Relevant for Risk Management.* European Commission, Brussels.

12

A Phytosanitary Risk Management Model

Opportunity and risk come in pairs.
Bangambiki Habyarimana

12.1. Introduction

This chapter introduces the risk management model that is proposed as a template for those national plant protection organization (NPPOs) with a desire to update and improve their approach to risk management. The model is introduced in its entirety in this chapter. The next two chapters unpack the model's contents and explain the risk management task groupings in details. Please feel free to adopt and adapt this model as you require. Use parts that improve your process and ignore those parts that make no sense for your organization.

12.2. ISPM 2 Framework for Pest Risk Analysis

Fig. 9.5 (see p. 111) presents framework for pest risk assessment from ISPM 2 (IPPC, 2006) of the International Plant Protection Convention (IPPC). Nothing in the risk management model proposed herein will interfere in any way with the workings of the ISPM 2 framework. That process is divided into three stages, as shown. The model introduced in this chapter simply pro-

vides a more detailed description of what the phytosanitary risk manager might productively be doing during each of the three stages of this framework.

12.3. Proposed Risk Management Model

The phytosanitary risk management model proposed for use in this handbook is presented in Fig. 10.2 (see p. 120). During Stage 1 of the PRA "process", initiation, the risk manager has responsibilities to plan the risk assessment. During Stage 2, "pest risk assessment", risk assessors are solely responsible for conducting the risk assessment. The risk manager has responsibilities to support, review, and issue the risk assessment. Stage 3, "risk management," has been broken down into the following responsibilities:

- traditional risk management
- formulation
- evaluation
- selection
- implementation.

Chapter 13 provides a detailed description of the risk manager's responsibilities for Stages 1 and 2 of the ISPM 2 PRA framework. Chapter 14 provides a description of the risk manager's Stage 3 risk management responsibilities.

12.4. Summary and Look Forward

Here are five things to remember from this chapter.

1. The phytosanitary risk management model presented is a prototype for NPPOs to adopt or adapt.
2. The three stage ISPM 2 PRA framework is supplemented by this model.
3. Risk managers have risk assessment planning responsibilities in Stage 1 of the ISPM 2 PRA framework.

4. Risk managers have the responsibility to support, review, and issue the risk assessment in Stage 2 of the ISPM 2 PRA framework.
5. Risk managers have traditional risk management, formulation, evaluation, selection, and implementation responsibilities in Stage 3 of the ISPM 2 PRA framework.

The next chapter unpacks the details of the risk manager's responsibilities in Stages 1 and 2 of the ISPM 2 PRA framework.

12.5. Reference

IPPC (2006) International Standards for Phytosanitary Measures, Publication No. 2: *Framework for Pest Risk Analysis*. Secretariat of the International Plant Protection Convention (IPPC), Food and Agriculture Organization of the United Nations, Rome.

13

Pest Risk Management
Through Stages 1 and 2

Good risk management fosters vigilance in times
of calm and instils discipline in times of crisis.

Dr. Michael Ong

13.1. Introduction

This chapter and the next one provide a detailed
description of the risk management model introduced in the preceding chapter as a proposed
template for those national plant protection organizations (NPPOs) with a desire to update and
improve their approach to risk management.
The risk analysis framework presented in ISPM 2
and depicted graphically as a flow chart in Fig.
9.5 (see p. 111) identifies three stages of risk analysis. Pest risk management is identified as Stage
3 in that process. Risk management, however,
has a role in all three stages of pest risk analysis.
It begins in the Initiation (Stage 1), well before risk
assessment begins. In Stage 1, risk managers
determine the need for a risk assessment and
provide the risk assessment team with the scope
of the risk assessment to be developed. Initial
problem formulation and boundary conditions
are to be established by risk managers during
scoping in the initiation stage with risk assessors. Risk managers should collaborate with risk
assessors to define clear, scientifically defensible
"risk questions" before the risk assessment is begun.

Risk assessment (Stage 2) is not the risk
manager's job but it is, to a great extent, the risk
manager's responsibility to see that it is done
well. The top half of Figure 10.2 (see p. 120) outlines the risk manager's role in the initiation and
risk assessment stages of the pest risk assessment
process. Risk assessors remain wholly responsible for the conduct of an independent science-based risk assessment. The risk manager's
responsibilities are organized into four phases
over the first two stages of the ISPM 2 framework. These are: planning the risk assessment; supporting the conduct of the risk assessment;
reviewing the assessment; and finally, issuing
the risk assessment. This chapter is organized
into four main sections corresponding to these
four phases of risk management responsibilities.

Most descriptions of the PRA process begin
with the initiation stage. Clearly, there is more to
the process than identifying pests of quarantine
concern. When an exporting country requests
market access for one of its plant products, the
importing country usually explores the kind of
response necessary. That is, does the importing
country's NPPO need to intervene in this trade
activity? This is often the first phytosanitary risk
management decision. If some form of regulation is deemed to be potentially necessary, a specific
risk management activity is initiated, usually by
starting a PRA, which begins with the initiation
stage. The risk management model of Fig. 10.2
begins with the risk management activities that
accompany the initiation stage. The risk assessment planning activities can be considered part

of Stage 1 of the ISPM 2 framework of Fig. 9.5. Likewise, risk managers have a role in the conduct, review, and issuing of the risk assessment. All of these activities coincide with Stage 2 of the ISPM 2 PRA process.

The risk manager's Stage 3 responsibilities are detailed in Chapter 14.

13.2. Plan the Risk Assessment

Some NPPOs tend to see the roles of risk managers and risk assessors as completely segregated, ceding the entire risk assessment process to risk assessors. This is both unrealistic and less effective than having risk managers play an active role in preparing for and helping to plan the risk assessment process. This involvement better assures that the risk assessment will meet the information needs of risk managers. The risk manager's role in planning the risk assessment takes place during the initiation stage of the PRA. It comprises the following tasks, which are developed in the text that follows:

1. Set priorities.
2. Identify risks.
3. Establish need.
4. Participants.
5. Scope of risk assessment.
6. Charge for risk assessment.
7. Resources.
8. Schedule.
9. Communicate.
10. Initiate.

13.2.1. Set priorities

No NPPO is likely to have all the resources required to act upon the extant trade requests and plant protection issues before them at any given time. Consequently, risk managers must prioritize risks for effective and efficient use of available resources. Factors that may be taken into account include program urgency, funding, staffing, expertise, time, and other factors. "Other factors" could include a media outcry about a perceived risk that has the potential to attract enough public attention to become a delicate public relations issue for the NPPO. Existing

international agreements that might be affected by priority-setting could also be a factor. Setting priorities is the act of analyzing and defining NPPO priorities based on importance and capability. Risk managers must decide which work gets done and in what order. This is presumably done by weighing the relative importance of different market access requests and considering the NPPO's capacity to accomplish specific activities within a specified time period with the available funding, staff, and expertise. This priority-setting step results in a go/no-go decision for each possible action an NPPO could take. Risk managers decide whether and when to address a risk issue during this task.

13.2.2. Identify risks

Once an issue is made a priority and the risk management team has decided to devote resources to it, a risk management activity is triggered and begins in earnest. The usual trigger for a pest risk management activity is a market access proposal. It may be a new product from an existing trading partner or new products from new trading partners. New trade simultaneously presents potential losses and opportunities for gain. The NPPO of the importing country will normally focus on the potential loss of crops or plant health that could result when quarantine pests are present. Businesses in the importing

A PRA process may be triggered in the following situations (ISPM 1):

- a request is made to consider a pathway that may require phytosanitary measures
- a pest is identified that may justify phytosanitary measures
- a decision is made to review or revise phytosanitary measures or policies
- a request is made to determine whether an organism is a pest

Similar language is found in ISPM 11 where the PRA process may be initiated as a result of:

- the identification of a pathway that presents a potential pest hazard
- the identification of a pest that may require phytosanitary measures
- the review or revision of phytosanitary policies and priorities.

and exporting countries will focus on the potential gains from trade.

The complete set of presenting risks is usually bigger than any one risk manager focuses on. Einstein purportedly said, "If I had one hour to save the world, I would spend 55 minutes defining the problem." Establishing a decision context that carefully identifies all the relevant risks from the outset helps to distinguish risk management from other decision-making paradigms. The purpose of risk management is to find the right problem and to solve it or to find the right opportunity and pursue it.

Identifying the decision-relevant risks is the risk manager's first major responsibility for a given risk management activity. Fig. 13.1 shows risk identification as a four-part process consisting of identifying the hazard or opportunity that initiates the process, identifying the undesirable consequence of concern, specifying the sequence of events that is necessary for the hazard/opportunity to result in the harm/gain, and characterizing the relevant uncertainty in all of this. The first iteration of risk identification is often speculative—actual risks may not be confirmed until after the risk assessment is complete. Nonetheless, a risk management activity begins with the preliminary identification of risks.

Phytosanitary risk identification begins by identifying the pest or thing that could cause harm. Uncertainty at the outset may restrict this identification to "pests of potential quarantine concern" pending confirmation of the identity of any pests via the PRA. Initial identification of the harm or harms these potential pests may cause may also remain speculative at the outset. As the risk assessment progresses, specific species and harms may be identified. The best risk identification will identify harms beyond those to plants, when they exist.

The first task for international trade advocates who deal with opportunity risks is to identify the opportunity for gain. These are usually gains from trade that accrue to exporters and importers. These may include gains in terms of sales, net income, jobs, increased product variety, lower prices, and the like.

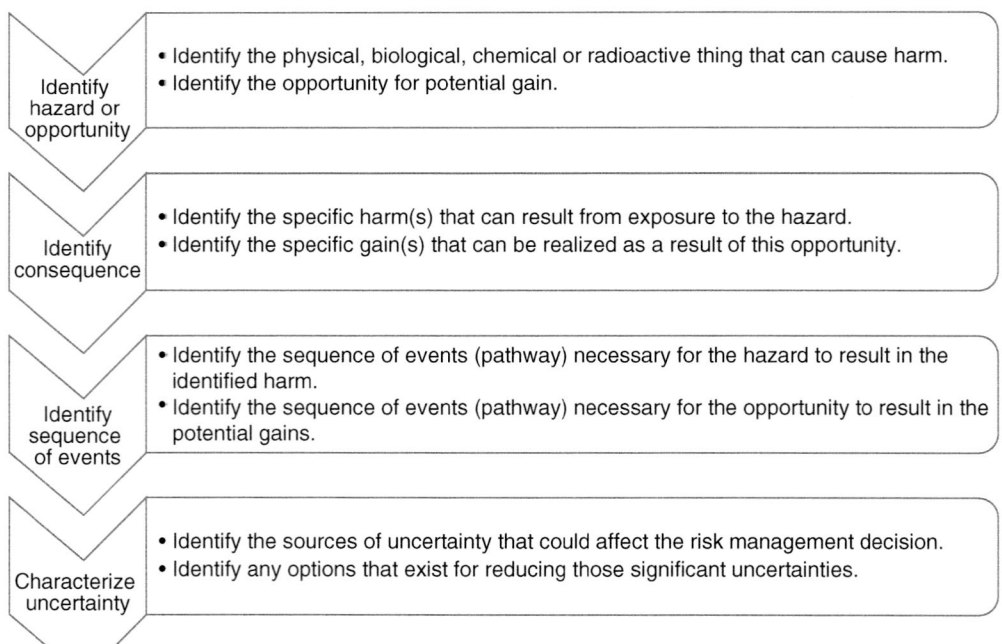

Identify hazard or opportunity
- Identify the physical, biological, chemical or radioactive thing that can cause harm.
- Identify the opportunity for potential gain.

Identify consequence
- Identify the specific harm(s) that can result from exposure to the hazard.
- Identify the specific gain(s) that can be realized as a result of this opportunity.

Identify sequence of events
- Identify the sequence of events (pathway) necessary for the hazard to result in the identified harm.
- Identify the sequence of events (pathway) necessary for the opportunity to result in the potential gains.

Characterize uncertainty
- Identify the sources of uncertainty that could affect the risk management decision.
- Identify any options that exist for reducing those significant uncertainties.

Fig. 13.1. Pest risk analysis framework adapted from ISPM 2.

Phytosanitary risk managers and assessors may not have much experience with the concept of a hazard. In some communities of practice hazard is a term reserved for human health risks. Hazard is used in a much broader and more general sense throughout this handbook.

The Ontario Ministry of Agriculture and Food (OMAF) defines a hazard as:

A thing or action that can cause adverse effects, e.g., a human pathogen, a plant pest, an animal disease agent, the introduction of a specific commodity or product. In the OMAF framework, the term "hazard" is used to imply the cause of an adverse event and not the negative consequences. When referring to the negative consequences of a causal hazard, we encourage the use of terms such as negative outcome, consequence, or impact.

(OMAF, 1997)

Harm is not automatic, given the existence of a hazard. There is usually a sequence of events that is necessary for the hazard to result in the identified harm(s). For example, one sequence of events might be that the hazard is present on the trade commodity and it survives transit to the importing country where it then finds sufficient life requisites to establish a sustainable breeding colony that, then, spreads to vulnerable plants and damages them. This is a common sequence of events for a PRA. However, when new or unique risks arise, these details may not become known until a risk assessment is completed.

By a similar logic, gains are not automatic given an opportunity. There will be a sequence of necessary events required for gains to be realized. This sequence ought to be identified. For long-term trade gains, for example, a product must enter the importing country in sufficient quantity and quality to have a market impact; it must also be free of quarantine pests if long-term gains are to be realized.

Pest risks are a global problem. NPPOs exist because countries recognize the problems associated with pests of quarantine concerns. The very existence of an NPPO is evidence of a country's willingness to devote resources to pest risk problems. Thus, every pest risk management activity begins with a preliminary understanding of the nature of the risks it will address. The final risk identification cannot be formulated, however, until the risk assessment is completed and the

potential pests and their consequences are identified.

Most organizations take a narrow view of risk when they establish the context of their decision problem. This makes sense. NPPOs have been charged with protecting plant life and health in their countries, private companies are generally motivated by profit. Each of these stakeholders owns a separate piece of the overall risk and neither typically takes an interest in broadening its perspective of the risk. Yet, that is what an optimal risk management strategy would do.

The party with the primary responsibility for stewardship of a country's plant health is the NPPO. By the powers vested in the NPPO to protect plant health, the NPPO also turns out to be the party with the greatest power over and impact on international trade outcomes. Thus, it behooves the NPPO risk manager to identify risks as broadly as possible from the outset. Even though the risk assessment may be restricted to plant pest risks it is important to be keenly aware of the risk–risk tradeoffs entailed by a risk management decision. Such an awareness, even in a general sense, is not possible unless the risk manager has given careful thought to identifying all of the risks relevant to a specific risk management activity. The desired outcome of this process is a written risk identification statement. This statement is a single piece of paper that succinctly identifies the reasons for a specific risk management activity. Risk management is an iterative process. You can expect to revise and refine this statement several times before the matter is decided. That is why it is important to keep it up to date.

Sample preliminary risk identification statement

1) Importing raw timbers from Fredonia carries a risk of plant pests of quarantine concern.
2) These timbers could increase jobs and income in the domestic furniture industry.
3) Increasing trade with Fredonia is a high priority of the current administration.
4) Fredonia seeks to establish an export industry that begins with our country.

The box presents a hypothetical first iteration of a risk identification statement. As the risk management activity progresses, this statement will be iterated and become more specific as the initial uncertainty is reduced via risk assessment and other analyses. Subsequent iterations can be expected to identify specific pests of quarantine concern. In most cases, the NPPO will restrict its risk management actions to matters of plant health using knowledge of the other risks as possible factors to consider when selecting a risk management option.

13.2.3. Establish need for risk assessment

Once a decision has been made to initiate a risk management activity, e.g., to address a market access request, and risks have been identified, risk managers decide whether risk assessment is needed or not. This decision begins with an initiating event such as a trade request. Risk managers consider the name and affiliation of the requestor and the plant species to be imported as well as seasonality for the export and pest risk assessment areas. Risk managers may decide, with the help of risk assessors, that a risk assessment is needed for any one of several reasons including:

- the requested market access is currently not authorized
- the commodity is currently permitted into a portion of the importing country, but the requestor wishes to expand the market and importing area
- the commodity is currently permitted from a portion of the exporting country, but the requestor wishes to expand the exporting area
- importation of the commodity with additional plant parts is requested
- the pest risk associated with the commodity has changed or is expected to change—for example:
  - o new mitigation measures for the commodity are proposed or implemented, e.g., irradiation
  - o there is reason to believe, based on scientific studies or port-of-entry interceptions, that a new pest associated with the commodity has become established in the exporting area
  - o new information indicates that a pest is likely to be more damaging in the PRA area than originally determined (USDA/APHIS/PPQ (2012)).

Most of the time, risk assessment is initiated to evaluate the pest risk associated with a market access request. Risk assessors are likely to make the recommendation but the need for risk assessment should be confirmed by the risk manager. The potential need for a weed risk assessment may be considered at the same time. The ultimate decision on whether one or more risk assessments are needed should rest with the risk manager. The aim of the initiation stage of PRA is to identify the pest(s) and pathways that are of quarantine concern in relation to the identified PRA area. Other information will be gathered as required to reach necessary decisions as the PRA continues.

13.2.4. Participants

Once a decision to pursue a risk assessment is made, the risk assessment team must be assembled. This is a risk management responsibility and the process begins with NPPO personnel. Risk managers must consider the type of expertise needed, the estimated duration of the project, the time per week that individuals will be needed, peak periods of activity, and the like.

If external experts may be required, this should be identified during the planning of the risk assessment. External expertise could include consultants, advisory committees, research personnel, industry representatives, or others.

Roles and responsibilities ought to be clarified at this time if they have not been previously clarified in the risk assessment policy. It is common to appoint a risk assessment team leader when the assessment involves numerous people. Coordinators, chief science officers, project managers, and other roles may be required for some risk assessments. Risk management activities that cross branches or divisions must carefully define the party with primary decision authority.

Few individuals possess sufficient knowledge to do a risk assessment alone, consequently, risk assessment is frequently a team effort. On a good team, diverse members of the team integrate their knowledge and skills in a synergistic fashion that results in an assessment that exceeds what independently functioning experts could have attained on their own. A multidisciplinary team has the requisite specialties available but on an interdisciplinary team those specialists work closely together, integrating the knowledge of their disciplines.

13.2.5. Scope of risk assessment

Determining the scope of the risk assessment is the risk manager's single most important responsibility in planning a risk assessment. Scoping a risk assessment brings the purpose of the assessment and the approach(es) to be used in it into focus. A clearly articulated scope is necessary to prepare an effective risk characterization and to eventually judge the success of the risk assessment. A good scope provides a "road map" for how the risk assessment will be accomplished (EPA, 2012). The road map helps all parties involved in the risk assessment understand the role of the assessment in the overall decision-making process. During the scoping process, risk managers, with the help of risk assessors, decide what will and will not be included in the risk assessment. This determines the complexity and focus of the assessment. A scope may be prepared by risk assessors but it must be approved by risk managers.

A good scope will define the purpose of the assessment, it will establish the depth and breadth of analysis, and it will identify the products needed from the risk assessment. The scope helps identify resource needs and schedules.

Import production

What is the risk assessment to assume about the conditions under which an import will be produced? Will it assume mangos for import will be washed or not? All mangos are routinely washed, so an argument can be made for describing the process as realistically as possible. A conservative risk manager might argue that washing, though a common process, cannot be assumed.

If the task is to assess the import of wooden 2 x 4s, risk assessors need not assess the logs from which the boards are cut because things happen during preparation that risk managers can count on. Timbers are debarked, for example, before the wood is cut. Things that happen routinely are part of the "without" condition. Things that risk managers do to reduce risk are part of the "with" condition.

Scoping often includes the enumeration of assumptions about production in pest risk management. Assumptions that affect the risk assessment may be made by risk assessors, risk managers, or a combination of the two. In pest risk management it is not uncommon for risk managers to handle the task of enumerating product production assumptions.

Scoping, in a phytosanitary risk management context, is an activity that occurs over time. It typically begins when the importing country requests "information prerequisites" from the exporting country. This may be done through the use of a questionnaire. When the requested information is obtained, it is reviewed by assessors and managers for clarity. Once this information is deemed sufficient, the scope of the analysis that will be done during the assessment is defined. Typically, risk assessors will decide if the pest list developed by the prerequisite information is adequate or not. Meanwhile, risk managers identify the assumptions that will frame the risk assessment. For example, will the risk assessment assume that mangos are washed or not? What other treatment and handling assumptions about the point of origin will be made?

Risk management needs to be intricately involved at this point in the PRA. Many market access requests begin from a place of wariness, if not distrust. Neither country wants to say too much in a process that stands or falls on the quality of the available evidence. Countries may "cherry pick" the evidence they provide, especially on the pest list. It is common practice for countries not to provide all of the available information. This practice is sometimes due to limited capacity, sometimes due to a strategic desire to control the flow of information, and sometimes due to deceit. The vertical team, which includes managers and assessors, needs to be more consistently involved in the evidence discovery part of the scoping process.

The scope of a risk assessment provides the first formal formulation of the problem to be solved. That formulation includes the commodity(ies) to be imported, their intended use, identification of the export area(s) and PRA area, field and/or harvest procedures to be considered when assessing the pest risk, preliminary identification of risk management options, and other details that provide the basis for the risk assessment. It is essential that risk managers and assessors agree on the scope of a risk assessment.

A well-executed scoping process provides the information needed for most of the remaining tasks in planning in the risk assessment, i.e., the charge, resources, schedules, and the like. In addition, an effective scoping process can reduce conflict between managers and assessors. When both parties participate in the process, disagreements can be surfaced more quickly and dealt with swiftly, before the assessment begins. An agreement among principal parties on the scope of realistic expectations about risk assessment outputs, resource commitment, and time frame is an additional benefit of scoping. A commonly shared understanding of the degree of complexity of the risk assessment required to adequately inform the decision at hand will support more effective and efficient decision-making (EPA, 2012).

Risk management objectives and constraints

Once the relevant risks have been scoped, it is time to identify the risk management objectives. What would successful phytosanitary risk management look like? What do risk managers want to see happen? These results are *risk management objectives*. What do they want to be sure does not happen? These are *risk management constraints*. Objectives and constraints define what successful risk management looks like. They provide a way to define and measure success.

Risk management objectives and constraints express the risk management organization's values. As such, these objectives are a transparent declaration of the desired outcomes of the risk management activity.

Objectives do not identify specific ways to manage a phytosanitary risk, they simply identify the desired outcomes of a risk management action. An objective is not a risk management measure, rather it is a statement of what risk management measures are intended to achieve. Good objectives are precise, practical, and measurable.

Sometimes there are things that are important *not* to do. Call these things constraints. Think of them as negative objectives. If a risk management activity tries to maximize its objectives, it also tries to avoid or at least minimize its constraints.

The desired outcome of this process is a written risk management objectives and constraints statement (hereafter called an "objectives statement" for simplicity), a single piece of paper to accompany the risk identification statement. Where risk identification identifies the reasons for the risk management activity, the objectives statement identifies the desired outcomes. Ideally, this statement is vetted and published as appropriate.

Sample preliminary risk management objectives and constraints statement

1) Prevent damage to domestic plants as a result of pests associated with raw timbers from Fredonia.
2) Minimize adverse economic impacts on the domestic furniture industry.
3) Implement the least trade restrictive measures if risk management is required.
4) Minimize environmental impacts associated with any treatment.

Notice that preventing damage to domestic plants is dramatically different from preventing entry of a pest, which might be a fairly typical unspoken objective of phytosanitary risk managers. Prevention of entry points strongly toward prohibition as a risk management measure. Prevention of damage opens risk management to consider a wider array of phytosanitary measures.

The box presents a hypothetical first iteration of a statement of risk management objectives in a scoping process. Notice that this statement tracks closely with the risk identification statement presented in an earlier box. It is not uncommon to see a one-to-one mapping between identified risks and risk management objectives. It is good practice to comprehensively identify and be aware of all the risks relevant to a given decision problem, but risk managers may not "own" all of those risks and so some may not be reflected in the objectives statement. The objectives statement prepared during scoping can be expected to change and evolve through the iterations of the risk assessment and the risk management process that engulfs it. In particular, both risks and the related risk management objectives become clearer as uncertainty is reduced. The success of a risk management activity is defined by the extent to which objectives are met and constraints are avoided. The objectives and constraints statement is a crucial output of the scoping process and an indispensable input for the charge for the risk assessment.

Consider diversion from intended usage while scoping

Some commodities have multiple intended uses. A commodity's intended use can be processing, ornamental, propagation, or consumption. Each of these uses may have a different risk profile. When market access is requested the intended use is specified. The intended use indicated for a commodity may, however, change. Diversion from intended use (DFIU) occurs when regulated articles are used for a purpose other than their originally declared purpose after the commodity is imported. Such unintended uses of a commodity may result in a higher probability of introduction than the declared intended use, possibly with significant negative consequences.

Objectives and constraints are not:

- absolute targets: they do not specify a particular level of achievement
- management options: objectives and constraints do not prescribe a specific course of action
- government goals: objectives and constraints are not a political or governmental objective
- risk assessment tasks: e.g. estimating a probability of introduction is not an objective
- resource constraints: it does not address time, money, or expertise.

Examples of commodities diverted from their intended use include:

- potato tubers imported for human consumption may be used as vegetative seed potatoes
- bulbs/corms/tubers like garlic, onion, and taro, imported for human consumption, may be planted for propagation
- fresh fruits and vegetables like avocado, pepper, and tomato, imported for human consumption, may be used as the source of seeds for propagation
- grain or seeds imported for industrial processing may be used as seeds for planting
- ornamental cut flowers imported for decorative purposes may be propagated
- seed imported for destructive laboratory testing may be used for planting
- wood chips imported as fuel may be used in landscaping.

An importing country may want to establish risk management measures for DFIU. This can only be done if technically justified evidence to support the application of measures is produced in the PRA process. Thus, risk managers must address the DFIU issue during the scoping process.

13.2.6. Charge for risk assessment

Risk managers are tasked with making a decision and decisions require information. Risk assessment supports decision-making and every good risk assessment requires a charge from the risk managers. The "charge" provides official instructions for conducting a risk assessment. It defines and formalizes selected important aspects of the overall scope of the risk assessment effort. The charge for a risk assessment includes a list of questions prepared by the risk managers that are to be answered by the risk assessment. The risk manager's questions should identify the technical and scientific issues that need to be addressed. The questions should contain enough detail to guide the risk assessment efforts. These questions cannot be formulated without significant interaction between risk managers and risk assessors and, at times, that may need to include dialogue with representatives of industry and the exporting country.

It is the risk manager's responsibility to generate questions that focus the risk assessment so it provides the science needed to inform the risk management decision at an appropriate level of detail (EPA, 2012). It is, then, the risk assessor's responsibility to design an assessment that addresses and answers the risk management questions as completely as possible. Getting the questions right is essential to the risk assessment's ability to provide the information needed for risk management.

Much of the current phytosanitary risk management process has been standardized by the guidance from the IPPC, in particular the ISPMs that pertain to the PRA process. These are routine and commonly conducted analyses. A one-time well-developed charge may suffice for the recurring aspects of PRAs. Many risk management activities are unique in some fashion, however, and this uniqueness should be well reflected in both the scoping process and the charge to the risk assessors. When decision-making requires more than the standard PRA information, risk managers must identify the unique information they need to make a decision.

This is best done through a series of specific questions. One might imagine recurring questions to include such things as:

- What are the species of quarantine concern?
- What is the probability that each pest will become established?
- What is the overall assessment of the risk of introduction for each pest?
- What are the economic, environmental, political, and other consequences associated with the introduction of each pest?

Most NPPOs have structured their risk assessment process in such a way as to produce the most frequently required data. There may be additional questions that need to be asked to support risk management decision-making. These may pertain to the efficacy of different risk management options such as:

- What is the probability that a pest will become established if risk management option A is implemented? In other words, what is the residual risk associated with risk management option A? B? etc.
- What is the cost of implementing option A? B? etc.

If there are other atypical questions of concern to risk managers it is important that they articulate and include them in the charge for risk assessment. The desired outcome of this process is a written charge for the risk assessment prepared by risk managers.

13.2.7. Resources

Adopting a view of risk managers as NPPO managers and administrators, it is the risk manager's responsibility to provide adequate resources for the risk assessment. This includes the personnel, as discussed above, but also financial and technical resources as needed. Risk assessors need a budget adequate for their assessment. When an assessment requires access to computerized databases, aerial photography, base mapping, or special equipment it is generally considered the risk manager's responsibility to provide resources sufficient to acquire these materials.

13.2.8. Schedule

It may be helpful to complete the preceding risk assessment planning steps before establishing the schedule. A good schedule identifies tasks and the tasks' deliverables. These might include identification of organisms of concern, pathway identification, risk estimates for pests, or residual risk estimates for the risk management options. A schedule can include a timeline for the accomplishment of the identified tasks and it may single a few of these tasks out as milestones. Milestones are used to mark specific points along a risk assessment's timeline. These points may signal anchors such as the start and end date, a need for risk manager input, external review or budget checks, among other purposes. Some risk assessments may have a financial schedule as well to control the obligation and expenditure of funds. A good schedule shows everyone what is going to happen and when as well as where the risk assessment stands in relation to all the planned work.

13.2.9. Communicate

Risk assessment is iterative by its nature and there may be a need to revise, refocus, and redirect the risk assessment from time to time. To monitor the progress of the risk assessment against its schedule requires effective internal risk communications. That means determining how often and for what purposes the risk assessment and risk management teams will communicate. The performance schedule should provide sufficient touchpoints between risk managers and risk assessors to meet all the communication needs of both parties. The nature of the interactions between managers and assessors is discussed at length in Chapter 11.

External risk communication may also begin with these risk assessment planning tasks. Although the extent of external risk communication will vary widely from case to case, risk communication should be intentionally planned, even if it is to be limited in scope. A plan should consider the following elements:

- Clearly identified stakeholders and collaborators.
- Specific communication goals.
- Audience assessments.
- Communication tailored toward each audience.

Risk communication specialists should be involved with the risk assessment effort from planning and scoping onward. Risk communication is addressed in more detail in Chapter 16.

13.2.10. Initiate

When planning for the risk assessment is complete, it is time to initiate the assessment. This occurs when the risk manager gives the order to begin the assessment. Initiation also requires the resources needed to complete the assessment. This initiation task will coincide with the Stage 1 initiation of a risk assessment when an assessment is needed.

13.3. Support for risk assessment

It bears repeating that risk managers are not to interfere in any way with the independent conduct of an objective, science-based risk assessment. There are, however, several ways for risk managers to support the conduct of a risk assessment. This set of tasks includes:

1. Review and refine the process.
2. Data and evidence.
3. Develop model.
4. Review results.
5. Residual risk.
6. Documentation.

Taken as a whole, risk managers support risk assessors in their effort to conduct a science-based risk assessment. This marks the beginning of Stage 2 of the ISPM 2 PRA process.

13.3.1. Review and refine the process

The risk assessment process may need to be reviewed and refined once it gets underway. It is an iterative process, so expect the risk assessment to change and evolve as it is conducted. That means, at times, risk managers may be called upon to review suggested changes to the risk assessment scope, its schedule, resource allocations, or the team's composition. Other times, risk managers may choose to refine the charge to reflect changes in information needs or risk management objectives.

A pre-existing completed risk assessment may also be reviewed and refined. ISPM 2 suggests the need for a new or revised PRA may arise from situations such as when a national review of phytosanitary regulations, requirements, or operations is undertaken; the regulatory proposal of another country or international organization is evaluated; or a new system, process, or procedure is introduced or new information that could influence a previous decision becomes available. An international dispute on phytosanitary measures, a change in another country's political boundaries or a change in a country's phytosanitary situation could be additional reasons for reviewing and revising an existing risk assessment. This review may or may not result in the need for new iterations of the risk assessment.

13.3.2. Data and evidence

Data and evidence are the primary responsibility of the risk assessors, but risk managers have some data and evidence responsibilities. These include establishing data quality objectives as part of the organization's risk assessment policy. Such objectives facilitate transparent documentation of the justification of why data are included or excluded (IRAC, 2000).

In actual practice, it is not uncommon for risk managers to request information from another country or a company. Peer-to-peer communication between importing and exporting countries can be relatively rare because scientists may be naïve about the sensitivity of trade negotiations to exchanges of unofficial information. Hence, that task may fall to risk management.

Risk managers are responsible for seeing that uncertainty is adequately addressed by risk assessors. Incomplete information and data gaps are a significant challenge throughout all aspects of risk analysis. Risk managers should require assessors to identify significant sources of uncertainty and to determine whether the available data are representative of the actual conditions of any given risk assessment.

The Office of Science and Technology Policy (OSTP) (2010), speaking in a memo on scientific integrity, said,

> The accurate presentation of scientific and technological information is critical to informed decision making by the public and policymakers. Agencies should communicate scientific and technological findings by including a clear explication of underlying assumptions; accurate contextualization of uncertainties; and a

description of the probabilities associated with both optimistic and pessimistic projections, including best-case and worst-case scenarios where appropriate.

Prescribing the manner in which uncertainty is to be handled is an important area of collaboration between managers and assessors.

There will be times when the question arises, "Are the data good enough to support risk management decision-making?" It is generally the risk assessor's responsibility to determine the extent to which a data gap exists and to reflect that judgment when determining one's confidence level in the available data and the risk assessment produced with it. These determinations are ultimately matters of scientific judgment rather than pure science. It is to be expected that different people may have different comfort levels for making decisions based on a given set of data. What is convincing evidence for one may be scant evidence for another. Realistically, such issues can be expected to arise from time to time during the conduct of a risk assessment and ultimately managers and assessors will have to reach agreement on the conflict. Situations like these can stall a risk assessment. When managers and assessors differ in these ways, careful dialogue is required. To minimize disagreement on what constitutes a significant data gap, managers and assessors should agree on criteria for evaluating data and setting a threshold for data sufficiency prior to initiating the risk assessment if this is not already addressed in the organization's risk assessment policy.

Risk managers, like risk assessors, are heavily invested in the quality of the data used for risk assessment. They should establish a policy for evaluating the results of the risk assessment. For example, they may establish minimum standards for conducting a sensitivity analysis on all risk assessments to determine whether a data gap is significant for decision-making. They may require risk assessors to identify the uncertain risk assessment inputs with the greatest impact on the risk characterization. Risk managers may prescribe the form by which results of quantitative risk characterizations are reported. For example, they may require that quantitative risk estimates be specified by a five-number summary or a credible interval consisting of expected value, 5th and 95th percentiles. Some risk managers may insist on seeing the data displayed in

probability density functions, cumulative distribution functions, box plots, or other specific displays.

Expert elicitation provides a useful methodology for characterizing uncertainty when significant data gaps exist. Risk managers may set policy detailing how and when expert opinion or judgment may be used. The circumstances under which expert elicitation may be employed as a means of characterizing uncertainty may be another topic of manager/assessor cooperation and coordination. For an excellent overview of methods for eliciting expert judgment see O'Hagan *et al.* (2006).

13.3.3. Develop model

A conceptual model is a representation of predicted relationships between the hazard and potential harm(s) in exposed populations. The conceptual model depicts the movement of the hazardous pest from the source to the new host. More sophisticated conceptual models may identify variables and data needed to conduct the PRA. To a great extent, NPPOs follow the conceptual model described in ISPMs 2 and 11. This framework and guidance has been widely accepted by risk managers and assessors alike and qualitative assessment models are far more common than quantitative ones.

Quantitative risk assessment may require several iterations to develop a sound and useful conceptual model. The conceptual model should describe or visualize the relationships among the assessment and measurement endpoints, the data required, and the methodologies that will be used to analyze the data. Risk managers may elect to be kept informed of the progress on a conceptual model in order to ensure that once that model is specified and built as a computational one, it will yield the desired information in a usable form.

Risk managers will generally have no role in the development of a specific risk assessment model. Certainly, they will have no direct role in the specification and build-out of a conceptual model. However, they may become involved in prescribing a model-building process as a matter of policy. Such policy might establish standards for verifying and validating models as well as establishing administrative requirements for documenting the results of the model. Risk managers must have sufficient knowledge of and comfort with the conceptual model to trust the results of the risk assessment and, when necessary, to explain and defend it in bilateral negotiations.

A quantitative model-building process

1. Get the question(s) right.
2. Know the uses of the model.
3. Build a conceptual model.
4. Specify the model.
5. Build a computational model.
6. Verify the model.
7. Validate the model.
8. Design simulation experiments.
9. Make production runs.
10. Analyze simulation results.
11. Organize and present results.
12. Answer the question(s).
13. Document the model and results.

(Yoe, 2019)

13.3.4. Residual risk and risk reduction

Chapter 2 introduced two dimensions of risk that are extremely important to risk managers, i.e., risk reduction and residual risk. A risk reduction is the extent to which an existing, future or historical risk is or might be reduced by a risk management option. Different risk management options offer different levels of risk reduction as depicted in the left panel of Fig. 13.2. There, option A is more effective in reducing an introduction risk than option B is. No risk management option, short of a total prohibition of trade is likely, at least in theory, to reduce the risk of a course of action to zero. Even with a total restriction on trade there may be other pathways, which could introduce some probability of introduction, such as smuggling.

If we call the risk with trade and no risk management the "without condition," signifying a condition without specific additional risk management controls in place, and the risk with trade and a risk management option in place the "with condition," signifying the presence of one or more new risk management measures, then measurements of risk reduction are obtained by comparing the without and with condition scenarios. Current best practice risk assessment routinely produces the without condition risk estimate. It is the risk manager's responsibility to request estimates of risk reduction for each risk management option considered.

A residual risk is the amount of existing, future or historical risk that remains or might remain after a risk management option has been implemented. Residual risks are shown in the right panel of Fig. 13.2. Residual risk also varies with the risk management option evaluated.

Residual risk is sometimes overlooked by risk managers, especially if it is not explicitly estimated in the PRA. Imagine that the risk with trade and no risk management shown in the figure is qualitatively estimated to be a high risk with moderate amounts of uncertainty. Next, imagine a set of measures, risk management option A, which reduces the risk to a low level, i.e., the residual risk of option A is low. Now imagine another set of measures, risk management option B, that reduces the risk from high to medium. The residual risk of option B is medium. Risk managers need to make sure that risk assessors know that they will need without condition

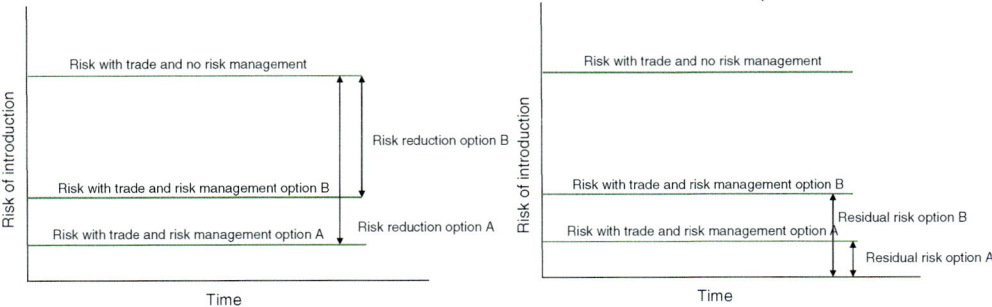

Fig. 13.2. Four phases of risk manager involvement in the initiation and risk assessment stages of pest risk analysis.

estimates of risk and with condition estimates of risk as well as estimates of residual risk for each risk management option under consideration. This request should be included in the charge.

It is somewhere between naïve to dereliction of duty to assume that prescribing a set of risk management measures will reduce the residual risk of introduction to zero. Risk managers should be universally suspicious of claims of zero residual risk. Risk control measures are not always implemented with the desired care and rigor and, even when they are, they may not be totally effective. Thus, residual risk becomes a very important risk estimate.

Establishment sub-elements

Pest prevalence on the harvested plant part(s)

- Likelihood of surviving post-harvest processing before shipment.
- Likelihood of surviving transport and storage conditions of the consignment.
- Likelihood of coming into contact with host material in the endangered area.
- Likelihood of arriving in the endangered area.

Potential direct and trade impacts

- Damage potential in the endangered area.
- Spread potential.
- Determining export markets at risk.
- Likelihood of trading partners imposing additional phytosanitary requirements.

Adapted from APHIS (2012)

It can be difficult to make fine distinctions in risk reductions, and consequently residual risk, when relying on qualitative risk estimates. This is when it may be very useful to consider the with condition ratings of sub-elements (see box above) used to arrive at an overall risk rating. Descriptions of uncertainty may be useful for making distinctions in the effectiveness of risk management options. If two options reduce a risk from high to low but one option has more uncertainty associated with the implementation and efficacy of a measure, choosing the less uncertain measure may be a viable means of distinguishing between options.

13.3.5. Review results

This review is primarily by NPPO risk managers only. When the first draft of the risk assessment

is completed, risk managers must review the results to ensure that the questions, provided in the charge for the risk assessment, have been answered. They must also make sure they understand the remaining uncertainty in the assessment and how it could affect the answers to the charge questions, as well as other significant results of the assessment. Risk managers need a keen awareness of the strengths and weaknesses of a risk assessment as well as a basic understanding of the conceptual risk assessment model.

A well-documented risk assessment will usually meet these needs. A complex assessment may require face-to-face dialogue between risk assessors and risk managers, to ask questions and gain the required understanding of the assessment and its results.

13.3.6. Documentation

Risk assessments and the risk management decisions based upon them must be documented. The IPPC's transparency principle (IPPC, 2016) requires countries to make the rationale for their phytosanitary requirements available upon request. ISPM 1 also requires contracting parties to provide technical justification of phytosanitary measures implemented as a result of the pest risk management stage of PRA, if requested. This technical justification of the identified or selected risk management measures must be documented using the best available and most reliable scientific information (IPPC, 2016). All of this establishes a need to document the results of both the risk assessment and any decisions to require phytosanitary

---

The main elements of documentation are:

- Purpose for the PRA.
- Pest, pest list, pathways, PRA area, endangered area.
- Sources of information.
- Categorized pest list.
- Conclusions of risk assessment:
  - probability
  - consequences.
- Risk management options identified.
- Options selected.

Source: ISPM 11 Pest risk analysis for quarantine pests, including analysis of environmental risks and living modified organisms

---

risk measures as a result of the risk assessment. This documentation is necessary to facilitate review and dispute resolution. It must plainly reveal the sources of information and the rationale used to reach the risk management decision.

Documentation of the pest risk management stage should include a discussion of all instrumental uncertainties considered during the part of the risk assessment used to identify and select the recommended pest risk management measures. It is important to clearly inform decision-makers about the level of uncertainty regarding the scientific evidence that formed the basis for the selection of risk management options in the risk management documentation (NAPPO, 2012). It is the risk manager's responsibility to see that the risk assessment and any and all subsequent risk management decision making are adequately documented.

Technology advances make it possible to envision alternative forms of documenting risk management decisions. A traditional written risk assessment document is likely to be the mainstay form of documentation for the foreseeable future. More adventurous risk managers might be interested in alternative forms of complying with the transparency principle. Video documentation is an obvious alternative. Showing people the pests, the problems and challenges they present, illustrating the risk management options, and showing the science are as yet unexploited alternative ways of telling a story. The use of social media and virtual or augmented reality are also untapped potential ways of documenting risk assessments.

Eight good documentation ideas

1. Tell a story: use an engaging beginning, an interesting middle, and a satisfying ending.
2. Narrative quality: no data dump, no default formats, no "take the last report and add a little to it."
3. Chronology is your friend.
4. No geek-speak.
5. Cut to the chase: write it the way you would say it, nouns and verbs beat adjectives and adverbs.
6. User-friendly and informative figures and features (let people access data/info in the ways they like)—write on a map or a picture.
7. Tie decisions and judgments to the evidence: say why you did what you did.
8. Tell the truth—no matter where it takes you.

13.4. Review the Risk Assessment

This review follows the risk manager's review of the risk assessment. The goal of the following set of tasks is to produce a critical review of the risk assessment that results in its approval followed by organizational clearance:

1. Team review.
2. Peer review.
3. Stakeholder input.
4. Revise risk assessment.
5. Clearance/approval.

The outcome of these tasks is a risk assessment document that is ready for release to stakeholders and interested publics. Once the risk assessment has been documented, it is time to submit it to a more thorough and more formal review process to prepare it for public release. That review begins by having the completed risk assessment, including documentation of the risk management decision, reviewed by the entire risk analysis team. This may be followed by an external technical (peer) review, stakeholder input, and final clearance and approval of the risk assessment document by the organization that produced it.

13.4.1. Team review

Ideally, the risk assessment team actively coordinates with the risk management team throughout the risk assessment process by providing opportunities for input while seeking feedback as needed. Team review need not be a one-time end-of-process event. The communication plan, coupled with scheduled milestones and other touchpoints of interaction should provide ample opportunities for the assessment team to brief risk managers on the progress of the risk assessment. Consequently, by the time the risk managers review the draft document, they should be fairly well acquainted, if not thoroughly knowledgeable, about the methodology used, the answers to the risk management questions, and the general conclusions of the assessment. Nonetheless, the entire risk analysis team, however it is defined, ought to be provided with an opportunity to review the documented draft risk assessment before it is released to external review.

13.4.2. Peer review

Peer review is important for quality control. It helps lend credibility, transparency, and technical accuracy to any scientific work product. Peer review is not required by the IPPC or any other international body, however, it is broadly considered part of best practice. It is a process routinely used for risk assessments prepared by the European Food Safety Authority (EFSA), the International Life Sciences Institute (ILSI), the US Environmental Protection Agency, Food Standards Australia New Zealand, and many other food-related organizations. There is widespread belief that external peer review promotes good science, which leads to better policy, and that better science increases the legitimacy of decisions (Guston, 2002).

Peer review of a risk assessment includes focused and formal charge questions covering:

- the completeness and strength of the data presented
- the defensibility of the assumptions
- the use of appropriate analyses and methods
- the strength and defensibility of the conclusions
- the strength and scientific defensibility of the rationales provided for choice of: study, effect, level, models, uncertainty factors, and the like
- more specific questions regarding key pest or document-specific issues.

(Patterson *et al.*, 2007)

Patterson *et al.* (2007) describes peer review as a formal, external, and independent review of a work product intended to be final. It is undertaken to gain agreement from an external group of peer experts on a document's conclusions and the scientific basis for those conclusions. If that agreement cannot be reached the document must be revised until it can be consensually achieved.

Peer review is an essential part of science. It provides independent scientists with the opportunity to review the risk assessment team's work, check it, comment on it, and determine whether it is good enough to support the conclusions of the risk assessment and the risk management decisions based upon them. It is common to seek a peer review of an assessment's data quality as well as the methods and models used. Subject matter experts outside of the NPPO organization preparing the risk assessment are typically enlisted to conduct the peer review. Different organizations can be expected to pursue peer review in a variety of ways. Experts from within the originating organization, who had no involvement with the document, may be used at times, but that is not a preferred method.

13.4.3. Stakeholder input

Stakeholder input is an important component of the risk assessment review process and one that is critical for transparency. RSPM 40 (North American Plant Protection Organization, NAPPO, 2012) suggests that stakeholders affected by the pest(s) of concern should be consulted, where feasible, when risk management measures are being developed. Understanding stakeholder expectations, in terms of pest or damage tolerance, will assist risk managers in designing specific risk management measures. These consultations are also useful to identify stakeholders who should be included as part of risk communication or for information-sharing purposes.

Stakeholders can be informed and listened to during workshops, public meetings, informal meetings, and technical presentations, as required or requested. Comments may also be submitted to the public docket and through other official avenues as provided.

13.4.4. Revise risk assessment

Team review, peer review, and stakeholder input can result in the need to revise the risk assessment and other decision documentation. Revising the risk assessment may be a simple iteration of the documentation task or it may entail a re-iteration of a number of risk assessment tasks.

13.4.5. Clearance and approval

Each NPPO should have a process for clearing and approving documents that ensures materials released by them are scientifically and technically accurate. Approval may require tracking and documenting actions and clearances of responsible offices. Collaborative projects with other government agencies may involve consideration of the clearance process of those agencies. Risk managers should ensure that regulatory notices are developed and reviewed using appropriate, standard operating procedures.

13.5. Issue the Risk Assessment

Finalizing the risk assessment and making it publicly available is made possible by the fourth set of tasks shown below:

1. roll-out
2. address comments
3. release
4. debrief

These tasks complete Stage 2 of the PRA process.

13.5.1. Roll-out

A roll-out makes a new product, service, or system available for the first time. Thus, the roll-out of a risk assessment introduces the results and findings of the PRA and any risk management decisions based upon it to external stakeholders. The roll-out plan should be tailored to the specific needs of the risk management activity. Thus, each roll-out will be unique.

Example elements of the roll-out plan

- A press contact list.
- Talking points/press announcements/briefing materials/Q&As.
- Schedule and hold public meeting(s).
- Plan and hold stakeholder briefing(s).
- Plan and hold other briefings or presentations, as needed.
- Develop and publish any required notices and availability of document(s).
- Make and distribute printed copies of document(s) to stakeholders and public.
- Prepare and post risk assessment document(s) on web page.

(Adapted from CFSAN, 2002)

The roll-out plan should identify the audience for the PRA. Who will be impacted by the

assessment? Who needs to be informed about the assessment and its findings? Who is the target audience for each document produced? What other groups may be interested in the assessment? How will the different audiences be informed? What channels will be used for communication? When will the document be released? Are any meetings necessary? If so, when will they be held? What immediate actions need to be taken and by whom? The risk communication team should be heavily involved in developing and implementing the roll-out plan, which should include proactive communication and outreach.

13.5.2. Address comments

Some NPPOs will have public dockets that require that stakeholders be permitted time to submit comments on the draft risk assessment. In such cases, the documents provided and the comments submitted are made available for public scrutiny and review at the NPPO's public docket. Typically, a discussion of the comments and how they were addressed in the risk assessment is presented with the final version of the document. Thus, addressing the comments can range from a rather perfunctory function to a substantive refinement of the draft risk assessment.

13.5.3. Release final assessment

Following the public comment period (stakeholder involvement described above), the draft risk assessment is revised and issued as a final document. A notice of availability of the risk assessment is often published by the NPPO in a manner prescribed by national policy or practice. The major initiative at this point is to decide where and how the related documents will be made available. Printed copies may be provided along with electronic copies. The content and findings of the documents may be presented at public meetings. Briefer summaries may be prepared for certain stakeholder groups. Copies may be proactively provided to trade partners as well as made available upon request to others.

13.5.4. Debriefing

The entire risk analysis team should meet to evaluate the overall risk assessment process following the release of the assessment, for the purpose of improving the process. The team may consider questions like the following adapted from CFSAN (2002).

- Was the assessment completed on time?
- Was the risk manager's charge to the risk assessors sufficiently clear?
- Did the teams take appropriate steps to resolve any concerns over the charge in a timely manner?
- Were sufficient resources identified and provided?
- Were there any gaps identified that suggest the need for further capacity-building?
- Were key assumptions from the risk managers clearly provided?
- Did risk assessors/managers and communicators sufficiently discuss and understand the significance of such assumptions in a timely manner?
- Were data gaps or other risk assessment assumptions identified by risk assessors clearly communicated to the other team members?
- Were the impacts of these uncertainties clearly identified?
- Was a sensitivity analysis performed?
- Do risk assessors believe the results are scientifically sound and that the assumptions used followed best risk assessment practice?
- Were any conflicts between risk assessors and managers identified and, if so, how were they resolved?
- Do risk managers feel they got an assessment that answered their question(s)?
- If not, why not, and what could have been done differently?
- At what point was the risk communication team brought into the process?
- Were stakeholder concerns identified early on in the process?
- Did the team feel it had sufficient information to facilitate the communication of results?
- Were appropriate and sufficient project milestones identified?
- Did they facilitate the identification of problems and corrective actions?

- Were problems identified in a timely manner?
- Was there a clear process for resolving problems?
- What other steps (if necessary) would be needed to better resolve issues raised or problems identified?
- Did sufficient communication among all team members occur?
- Was communication ongoing?
- Did team leaders sufficiently communicate?
- Did the appropriate level of information filter down to all team members?

The tasks described in this chapter coincide with the description of Stages 1 and 2 of the ISPM 2 Framework for PRA. However, it should be noted that the work described here is iterative in nature and will not always proceed in a smooth, linear way. It is of particular note that some of the tasks described here, including documentation of risk management decisions in the risk assessment, cannot occur until significant parts of Stage 3 of the PRA process are completed. The risk management tasks coinciding with that stage are described in the next chapter.

13.6. Summary and Look Forward

Here are five things to remember from this chapter.

1. The conduct of a risk assessment is to be free from interference from risk managers.
2. Risk managers are primarily responsible for planning the risk assessment with input from risk assessors.
3. Risk managers and assessors must coordinate and cooperate throughout the conduct of the risk assessment.
4. Risk managers are responsible for an effective and constructive critical review of the risk assessment, which may include a peer review, that leads to approval and organizational clearance of a risk assessment document.
5. Risk managers are responsible for the release of a final risk assessment and a debrief of the process that produced it.

The next chapter describes the remainder of the generic risk management model presented in this handbook. It includes the tasks that coincide, conceptually, with Stage 3 of the PRA process.

13.7. References

CFSAN (Center for Food Safety and Applied Nutrition) (2002) *Initiation and Conduct of all "Major" Risk Assessments Within a Risk Analysis Framework*. CFSAN Risk Analysis Working Group. Available at: http://www.fda.gov/Food/FoodScienceResearch/RiskSafetyAssessment/ucm475127.htm (accessed February 1, 2017).

EPA (Environmental Protection Agency) (2012) *Microbial Risk Assessment Guideline Pathogenic Microorganisms with Focus on Food and Water*. Prepared by the Interagency Microbiological Risk Assessment Guideline Workgroup. Washington, District of Columbia, USA.

Guston, D.H. (2002). *Comments Made at Society for Risk Analysis Forum on "Conflict, consensus, and credibility: a forum on regulatory peer review."* Arlington, VA.

IRAC (Interagency Risk Assessment Consortium) (2000) *Public Meeting on Food Safety Risk Analysis Clearinghouse Data Quality Objectives*. Available at: http://foodrisk.org/IRAC/events/2000-12-05/index.cfm (accessed January 31, 2016).

IPPC (2016) International Standards for Phytosanitary Measures, Publication No. 1: *Phytosanitary Principles for the Protection of Plants and the Application of Phytosanitary Measures in International Trade*. Secretariat of the International Plant Protection Convention (IPPC), Food and Agriculture Organization of the United Nations, Rome.

North American Plant Protection Organization (NAPPO) (2012) *NAPPO Regional Standards for Phytosanitary Measures 40 Principles of Pest Risk Management for the Import of Commodities. Ontario, Canada*. Available at: http://www.nappo.org/files/8314/3889/6413/RSPM40-e.pdf (accessed February 1, 2017).

Office of Science and Technology Policy (OSTP) (2010) *Memorandum for the Heads of Executive Departments and Agencies, Subject: Scientific Integrity*. Available at: http://www.whitehouse.gov/sites/default/files/microsites/ostp/scientific-integrity-memo12172010.pdf (accessed January 31, 2016).

O'Hagan, A., Buck, C.E., Daneshkhah, A.J. Eiser, R. Garthwaite, P.H. *et al*. (2006) *Uncertain Judgements: Eliciting Experts' Probabilities*. John Wiley & Sons, West Sussex, UK.

OMAF (Ontario Ministry of Agriculture Food and Rural Affairs OMAF) (1997) *A General Risk Assessment Framework Food Safety Risk Assessment*. Available at: http://www.gov.on.ca/OMAFRA/english/research/risk/frameworks/as1.html#core (accessed January 24, 2005).

Patterson, J., Meek, M.E., Strawson, J.E. and Liteplo, R.G. (2007) Engaging expert peers in the development of risk assessments. *Risk Analysis,* 27, 6.

USDA/APHIS/PPQ (2012) *Guidelines for Plant Pest Risk Assessment of Imported Fruit and Vegetable Commodities*. Plant Epidemiology and Risk Analysis Laboratory, Center for Plant Health Science and Technology. United States Department of Agriculture, Animal and Plant Health Inspection Service.

Yoe, C. (2019) *Principles of Risk Analysis Decision Making Under Uncertainty*, 2nd edn. CRC Press, Boca Raton, FL, USA.

14

Pest Risk Management Through Stage 3

Some risks that are thought to be unknown, are not unknown. With some foresight and critical thought, some risks that at first glance may seem unforeseen, can in fact be foreseen. Armed with the right set of tools, procedures, knowledge and insight, light can be shed on variables that lead to risk, allowing us to manage them.

Daniel Wagner

14.1. Introduction

Continuing with the pest risk management model, the lower half of Fig. 10.2 (see p. 120) provides an illustrated outline for this chapter. Stage 3 of the ISPM 2 PRA framework consists of the work most consistently recognized as the responsibility of the risk manager. Fig. 10.2 groups the primary risk management tasks of Stage 3 into five categories: traditional risk management; formulating risk management options; evaluating individual options; selecting the best risk management option from among the available options; and implementing the option, which includes the monitoring and review responsibilities of risk management. Each of these categories is further subdivided into a sequence of tasks, which the risk manager must perform or for which the risk manager is responsible.

The risk management tasks described in this chapter may be quite different from common practice in any given national plant protection organization (NPPO). These tasks are described

to help improve common practice. Although presented in a sequential fashion they do not always occur in sequence. The tasks described here may comingle with or even precede some of the tasks presented in the previous chapter.

14.2. Traditional Risk Management

Traditional risk management is understood here to include the state of phytosanitary risk management this handbook seeks to improve. It is defined primarily by risk management guidance promulgated by the International Plant Protection Convention (IPPC), especially as found in ISPMs 2 and 11. The traditional risk management ideas of particular interest in this first task of the pest risk management model are:

1. IPPC guidance.
2. Acceptable risk.
3. Principles of plant quarantine (ISPM 1).

These traditional ideas form the foundation for the risk management model offered here. Each is developed in the following sections.

14.2.1. IPPC risk management

IPPC risk management guidance is an integral and indispensable part of the pest risk management model developed here and in Chapter 13. This model begins with traditional phytosanitary risk management, as described by the IPPC. The IPPC, as an international organization representing the interests of the world's plant health community, has had to walk the tightrope of harmonizing the world's phytosanitary risk management practice while not being too prescriptive.

In ISPM 1 (IPPC, 2016c), the IPPC says the concept of mitigating pest risk to an acceptable level should be the guiding principle of managing risk; aiming for zero risk is not a reasonable option. The pest risk management task is outlined in ISPM 2 (IPPC, 2019a), which describes the conclusion of the pest risk management stage occurring when risk managers can determine whether or not appropriate phytosanitary measures adequate to reduce the pest risk to an acceptable level are available, cost-effective, and feasible. Risk management is defined in ISPM 5 (2019b) as the "evaluation and selection of options to reduce the risk of introduction and spread of a pest."

ISPM 11 (2019d) describes pest risk management as the process of identifying ways to react to a perceived risk, evaluating the efficacy of these actions, and identifying the most appropriate options. Risk managers use the conclusions from pest risk assessment to decide whether risk management is required and, if so, the strength of measures to be used. A second guiding principle is introduced in ISPM 11 as managing risk to achieve the required degree of safety that is both justified and feasible given the options and limited resources available. The document also says uncertainty in the assessments of economic consequences and probability of introduction should also be considered during decision-making.

To many if not most people, phytosanitary risk management means choosing appropriate risk management measures. But that is only part of the analytic-deliberative process of risk management. Risk managers must consider and weigh various options for mitigating unacceptable risks to tolerable or acceptable levels. The efficacy, technical and economic feasibility of each option should be analyzed and considered. This analysis requires a close interaction with risk assessors as the risk assessment should provide relevant information and analysis about aspects of an option's efficacy, including, for example, a

particular pathway(s), control points, uncertainties identified in the assessment, pests likely to be in a pathway, and important information about the biology of the pest that may factor into the selection of risk management options (Devorshak, 2012). Risk managers are, therefore, dependent on risk assessors to provide information about how effective different mitigation options may be in reducing the overall risk. The deliberative aspect of risk management is selecting and applying mitigation options that are identified based on policy considerations and analytical components of pest risk management. These traditional responsibilities are considered in subsequent categories of risk management tasks. One of the most critical aspects of these deliberations is determining what constitutes an acceptable risk, the next topic.

14.2.2. Acceptable risk

The risk manager's first substantive decision following completion of a risk assessment is whether the assessed risk requires risk management measures. Managing pest risk to an acceptable level is identified as the guiding principle of managing risk in ISPM 1. A risk is acceptable when its probability of occurrence is so small, its consequences are so slight or its benefits (perceived or real) are so great, that individuals or groups in society are willing to take or be subjected to the risk that the event might occur (Yoe, 2019). Determining whether an assessed risk is acceptable or not is the responsibility of the phytosanitary risk manager. It is a matter of subjective judgment, not a scientific determination. A phytosanitary risk that is judged acceptable requires no additional risk management. A risk that is not acceptable is by definition unacceptable and must be managed. It is conceptually possible to take steps to reduce an unacceptable level of risk to an acceptable level. More often than not however, unacceptable risks are managed to tolerable levels, as zero risk is not considered a reasonable option (Specification 63).

"Tolerable risk" is a term that is widely accepted and used in the international food safety community. It has been less commonly used in phytosanitary risk management, but it is a useful term to introduce. A tolerable risk is not an acceptable risk. It is a non-negligible risk that has not yet been reduced to an acceptable level. Such a risk is tolerated for one of three general reasons: it may be impossible to reduce the risk further; the costs of further reduction are considered excessive; or, the magnitude of the benefits associated with the risky activity are too great to reduce it further. A tolerable risk is an unacceptable risk whose severity or chances of occurring have been reduced to a point where it is tolerated. There is nothing in the Sanitary and Phytosanitary Agreement (SPS Agreement) or other IPPC policy that prevents a tolerable level of risk being identified as the appropriate level of protection (ALOP). The SPS Agreement does not use the nuanced refinements found in the concept of tolerable risk. In SPS Agreement parlance, an ALOP results in an acceptable level of risk. The ALOP is described in more detail in Chapter 18.

14.2.3. Principles of plant quarantine

The general principles of plant quarantine are to be considered when risk management options are formulated. These principles, as related to international trade, articulated in ISPM 1, are:

- sovereignty—countries may exercise the sovereign right to utilize phytosanitary measures to regulate the entry of plants and plant products and other materials capable of harbouring plant pests
- necessity—countries shall institute restrictive measures only when necessary to prevent the introduction of quarantine pests
- minimal impact—phytosanitary measures shall represent the least restrictive measures available which result in the minimum impediment to the international movement of people, commodities and conveyances
- modification—as conditions change phytosanitary measures shall be modified promptly
- transparency—countries shall publish and disseminate phytosanitary prohibitions, restrictions and requirements and make the rationale for such measures available upon request
- harmonization—whenever possible phytosanitary measures shall be based on inter-

national standards, guidelines and recommendations, developed within the IPPC framework
- equivalence—phytosanitary measures that are not identical but which have the same effect shall be recognized as equivalent
- dispute settlement—disputes between two countries regarding phytosanitary measures will, preferably, be resolved at a technical bilateral level.

Risk management options are alternative means and measures that can be taken to reduce unacceptable phytosanitary risks to acceptable or tolerable levels. These measures are developed iteratively throughout a risk management activity and collaboratively by risk managers, risk assessors, contractors to trade, industry, academia, and others. The primary responsibility for developing risk management options rests with risk managers.

An option is a thing that is or may be chosen. The working definition of a risk management option (also known as, "solution," "option," or "alternative") is that it is a set of one or more risk management measures functioning together to address one or more risk management objectives. Many risk management options have more than one measure, especially those formulated in a systems approach. Different options have different measures or they combine the same measures in significantly different ways.

Good risk management requires more than one option from which to choose a course of action. It is difficult to establish a chosen course of action as the best course of action if it was the only option considered. A best option ought to be chosen from among several alternatives. Risk management options are, ideally, formulated to meet the full range of risk management objectives, chief among which is reducing unacceptable phytosanitary risks.

The above principles describe things that risk management options ought to do (sovereignty, modification, transparency, harmonization, equivalence, and dispute settlement) as well as things options ought not to do (necessity and minimal impact). Other risk management objectives identified during the scoping process, such as encouraging new trading partners or trade in new commodities, for example, should also be considered during the process of formulating risk management options consistent with these principles.

14.2.4. A decision example

The practice of pest risk management is too varied to suggest there is a typical example of how decisions are made. The SPS Agreement and ISPMs developed by the IPPC do, however, structure the process to a significant extent. Thus, it is possible to offer an example of a not uncommon decision process. If a risk assessment identifies a pest of quarantine concern, that is often considered sufficient evidence of "potential concern." There may be no explicit consideration of consequences. If there are one or more pests of quarantine concern, the risk is unacceptable and some additional risk management measures are warranted.

The extent of adverse consequences becomes relevant only when the strength of measures recommended is contested. At that point, additional evidence may be required to establish a relative level of consequences. Chapter 29 demonstrates how economic data can be used to strengthen and improve this decision point.

14.3. Formulation

Once it is determined that additional risk management measures are warranted, the formulation process begins. Formulation is the process of creating risk management options that meet risk managers' objectives. It is a set of tasks that requires close cooperation and collaboration between risk managers and risk assessors. The tasks in this step include:

1. Risk management objectives.
2. Identification of measures.
3. Formulation of strategies.

Risk management objectives are used to identify measures that can subsequently be combined via formulation strategies into risk management options. Examples of formulation strategies are found in Chapter 18.

14.3.1. Objectives

Risk management objectives and constraints should be identified in Stage 1 of the pest risk assessment process during the risk assessment planning tasks. Together these define what a successful risk management activity will look like. Objectives describe what risk managers hope to accomplish, e.g., protecting plant life and promoting international trade. Constraints describe the things risk managers hope to avoid doing, e.g., do not violate any environmental statutes.

The first thing that must be done to formulate risk management options is to confirm and affirm the risk management objectives and constraints. If the risk assessment produced information and insights that can be used to update the objectives, they should be modified to reflect the current goals of risk managers. These objectives will be used extensively throughout the formulation, evaluation, and selection tasks.

14.3.2. Measures

Risk management options are formulated from measures. A measure is a means to an end; an act, step, or process designed to accomplish an objective. The definition of a risk management measure is a feature or activity, that can be implemented at a specific geographic site to address one or more risk management objectives. Measures are the building blocks of which risk management options are made. Measures become more specific and better defined as the risk management activity progresses. In the early stages of risk management option formulation, a heat treatment may be envisioned. In later stages, the temperature and duration of heating will be identified.

A feature is a physical change in a process, e.g., developing new cultivars, heat or cold treatments. An activity is an action that is taken, like culling, inspection, or restricting end use. An activity can be a one-time occurrence, or it can be continuing or periodic occurrences. Measures to address objectives other than plant protection must be uniquely identified. Examples of risk management measures commonly used to protect plant health are provided in Table 14.1. These measures have been compiled from ISPM 14 and USDA/APHIS/PPQ (2012).

An alternative list of measures is provided in Table 14.2 (UNCTAD, 2012).

Risk management measures identified as potential components of a risk management

Table 14.1. Commonly used phytosanitary risk management measures identified by stage of production pathway.

Pre-planting
- Peaty planting material.
- Pest free areas, pest free places of production or pest free production sites.
- Producer registration and training.
- Resistant or less susceptible cultivars.

Pre-harvest
- Field certification/management, e.g. inspection, pre-harvest treatments, pesticides, biological control, and so on.
- Healthy planting material.
- Low pest prevalence (continuous or at specific times).
- Pest-free areas, pest-free places of production, or pest-free production sites.
- Pest mating or development disruption.
- Protected conditions, e.g. glasshouse, fruit bagging, and so on.
- Pesistant cultivars.
- Sanitation and cultural controls.
- Testing.
- Cultural controls, e.g. sanitation/weed control.

Harvest
- Culling infested fruit.
- Field sanitation.
- Field surveillance.
- Harvest technique, e.g. handling.
- Harvesting plants at a specific stage of development or time of year.
- In-field chemical treatment.
- Removal of infested products, inspection for selection.
- Sanitation, e.g. removal of contaminants, "trash."
- Stage of ripeness/maturity.
- Tarping.

Post-harvest treatment and handling
- Certification of packing facilities.
- Culling.
- Inspection and grading (including selection for certain maturity stages).
- Method of packing.
- Packinghouse inspection.
- Processing degree and type.
- Sampling.
- Sanitation (including removal of parts of the host plant).
- Screening of storage areas.
- Testing.
- Treatment, e.g. fumigation, irradiation, cold storage, controlled atmosphere, washing, culling, brushing, waxing, dipping, heat, and so on.

Shipping and distribution
- Inspection and/or testing.
- Method of packing.
- Post-entry quarantine.
- Pre-shipment inspection.
- Restrictions or end use, distribution and points of entry.
- Restrictions or the period of import due to difference in seasons between origin and destination.
- Sanitation (freedom from contamination of conveyances).
- Speed and type of transport.
- Testing.
- Treatment or processing during transport: treatment or processing on arrival.

End use
- Packaging.
- Post-entry processing.
- Restrictions on end-use.

Table 14.2. United Nations Conference on Trade and Development Classification of SPS Measures (UNCTAD, 2012).

UNCTAD classification of SPS measures

A1 Prohibitions/restriction of imports for SPS reasons.
A11 Temporary geographic prohibitions for SPS reasons.
A12 Geographic restrictions on eligibility.
A13 Systems approaches.
A14 Special authorization requirement for SPS reasons.
A15 Registration requirement for imports.
A19 Prohibitions/restrictions of imports for SPS reasons, not elsewhere specified (NES).

A2 Tolerance limits for residues and restricted use of substances.
A21 Tolerance limits for residues of or contamination by (non-microbiological).
substances.
A22 Restricted use of substances in food and feed and their contact materials.

A3 Labeling, marking and packaging requirements.
A31 Labeling requirements.
A32 Marking requirements.
A33 Packaging requirements.

A4 Hygienic requirements.
A41 Microbiological criteria on final products.
A42 Hygienic practices during production.
A43 Hygienic requirements, NES.

A5 Treatments for elimination of plant and animal pests and disease-causing organisms.
A51 Cold/heat treatment.
A52 Irradiation.
A53 Fumigation.
A50 Treatments for elimination of plant and animal pests, NES.

A6 Other requirements on production or post-production processes.
A61 Plant-growth processes.
A62 Animal raising or catching processes.
A63 Food and feed processing.
A64 storage and transport conditions.
A69 Other requirements on production or post-production processes, NES.

A8 Conformity assessment related to SPS.
A81 Production registration requirement.
A82 Testing requirement.
A83 Certification requirement.
A84 Inspection requirement.
A85 Traceability requirement.
A86 Quarantine requirement.

option should be appropriate to the risk identified by the risk assessment. The choice of measures will vary by consignment type (taxa, parts of plants) and origin. Combinations of measures ought to be considered to achieve an acceptable level of phytosanitary security. The measures listed above represent a continuum of options that may be applied throughout the production and international movement of plant and plant products. Descriptions of some of these measures adapted from RSPM 40 and ISPM 11 follow.

Inspection, a visual examination of plants, plant products or other regulated articles (ISPM 5),

is the most widely used phytosanitary procedure for imported and exported commodities. It is appropriate as a risk management measure where the target pest(s) are easily detected, produce characteristic signs or symptoms, and present a low risk of introduction. Inspection may not be suitable for pests when several individuals are likely to be in or on one fruit, are difficult to detect, or are likely to survive in the commodity and are mobile enough to leave, find suitable hosts, and mate. The most common inspections are (1) post-harvest, (2) in packinghouses, (3) pre-shipment, (4) fruit cutting in orchard, post-harvest, or (5) at point of entry.

It is simply not feasible to inspect every article in an entire consignment so inspection to detect pests is almost always based on some type of sampling. The quality of sampling programs varies among countries. It may be based on statistics or operational feasibility. If the level of risk (i.e., threshold level of infection, infestation or contamination, tolerance) is fixed, the required sample size will vary according to the size of the consignment. When the inspection sample size is a fixed percentage of each consignment the inspection results in risk rates that vary depending on the variable lot size.

When visual inspection is not adequate to detect the pest, destructive sampling methods like fruit cutting may be required. Certain pests are not detectable by simple visual inspection and may require microscopic examination. Laboratory testing may be required for pests such as viruses that can infect a plant without producing symptoms or that produce masked symptoms. When inspection is not effective, other forms of examination may be appropriate.

Certification procedures may be considered in developing risk management protocols. They can include export certification (ISPM 7) or phytosanitary certificates (ISPM 12) that confirm that the specified risk management options have been followed. Other examples of certification measures include:

- oversight by the NPPO of the importing country
- registered or approved packinghouses
- label requirements for limited distribution
- compliance agreements with packinghouses
- bilateral work plans
- production of plants or plant parts in a certification program.

ISPM 5 says treatments including chemical, thermal, irradiation or other physical methods may be applied pre-harvest to the crop, field, or place of production for suppression, containment, or eradication of pests. Post-harvest treatments may also be applied to the consignment for a number of desired effects including killing, inactivating, or removing pests, or rendering pests infertile. A combination of treatments may also be used. Examples of treatments include:

- chemical, e.g., fumigants, aerosols, mists, fogs, dusts, dips, granules, and sprays

- thermal, e.g., hot water dip, hot air, vapor heat, and cold
- drying
- controlled atmospheres
- physical methods; culling and grading, e.g., at harvest or during packing, or bagging, e.g., fruit bagged on the tree to prevent infection/infestation
- brushing and washing
- irradiation, e.g., gamma, X-ray, microwave.

The ranges of responses to the treatments include mortality, sterilization, inactivity, or devitalization. Measures can be taken to prevent or reduce infestation in the crop. These measures include:

- treatment of the crop, field, or place of production
- restricting the composition of a consignment so that it is composed of plants belonging to resistant or less susceptible species
- growing plants under specially protected conditions, e.g., in a glasshouse or in isolation
- harvesting plants at a certain age or a specified time of year
- production in a certification scheme that could involve a number of carefully controlled generations, beginning with nuclear stock plants of high health status.

Surveillance and monitoring are used during the production phase of a commodity to detect pest populations. If specified pest population levels are exceeded, additional control measures may be required. Pest-free areas and areas of low pest prevalence require surveillance and monitoring as pest risk management components.

Sanitation can be an important component of a pest risk management option. In the field, sanitation can include removing fallen fruit from orchards, destroying or plowing under crop residues, or similar activities designed to remove materials that may attract or harbor regulated pests. Sanitation can also be applied during or after harvest in packinghouses or before shipping. Removing damaged fruit or contaminants such as leaves or soil, and ensuring facilities are properly maintained and cleaned by removing residues and trash that may attract regulated pests are common practices.

Pest-free concepts are based on the exclusion of pests from an area as a result of any combination of pest management measures, cultural practices, climate, pest biology, physical geography, and other factors. Some related concepts include:

- pest-free area (PFA)
- pest-free place of production or pest-free production site
- pest-free growing period
- harvest and shipping windows
- shipment free from pests.

Some commodities may be processed after harvest in a way that reduces the risk associated with certain pests. Such handling may include peeling, dicing, slicing or chopping and other activities like washing that can effectively remove pests from a commodity.

The importing country may require post-entry measures to be applied to the commodity. These measures can be stand-alone measures or used as a component of a systems approach. Post-entry measures include such things as:

- post-entry quarantine—used for plants for planting when certain pests are not detectable on entry
- limits on the end use of commodity, e.g., allowing grain imports only for milling
- limited distribution—removing portions of the pest risk area from the area at risk.

Pests can spread via non-trade pathways and other risk management measures may be viable treatments along these pathways. For natural spread, traveler-mediated, and machinery/transport pathways, special measures may be considered. In natural spread, pests may move by flight, wind dispersal, or transport by vectors such as insects or birds and natural migration. Phytosanitary measures may have little effect if the pest enters the PRA area by natural spread. Control measures applied in the area of origin could be considered as could containment or eradication, supported by suppression and surveillance, in the PRA area after entry of the pest. Targeted inspections, publicity and fines or incentives could be effective measures for human travelers and their baggage. In some cases, treatments may be possible. Ships, trains, planes, road transport, and contaminated machinery could be cleaned or disinfested.

If a satisfactory means to reduce risk to an acceptable level cannot be found, the commodity may be prohibited. This should be considered a last resort.

14.3.3. Formulation strategy

Formulation is the process of building alternative risk management options from risk management measures like those described above. Option formulation has three essential tasks. They are:

- identify risk management measures that meet the risk management objectives
- combine these measures in compatible ways to build options
- reformulate the options as necessary.

Risk managers and risk assessors can be expected to collaborate in accomplishing these three tasks. Countries will, at times, cooperate in formulating options. There will be times when assessors will be in the best position to advise risk managers on the most viable combinations of measures. Other times, risk managers may be in the best position to say what measures will be most effective and practical in a given situation.

These three tasks will likely overlap and be repeated in an iterative fashion as new evidence is compiled and uncertainty is reduced. Formulation proceeds by taking each risk management objective specified, one by one, and identifying individual measures that could contribute to achieving each objective. One easily anticipated objective will be to reduce the risk of pests of quarantine concern. Depending on the size and nature of the consignment and the life history of the pest(s), among other things, one would identify measures from a list, such as presented in Tables 14.1 and 14.2, that could be useful in reducing the risk associated with the identified pests. Candidate measures identified in this way are the building blocks that will be put together to form alternative risk management options.

This process would be repeated for each of the other objectives as well. Some objectives, like promoting international trade or complying with environmental statutes, can be used to screen some of the measures originally identified to reduce pest

> The most basic method of developing risk management options is to take each risk management objective, one at a time, and identify all the measures that could contribute to its achievement. These are the building blocks for risk management options. These measures are then assembled into coherent options that address the full range of objectives.

risk. Other objectives may require measures quite different from those of Tables 14.1 and 14.2.

The next formulation task requires matching and mixing management measures identified in the above task into alternative risk management options. This process, which is especially relevant to an integrated approach to risk management, is best served by observing the realistic combinability and dependency of measures.

Combinability and dependency

Risk management measures may or may not be mutually exclusive. Measures that are not mutually exclusive are combinable. Combinability allows us to mix and match measures into different plans. Conversely, some measures may preclude others. When building risk management options, consider whether two measures may be mutually exclusive because of location, function, or overlapping.

Some measures may be dependent on other measures in order to be implemented. The dependency of two measures can exist for several reasons. First, one measure may be necessary to the function of another measure. Second, dependencies may serve to reduce risk or uncertainty in project performance through redundancy.

Reformulation, the third task, is a special type of iteration during which alternative options previously formulated are changed or refined for one or more reasons. Measures may be added, dropped, rescaled, or otherwise modified such that the reformulated option will better achieve a risk management objective or stay within the limits of a constraint. Measures can be modified to develop a reformulated option that is less costly, i.e., more cost-effective. Reformulation may be required to ensure an option includes everything that it needs to function successfully. "Mitigation" could be a reason to reformulate a risk management option that would produce adverse impacts. Sometimes reformulation is just the way a good option evolves, other times it is what you do to an option that is not good enough to warrant further consideration as originally formulated.

Consider a systems approach to formulating risk management options. The IPPC (2019b) defines a "systems approach" as "the integration of different risk management measures, at least two of which act independently, and which cumulatively achieve the ALOP against regulated pests." Systems approach thinking arose from situations where a single phytosanitary treatment was either not available or those that were did not provide the appropriate level of protection. Faced with prohibition as an alternative, risk managers began to consider combinations of measures that safeguarded plant health through redundancy. Combining measures that act independently of one another prevents a "one fails all fail" situation and provides a higher level of efficacy than a single measure, through redundancy. It can have the added incentive of contributing to the objective of promoting international trade.

To understand how the systems approach works and how to formulate systems approach risk management options it helps to think about pathways and consequences. In order for a risk to exist there must be a hazard or pest that can cause specific harms. Then there is a sequence of events that must occur to link the hazard to its harmful consequences. Thus, knowledge of the pathway and the consequences of introduction provide a strategy for formulating risk management options from a systems approach. Take the simple stylized pathway below, which could have been specified when risks were initially identified during the risk assessment planning task of risk management executed during Stage 1 of the PRA process:

Pest present $\Rightarrow$ Harvest $\Rightarrow$ Handling $\Rightarrow$ Shipment $\Rightarrow$ Entry $\Rightarrow$ Establishment $\Rightarrow$ Spread $\Rightarrow$ Consequences

There could be a separate measure at each step in the pathway, some in the exporting country and some in the importing country, which cumulatively provide the phytosanitary risk reduction that no one measure alone could provide. It might also be possible to have redundant or overlapping measures at a single point in the pathway. That way if any one measure at a point fails there will be other measures to assure the risk is mitigated. To formulate a systems approach, identify all the measures that can be applied to prevent the occurrence of the pest in the region of origin. This would include measures anywhere in the production chain from pre-planting to pre-harvest. Then risk managers would identify all the measures that can be applied during harvest, then all the measures that can be applied during handling, shipment, and so on. Armed with this list of potential measures, systems approach options can be built by assembling independent measures that, taken together, provide the desired level of risk reduction.

The basic formulation strategy is to look for control points in the pathway. These are points in the pathway or production chain where measures can be effectively applied to affect some aspect of the risk in a desirable way. With a systems approach, it may or may not be possible to measure the exact effect of the risk control, but risk managers expect the control would have some level of efficacy. It is not necessary to prescribe

risk measures for each control point. A critical control point is a point, step, or procedure in the plant product supply chain at which risk controls can be applied and the pest of interest can be prevented, eliminated or reduced to acceptable levels.

The USDA *Guidelines for plant pest risk assessment of imported fruit and vegetable commodities* (2012) provides an excellent example of a list of measures that could be combined for formulating systems approach risk management options. Mitigation measures would be selected from the middle column of Table 14.3 for a variety of control points sufficient to achieve the anticipated cumulative effect.

Table 14.3. Examples of risk management measures of potential use in formulating systems approach risk management options for pest risk management (adapted from Table 2.5, USDA/APHIS/PPQ, 2012).

Control point	Potential mitigations	Anticipated effect or impact
Production areas	Protected cultivation, pest-free areas, or areas of low pest prevalence	Prevention of infestation
Planting and harvest time	Pest-free periods/reduced host susceptibility	Reduce or prevent infestation
Cultivar selection	Host status/resistance	Reduce or prevent infestation
Post-harvest handling	Washing, waxing, brushing, bathing, culling, and the like	Reduce or prevent infestation
Processing	Peeling, baling, heating, cooling, cooking, milling, and the like	Reduce or prevent infestation
Packing and safeguarding	Prevention of reinfestation	Prevention of reinfestation
Treatments	Commodity tolerates various treatments	High mortality
Shipment	Mode of shipments includes treatment potential, e.g., cold transit	Potential mortality
Inspection	Commodity is easily inspected/tested	Detect infestation
Intended use	Devitalization	Prevent establishment
Volume of exports	Note whether or not increased volumes beyond originally requested volume will affect overall risk	Potentially increase risk
Pest biology (feeding, reproduction, development)	In-field pest management practices	Reduce or prevent infestation
Seasonality	Time of harvest or shipping, e.g., when pest is not active	Reduce or prevent infestation
Host specificity	Resistant cultivars, non-host status	Reduce or prevent infestation
Susceptibility or resistance to treatments	Pre-shipment, in-transit, or post-entry treatments	Reduce or eliminate infestation
Geographic and temporal distribution	Restrictions on time of year or destination of imports	Prevent establishment
Climate suitability	Restrictions on time of year or destination of imports	Prevent establishment
Environmental resistance or susceptibility	Various types of shipping conditions or post-harvest handling	Reduce infestation
Detectable/testable	Inspection	Detect infestation
Role of vectors	Determine presence or absence of vectors by surveillance	Reduce establishment

The mix of measures employed in a systems approach has the added advantage of mitigating the risk of pests that are not normally associated with the commodity and pests that may not have been accounted for in the risk assessment, such as unknown pests (MAF, 2008). Systems approach risk management options are more promising the better the risk analysis team's knowledge of production practices, biology of the pest(s), its relationship to the host(s), and post-harvest practices is. The complexity of a systems approach option will depend on the:

- level of risk involved
- cost, feasibility, and efficacy of possible measures
- suitability of any given management option for managing that risk
- availability of information for the pest(s) and associated commodity
- level of uncertainty
- ALOP or acceptable/tolerable level of risk (USDA/APHIS/PPQ, 2012).

When knowledge uncertainty and natural variability present challenges to risk assessors and managers, the redundancy of systems approaches provides a hedge against uncertainty by varying the number and strength of measures.

> A control point that might be difficult to measure is field sanitation and the removal of fallen fruit from orchards to manage risks associated with fruit flies. The sanitation would reduce risk, but the exact level of efficacy would be difficult to measure.
>
> (USDA/APHIS/PPQ, 2012)

14.4. Evaluation

Not every idea a team has for managing a risk is going to be a good one. Risk managers need a way to separate the worthy ideas, i.e., the ones worth serious consideration, from the others. Evaluation provides that way. The components of evaluation of interest here are:

1. Options.
2. Criteria.
3. Judgment.

The purpose of evaluation is to find the value or worth of the risk management options that have been formulated. Once an option is formulated, it is time to begin to understand how it is likely to perform if implemented. Each option should be evaluated. Evaluation is a two-part process that includes assessment (measurement) and appraisal (judgment). In this task, risk management turns from the creative, divergent thinking of formulation to the critical, convergent thinking needed to figure out how best to meet the risk management objectives and reduce an unacceptable risk to a tolerable or acceptable level.

Evaluation is a qualifying step. It consists of a qualitative or quantitative analysis of the effects of a risk management option followed by an appraisal of those effects. The result of the evaluation task is to decide whether an individual risk management option is good enough to consider as a viable solution to the phytosanitary risk management problem. If it is, the option is included among the set of viable solutions. The goal of the evaluation step is to produce a set of viable solutions, which will subsequently be compared and from which a risk management strategy that provides an ALOP will be selected and implemented. Evaluation requires close collaboration with risk assessors, who will be called upon to assess the risk reduction and other significant impacts of each risk management option.

This evaluation task can begin as soon as risk management options have been formulated. Risk managers must identify the criteria they will use to judge the value of an option. A key question for risk managers to address is, does the individual formulated option make a difference among the criteria that matter most to risk managers? The answer to this question requires a comparison of two distinct scenarios and a subjective judgment of the value of the differences. This process is described in the pages that follow.

14.4.1. Options

Risk management options are an output of the formulation risk management task and they are an input to the evaluation task. An option was described earlier as a set of one or more risk management measures functioning together to address one or more risk management objectives. An option may consist of any one of the risk management measures described above or it could be a combination of measures developed in a systems approach to managing risk. Some specific commodity–pest pairs may have a limited range of options available. Other pairs may have many feasible options for managing the risk to an acceptable level. The evaluation task first requires options to consider.

Every risk management activity begins with a list of risk management measures like those in Tables 14.1 and 14.2 as possible solutions. Experienced risk managers may quickly zero in on a subset of these measures that may be applicable in any given situation. Experience may enable the risk manager to evaluate the measures almost intuitively. As stewards of the health of plants for their country, however, risk managers have an obligation to evaluate options objectively and transparently.

There may be rare instances where multiple options are not formulated. If a pest risk assessment reveals the risk to be very similar to an already managed pest risk, it may be judged sufficient to treat the new risk in a similar fashion. This could enable the risk manager to quickly focus in on a single specific risk management option. In the vast majority of risk management activities, risk managers ought to produce an array of viable options from which the best risk management option will be selected. This is so for a simple and compelling reason. One cannot be sure one has the best solution unless several possible solutions are considered. If a risk manager keeps going to the same set of risk management measures out of force of habit there is a very real danger that the risk manager may have ceased to think critically about how best to manage each risk. Even the same pest–commodity pair may warrant different measures at times. It is always better to consider a few alternatives, if only to confirm one's instinct about what is best. How does one know one has identified an option with minimal impact without considering others?

Once a risk manager begins to identify specific risk management options, it is time to evaluate those options. Evaluation is an iterative process. The first iteration may occur almost organically when an experienced risk manager narrows the large set of potential solutions to a subset of viable options. Options may be evaluated as they are formulated or evaluation may be delayed until all the formulated options are available. Not all risk management options will

be developed to the same level of detail. Some options may remain conceptual and never progress to a design level of detail. Other options may be expressed in impeccable detail. Options need not be at the same level of detail to be evaluated.

To progress through the evaluation task, risk managers need one or more risk management options in sufficient detail for evaluation. Options are evaluated individually and not in comparison to one another. The risk manager, in effect, picks up an option, examines it on its own merits and then pronounces the option a potentially viable solution or not. Options that are found non-viable are either reformulated, i.e., fixed to make them viable, or they are dropped from further consideration. Evaluation requires assessment and appraisal and specific criteria are needed to complete these tasks. Criteria are the subject of the next section.

14.4.2. Evaluation criteria

The evaluation of risk management options begins as an analytical task. Good evaluation requires decision criteria or things to measure. There are things one wants a risk management option to do and things one wants it not to do. The evaluation process can be qualitative or quantitative and it should be based on very basic principles and criteria like:

- feasibility
- efficacy
- impacts
- minimal impact
- non-discrimination
- equivalence
- residual risk, and
- other criteria as needed.

"Feasibility" means the option will work because it is well thought out. A feasible option is complete. All the necessary measures and implementation actions needed to make the plan successful have been accounted for in the formulation process. The negative effects of the option are limited and tolerable. Ask yourself, is everything present that is needed to make this option work as desired? Will this option produce the desired outcomes? If so, it is feasible.

"Impacts" mean the impacts of implementing risk management measures (or not), which is distinguished from the consequences of pest introduction. For example, a fumigation may be very effective for eliminating pests, but have environmental impacts from the use of poison gas.

Guidance on feasibility can be found in ISPM 14, ISPM 11, ISPM 2 and ISPM 28. The phytosanitary treatments must be feasible. That means considering the negative effects of treatments on the commodity like phytotoxicity, physical damage, shelf life, and cost. Will the facilities and equipment required be available? Is the option practical?

Does it produce unacceptable outcomes? If so, it needs to be reformulated or rejected from further consideration. A feasible plan is implementable. During the evaluation task consideration of feasibility tends to remain qualitative.

"Efficacy" means an option is effective and efficient, it will make progress. An effective plan reduces pest risk and is responsive to the wants and needs of people as revealed in the risk management objectives. The most effective solutions make significant contributions to these objectives. Generally, efficacy is considered to apply primarily to an option's ability to reduce pest risk and this is generally a qualitative judgment in the evaluation task. That judgment of whether an option is "efficacious enough" should consider things like:

- the level of phytosanitary risk present
- the ALOP
- the nature of the phytosanitary risk being addressed
- the biology of the organism(s) being managed
- the tolerance of the commodity to the applied risk management measure
- the operational and technical considerations like practicality, cost, timing, available technologies, infrastructure, and so on.

"Impacts" include the economic, social, and environmental impacts of phytosanitary measures. These impacts should be identified and considered, generally in qualitative terms, during the evaluation process. The nature and magnitude of an option's impacts will influence its acceptability. An acceptable option is compatible with existing legislation, policy and authorities. There is no clear reason the option will be stopped, it is doable. There are impacts that can render an option infeasible and they will vary by type and magnitude from one case to the next. Opposition to an option does not make it unacceptable, that simply makes it unpopular. Nonetheless, there is no sense to give serious consideration to a plan that has unacceptable impacts.

There is no reason to further evaluate an option if it fails to meet any one of these criteria. A solution that is not feasible won't work. There is no point to an ineffective plan. A plan that produces unacceptable impacts is a non-starter. The real evaluation challenge is in determining thresholds for these criteria. They are usually assessed qualitatively, if not totally subjectively, at this stage of the risk management process. When risk managers judge that a plan does meet these criteria, it is wise to document the rationale and the evidence used. Options that fail to meet these criteria should be fixed by reformulation or dropped from further consideration.

There are also a few constraints a risk management option must not violate. These include the constraining principles of minimum impact, non-discrimination, and equivalence. An option that goes beyond the least restrictive measures available, resulting in more than the minimum impediment to the international movement of people, commodities, and conveyances violates the "minimum impact" principle and should be disqualified from further consideration as a viable solution. An option that discriminates by imposing different requirements on a contracting party that has the same phytosanitary status violates the "non-discrimination" principle and should be disqualified from further consideration. It is discriminatory to require a contracting party to use more stringent measures than the importing party uses on the same pest if it is already present in the importing country. Required pest risk management measures should be consistent across countries with the same phytosanitary status and comparable domestic phytosanitary situations. There may be times when the exporting country proposes an alternative option. If an importing country fails to recognize risk management measures demonstrated to achieve the desired level of stringency and ALOP by alternative phytosanitary measures it violates the principle of "equivalence."

An option may be judged against other criteria as well. It is likely that an option's risk reduction will be part of its effectiveness and its "residual risk" may go a long way toward determining if an option is acceptable or not. For some of these criteria, a more formal assessment of the differences may be warranted. Measuring those differences is the subject of the next section.

14.4.3. Judgment

The purpose of this judgment task is to identify a set of risk management options that would meet the risk management objectives of a particular risk management activity. Think of this as a screening step in the risk management option selection process. Using the options identified during the formulation process and the evaluation criteria of the risk managers, with or without the assistance of risk assessors, will identify risk management options worthy of proceeding to the more detailed selection process. The evaluation criteria are likely to include some measures of efficacy, feasibility, and impacts. Evaluation may or may not be quantitative. It does, however, consist of considering each option on its own merits and it involves the comparison of scenarios.

Comparing scenarios

Comparing scenarios is an important analytical process. Comparing without and with condition scenarios is the essence of the analytical part of the evaluation process. The effectiveness of a risk management option is judged based on changes in risk and other decision criteria that are observed through scenario comparisons. Table 14.4 presents results of a hypothetical evaluation scenario comparison. In actual fact,

evaluation is likely to rely on the appraisal of qualitatively assessed impacts. A quantitative example is used here to enhance the transparency and explanatory value of the example. It shows that if trade is allowed without any risk management measures taken to control the risk that is expected to result from the market access, the probability of introduction of the identified pest is high. Annual economic damages are estimated to total C15 million. (The label "C" is used to represent hypothetical currency units. If no action is taken to control risk then control costs are zero).

Imagine a risk management option "A" is developed to help control the risk. The measures that comprise Option A would reduce the probability of introduction to medium. The expected annual economic damages with this option in place are C4 million. The option would impose modest annual costs of C2.5 million. Risk assessors and other staff analysts would be required to estimate and provide these results if requested to do so by the risk manager.

Now it is time to compare the two scenarios. Option A reduces P(introduction) from high to medium; it reduces expected annual damages by C11 million at a cost of C2.5 million annually. These differences must be appraised by risk managers who will decide if it is worth considering Option A further. Although it reduces damages by C11 million at a cost of C2.5 million it leaves a residual risk of introduction that is medium with residual damages of C4 million. Note, the risk manager is not deciding to select this option or not, the risk manager only has to decide if this option is good enough to continue considering as a potential solution, on its own merits, without respect to any other option.

This is a subjective decision; it is not science. This is where the risk manager's judgment must be exercised. If the economic gain, spending C2.5 million to prevent C11 million in damages,

Table 14.4. Example of a without and with condition comparison for habitat units.

	P(introduction)	Adverse economic impacts (Annual)	Control costs (annual)
Trade is allowed without additional risk controls	High	C15,000,000	C0
Risk management option: A	Medium	C4,000,000	C2,500,000
Differences attributable to RMO A	Reduction	−C11,000,000	+C2,500,000

dominates Option A will be retained for further consideration. If residual risk dominates, it may be judged that a medium probability of introduction with remaining damage of C4 million is too great to bear and the plan would be rejected or returned to reformulation to see if there is anything that can be done with this option to further reduce the residual risk enough to make it a viable solution.

Fig. 14.1 illustrates three different methods of scenario comparison for a pest risk decision criterion. Imagine a criterion like risk. It might be measured as the expected annual damages measured in currency units C, as annual P(introduction), or by some other metric. The simplest comparison is the before and after comparison. This compares a baseline or "before" estimate of a criterion to the "after" value of that same criterion with an option in place and functioning. The difference between these two values is calculated in a before and after comparison. This approximates the most common current means of estimating impacts.

This method does not take into account changes in the risk impacts that would naturally occur over time. The true impacts of accelerating or self-attenuating problems are not seen in a before and after comparison. To account for these kinds of changes the without and with comparison should be used. The figure shows a risk that grows worse over time under the "without condition." The "with condition," shows Option A's effect. In this case, the option did not reduce the risk to its background level and there is some elevation of risk with Option A in effect. The without and with condition comparison can yield a significantly different view of the risk reductions produced by an option when compared to the before and after analysis. The best analysis would estimate the changes in the risk over time.

Gap analysis is a third kind of comparison. This is predicated on risk managers or some higher authority establishing a target level for the risk (or other effect of interest). No increase in risk above the background level might be a common target, a specific ALOP would be another. Once a target is established, risk managers try to hit the target by creating an option that yields a with condition scenario that meets or exceeds the target. When the target is ambitious or when risk reduction is traded-off against other values, like product variety, price, or jobs and income, some solutions may fall short of the target, establishing a gap between the desired level of performance of the option and its actual performance. Gap analysis is a comparison technique that focuses on the distance between the desired target and the actual performance of a

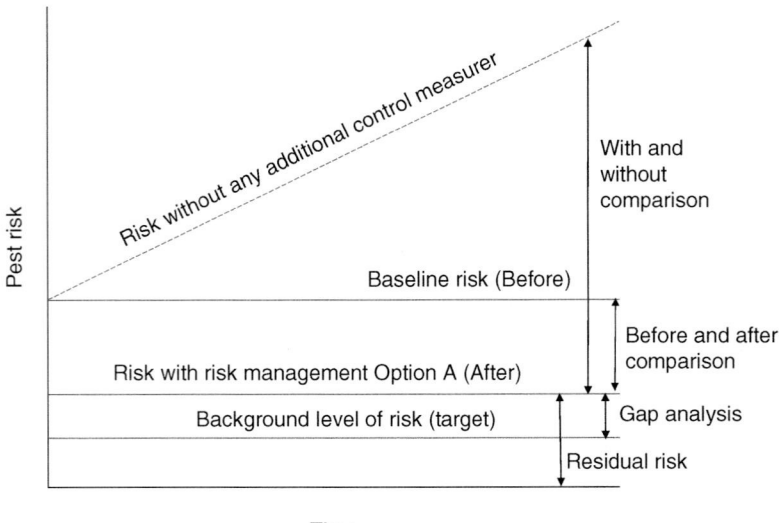

Fig. 14.1. Three scenario comparison methods that can be used in the appraisal or judgment stage of the risk management option evaluation task.

risk management option. Fig. 14.1 also illustrates the residual risk associated with Option A. The example of Table 14.4 uses with and without condition comparison.

A critical judgment to be made in every PRA is whether the without condition risk of a pest is an acceptable risk or not. A critical evaluation judgment to be made about every risk management option formulated is whether the with condition risk of that same pest is now acceptable, i.e., does it provide an ALOP?

Without and with condition scenarios compared

If scenario comparisons are to be useful for decision-makers, they must highlight differences in scenarios that make a difference, i.e., differences that are important and that matter to decision-makers. In best practice, these metrics will reflect some or all of the risk management objectives as well as other decision criteria. It is recommended that risk managers compare scenarios with the help of risk assessors. Specifically, they should consider the risks and other relevant effects that would result if unmitigated trade was allowed and compare them to the risks and other effects that would occur with different risk management options in effect.

The image of a risk assessment stage providing all the analysis followed by a risk management stage providing all the deliberation is a limited depiction of the process. It is more likely, more rational, we would argue, that the risk analysis process proceeds in an iterative fashion with cycles of analysis and deliberation. Risk analysis is an analytic-deliberative process. There is analysis followed by deliberation followed by more analysis and more deliberation, etc.

Once the risk assessment establishes that the risk associated with importing a plant or plant product would result in an unacceptable risk, it is necessary to consider risk management options that could reduce the risk to an acceptable level of risk by providing an ALOP. The risk assessment may identify and evaluate some of the more obvious risk management options but risk managers can be expected to identify and add options of their own. These options may require additional analysis by the risk assessors. As risk management options are added, adjusted and deleted, there could be several rounds of

collaborative work between risk managers and risk assessors to ensure that the relevant effects of a risk management option have been assessed.

The presenting problem at this stage of the risk management process, however, is for risk managers to identify those options that represent potentially viable solutions to the risk management problem that was scoped. Feasibility, efficacy, and impacts are criteria that have been suggested (ISPM 2, 11, 14 and RSPM 40) for evaluating risk management objectives. A good evaluation process assesses the differences between the metrics used to represent these criteria. Risk managers must, then, by some means, determine whether the differences are favorable enough to consider the option a viable risk management solution.

The ultimate goal is to identify and select the best option for providing an ALOP, consistent with the criteria identified by the SPS Agreement and the IPPC. Evaluating options is an intermediate screening step that requires risk managers to identify a subset of viable options from among that hypothetical universe of all possible options. The output of the evaluation task is a short list of risk management options, each of which represents a potentially viable solution to the presenting problem. Formulation, a divergent thinking process, identifies as many ways to meet the risk management objectives as possible. Evaluation is the first step in risk management's convergent thinking process. It identifies a select few of the many possible solutions as viable solutions.

This process is subjective and it is usually qualitative. Quantitative work does not normally begin until a set of viable options have been identified, if it is done at all. As a first iteration of decision-making, the process tends to be informal. It could be as simple as identifying the options that exceed the subjective thresholds of feasibility, efficacy, and impacts. More formal consideration of IPPC decision criteria is reserved for the next iteration of decision-making. Evaluation enables risk managers to deploy analytical resources more efficiently by assuring that options that are subjected to more detailed risk assessment, cost estimates, or other analysis are viable. Evaluation also limits the risk manager's own deliberations to a smaller set of viable options.

The judgment or appraisal step of the evaluation task results in a short list of well identified potential risk management options. Given this small set of viable options, it is now time to move forward to identify and select the best option. That is part of the selection process.

14.5. Selection

The selection process requires risk managers to move from the list of viable options produced by the evaluation process to a single best risk management option to be implemented. That option can include granting market access without any risk management controls. The selection process identified here consists of three major tasks:

1. Comparison.
2. Decision.
3. Approval.

Now that risk managers have individually evaluated each risk management option, it is time to compare those that remain viable. If any one option is more desirable on every criterion used for decision-making, the decision is simple: chose the option that dominates all others. However, when comparing options to one another it is common to find that a plan may be better than others on some criteria and worse than others on other criteria. Thus, identifying the tradeoffs among plans during the comparison process is often an important part of the selection process.

Once tradeoffs have been identified, it is time to make a decision and select the best risk management option for the given situation. Following this decision, the chosen option must be vetted and approved according to the policies and procedures of the NPPO.

14.5.1. Comparison

Evaluation requires risk managers to assess the differences an option makes when compared to allowing market access without any additional risk control. They must, then, decide if those differences are desirable enough for an option to warrant serious consideration as a viable option for addressing the risk management objectives and determining an ALOP. Given a set of potentially

viable plans, risk managers need a means to compare the viable plans to one another for the purpose of identifying the best plan.

> RSPM 40 says the first step in selecting risk management measures is to compare identified options against criteria such as efficacy and cost. This can be a simple list of options with expected efficacy, predicted effect(s) on pest risk, potential costs, and impacts of each measure.

Risk management option effects are identified by means of scenario comparisons, preferably without and with condition comparisons. Some subset of potential effects must be identified by risk managers as the decision criteria. The decision criteria will normally include some combination of efficacy, feasibility, and impacts metrics. The metrics may be exactly the same ones used to evaluate plans or they may differ, representing another iterated level of detail or uncertainty reduction.

Let us continue the hypothetical example introduced earlier. Table 14.5 presents hypothetical metrics for five decision criteria. Unrestricted market access is expected to adversely impact the crops of farmers by C15 million but it will provide benefits to consumers and importers valued at C20 million annually. There are no risk control costs associated with this option. Option A relies primarily on inspection and sampling. It reduces the P(introduction) from high to medium, reduces impacts on farmers to C4 million and reduces benefits to consumers by C10 million to C10 million at a control cost of C2.5 million. Option B relies on inspection, sampling, and a treatment. It reduces risk to low and results in C16 million in benefits to consumers and importers at a control cost of C23 million. Impacts on domestic farms falls to C2 million. Option C is a prohibition on market access with the noted impacts. It costs C1 million annually to enforce the ban, importers and consumers do not benefit at all.

Table 14.6 translates the values in Table 14.5 into relative ranks for each decision criterion, where lower numbers are best. Option C is best on reducing P(introduction), it is also best on losses to farmers and adverse environmental impacts. Doing nothing is best on impacts to importers and consumers and on control costs. No plan is best on all criteria. No option has all 1s, so no option dominates all others. No option has all

Table 14.5. Hypothetical comparison of the annual effects of three risk management options compared to the no controls option.

Course of action	P(introduction)	Adverse environmental impacts	Domestic losses to farmers (annual)	Domestic impacts on importers & consumers (annual)	Control costs (annual)
Market access is allowed without additional risk controls	High	Medium	C15,000,000	C20,000,000	C0
Risk management Option A	Medium	Low	C4,000,000	C10,000,000	C2,500,000
Risk management Option B	Low	Low	C2,000,000	C16,000,000	C23,000,000
Risk management Option C	None	None	C0	C20,000,000	C1,000,000

Table 14.6. Hypothetical comparison of the ranks of effects of three risk management options compared to the no controls option.

Course of action	P(introduction)	Adverse environmental Impacts	Domestic losses to farmers (annual)	Domestic impacts on importers & consumers (annual)	Control costs (annual)
Market access is allowed without additional risk controls	4	4	4	1	1
Risk management Option A	3	3	3	3	3
Risk management Option B	2	3	2	2	4
Risk management Option C	1	1	1	4	2

4s, so no option is dominated by all others. Option A, with all 3s is better than each other option on at least one criterion, so it cannot be logically eliminated either.

If we presume these are the four criteria upon which a decision will be based, then Table 14.6 indicates the tradeoffs among the options. Clear displays of the tradeoffs among alternative risk management options is the desired output of the comparison step of the selection task.

Prescriptions of the SPS Agreement may circumscribe the criteria set upon which a decision may be based, but neither the SPS Agreement nor any other guidance can or should limit what risk managers consider in their deliberations leading up to decision-making. If there are likely to be adverse environmental impacts along with impacts on importers and consumers, only a foolish risk manager would fail to be aware of such impacts. Information, in addition to that derived from a risk assessment, may be considered in determining an ALOP for decision-making as long as risk managers do not violate the principle of consistency (SPS Agreement, Article 5.5).

> In actual practice the evaluation, selection and decision steps may be a great deal more intermingled than presented here. They may happen indistinctively all at once in a single sitting or in discrete steps over a number of months.

14.5.2. Decision

The efficacy of the risk management option is likely to be the primary selection criterion, given that a quarantine pest has been identified, but it need not be the only criterion. ISPM 11 takes a rather narrow view of the selection process when it says "appropriate measures should be chosen based on their effectiveness in reducing the probability of introduction of the pest." It does go on, however, to open the door to considering other criteria when it says options should consider cost-effectiveness and feasibility. The ISPM identifies the benefit from phytosanitary measures as preventing the introduction of the pest and, consequently, the potential economic consequences of introduction. It goes on to suggest a cost–benefit analysis for each of the minimum measures found to provide acceptable security.

Emergency and provisional measures

Pest risk management measures may be applied in the absence of a PRA for emergency measures established as a matter of urgency in a new or unexpected phytosanitary situation (ISPM 5). Such an emergency measure may or may not be provisional, i.e., subject to review and full justification as soon as possible. In general, emergency actions for new or unexpected phytosanitary situations are temporary and are to be the subject of a detailed PRA as soon as possible (ISPM 5).

Several evaluation/comparison/selection criteria may be discerned from ISPM 11. In addition to risk reduction and the economic efficiency implicit in a cost–benefit analysis, these include:

- principle of "minimal impact"—measures should not be more trade restrictive than necessary and they should be applied to the minimum area necessary to protect the endangered area
- reassessment of previous requirements—if existing measures are effective, no additional measures will be necessary
- principle of "equivalence"—different measures with the same identified effects should be accepted as alternative controls
- principle of "non-discrimination"—requirements for imports should be no more stringent than domestic controls.

Explicitly including these criteria in the example would expand the comparison table, but it could readily be done. Now, it is time to revisit some of the evaluation criteria in their role as decision criteria.

Efficacy

Efficacy is important enough to warrant special consideration and definition as a selection criterion. Determining the efficacy of each risk management option is essential to evaluate the extent to which an option reduces pest risk. Describing the efficacy of an option requires specification and measurement of a desired outcome or endpoint. Mortality for with condition estimates of consequences, probability of introduction, or risk ratings are some examples of such endpoints.

It is desirable to measure efficacy quantitatively, whenever possible. Quantitative estimates ought to be designed to appropriately convey the uncertainty present in the estimate. This can be done through the use of confidence levels or expressing values as a five-number summary. When quantitative estimates of efficacy are not possible, qualitative estimates, e.g., high, medium and low, may be used.

RSPM 40 suggests several factors to consider when estimating the efficacy of a measure or the more difficult task of estimating the efficacy of several combined measures that may have different metrics. These include the nature and magnitude of the pest risk and the ALOP as well as the biology of the organism(s) being managed, the tolerance of the commodity to the proposed risk management measure(s), and such things as practicality, cost, timing, available technologies, infrastructure, and the like. These factors support a characterization of efficacy rather than an estimate of efficacy.

Treatment efficacy can be specified as mortality but that may not always be the most appropriate endpoint. Low-risk pests, for example, may not require or be well served by high mortality treatments. Likewise, low infestation rates or a pathway that precludes introduction of a pest will not be best served by a mortality endpoint. Alternative treatment efficacy measures include sterility (including sterility of F1 generation), inactivation, altered behavior, prevalence of pests in a consignment, proportion of pests removed, pest approach rates, probability of entry, probability of introduction, probability of pest outbreaks, probit analysis, or population counts. Risk managers and risk assessors should cooperate on the choice of the most appropriate endpoint.

Feasibility

Can the option be easily or conveniently done? If so, it is feasible and applicable. ISPM 28 suggests that feasibility begins by considering the procedure for carrying out the risk controls. For treatments, for example, this could include ease of use, risks to operators, technical complexity, training and equipment required, and facilities needed. The costs of risk controls will be important. Life cycle costs should be considered. The extent to which other NPPOs recognize and approve the controls will matter as will having the necessary expertise available to apply the controls.

The most feasible controls are versatile in the sense that they are applicable to a wide range of countries, pests, and commodities. The extent to which a control complements other phytosanitary measures will affect its potential use as part of a systems approach. The availability of information on potential undesirable side effects, e.g. impacts on the environment, non-target organisms, human and animal health, also affect feasibility. Then, of course, there is the technical viability of the control and consideration of the risk of the target organism developing resistance to a treatment.

Impacts

The impacts of a risk management option provide the greatest potential for diversity and variation in decision criteria. Generally, the economic, social, and environmental impacts of phytosanitary measures should be identified and considered during evaluation and selection. In some risk management actions these impacts will be considered informally. RSPM 40 suggests a more formal assessment of a wider variety of impact assessments may be warranted when:

- significant unintended social or environmental impacts could accompany a risk management option, as when the magnitude and scope of environmental impacts are unclear or when there are public health sensitivities about certain control technologies
- different groups in society are affected differentially by economic impacts, as when producers in one region benefit while others are harmed

- risk management options pose challenges that subject them to public scrutiny, as when irradiation facilities are constructed or an option is expensive.

Making decisions

ISPM 11 says the decisions to be made in the pest risk management process will be based on the information collected during the PRA. That information will usually include an estimate of the probability of introduction and an evaluation of potential economic consequences in the PRA area. In many cases, identification of a pest as a quarantine pest is considered sufficient evidence of potential economic damage. It is, unfortunately, rare for risk managers to have explicit estimates of economic damages, much less a broader array of economic consequences or other impacts.

Requiring no new or additional risk management controls, i.e., doing nothing, is always a viable result of the selection process. The alternative outcome will be the selection of a risk management option that consists of one or more management measures that are expected to lower the risk associated with the pest(s) to an acceptable level. This risk management option forms the basis of phytosanitary regulations or requirements.

RSPM 40 identifies the following factors to consider in a cost–benefit analysis of risk management options.

- Monetary estimates of the impacts of each option on existing regulations.
- All commercial, environmental, and social costs that result from each option.
- Benefits, usually losses averted, associated with each option.
- Appropriate discounting of costs and benefits that accrue in different time periods.
- Benefit of each measure's efficacy against other quarantine pests.

Viable decision choices must survive the evaluation process, where their efficacy and feasibility are established. When a decision involves a large monetary impact or is politically or socially controversial, a cost–benefit analysis may be conducted to measure the economic efficiency of the alternative risk management options. cost–benefit analysis measures the benefits of a risk management option and compares them to the costs of that option. The risk management option with the largest net benefits (present value of all benefits minus present value of all costs) maximizes economic efficiency. Cost–benefit analysis is discussed at greater length in Chapter 29.

There are four basic methods of decision-making from which risk managers can choose. Fig. 14.2 presents them. First, the methods are separated into formal and informal decision theory methods. Next, decision makers can consider all or just some of the decision criteria. These two decision-making dimensions create the four quadrants in the figure.

Formal methods that use a subset of the decision criteria or a single criterion, like maximizing risk reduction efficacy, are commonly used but it offers great potential for analytical decision-making. The formality emerges from an intentional effort to minimize risk. Multi-criteria decision analysis, a formal method that considers all of the decision criteria has seldomly been used but it offers great potential for analytical decision-making. Informal methods that use a subset of the decision criteria are satisficing methods that seek to assure a risk management option is "good enough" based on the subset of criteria favored by the risk managers. The remaining method relies on the use of a decision matrix like that shown in Table 14.4 that displays all of the decision criteria. The risk manager then considers all of the data in an often ad-hoc manner to arrive at a decision. For most decision making, one or a few criteria will emerge as most important to risk managers. Best practice decision making considers all the decision criteria, but it is discriminating in identifying these criteria.

14.5.3. Approval

Chapter 5 differentiated three tiers of pest risk manager, the first of these were the analytical risk managers who direct and support the risk assessment as well as make phytosanitary risk management recommendations. The analytical risk managers are not always the final decision-makers in a risk management activity. This is a responsibility that is usually reserved for the policy risk managers, whose job it is to make the final risk management decision. There were 183 contracting parties to the IPPC at the time of this writing. That means there are 183 unique ways for a sovereign nation to finalize a regulatory decision. Each risk management activity initiated must clear the vetting and approval process of its nation before it can be implemented. The approval process is the proprietary property of the sovereign nations engaged in trade that must be respected as part of the overall risk management process.

Approval of a risk management decision establishes the appropriate level of phytosanitary protection for a nation. The SPS Agreement defines this as, "The level of protection deemed appropriate by the Member establishing a sanitary or phytosanitary measure to protect human, animal or plant life or health within its territory." This concept may also be referred to as the "acceptable level of risk."

14.6. Implementation

Once a risk management option has been selected and approved, it remains to be implemented. The relevant tasks are:

1. Verification.
2. Monitoring.
3. Review.

Risk managers must verify that the option is, in fact, being enacted by all parties with a responsibility for risk management. In other words, risk

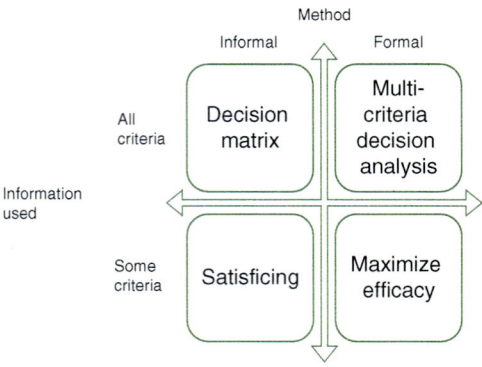

Fig. 14.2. Four decision-making methods for risk managers.

managers must make sure people are doing what they are supposed to do to manage the phytosanitary pest(s). Then the risk manager has a responsibility to see if the risk control measures are working. This is begun through the monitoring task. Finally, risk managers must evaluate the evidence collected in the monitoring process to see if the risk management goals are being achieved. If they are not, the risk management option needs to be modified.

14.6.1. Verification

Verifying actual implementation of the risk management option is likely to be important in the near term after an NPPO decides to implement a specific risk management option. Are people doing what they are supposed to be doing? Audits can answer questions about processes under the direct control of an organization, but when implementation requires large groups of stakeholders or the public to take or avoid certain activities, other methods of verification will be required. Once risk managers confirm that everyone is doing what they need to do to get the risk management option to work, then it is time to begin monitoring outcomes.

14.6.2. Monitoring

Risk management decisions are made under uncertain conditions. There is no way to know with certainty if a risk management option will be effective in advance of its implementation. ISPM 11 says the success of the risk management option should be determined by monitoring. Monitoring is the ongoing gathering, analysis, and interpretation of data to determine whether the desired risk reductions are being realized. Risk managers are responsible for monitoring the outcomes of their decisions to see if they are working. Monitoring reveals how well the implemented risk management option has performed. Review determines what steps, if any, need to be taken to improve the risk management outcomes. Monitoring is an essential part of risk management. It is a continuous process that is normally the responsibility of competent national authorities.

Examples of monitoring data include inspection data, intercepts, and outbreaks. Notifications are, here, considered part of the monitoring task. National monitoring and surveillance activities should be designed to collect information of high utility to future risk management decisions, including information for undertaking pest risk assessments, e.g., concentration and prevalence data for pests in commodities.

Inspection

Risk managers are responsible for the process used to inspect regulated commodity shipments. Inspection is defined in ISPM 5 as the official visual examination of plants, plant products, or other regulated articles to determine if pests are present and/or to determine compliance with phytosanitary regulations. Interception of a pest is the detection of a pest during inspection or testing of an imported consignment. Interception of a consignment can lead to its refusal or controlled entry due to failure to comply with phytosanitary regulation.

> The use of inspection as a means to detect the presence of pests in, or to determine or verify the pest level of a consignment is based on the following assumptions (ISPM 23):
>
> - pests of concern, or the signs or symptoms they cause, are visually detectable
> - inspection is operationally practical
> - some probability of pests being undetected is recognized.

Consignments are inspected to confirm compliance with import or export requirements related to quarantine pests or regulated nonquarantine pests (ISPM 23). Consequently, inspection may serve to verify the effectiveness of phytosanitary measures taken at a previous point in time. Export inspections ensure the consignment meets the specified phytosanitary requirements of the importing country. Import inspections verify compliance with phytosanitary import requirements. Inspection may also be used to detect organisms that have not yet been determined as phytosanitary risks. Inspections are often used as a risk management measure, so they can do double duty as a monitoring technique as well.

It is not often feasible to inspect an entire consignment. Inspection usually relies on sampling, and that means there is some probability that pests that are present will go undetected. Samples do not cover 100% of the consignment and inspection is not 100% effective. The inspection sample size is normally based on a specified regulated pest associated with a specific commodity. Sample size may be more difficult to determine where inspections of consignments target several regulated pests (ISPM 23). Risk-based sampling, a topic of Chapter 29, offers promise as a more efficient new approach to consignment sampling.

Notifications

When an exporting country fails to comply with specified phytosanitary requirements, the importing country is to notify them of the failed consignments. Likewise, the exporting country is notified when the importing country must take emergency action upon detection of a pest posing a potential threat. Voluntary notification may be used for other purposes with the aim of international cooperation to prevent the introduction and/or spread of regulated pests (IPPC Articles I and VIII).

Non-compliance notifications are intended to aid the investigation of the cause of the non-compliance so that steps may be taken to avoid recurrence. The importing country is expected to first consult with the exporting country and provide them with the opportunity to investigate the apparent non-compliance and correct it as necessary. Changes in the phytosanitary status of a commodity or area, or failures of phytosanitary systems in the exporting country, should not be reported until the notification is addressed bilaterally (ISPM 13).

Risk managers should provide prompt notification once non-compliance has been confirmed or emergency phytosanitary actions have been taken. Notifications should use a consistent format and include certain minimum information which includes a description of the phytosanitary actions taken and the parts of the consignment affected by those actions (ISPM 13).

Uncertainty

Many risk assessments are conducted under conditions of considerable uncertainty. Data

gaps can affect the structure of a risk management option. It is the risk manager's responsibility to see that the constantly expanding body of evidence is monitored to reveal when new data become available, that could substantively alter the conclusions of the risk assessment or its associated degree of uncertainty.

> Notification is provided when regulated pests are detected in imported consignments or in other significant non-compliance instances such as:
>
> - failure to comply with phytosanitary requirements
> - detection of regulated pests
> - absence of phytosanitary certificates
> - uncertified alterations or erasures to phytosanitary certificates
> - serious deficiencies in information on phytosanitary certificates
> - fraudulent phytosanitary certificates
> - prohibited consignments
> - prohibited articles in consignments, e.g., soil
> - evidence of failure of specified treatments.
>
> (ISPM 13)

14.6.3. Review

Once the monitoring information is gathered it must be evaluated. This process should compare risk management results to risk management objectives in order to decide whether the objectives are being met and the risk management option is successful. This means looking at the results of the risk management strategy and judging them as satisfactory or not. One way to do this is to compare them with risk management expectations based on the risk assessment and other data. Are the desired/expected risk reductions being achieved? Have the potential benefits from opportunities been attained? Alternatively, this evaluation could contrast the results to date with what you believe is possible from other actions. The information supporting the PRA should be periodically reviewed to ensure that any new information that becomes available does not invalidate the previous decision taken. This sort of evaluation is part of the risk manager's post-implementation responsibility, which may be assigned to risk assessors.

If the evaluation of monitored results reveals unsatisfactory outcomes, the risk management decision should be modified. That modification most often will take the form of a new iteration of some or all of the phytosanitary risk analysis process. It could vary from a completely new risk assessment and decision-making round to a slight adjustment of the decision or its implementation strategy. The public and stakeholders should be kept informed of any and all post-implementation findings and changes in the risk management option.

> The principle of "modification" states: "As conditions change, and as new facts become available, phytosanitary measures shall be modified promptly, either by inclusion of prohibitions, restrictions or requirements necessary for their success, or by removal of those found to be unnecessary."
>
> (ISPM 1)

ISPM 11 cautions that the "implementation of particular phytosanitary measures should not be considered to be permanent." The review can consider the risk management strategy as a whole or a particular risk management measure. A review of monitoring data may indicate that current risk management activities are not producing appropriate risk reductions. More stringent measures may be required to reach an acceptable or tolerable level of risk. Targeted monitoring may indicate the need to review a particular phytosanitary

risk management measure. Risk management decisions should also be opened to review when new data or new risk management options become available.

14.7. Summary and Look Forward

Here are five things to remember from this chapter.

1. The risk management responsibilities of Stage 3 of the PRA process can be expanded to a more rigorous risk management model.
2. Formulation is the process of devising solutions to risk management problems and opportunities.
3. Evaluation is a screening process used to determine whether a formulated risk management option is a viable solution in a given phytosanitary risk management activity.
4. Selection begins with the risk management options that survive evaluation, it compares them across a set of decision criteria and identifies the best option from among the viable alternatives.
5. The implemented options must be monitored and occasionally reviewed to ensure that the desired risk management outcomes are being achieved.

The next chapter considers how risk managers ought to handle the uncertainty that accompanies the PRA process.

14.8. References

Devorshak, C. (ed.) (2012) *Plant Pest Risk Analysis: Concepts and Application.* CAB International, Wallingford, UK.

IPPC (2015) *Specification 63: Guidance on Pest Risk Management.* Secretariat of the International Plant Protection Convention (IPPC), Food and Agriculture Organization of the United Nations, Rome, Italy.

IPPC (2016a) International Standards for Phytosanitary Measures, Publication No. 1: *Phytosanitary Principles for the Protection of Plants and the Application of Phytosanitary Measures in International Trade.* Secretariat of the International Plant Protection Convention (IPPC), Food and Agriculture Organization of the United Nations, Rome, Italy.

IPPC (2016b) International Standards for Phytosanitary Measures, Publication No. 13: *Guidelines for the Notification of Non-compliance and Emergency Action.* Secretariat of the International Plant Protection Convention (IPPC), Food and Agriculture Organization of the United Nations, Rome, Italy.

IPPC (2016c) International Standards for Phytosanitary Measures, Publication No. 28: *Phytosanitary Treatments for Regulated Pests.* Secretariat of the International Plant Protection Convention (IPPC), Food and Agriculture Organization of the United Nations, Rome, Italy.

IPPC (2016d) International Standards for Phytosanitary Measures, Publication No. 7: *Export Certification System.* Secretariat of the International Plant Protection Convention (IPPC), Food and Agriculture Organization of the United Nations, Rome, Italy.

IPPC (2017a) International Standards for Phytosanitary Measures, Publication No. 12: *Phytosanitary Certificates*. Secretariat of the International Plant Protection Convention (IPPC), Food and Agriculture Organization of the United Nations, Rome, Italy.

IPPC (2017b) International Standards for Phytosanitary Measures, Publication No. 23: *Guidelines for Inspection. Secretariat of the International Plant Protection Convention (IPPC)*, Food and Agriculture Organization of the United Nations, Rome, Italy.

IPPC (2019a) International Standards for Phytosanitary Measures, Publication No. 2: *Framework for Pest Risk Analysis*. Secretariat of the International Plant Protection Convention (IPPC), Food and Agriculture Organization of the United Nations, Rome, Italy.

IPPC (2019b) International Standards for Phytosanitary Measures, Publication No. 5: *Glossary of Phytosanitary Terms*. Secretariat of the International Plant Protection Convention (IPPC), Food and Agriculture Organization of the United Nations, Rome, Italy.

IPPC (2019c) International Standards for Phytosanitary Measures, Publication No. 14: *The Use of Integrated Measures in a Systems Approach for Pest Risk Management*. Secretariat of the International Plant Protection Convention (IPPC), Food and Agriculture Organization of the United Nations, Rome, Italy.

IPPC (2019d) International Standards for Phytosanitary Measures, Publication No. 11: *Pest Risk Analysis for Quarantine Pests*. Secretariat of the International Plant Protection Convention (IPPC), Food and Agriculture Organization of the United Nations, Rome, Italy.

MAF (Ministry of Agriculture and Forestry) (2008) *Import Risk Analysis: Fresh coconut (Cocos nucifera) from Tuvalu. Draft for Public Consultation*. MAF Biosecurity, Wellington, New Zealand.

North American Plant Protection Organization (2012) NAPPO Regional Standards for Phytosanitary Measures. RSPM 40: Principles of Pest Risk Management for the Import of Commodities. Ontario, Canada. Available at: http://www.nappo.org/files/8314/3889/6413/RSPM40-e.pdf (accessed February 1, 2017).

UNCTAD (United Nations Conference on Trade and Development) (UNCTAD) (2012) *Classification of Nontariff Measures, February 2012 version* (UNCTAD/DITC/TAB/2012/2). New York, Geneva: United Nations.

USDA/APHIS/PPQ (2012) *Guidelines for Plant Pest Risk Assessment of Imported Fruit and Vegetable Commodities*. Plant Epidemiology and Risk Analysis Laboratory, Center for Plant Health Science and Technology.

Yoe, C. (2019) *Principles of Risk Analysis Decision Making Under Uncertainty*, 2nd edn. CRC Press, Boca Raton, Florida, USA.

15

Uncertainty and Pest Risk Management

Uncertainty is a permanent part of the
leadership landscape. It never goes away.

Andy Stanley

15.1. Introduction

Uncertainty is an inherent part of pest risk
assessment, in fact, without uncertainty
there is no risk. Risk management, then, is
decision-making under uncertainty. Risk managers must be aware of the uncertainty that attends the risk assessment and other evaluations
they use to support their decision-making. This
uncertainty can arise from knowledge uncertainty or natural variability, as described in
Chapter 3. Good risk management requires the
most complete and objective view of the risk possible. This places a burden on the risk assessors
to carefully consider and communicate the significance of the uncertainty encountered in the
pest risk assessment to risk managers.

It is the risk assessor's responsibility to address the uncertainty encountered in the risk assessment. Likewise, analysts must address the
decision-making in other evaluations that are
done to support risk management decision-
making. It is the risk manager's responsibility to
address decision-making. This chapter focuses on
the role of decision-making in risk management
decision-making. It proceeds with a brief review of
some of the sources of uncertainty confronted

in a pest risk assessment. Next, it addresses the
need for a systematic method of characterizing
uncertainty. The chapter concludes by examining some options for addressing uncertainty in
the risk management process.

15.2. Sources of Uncertainty

The European Food Safety Authority (EFSA, n.d.)
defines uncertainty "as referring to all types of
limitations in the knowledge available to assessors at the time an assessment is conducted and
within the time and resources available for the
assessment." EFSA offers the general principle
that assessors are responsible for characterizing
uncertainty, while decision-makers are responsible for resolving the impact of uncertainty on
decisions. In best practice risk management, risk
assessors are charged with the responsibility of
answering a specific set of risk questions prepared by the risk managers.

> **Practical advice to risk assessors**
>
> Say what is uncertain. Say why it is uncertain.
> Say how uncertain it is. Say why the uncertainty
> is important.

Toward this end, scientific uncertainties should
be explicitly considered at each step in the risk
assessment and documented in a transparent
manner. The expression of uncertainty may be

qualitative or quantitative, but it should be quantified to the extent that is scientifically achievable. When addressing the uncertainty in the answers to the risk manager's risk questions, say how different the outcome might be. Then, say how likely the outcomes of interest to risk managers are. Quantify as much of the uncertainty as possible, then leave risk management to risk managers. Critical areas of concern to NPPO risk managers include the existence, nature and magnitude of a risk, and when possible, how effective risk management options will be in reducing the risk that exists. Areas of concern to private sector risk managers include the costs of complying with phytosanitary measures, if any, and the magnitude, duration, and likelihood of economic or other expected benefits.

C-U-U

A practical approach for risk assessors is to categorize their evidence, reasoning and information as confirmed, unconfirmed, and unknown (C-U-U). Information that is confirmed is well established and unlikely to change. The truth or accuracy of unconfirmed information is not reliable. Unknown information is speculative as to its existence. A modest goal for risk assessors is to move as much information into the confirmed column as possible. Risk managers should try to base decisions on confirmed information, they should tread very lightly around unconfirmed information and should avoid relying on unknown information.

Some of the most common sources of uncertainty include:

- inherently unobservable phenomena of concern
- missing information
- incomplete information
- old and outdated information
- contradictory information
- alternative explanations with different consequences
- misinterpretation of information
- incorrect information
- indirect information (extrapolations)
- incorrect models or assumptions
- natural variability in climate, geographic factors, consignment attributes, and the like

- reliance on expert judgment, especially vague opinions.

Assessors should describe uncertainty encountered in characterizing the pest, its probability of introduction, and the consequences of that introduction, as well as uncertainty in the feasibility, efficacy, and impacts of proposed risk management options. Uncertainty that affects the answers to the risk manager's questions is instrumental uncertainty and it must be conveyed. Instrumental uncertainties are those that have the capacity to affect the risk manager's decisions or the outcomes of their decisions.

15.3. Addressing Uncertainty for Risk Management

Conveying the nature and significance of the scientific uncertainty cannot be a haphazard effort, neither should it be an ad hoc process. Risk managers should prescribe or, at least, provide guidance on how they would like to have scientific uncertainty expressed and communicated in risk assessment in the NPPOs risk assessment policy. Fig. 15.1 summarizes a range of qualitative and quantitative methods that can be used to analyze uncertainty.

Notice, from the figure, the increasing levels of refinement in the manner uncertainty can be characterized. The second level of refinement after identifying the uncertainties is qualitative characterization of the uncertainties. The third level of refinement is to quantify the characterizations with bounds or ranges. The fourth level of refinement is to provide probabilistic quantification of the uncertainty. Descriptions of the methods listed are available from the *Draft guidance on uncertainty* (EFSA) (see http://www.efsa.europa.eu/sites/default/files/160321DraftG-DUncertaintyInScientificAssessment.pdf, accessed November 19, 2019). It may be desirable to identify one or more of these methods for use in the NPPOs' risk assessment policy document.

Although an overall probabilistic quantification of the uncertainty may be an ideal, the EFSA authors recognize it may not always be possible to quantify the overall uncertainty. In this case, they recommend that assessors:

- report that the overall uncertainty cannot be quantified

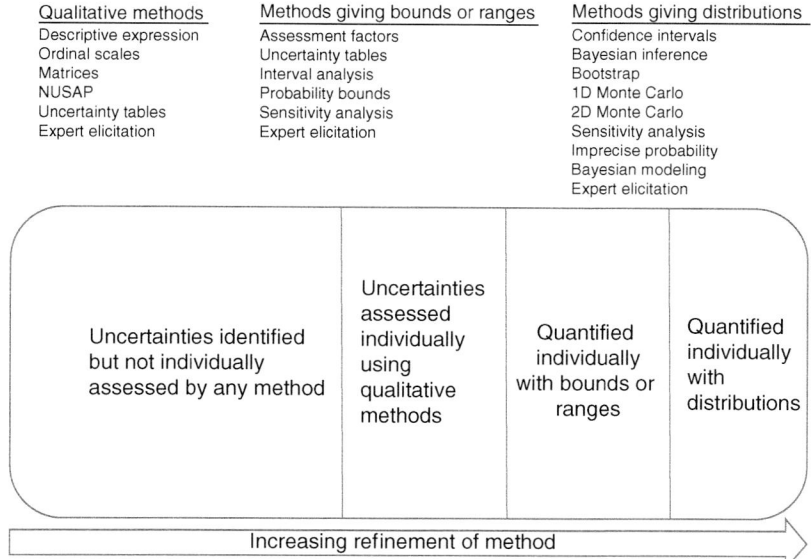

Qualitative methods	Methods giving bounds or ranges	Methods giving distributions
Descriptive expression	Assessment factors	Confidence intervals
Ordinal scales	Uncertainty tables	Bayesian inference
Matrices	Interval analysis	Bootstrap
NUSAP	Probability bounds	1D Monte Carlo
Uncertainty tables	Sensitivity analysis	2D Monte Carlo
Expert elicitation	Expert elicitation	Sensitivity analysis
		Imprecise probability
		Bayesian modeling
		Expert elicitation

Uncertainties identified but not individually assessed by any method	Uncertainties assessed individually using qualitative methods	Quantified individually with bounds or ranges	Quantified individually with distributions

Increasing refinement of method

Fig. 15.1. Increasingly refined methods for characterizing uncertainty (adapted from Figure S.2, draft EFSA guidance).

- consider partial quantification, conditioned on assumptions about the unquantified uncertainties
- highlight and describe the unquantified uncertainties.

In a system's approach to pest risk management there may be differing levels of uncertainty about the efficacy of different measures. There may be no uncertainty about some measures and high uncertainty about others. The uncertainty about one key aspect of a risk management option could dwarf other uncertainties. Such an uncertainty is dominant and it deserves special attention by risk managers.

In other instances, a great deal of uncertainty about an option can be eliminated by a measure with no uncertainty. A risk management option that includes an irradiation measure can eliminate concern about a lot of other uncertainties. Consequently, it is wise to focus on the most significant uncertainties.

When the characterization is wholly qualitative, assessors should not use qualitative expressions that imply quantitative judgments.

Consider an example of an effort to characterize the consequences of an infestation of the hypothetical Ferocies spp. Consistent with Fig. 15.1, the process begins by identifying the uncertainty impact as the monetary losses of affected crops. A qualitative characterization could be as simple as: the consequences are uncertain but are believed to be moderate to high. The next step could be to estimate that damages will fall between 1,000,000C and 100,000,000C. The most difficult step is the last one. Fig. 15.2 demonstrates the value of this step. The preceding step specifying a bound or interval is a useful step forward but it reveals nothing about the relative likelihood of different values in that interval occurring. The alternate distributions in Fig. 15.2 reveal a great deal.

Part I, the distribution on the left, reveals that large amounts of damage are far more likely than small damage amounts. For contrast, Part II, on the right, shows just the opposite pattern. These two different characterizations of uncertainty might well elicit entirely different risk management responses. The five-number summary for each distribution appears beneath its distribution.

EFSA has proposed a probability judgment scale to express the probability of uncertain outcomes. Ideally, the probability of an uncertain outcome would be calculated by combining the

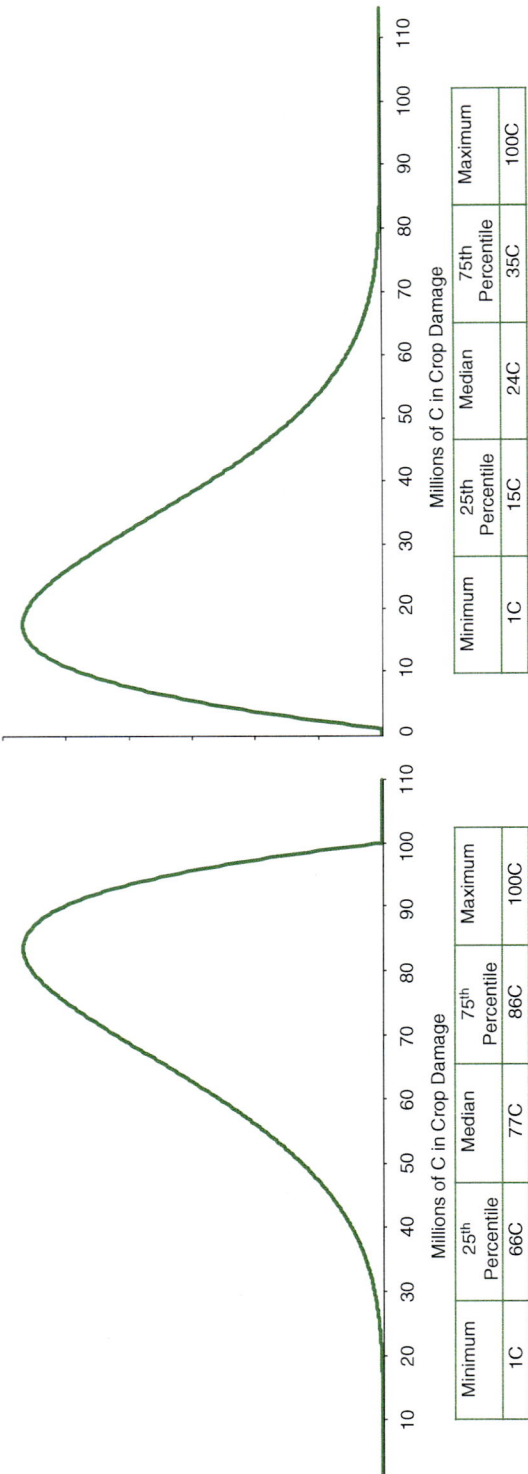

Fig. 15.2. Hypothetical demonstration of alternative quantified characterizations of uncertainty using alternate distributions.

contributions of individually quantified uncertainty quantities when possible. Probabilities of uncertain outcomes can also be generated by eliciting a judgment of the probability of a specified outcome that is important to risk managers from an expert panel.

Assessors may find it difficult to express a precise probability of a risk's existence or its probability of exceeding an acceptable limit, but a probability bound—it is no less than x% or no more than y%—might be easier to express. Interval estimates that specify a minimum and maximum value of such a probability are also easier to define than point estimates.

Table 15.1 provides a convenient default set of definitions for expressing the probability of an uncertain outcome. The quantitative ranges serve the purpose of approximately anchoring the words used in natural language descriptions of probabilities. Whatever use is made of the table, it is important to record the rationale for the judgments used.

A more suitable method for qualitatively characterizing uncertainty focuses on the quality of the evidence. The US Army Corps of Engineers has defined confidence levels (the obverse to uncertainty ratings) to be used for situational awareness of risk assessment judgments and results. When making key judgments, defined as those that may materially affect decisions or understanding, they ascribe *high, medium,* or *low* levels of confidence to the assessments, based on the nature of the evidence, as described below.

High: indicates that the nature of the issue is knowable and there is consistent evidence supporting the judgment from a variety of credible, independent sources or that the evidence, though limited, is measurable and consistent with accepted physical principles and scientific

Table 15.1. Scale proposed by EFSA to harmonize expressions of the probability of uncertain outcomes.

Probability term	Subjective probability range
Extremely likely	99–100%
Very likely	90–99%
Likely	66–90%
As likely as not	33–66%
Unlikely	10–33%
Very unlikely	1–10%
Extremely unlikely	0–1%

methods. Inferences are few and there are no apparent plausible alternatives deemed as likely as the judgment. A "high confidence" judgment is not a certainty.

Medium: indicates that the nature of the issue is either knowable or sufficient evidence minimizes uncertainty. Although there may be minor inconsistencies, gaps, or questions, credible evidence largely supports the judgment. There may be some foundational assumptions or inferences, and a plausible alternative deemed nearly as likely as the judgment.

Low: indicates that the nature of the issue may not be knowable, e.g., complex future-oriented judgments, or there are significant inconsistencies, gaps or questions regarding the evidence. There may be several foundational assumptions or inferences, and a plausible alternative(s) deemed as likely as the judgment.

Methods like this can be used to describe the confidence or uncertainty attending a risk assessment result. Other methods examine the evidence more carefully. Pest risk assessments and pest risk management decisions should be based on the available evidence. Uncertainty may, therefore, best be associated with the available evidence. USDA/APHIS/PPQ (2012) has developed a method for evaluating the evidence used to make an assessment judgment that considers both the reliability and applicability of the available scientific evidence.

Reliability refers to how trustworthy the evidence is. The reliability of the evidence is determined by the quality of the source, its recency (year of publication), methodology used, and the degree of consensus over the methods used or the interpretation of the results. The quantity of available information can also be a reliability factor. Many resources from lower quality sources that have the same conclusion may be more powerful evidence in terms of certainty than a single reference from a higher quality publication source.

Applicability refers to how germane the evidence is to a specific assessment or situation. Ideally, the risk assessment findings should be based on evidence of the pest's current behavior under conditions corresponding to the PRA area. The most applicable data requires no extrapolation.

The reliability and applicability of the evidence, together, provide an estimate of the level of uncertainty (or confidence) about the available

Table 15.2. Reliability of evidence rating (Regional Standards for Phytosanitary Measures, RSMP 40).

Publication source	Reliability	Examples
Well-known peer-reviewed journal	Low	None
	Moderately low	Few or no original research papers; any found do not describe methodology OR methodology used is not widely accepted.
	Moderately high	At least one original research paper with detailed description of methodological approach. Several original research papers without specified methodology. Multiple published review articles; articles cite independent (separate) sources of information.
	High	Multiple original research papers with detailed description of the methodological approach(es) used; approaches are widely accepted.
Obscure or less-known peer-reviewed journals	Low	No original research.Few or no review or summary articles.
	Moderately low	Few or no original research papers; methodology may or may not be described. Multiple published review articles which may or may not cite independent (separate) sources of information.
	Moderately high	Multiple original research papers (with specified methodology).
	High	Many original research papers (by multiple authors) that include a detailed description of the methodological approach(es) used; approaches are widely accepted; supported by other evidence.
Other expert sources that are not peer-reviewed, e.g., universities, subject matter experts, scientific societies— may include extension reports, non-journal articles, bulletins, etc.	Low	Single reports; if more than one report, those that are found may or may not be based on independent (different) information sources. No supporting evidence found.
	Moderately low	A few articles and reports that may or may not have each been based on independent (different) information sources.
	Moderately high	Several independent articles/reports based on independent information; methodology is described.
	High	Many reports from independent sources; well understood methodology; general consensus between information source.
Information from trading partner or developed by NPPO itself	Low	Evidence not well-documented or inconsistent with other sources; methodology not verified, not reported, or is not widely accepted.
	Moderately low	Evidence is well-documented and consistent with other sources; methodology has either not been verified or it is not widely accepted.
	Moderately high	Well-documented evidence that is generally consistent with other information; methodology verified and widely accepted.
	High	Well-documented evidence; methodology verified and widely accepted; supported several other sources.
Scientific consensus—for instance, the degree to which there is agreement in the literature or among experts on methods or interpretation of results, despite some contradictory views or report	Low	May or may not have scientific consensus. Methodology may or may not be generally accepted. NPPO may or may not have experience with pest; if so, the evidence may or may not be consistent with NPPO experience.
	Moderately low	May or may not have scientific consensus. Any methodology described may or may not be generally accepted. NPPO may or may not have experience with pest; if so, the evidence is generally consistent with NPPO experience.
	Moderately high	General consensus in scientific literature and other sources (but may include a few contradictory reports). Methodological approaches used are generally accepted. Evidence consistent with NPPO experience with pest.
	High	High consensus in scientific literature and other sources (no or practically no contradictory evidence found). Methodological approaches are widely accepted. Evidence generally consistent with NPPO experience with pest.

scientific evidence. Evidence that is highly reliable and applicable leads to a higher level of confidence in the assessment findings. Table 15.2 presents reliability ratings for five different publication sources. Table 15.3 defines applicability for species and environment.

A reliability/applicability matrix, as shown in Table 15.4, can be constructed to identify overall levels of uncertainty in the evidence used to support a key risk assessment finding.

Let us use the hypothetical pest Ferocies spp. once more. Imagine that the scientific evidence for the probability of introduction, which has been rated high, is judged by assessors to be moderately high in reliability and low in applicability for an overall evidence uncertainty rating of moderately uncertain. The consequences of introduction have been rated medium, with moderately low reliability but high applicability, yielding a consequence uncertainty rating of moderately certain. This produces the summary of Table 15.5.

The probability of introduction is more uncertain than the consequences, which are, themselves, only moderately certain. Thus, despite the seeming definitiveness of the risk ratings, when considered alone, we learn the evidence for these judgments was not strong.

In an actual risk assessment, this is the point at which the risk assessor would say what is uncertain in the assessment of P(introduction) and why it is uncertain. The rating of evidence says how uncertain the evidence for P (introduction) is. Next, the risk assessor must say why the uncertainty is important. The risk has been rated as high, so resolving the uncertainty will not produce a higher estimate of the risk, although it could confirm the high risk. It is more likely that resolution of the uncertainty would reduce the P(introduction) risk rating. Let us suppose that such resolution could not reduce the risk rating below a medium risk and a high end of medium at that. This is the kind of information that ought to be conveyed to the risk manager. It is helpful to know that resolution of the uncertainty will not lead to low or no risk when that is the case. Likewise, consequences must be addressed in a similar fashion.

15.4. Managing Risk with Uncertainty

Assessing risk and characterizing uncertainty are two independent risk assessor tasks. Risks

Table 15.3. Applicability of evidence ratings (RSPM 40).

Publication Information	Applicability	Examples
Species-specific data	Low	Species-specific data were limited; most of the species data were approximated or extrapolated from congeneric species, or other similar species.
	Moderately low	Species-specific data were used; some of the species data were approximated or extrapolated from congeneric (or other) species known to behave similarly.
	Moderately high	Species-specific data were used.
	High	Data for both pest and host(s) species were used.
Environment-specific data	Low	Environment-specific data were limited; no close proxy data were available; extrapolations were based on situations that may or may not be applicable.
	Moderately low	Some environmental-specific data were used, but most were approximated or extrapolated from similar situations (i.e., research conducted in the areas of comparable climate, on a closely related host).
	Moderately high	Some environment-specific data were used, but at least some data were approximated or extrapolated from similar situations (i.e., research conducted in the areas of comparable climate, on a closely related host).
	High	Environment-specific data were used.

Table 15.4. Evidence uncertainty rating matrix (RSPM 40).

		Reliability			
		Low	Moderately low	Moderately high	High
Applicability	High	Moderately uncertain	Moderately certain	Certain	Certain
	Moderately high	Moderately uncertain	Moderately certain	Certain	Certain
	Moderately low	Uncertain	Moderately uncertain	Moderately certain	Moderately certain
	Low	Uncertain	Uncertain	Moderately uncertain	Moderately uncertain

Table 15.5. Example summary risk and uncertainty ratings for hypothetical Ferocies spp.

Item	Risk rating	Evidence uncertainty
P(introduction)	High	Moderately uncertain
Consequence of introduction	Medium	Moderately certain

are to be assessed on the strength of the best available scientific evidence. The uncertainty attending that risk is to be characterized quantitatively or qualitatively, independent of the assessment of any risk element. It is the risk assessor's responsibility to be objective in assessing all risk elements. It is the risk assessor's responsibility to characterize the uncertainty about the evidence used to rate a risk element. It is, then, the risk manager's responsibility to decide how to regard the uncertainty in the decision-making process.

Let us imagine a risk assessment with substantial uncertainty. What are the risk manager's options for responding to that uncertainty? Here are some potential responses:

Some risk assessors have been mistakenly tempted to adjust the qualitative rating of a risk element in response to the uncertainty surrounding that element. For example, a risk element that might be qualitatively rated low based on the available scientific evidence is subjectively raised to a medium rating to reflect the substantial uncertainty about that risk element. This is never an acceptable method for addressing uncertainty in a risk assessment. The risk and uncertainty ratings are to be independent of one another.

- reduce the uncertainty
- conservative risk management
- precaution
- redundancy
- monitoring
- adaptive management
- provisional measures.

The first option is often to reduce the uncertainty as much as possible. Options for doing so include gathering more data, employing more sophisticated or critical analysis of the available data, employing more experienced assessors, or using expert elicitation to characterize the uncertainty. When these options are impractical, research may be an option for reducing instrumental uncertainties, although research can be time consuming and expensive.

The risk manager can adopt a conservative approach to risk management when there is significant uncertainty. When adopting a conservative approach to uncertainty, the risk manager errs in favor of giving more credence to the riskier possibilities in the assessment. Such a conservative bias must still be based on the best available scientific information. Risk management measures adopted in response to this conservative bias may well be too stringent for the actual facts of the situation so they should be regarded as temporary measures. These measures should be accepted as provisional and reviewed as soon as additional information becomes available.

Precaution is a valued and controversial risk management response to uncertainty. Unlike a one-off conservative approach to risk management, precaution represents a more philosophical state of mind. Uncertainty and precaution have a direct relationship in this philosophy: the more uncertainty there is the greater the need for additional precaution. It is useful to note that precaution is not incompatible with the Sanitary and Phytosanitary Agreement (SPS Agreement) or the IPPC. Risk assessment is required to evaluate the evidence and identify the uncertainty that gives rise to precaution. Precaution cannot be supported as a legitimate response to uncertainty when the adequacy of the available information is determined before risk analysis is undertaken or completed. If the risk analysis is completed and becomes the basis for identifying the uncertainty, precaution is a valid response. As with a conservative approach, the measures should be considered temporary and provisional, to be reviewed as uncertainty is reduced. A presumption of due diligence in reducing uncertainty is made.

Uncertainty in risk may be met by redundancy. Redundancy adds risk management measures or extra strength to measures as a means to compensate for uncertainty. Redundancy may be part of a conservative or precautionary response or it may be a legitimate response to uncertainty. As a type of provisional measure, it requires technical justification to be continued. Measures should be adjusted as appropriate as uncertainty is reduced. In addition to compensating for uncertainty, redundancy can also be used:

- as a safeguard for lack of experience
- when no less stringent measure is available
- when no single measure is available
- as an alternative to a single more stringent measure (RSPM 40).

Monitoring is a normal part of the risk management process. Collecting and evaluating information to assure that risk reductions are being achieved is expected for every risk management option. Monitoring also includes monitoring the state of knowledge about risky situations for the expressed purpose of reducing uncertainty. Either of these two situations may lead to an opportunity to adjust the risk management measures initially imposed as a hedge against uncertainty.

Adaptive management strategies may be invoked by risk managers as a reasonable approach to uncertainty. The National Research Council (2004) defines adaptive management as a decision process with "flexible decision making that can be adjusted in the face of uncertainties as outcomes from management actions and other events become better understood. Careful monitoring of these outcomes both advances scientific understanding and helps adjust policies or operations as part of an iterative learning process."

Adaptive management is an evolving process that involves learning, i.e., the accumulation of understanding over time through the reduction of uncertainty, and adaptation, i.e., the adjustment of management over time. This sequential cycle of learning and adaptation leads to two beneficial consequences: better understanding of the phytosanitary risk, and better management based on that understanding.

Article 5.7 of the SPS Agreement provides for the use of provisional measures in cases where relevant scientific evidence is insufficient to complete an adequate risk assessment. Risk management measures adopted in such a case must be identified on the basis of available pertinent information. This includes information from relevant international organizations as well as from phytosanitary measures applied by other countries. A country that uses provisional measures is further obligated to attempt to reduce the uncertainty by obtaining the additional information necessary for a more objective risk assessment. The SPS Agreement further obligates the country to review the phytosanitary measure(s) used within a reasonable period of time.

15.5. Summary and Look Forward

Here are five things to remember from this chapter.

1. It is the risk assessor's job to characterize and communicate the uncertainty in risk assessment.
2. It is the risk manager's job to decide how to handle uncertainty in decision-making.
3. Increasingly refined methods for characterizing uncertainty are available.
4. The reliability and applicability of the evidence, together, provide an estimate of the level of uncertainty about the available scientific evidence.
5. Potential responses to uncertainty include: reduce the uncertainty, conservative risk management,

precaution, redundancy, monitoring, adaptive management, and provisional measures.

The next chapter addresses the risk communication task in PRA. Risk communication presents a unique challenge in the world of phytosanitary risk management because much of the best practice theory is precluded by the rule-making processes of nations and the political sensitivity of trade negotiations.

15.6. References

EFSA Scientific Committee (n.d.) *Guidance on Uncertainty in EFSA Scientific Assessment*. European Food Safety Authority (EFSA), Parma, Italy. Available at: https://www.efsa.europa.eu/sites/default/files/160321DraftGDUncertaintyInScientificAssessment.pdf (accessed February 1, 2017).

National Research Council (2004) *Adaptive Management for Water Resources Project Planning*. The National Academies Press, Washington, District of Columbia, USA, *doi*: https://doi.org/10.17226/10972. Available at: https://www.nap.edu/catalog/10972/adaptive-management-for-water-resources-project-planning (accessed November 19, 2019).

USDA/APHIS/PPQ (2012) *Guidelines for Plant Pest Risk Assessment of Imported Fruit and Vegetable Commodities*. Plant Epidemiology and Risk Analysis Laboratory, Center for Plant Health Science and Technology.

16

Stakeholders and Risk Communication

_____

To keep everyone invested in your vision, you have to back up a little bit and really analyze who the different stakeholders are and what they individually respond to.

Alan Stern

16.1. Introduction

Risk communication is widely accepted as one of three components comprising risk analysis. A 1998 Expert Consultation to the Codex Alimentarius Commission recommended adopting the following definition of risk communication: "Risk communication is the exchange of information and opinions concerning risk and risk-related factors among risk assessors, risk managers, consumers and other interested parties." The WTO SPS Committee defines risk communication as: "the exchange of real-time information, advice and opinions between experts and people facing threats to their health, economic or social well-being. The ultimate purpose of risk communication is to enable people at risk to take informed decisions to protect themselves and their loved ones" (WHO, http://www.fao.org/fao-who-codexalimentarius/roster/detail/en/c/308434/, accessed November 19, 2019).

Devorshak (2012) reports that currently there are no international standards that explicitly describe how countries should incorporate risk communication into their pest risk assessment process. It is further noted that the term "risk

communication" is not yet included in ISPM 5 (IPPC, 2019).

Risk communication can be described as an "interactive process" where information is exchanged between the national plant protection organization (NPPO) and its stakeholders. Rather than a distinct step in the PRA process, risk communication is more profitably considered an ongoing process conducted continually throughout the entire PRA process. Ideally, the PRA process would involve the constant exchange of information and opinion on plant trade issues and related risks among risk assessors, risk managers and policymakers, trading partners, industry, special interest groups, and all other interested parties from the beginning of a new trade initiative through risk assessment and risk management. Realistically, NPPOs may face institutional and cultural barriers and other constraints that make this level of collaboration difficult to achieve (Lundgren and McMakin, 2009).

An important additional factor for phytosanitary risk managers is that risk communication is conducted within a legal framework for trade. Many PRA processes lead to rule-making. This can involve peer review of the draft analyses and a public comment process on proposed rules and these processes can have relatively rigid bureaucratic designs that are sometimes inconsistent with the objectives of "open, two-way communication." For instance, in the United States, once a regulatory proposal is drafted, the risk assessors and risk managers who did the work are prohibited from offering any additional information beyond clarification of what is in the proposal. This is not consistent with the conceptual best practice of complete transparency.

Transparency can be further impaired by the strategic and politically sensitive nature of trade negotiations. Although the aspiration to transparency may be desirable it is not yet reality in a practical world where there can often be strategic cat and mouse struggles over information, as described later in this chapter. As a result, there is a great deal of potential room for improvement in the involvement of stakeholders and risk communication in the PRA process.

This chapter leans heavily on the risk communication lessons learned by the international food safety community of practice, while offering numerous original suggestions for improving the process. The chapter proceeds by examining who the primary PRA stakeholders are. Next it considers risk communication and the Sanitary and Phytosanitary Agreement (SPS Agreement) and some PRA goals for risk communicators. From there, the chapter proceeds to consider some specific risk communication shortcomings in PRA as well as some specific opportunities for improving pest risk communication. Elements of effective pest risk communication are considered before three types of risk communication strategy are introduced. At that point the chapter begins to draw most heavily on the food safety risk communication literature by presenting principles of risk communication, barriers to effective risk communication, and strategies for effective risk communication.

16.2. Stakeholders

Stakeholders are people who are affected by or who can affect a trade decision. The main stakeholders to an international plant and plant product trade proposal and its ramifications can be categorized as:

- governments
- industry
- consumers and consumer organizations
- academic and research institutions
- media
- World Trade Organization (WTO).

16.2.1. Governments

Governments have assumed the responsibility for managing pest risks associated with international trade proposals. With the responsibility for managing pest risks comes the responsibility to communicate information about pest risks to all interested parties to an acceptable level of understanding. Risk managers in the NPPOs are obligated to ensure effective communication with interested parties and "appropriate involvement" of the public in the risk analysis process. The definition of "appropriate involvement" can impose limitations on the risk communication process.

16.2.2. Industry

Industry is responsible for the quality and safety of the plants and plant products it produces. Industry often has the best understanding of the specific processes used to produce and handle plants and plant products. Consequently, industry's participation in PRA provides an important source of information for risk assessment and risk management that is essential to effective decision-making. In addition, industry has an obligation to communicate information regarding risks to affected stakeholders. To do that, industry must understand the relevant risks as well as the means by which they will be managed.

16.2.3. Consumers and consumer organizations

Risk communication includes the public's ability to participate in the PRA process when public health issues or other issues affecting consumer welfare arise. Involving the public early in the process helps to ensure that consumer concerns are addressed. Their continued involvement can produce better public understanding of the risk assessment process and how risk management decisions are made. Consumers and consumer organizations have a responsibility to make their concerns and opinions on risks known to risk managers when given the opportunity to do so. Consumer organizations can work with government and industry to make sure that risk messages addressed to consumers are appropriately formulated and delivered.

16.2.4. Academia and research institutions

The academic and research communities may be called upon to serve as subject matter experts on matters of importance to PRA. They may also be asked by the media or other interested parties to comment on government decisions. These communities may also provide helpful advice on risk communication approaches and strategies to risk managers.

16.2.5. Media

The media can play a critical role in risk communication when pest risks rise to their attention. Much of the information that the public receives on PRA comes to them through the media. Media can be used to help transmit a message, or they can create or interpret a message. The media are not limited to official sources of information and their messages can reflect the concerns of the public and other sectors of society, thus making risk managers aware of concerns they may have missed.

16.2.6. WTO

The WTO SPS Agreement encourages harmonization, which includes the establishment, recognition, and application of common sanitary measures by different member countries. It places a strong emphasis on the principles of transparency and consistency in the development and application of phytosanitary measures. These tasks are clearly dependent on effective risk communication among nations. The notification procedure required by the SPS Agreement communicates risk management decisions among affected member countries.

16.3. Risk Communication and the SPS Agreement

You will not find the term "risk communication" anywhere in the SPS Agreement. What you will find are repeated implicit references to the need for risk communication. These are found in the principles of harmonization, equivalence, and transparency as well as in the technical assistance, special and differential treatment, and consultations and disputes settlement provisions. Risk communication also comes to the fore in the SPS guidance on publication of regulations, the enquiry point and notification procedures.

Risk communication is fundamental to the development of the standards and guidelines that form the foundation for harmonization. A determination of equivalence depends on a clearly communicated understanding of the efficacy of alternative treatments. Technical assistance obviously requires close communication as do the consultations required to establish the need for special or differential treatment. Transparency is the subject of its own appendix to the SPS agreement. This is where the need for risk communication is most pronounced. It begins with the prompt publication of all regulations and proceeds with the establishment of a single point of enquiry to facilitate discussion of all communication relative to any risk management decision. It culminates in notification procedures that can require extensive risk communication.

The International Plant Protection Convention (IPPC) has little to say on the subject of risk communication as found in the ISPMs. The most substantive discussion is found in ISPM 7, which comments on the need for communication within and outside the exporting country. Within the country it says the NPPO should have procedures in place for timely communication to relevant personnel and to industry. Outside the exporting country, the NPPO should:

- liaise with the nominated representatives of relevant NPPOs to discuss phytosanitary requirements
- make available a contact point for importing country NPPOs to report cases of noncompliance
- liaise with the relevant regional plant protection organizations and other international organizations in order to facilitate the harmonization of phytosanitary measures and the dissemination of technical and regulatory information.

Goals of risk communication

1. Promote awareness and understanding of the specific issues under consideration during the risk analysis process, by all participants.
2. Promote consistency and transparency in arriving at and implementing risk management decisions.
3. Provide a sound basis for understanding the risk management decisions proposed or implemented.
4. Improve the overall effectiveness and efficiency of the risk analysis process.
5. Contribute to the development and delivery of effective information and education programmes, when they are selected as risk management options.
6. Foster public trust and confidence in the safety of *plant product imports*.
7. Strengthen the working relationships and mutual respect among all participants.
8. Promote the appropriate involvement of all interested parties in the risk communication process.
9. Exchange information on the knowledge, attitudes, values, practices and perceptions of interested parties concerning risks associated with *international trade of plants and plant products*.

(Adapted from FAO, 1998, italics indicate changes from original)

16.4. Pest Risk Analysis Goals of Risk Communication

The primary goal of risk communication in a pest risk assessment context is to provide timely, meaningful, relevant, and accurate information to all stakeholders to the pest risk assessment. Risk communication should effectively gather data from stakeholders, lead to a clear understanding of the findings of the risk assessment and the rationale behind them, as well as the details of the risk management decision and the measures required as a result of the PRA.

Paraiso *et al.* (2012) conducted a survey of 105 decision-makers working in the PRA field. Five goals of risk communication were identified in the following order:

1. Explain risks.
2. Explain decisions.
3. Encourage good practices among biologic control practitioners.
4. Respond to external and peer review recommendations.
5. Explain petition requirements.

To this list we add obtaining useful information from stakeholders and good practice of the SPS Agreement communication responsibilities.

16.5. Pest Risk Analysis Shortcomings

Covello *et al.* (2004) and Walls *et al.* (2004) found that although risk communication is an important component of the risk analysis process, it is still an ambiguous concept to many regulatory professionals and their stakeholders. Several studies by APHIS (1996, 2006, 2007) showed that stakeholders had little knowledge of the PRA framework, which limits their ability to participate in the process. There would seem to be little expectation or opportunity for stakeholder involvement. Paraiso *et al.* (2012) have suggested that development and use of practical mechanisms, such as public notifications in newspapers, direct mail, or email alerts and by presenting information in different formats including the internet, brochures, newspapers, and relevant guidelines/standards will improve understanding of the process. The study also showed that stakeholder access to risk-related information could be increased via distribution of risk messages through television, radio, and/or newspaper announcements, as well as by more novel distribution modes like e-alerts, text messages, Facebook pages, blogs, and other social media.

Based on the findings of the Paraiso *et al.* study a set of recommendations are offered to overcome some of the more serious shortcomings in pest risk communication. They are:

- increase the transfer of information pertaining to the PRA
- better characterization of risk communication goals
- careful identification and development of risk communication messages specific to different types of stakeholders

- greater involvement of government agencies in educating stakeholders about PRAs
- develop PRA frameworks that increase stakeholder involvement in the decision-making process
- consistency in risk communication messages.

The following paragraphs contain a great deal of useful guidance that can be used to improve risk communication throughout phytosanitary risk management. NPPOs are encouraged to move as fully into this transparent world of risk communication as possible. However, it is important to realize that will not be possible in every country because of the various ways that reality can constrain risk communication during phytosanitary risk management. The concept of risk communication and the reality usually diverge to varying extents.

In many nations, phytosanitary risk management can have regulatory rule-making potential. Each nation will have its own regulatory processes, which often influence communication. For example, in the United States, once a draft risk assessment is released for public comment, no further communication with the public is allowed. No matter how conceptually desirable communication may be during this time, it is not possible. Different regulatory processes may restrict or otherwise influence what information can be shared and when.

Many nations must cope with the challenges presented by the existence of confidential and proprietary information. Varying cultural mores and legal systems will necessitate restrictions on the handling of such information.

Peer-to-peer communication can be fraught with trouble in part due to when and how information becomes "official information." Restrictions on the exchange of unofficial information can be routinely expected. Scientists cannot always speak to other scientists, for example, when the lines of communication are strategically restricted. Information is usually asymmetric in a phytosanitary risk management situation and maintaining some of that asymmetry can be strategically important.

Importers and exporters routinely have different views on what constitutes relevant information. Importing countries will typically want all of the available information. Exporting countries, by contrast, may prefer to limit the information provided on the commodity and its region of origin.

Feedback loops between importing and exporting countries are not especially common. A risk assessment may begin with the information provided by the exporting country, but it is rare for the importing country to provide feedback on what information proved useful and reliable or not. Likewise, it is rare for exporting countries to be provided with feedback obtained from monitoring data.

These are but some of the reasons that the reality of phytosanitary risk communication differs from the theory. Nonetheless, there is a growing body of evidence that suggests more and better risk communication leads to better risk management decisions and a less combative and contested decision-making process.

16.6. Opportunities to Improve Pest Risk Communication

Following the Paraiso *et al.* findings there are several opportunities to improve risk communication in the PRA process. Table 16.1 begins with opportunities for information transfer among stakeholders. The row entry identifies the stakeholder seeking information, the column identifies the stakeholder dispensing information. Government can obtain related risk assessments or information about regulatory approaches used by other governments. They can also obtain industry practice information and pathway data from industry. Thus, each cell describes examples of information the row stakeholder might obtain from the column stakeholder.

In the "Plan the risk assessment" activities in the generic risk management model presented in Chapter 13, there is a communication activity. Among the suggested tasks to accomplish during this activity were:

- Clearly identify stakeholders and collaborators.
- Specify communication goals.
- Assess audiences.
- Tailor communication messages for each audience.

Table 16.1. Opportunities for information transfer during a pest risk analysis activity.

	Government	Industry	Consumers	Academia	Media	WTO
Government	1. Risk assessments 2. Regulations	1. Industry practice 2. Pathway data	1. Consumer practice 2. Consumer concerns	1. Detailed science 2. Best practices	1. Public concerns 2. Publicity for issue	1. Guidance 2. Previous experience
Industry	1. Hazard information 2. Risk management options	1. Best practices	1. Consumer preferences	1. Detailed science 2. Best practices	1. Public concerns 2. Publicity for issue	NA
Consumers	1. Hazard information 2. Risk management options	1. Product availability 2. Market information	1. Consumer reviews	1. Unbiased information	1. Public concerns 2. Publicity for issue	NA
Academia	1. Hazard information 2. Risk management options	1. Industry practice	1. Consumer practice 2. Consumer concerns	1. Information exchange 2. Technology exchange	1. Public concerns 2. Publicity for issue	NA
Media	1. Hazard information 2. Risk management options	1. Product Availability 2. Industry practice	1. Consumer concerns 2. Consumer preferences	1. Scientific background	NA	NA
WTO	1. Dispute resolution information	1. Industry practice	1. Consumer concerns	1. Detailed science 2. Best practices	NA	NA

This handbook is one effort to educate the public about the PRA process, in particular the risk management function. Government agencies could do more to educate the public about their process, especially through development of their websites. There are a great variety of options for educating the public. The United States offers the public a toll-free hotline number, while New Zealand offered the opportunity to consult with a risk analyst before the submission of a permit application (Paraiso *et al.*, 2012).

Adopting a risk management model, such as the one presented in this handbook, is a practical way of increasing stakeholder involvement. As for consistent risk management messages, they become more reasonable and feasible once an NPPO commits to a risk communication plan.

16.7. Patterns of Risk Communication

The market access risk communication process begins with the exporter understanding the phytosanitary requirements of the importing country. These could be provided by the overseas buyer or they can be officially requested from the importing country by the exporter's agent or the exporting country's NPPO. In most cases, this will at least involve phytosanitary certification which requires the exporter to ask and pay the exporting NPPO for a phytosanitary inspection. This results in a certificate stating that the shipment in question meets the requirements of the importing country. Certificate requirements could be as simple as certification for pest freedom based on inspection, or a more complex certification of freedom from specific pests, treatment, or other conditions presumably designed to mitigate pest risk.

The risk communication process gets more complicated if there are no known requirements for the importing country. This could be because the commodity is not authorized, i.e., it is restricted, or has never been requested before, or the importing country simply doesn't have any interest in setting requirements for the article in question. It is also possible that there are requirements in place, but a phytosanitary certificate is not one of them, so the exporting NPPO may have no information in its export certification database.

> Governments tend to encourage agricultural exports through three different export strategies: to increase market access (new products); to expand market access (new markets); and to maintain market access (avoid loss).

If this market access request is for a new commodity that has not been authorized previously, then someone needs to raise the question of requirements officially with the NPPO of the importing country. This could be done by the importing industry (the buyer) or it could be done by the exporting NPPO, or more likely both. Where the exporting NPPO is concerned, the specific trade issue is likely to get mixed in with all the other import-export issues and trade agreements for the country in question. At some point, it will be officially raised. This could be in the context of bilateral trade talks or perhaps only through official correspondence. This is where the unwritten principle of quid pro quo arises and each country tries to leverage their trade interests to their advantage.

Assuming access will be considered, the next step is usually a PRA or some kind of evaluation process by the importing NPPO that results in proposed requirements that trigger further negotiation and perhaps finally some decision on the entry status and requirements. It is during this time that the importing NPPO can expect to involve domestic interests in the risk management process. It is also during these further negotiations that the exporting NPPO can disagree on behalf of the exporter. The astute exporter will have advised its NPPO in advance regarding its risk management capability and especially its "bottom line" for requirements it would accept. Otherwise, the process could end up with a deal the exporter can't accept and everybody will have wasted their time.

The analytical aspect of this process has some essential pieces and key principles, but it does not need to be overly complex or resource intensive to be effective. This is not a one-sided process and it is best when it involves collaboration by both trading partners with the shared objective of safe trade. Fig. 16.1 illustrates a potential generic pattern of risk communication among the stakeholders to this process. Exporting and importing interests represent all the stakeholders from the private side of an issue in

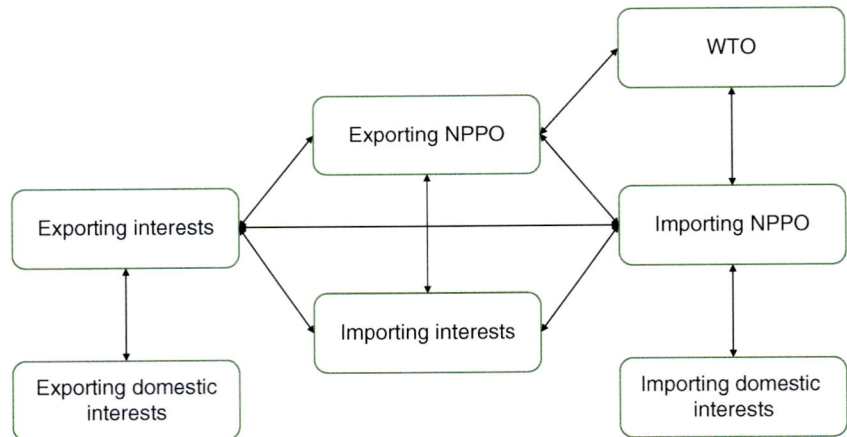

Fig. 16.1. A generic pattern of potential communication among stakeholder groups involved in a trade issue.

the countries of origin and destination. The exporting and importing country NPPOs represent the government interests in those nations. Domestic interests represent those stakeholders who do not have a direct financial interest in the trade issue. This would include consumers and consumer groups, academia, and the media.

16.8. Elements of Effective Pest Risk Communication

A Food and Agricultural Organization (FAO) expert consultation (1998) produced a wealth of insight into the elements of effective risk communication. These lists are reproduced below. Depending on the nature and purpose of the risk communication, messages may contain some or all of the following categories of information.

16.8.1. Nature of the risk

The nature of the risk will include the following kinds of information:

- the characteristics and importance of the hazard of concern
- the magnitude and severity of the risk
- the urgency of the situation
- whether the risk is becoming greater or smaller (trends)

- the probability of exposure to the hazard
- the distribution of exposure
- the amount of exposure that constitutes a significant risk
- the nature and size of the population at risk
- who is at the greatest risk?

16.8.2. Nature of the benefits

Information about the nature of the benefits of the risky activity includes such things as:

- the actual or expected benefits associated with each risk
- who benefits and in what ways?
- where the balance point is between risks and benefits
- the magnitude and importance of the benefits
- the total benefit to all affected populations combined.

16.8.3. Uncertainties in risk assessment

Critical uncertainties such as the following may be encountered in the risk assessment:

- the methods used to assess the risk
- the importance of each of the uncertainties
- the weaknesses of, or inaccuracies in, the available data

- the assumptions on which estimates are based
- the sensitivity of the estimates to changes in assumptions
- the effect of changes in the estimates on risk management decisions.

16.8.4. Risk management options

There will also be critical information about the risk management options such as:

- the action(s) taken to control or manage the risk
- the action individuals may take to reduce personal risk
- the justification for choosing a specific risk management option
- the effectiveness of a specific option
- the benefits of a specific option
- the cost of managing the risk and who pays for it
- the risks that remain after a risk management option is implemented.

16.9. Types of Risk Communication

Lundgren and McMakin (2009) identify three fundamentally different types of risk communication that lend themselves readily to PRA. They are:

1. Care communication
2. Consensus communication
3. Crisis communication.

Care communication is communication about risks that have been assessed. The risk and the way to manage it are well determined and accepted by most of the audience. An example would be describing the risk associated with a pest–commodity pair that has been successfully managed in multiple locations. An audience may sometimes be unaware of the risk, or unconcerned about it; then the primary goals of care communication are to raise awareness and motivate people to take action.

Consensus communication is used to reach a decision about how the risk will be managed. An example would be the importing NPPO and

importing and exporting industry working together to determine how best to avoid the introduction of a pest as a result of a trade proposal.

Crisis communication requires communication in the face of extreme, sudden danger. For example, following introduction of a pest like citrus canker that may have severe impacts on members of the public. This communication occurs both during and after the emergency. Care or consensus communication could be used when planning on how to deal with potential emergencies, depending on how much the audience is involved.

16.10. Principles of Risk Communication

The FAO (1998) offered a set of principles to help craft effective risk communication for food safety. These same principles provide valuable guidance for risk communicators dealing with phytosanitary issues. The principles include:

- know the audience
- involve scientific experts
- establish expertise in communication
- be a credible source of information
- share responsibility
- differentiate between science and value judgment
- assure transparency
- put the risk in perspective.

The discussion that follows draws heavily from the 1998 FAO document.

16.10.1. Know the audience(s)

There is no such thing as "the audience." Instead, any risk communication is going to involve multiple audiences. You cannot speak in the same ways to other plant pest regulators, industry and consumer groups and expect to be carefully understood by each. The vocabulary, experience and level of interest varies enough among groups that separate risk communications ought to be designed for separate audiences.

Effective risk communication requires knowing and understanding the concerns, beliefs, and feelings of each audience. That knowledge is best achieved through an analysis of the audience.

An analysis begins by identifying an external audience who are stakeholders for the proposed plant product trade. Define this audience's demographics, i.e., who they are, and psychographics, i.e., why they are interested in your PRA. Then use this information to identify risk communication strategies that depend on this information. These data can be gathered formally through primary research and available data or informally through conversations and even informed speculation.

An audience analysis helps one to understand the motivations and opinions of the audiences. Ideally, one will get to know the audiences as groups and individuals. This knowledge better enables one to open and maintain a channel of communication with them. Listening to all interested parties, i.e., audiences, is an important part of risk communication.

16.10.2. Involve the scientific experts

Risk assessors, in their capacity as scientific experts, must be able to explain the concepts and processes of risk assessment to the different audiences. This includes the results of the assessment, the scientific data, assumptions, subjective judgments that results are based upon, and significant remaining uncertainty necessary for any interested parties to clearly understand the risk. It is important that assessors be able to clearly communicate what they know and what they do not know, and to proclaim the uncertainty effectively. That does not mean the scientists themselves should be communicating directly with stakeholder audiences but their knowledge and expertise may be key parts of a risk communication message. The scientific efficacy of the risk management options considered and implemented also needs to be appropriately and adequately described for the various audiences.

16.10.3. Establish expertise in communication

Successful risk communication requires the expertise to convey understandable and usable information to all interested parties. Risk managers, assessors and other experts may lack the time or

skill to engage in complex risk communication tasks. Responding to the needs of various audiences and preparing effective messages requires people with risk communication expertise. These specialists should be involved as early as possible and the best way to do that is for an NPPO or a company to develop in-house risk communication expertise. Experts can be hired directly or developed through training and experience.

16.10.4. Be a credible source of information

Be first, be right, be credible (CDC, 2012). Although this advice was initially developed for crisis communication it is equally applicable for care and consensus communication risk messages as well. Getting the information right is fundamental. Credible sources are far more likely to influence the perception of a risk than is a less- or non-credible source. Credibility is accorded a source by a target audience and it may vary depending on the nature of the hazard, culture, social and economic status of the audience, and other factors. Timeliness is at times more important to credibility than the risk itself. Credibility is damaged in the long run by omissions, distortions and self-serving statements and it is enhanced by acknowledging current issues and problems.

Consistent messages received from multiple sources reinforce the credibility of the message. Credibility is determined based on recognized competence or expertise, trustworthiness, fairness, and lack of bias. Consumers associate high credibility with terms like factual, knowledgeable, expert, public welfare, responsible, truthful, and good "track record." Credibility and the trust that results must be nurtured. Trust can be eroded or lost through ineffective or inappropriate communication and once lost, it is difficult to re-establish. Consumers have indicated, in studies, that distrust results from exaggeration, distortion and perceived vested interest.

16.10.5. Share risk communication responsibility

NPPOs have a fundamental responsibility for risk communication. Stakeholders and the public in general expect them to play a leading role in managing phytosanitary risks. This is true whether the risk management decision involves regulatory or voluntary controls. It is expected even when the NPPO decides no risk controls are necessary for a particular trade issue. In this latter event, it is still essential to communicate the reasons why no action is the best option. Efficient and efficacious risk management decision-making requires risk managers to understand public concerns and to ensure that decisions respond appropriately to those concerns. This requires the NPPO to learn what the public knows about the risks and what it thinks of the alternative options being considered to manage those risks.

Industry has a responsibility for risk communication, especially when the risk is a result of their products or processes. They must be forthright and transparent about industry practices that affect the level of phytosanitary risk faced by an importing country. The media may at times share in these risk communication responsibilities, particularly when industry practices and/or NPPO decisions may affect the public, significant plant or other environmental resources.

When multiple parties, e.g. government, industry, and the media, are involved in the risk communication process they have a joint responsibility for the consistency and outcome of that communication even though their individual roles may differ.

16.10.6. Differentiate between science and value judgment

Risk analysis has been described, at times, as the confluence between science and social values. It is important to separate science from values when making a risk management decision or when explaining the basis for a risk management decision. As a practical matter, this means report the scientific facts used to support the risk management decision as well as the uncertainties that are instrumental to the risk management decision that has been made, then identify the specific criteria upon which risk management decisions were based.

Risk communicators are responsible for explaining what is known and what is not known. It is the risk communicator's responsibility to convey where scientific knowledge begins and ends. Value judgments enter the risk management decision-making process most prominently when deciding what level of phytosanitary risk is acceptable or not. Thus, risk communicators ought to be able to justify the level of acceptable risk and the level of residual risk to the public that will remain with risk management measures in place. Many people may expect that the residual risk with phytosanitary risk management measures in place means zero risk, but zero risk is often unattainable. In practice, phytosanitary risk management options produce a tolerable nonzero level of risk. Making this transparently clear is an important function of risk communication.

16.10.7. Assure transparency

The PRA process must be open and available for public scrutiny by interested stakeholders. This level of transparency is essential to the acceptability of the process. Effective two-way communication between NPPO risk managers and interested parties, including the public, is essential to good risk management and it is a key to achieving transparency. Transparency need not compromise legitimate concerns to preserve confidentiality, e.g. proprietary information or data.

16.10.8. Put the risk in perspective

A pest risk should be examined in the context of the benefits associated with the proposed market access that poses the risk. It can sometimes be helpful to compare the risk at issue with other similar, more familiar risks. Risk comparisons can backfire, however, if stakeholders perceive that the comparison has been intentionally chosen to make the risk in question seem more acceptable. Effective risk comparisons should assure that:

• both (or all) risk estimates are equally sound
• both (or all) risk estimates are relevant to the specific audience
• the degree of uncertainty in all risk estimates is similar

- the concerns of the audience are acknow-ledged and addressed
- the substances, products or activities them-selves are directly comparable, including the concept of voluntary and involuntary exposure (FAO, 1998).

16.11. Barriers to Effective Risk Communication

Effective risk communication requires commu-nicators to be aware of and to recognize the most common barriers to risk communication and to know how to overcome them. Much of risk communication is internal, in the form of it-erative exchanges between risk managers and risk assessors. Chapter 11 on the interactions between risk managers and assessors describes the situations where many of the internal bar-riers to risk communication will occur. The bar-riers described below draw heavily from the 1998 FAO document prepared for the food safety community of practice.

16.11.1. Access to information

Stakeholders with information vital to the risk analysis process may not always be willing to share that information. They may be reluctant to release proprietary information or to share their information with government agencies with regulatory authority. Access to all the relevant data about a pest risk may not exist for risk managers and assessors in every situ-ation. Risk communication is more difficult when there is a lack of access to critical risk data.

16.11.2. Participation in the process

At the present time, external parties do not participate a great deal in the PRA process. This can become a significant barrier to effect-ive risk communication. The participation of stakeholders increases the overall understand-ing of the process and its decisions, and makes it easier to communicate with the public about

those decisions. Stakeholders who are given the opportunity to become involved in the PRA process are less likely to challenge the out-come, especially if their concerns have been addressed. The PRA process does not yet in-volve industry and other interested parties ef-fectively in its processes.

16.11.3. Differences in perceptions

NPPOs of different countries and industry repre-sentatives can all perceive the same pest risk very differently. They may disagree with the risk assessors and managers about the pest itself or the relative magnitude or severity of the risks as-sociated with the pest. Efforts to gain an under-standing of how the various stakeholders per-ceive the risk can enhance the effectiveness of the risk communication task.

16.11.4. Differences in receptiveness

Stakeholders may vary in their receptiveness to a risk management decision. Individuals may believe the risks are not as great as they have been characterized. Some people may believe they are personally more knowledgable about the pest and the risks it presents than the regu-lators are. Communicating effectively with such unreceptive groups requires risk managers to understand their attitudes, beliefs and concerns, and to address those concerns in risk communi-cation messages.

16.11.5. Lack of understanding of the scientific process

The pest risk assessment's reliance on precise scientific terminology may obscure the nature of the risk for some interested parties. Messages that are not kept relatively simple may be misunderstood. It is especially important to ac-knowledge and put scientific uncertainties into context to assure that the public has an accurate perception of what is and is not known about the risk. Explicitly identifying the value judgments used to make the risk management decisions and distinguishing them from the scientific judgments of risk assessment is essential to aiding the public's understanding of the basis for the decisions that are made. Using natural language to the greatest extent possible, while explaining the technical terms that are used, can mitigate this problem.

16.11.6. Source credibility

Trust is a communicator's most important asset when laboring under conditions of uncertainty. Perceptions of bias or past failures to provide accurate information will be a source of distrust. To a very great extent, credibility depends on the extent to which the risk assessment and risk management processes are believed to be transparent and open to public scrutiny.

16.11.7. The media

The public relies on the media for their risk information and the media do not always accurately convey that information. Relatively few reporters are trained in the complex scientific and policy aspects of pest risk issues and it can be difficult for them to prepare a story on highly technical matters under deadline pressure. The media also have their own agenda. Risk communicators may lack the experience needed with the media to understand how to work with reporters to enhance the quality and accuracy of their reports. Risk communicators need training in media skills.

16.11.8. Societal characteristics

Societal barriers, such as language differences, cultural factors, religious dietary laws, illiteracy, poverty, a lack of legal, technical and policy resources, and a lack of infrastructures that support communication can make risk communication more difficult. To the extent possible, these and other cultural and social attributes that hamper risk communication need to be identified and addressed, as part of the process of designing messages for target audiences.

16.12. Strategies for Effective Risk Communication

This section continues to lean heavily on the strategies developed by the food safety community of practice by summarizing several strategies for risk communication originally presented in the FAO (1998) risk communication report. Recent experience suggests that different risk communication strategies are appropriate for the various contexts in which it occurs. These strategies are discussed in terms of general considerations, points to consider regarding public concerns, and non-crisis strategies.

16.12.1. General considerations for effective risk communication

General considerations are guided by the sequence of events encountered in a risk communication: gathering background information, preparing the message, disseminating the message, and the follow-up review and evaluation of its impact. Bullet lists of considerations are presented for each of these events.

Background information

- Understand the scientific basis of the risks and attendant uncertainties.
- Understand the public perception of the risk through such means as risk surveys, interviews, and focus groups.
- Find out what risk information people want.
- Be sensitive to related issues that may be more important to people than the risk itself.

Expect different people to see the risk differently.

Prepare messages

- Avoid comparisons between familiar risks and new risks, as they may seem flippant and insincere unless presented properly.
- Recognize and respond to the emotional aspects of risk perceptions. Speak with sympathy and never use logic alone to convince an audience characterized by emotion.
- Express risk in several different ways, making sure not to evade the risk question.

- Explain the uncertainty factors which are used in risk assessment and standard-setting.
- Maintain an openness, flexibility, and recognition of public responsibilities in all communication activities.
- Build an awareness of the benefits associated with a risk.

Disseminating the message

- Accept and involve the public as a legitimate partner by describing risk/benefit information and control measures in an understandable way.
- Share the public's concern rather than deny it as not legitimate or as unimportant. Be prepared to give people's concerns as much emphasis as the risk statistics.
- Be honest, frank, and open in discussing all issues.
- If explaining statistics derived from risk assessment, explain the risk assessment process before presenting the numbers.
- Coordinate and collaborate with other credible sources.
- Meet the needs of the media.

Review and evaluation

- Evaluate the effectiveness of risk messages and communication channels.
- Emphasize action to monitor, manage, and reduce risk.
- Plan carefully and evaluate efforts.

16.12.2. Points to consider regarding public concerns

Some risks concern the public more than other risks. Their characteristics are identified below. Strategies that can be used to mitigate public concern about risks are also identified.

Factors that raise public concern about risks

- Unknown, unfamiliar, or rare events as opposed to well-known or common hazards.
- Risks that are controlled by others, rather than those where the public or the individual is in control.

- Risks resulting from industry action or from new technology, rather than those perceived as natural.
- Risks where there is significant scientific uncertainty, or where there is open controversy among experts as to the probability and severity of the hazard.
- Risks that raise moral or ethical questions, such as the fairness of the distribution of risks and benefits, or the rights of one group in society to put others at risk.
- The decision-making process by which a risk is assessed is seen as being unresponsive or is unknown.

Strategies to mitigate public concerns about risk

- Make risks voluntary by giving consumers choices, whenever possible.
- Acknowledge uncertainty.
- Show that expert disagreement on an issue is merely uncertainty, by estimating risks as a range that includes estimates from both sides of the debate.
- Determine where control is and look to share it with interested parties.
- Treat all interested parties with courtesy.
- Always consider concerns and complaints seriously.

16.12.3. Strategies for non-crisis risk communication situations

Risk communication for non-crisis situations differs from other forms of risk communication so we revisit the sequence of events from Section 16.12.1 for these situations in the sections that follow.

Background Information

- Anticipate emerging public concerns before they become significant.
- Determine the public's perception of the hazard being considered and their knowledge and behavior regarding the risks involved.
- Analyze the target audience of a risk communication and understand their motivations. Try to determine the full range of the audience's concerns and their perceived importance.

- Analyze which information channels and messages are best to be used. Use the mass media and other appropriate channels to convey information.

Prepare messages

- Describe to concerned groups how risk is determined, how it can be monitored and how an individual can control or reduce risk.
- Identify shared values and help individuals identify an approach to meet their values.
- Make messages interesting and relevant by emphasizing the human rather than the statistical aspects of a story.
- Use extra care to make a message interesting enough for the media to publish. Claims of risk are usually considered to be more newsworthy by the media than claims of safety.

Disseminating the message

- Use the mass media where possible to address those consumer concerns which have been identified. For example, public forums with local opinion leaders can be televised.
- Sustain communication, thus enabling the public to make decisions based upon personal values and goals and to gain a greater understanding of the potential risks and benefits involved.
- Make risk communication multi-directional, not just from technical experts to the public, but from the public back to the experts.
- Use participation to sustain efforts. People have to feel that they are at the center of a risk management action or decision for the process to be effective.

Review and evaluation

- Add an evaluation component into any risk communication strategy.
- Test the clarity and understanding of the message with a representative segment of the target audience.

- Integrate risk communication with risk assessment and risk management activities, to increase the effectiveness of risk analysis and ensure proper utilization of resources.
- Educate and train risk assessors and risk managers in the principles and uses of risk communication.
- Effective risk communication can break through traditional boundaries within government sectors, between governmental and non-governmental organizations, and between the public and private sectors. Co-operation is essential and this requires the creation of equal partnerships between the different sectors at all levels of governance in societies.

16.13. Summary and Look Forward

Here are five things to remember from this chapter.

1. Risk communication is essential to a successful risk management process.

2. There are ample opportunities to improve the quality of risk communication in PRA.

3. Risk managers should seek to involve and gain input from all stakeholders to assure they consider valid issues and concerns other than science.

4. There are many valuable risk communication lessons to be learned from the food safety community of practice.

5. Risk communication requires the use of persons trained in risk communication.

Enterprise risk management (ERM) is a framework for decision-making under uncertainty that appeals broadly to private and public institutions alike. ERM is introduced in the next chapter where an overview of the COSO and ISO 31000 enterprise risk management frameworks are presented as examples.

16.14. References

APHIS (Animal Plant Health and Inspection Service) (1996) *Options for Changes in Biological Control Regulations and Guidelines in the United States: A Strawman for Comment*. National Biological Control Institute, Riverdale, Maryland, USA.

APHIS (Animal Plant Health and Inspection Service) (2006) *Plant Protection and Quarantine Permitting Review Highlights*, DA2006-04. Available at: http://ipm.ifas.ufl.edu/pdf/Organisms PermittingReview. pdf (accessed 15 December 2010).

APHIS (Animal Plant Health and Inspection Service) (2007) *Import and Export*. Available at: http://www. aphis.usda.gov/ import export/index.shtml (accessed 15 December 2010).

CDC (Centers for Disease Control) (2012) *Crisis and Emergency Risk Communication: 2012 Edition*. U.S. Department of Health and Human Services, Washington, District of Columbia.

Covello, V.T., McCallum, D.B. and Pavlova, M. (eds) (2004) *Effective Risk Communication: The Role of Government and Non-Governmental Organizations*. Plenum Press, New York, New York, USA, DOI 10.1007/978-1-4613-1569-8.

Devorshak, C. (ed.) (2012) *Plant Pest Risk Analysis, Concepts and Application*. CAB International, Wallingford, UK.

FAO (Food and Agricultural Organization of the United Nations) (1998) *The Application of Risk Communication to Food Standards and Safety Matters*. FAO Food and Nutrition, Paper 70. Report of a Joint FAO/WHO Expert Consultation Rome, M-82 ISBN 92-5-104260-8.

IPPC (2019) International Standards for Phytosanitary Measures, Publication No. 5: *Glossary of Phytosanitary Terms*. Secretariat of the International Plant Protection Convention (IPPC), Food and Agriculture Organization of the United Nations, Rome.

Lundgren, R.E. and McMakin, A.H. (2009) *Risk Communication: A Handbook for Communicating Environmental, Safety, and Health Risks*. Wiley, Hoboken, NJ.

Paraiso, O., Kairo, M. Leppla, N.C; Cuda, J.P. Owens, M., Olexa, M.T. and Hight, S.D. (2012) *Opportunities for Improving Risk Communication During the Permitting Process for Entomophagous Biological Control Agents: A Review Of Current Systems*. Publications from USDA-ARS / UNL Faculty Paper 1420. Available at: http://digitalcommons.unl.edu/usdaarsfacpub/1420 (accessed February 3, 2017).

Walls, J., Pidgeon, N., Weyman, A. and Horlick-Jones, T. (2004) Critical trust: understanding lay perceptions of health and safety risk regulation. *Health Risk Society* 6, 133–150.

17

Enterprise Risk Management

_____

Risk is what an entrepreneur eats for breakfast. It's what she slips into bed with at night. If you have no appetite for this stuff, or no ability to digest it, then get out of the game right now.

Heather Roberts

17.1. Introduction

Business leaders face a significantly magnified exposure to risk and uncertainty as they attempt to manage the ever-changing economic, environmental, political, and technological landscapes in which they operate. A growing number of organizations have embraced the concept of enterprise risk management (ERM) so that strategic risks can be managed proactively to increase the likelihood the organization will achieve its core objectives (Beasley *et al.*, 2019). Based on the insight of 2,415 risk management experts from 86 countries, business interruption including supply chain disruption is the number one most important global business risk (Allianz, 2019). Quarantine pests guarantee supply chain disruption.

The World Economic Forum (WEF) (2019) suggests the world is moving into a new phase of state-centered politics. The idea of nations "taking back control" weakens collective responses to emerging global challenges. Erosion of multilateral trading rules and agreements has been identified by the WEF as the short-term risk of

second greatest concern to its respondents, right behind economic confrontation and frictions between world powers.

Businesses involved in the plant production chain are increasingly threatened by phytosanitary risks. Risks of quarantine diseases and pests increase as a result of growing trade, travel, transportation, and tourism. The Dutch government, for example, desires a phytosanitary policy that involves the full plant production supply chain. This would entail a more optimal allocation of responsibilities between government and the private sector.

Stakeholders who do not take appropriate risk-reducing measures not only put their own business at stake, but also that of others in the production chain as well as threats of the economic, environmental and other damages that can accompany quarantine pests (Breukers *et al.*, 2009). ERM is a bit of critical infrastructure in the aspirational world of an integrated phytosanitary risk management program where government and industry partner for the good of all involved in and affected by international trade of plants and plant products. Stakeholders have to be better capable of taking responsibility in phytosanitary risk management if it is to prevail and survive these challenges. This chapter provides an overview of ERM.

Up until now the focus of Part 2 of this handbook has been on phytosanitary risk management and NPPO risk managers. Private industry, which is responsible for production risk management owns a substantial part of the phytosanitary and economic risks that are involved with international trade in plants and plant products, is turning more frequently to ERM. ERM is not so fundamentally different, in its principles, from the phytosanitary risk management process that has been described, but it looks different because the entity is protecting its own objectives rather than serving a stewardship role for society.

This chapter proceeds by previewing the value of ERM to organizations engaged in international trade. It then introduces and reviews two popular ERM frameworks, ISO 31000 Risk Management Standard and the Committee of Sponsoring Organizations' (COSO) ERM framework.

17.2. ERM's Value to Phytosanitary Risk Management

The underlying premise of enterprise risk management is that every entity exists to provide value for its stakeholders and every entity faces uncertainty. Uncertainty presents both risk of loss and opportunity for gain. This uncertainty affects an enterprise's ability to achieve their objectives. ERM provides a framework for management to effectively deal with uncertainty and its associated risks of loss and gain.

Companies involved in the export and import of plants and plant products operate in environments where globalization, climate change, technology, regulation, restructurings, changing markets, and competition are some of the factors that create uncertainty. They recognize opportunities for potential gain through international trade. Trade, whether successful or unsuccessful, has important implications for a company's profit, revenue streams, and labor market activities. Entities with a direct financial stake in the movement of plant commodities also have a direct stake in the outcomes of phytosanitary risk management decisions and measures taken. The burden of implementing many risk management measures falls directly on the shoulders of exporters and importers. Thus, these enterprises face risks from within and without their organizations.

This handbook has espoused an, admittedly aspirational, vertically integrated approach to phytosanitary risk management. The pursuit

and expansion of international trade in plants and plant products has undeniable benefits for exporting and importing countries. The opportunities for developing efficacious systems approaches for managing phytosanitary risk multiply when importing and exporting firms cooperate with the regulatory authority of NPPOs. An ERM approach by the private sector enables organizations to better anticipate risk so they can get ahead of it, with an understanding that change creates opportunities, not simply the potential for rising costs or crises.

The benefits of exporting and importing firms practicing ERM include the following (COSO, 2017; CGMA, 2019):

- greater awareness of the risks an organization faces and the ability to respond effectively to them
- enhancing confidence about the organization's achievement of its strategic objectives
- improving compliance with legal, regulatory, and reporting requirements
- increasing efficiency and effectiveness of operations
- a consideration of both positive and negative aspects of risk, enabling management to identify new opportunities and unique challenges associated with current opportunities
- the identification and managing of risk across the international trade enterprise—sometimes a risk can originate in one part of the enterprise and impact a different part
- increasing positive outcomes and advantages while reducing negative surprises
- reducing performance variability
- improving resource allocation: robust information on risk allows management to assess overall resource needs
- enhancing enterprise resilience and its ability to anticipate and respond to change, not only to survive but also to evolve and thrive.

ERM is an organization's culture, it is not a function or department. An organization manages risk to create, preserve, and realize value. ERM defines the capabilities and practices that organizations use to do that. ERM is more than a risk register. It is not a checklist. It is a set of principles on which processes can be built or integrated for a particular organization, and it is a system of monitoring, learning, and improving performance. It can be used by organizations of any size.

17.3. ERM Frameworks

There are numerous ERM frameworks that a company could follow, including several national standards. Among the more widely known frameworks are the ISO 31000 Risk Management Standard (2009a, 2009b, 2009c) and the COSO ERM framework. Brief descriptions of these models are offered for private sector stakeholders interested in learning more about the practice of ERM. Private industry stakeholders in the international trade of plants and plant products would be well advised to adopt an enterprise risk management framework, especially if they are interested in sharing responsibility for phytosanitary risk management.

The ISO 31000 Risk Management Standard was first published in 2009 and revised in 2018 by the International Organization for Standardization (ISO). It does not use the term ERM, instead it defines the risk management process as "coordinated activities to direct and control an organization with regard to risk." It also provides a definition of the risk management framework as a "set of components that provide the foundations and organizational arrangements for designing, implementing, monitoring, reviewing and continually improving risk management throughout the organization." As an International Standard, it enjoys wide acceptance in many countries and large corporations. It is practical and business oriented.

The COSO ERM framework was published in 2004 and revised in 2017 by the Committee of Sponsoring Organizations of the Treadway Commission. It defines ERM as "a process, effected by an entity's board of directors, management and other personnel, applied in strategy setting and across the enterprise, designed to identify potential events that may affect the entity, and manage risk to be within its risk appetite, to provide reasonable assurance regarding the achievement of entity objectives." It is popular in the United States. Each is summarized in sections that follow.

17.3.1. ISO 31000

Fig. 17.1 shows the relationship among the principles, framework, and process, the three components that comprise the ISO framework. Each component is summarized below.

ISO 31000: 2018 principles

The ISO principles, seen in the top part of Fig. 17.1 define an effective risk management organization. Risk management is an integral part of all of an organization's activities. The process is structured and comprehensive to ensure that it produces consistent and comparable results. Ideally, the framework is customized to meet the needs of the organization. It includes appropriate and timely two-way involvement of stakeholders to enable their knowledge, views, and perceptions to be considered. Risks emerge, change, or disappear as the organization's external and internal contexts change. Risk management should be based on the best available information and should include appropriate consideration of the limitations and uncertainties associated with that information. Human behavior and culture influence every aspect of risk management. Experience and learning gained through risk management ensure the continual improvement of the process.

The risk management framework (leftmost part of Fig. 17.1) provides the foundations and organizational arrangements for designing, implementing, monitoring, reviewing, and continually improving risk management throughout the organization. A customized framework can integrate risk management into the organization's activities and functions. The effectiveness of the ERM framework depends on the extent to which it is integrated into the organization's governance. The organization's risk management leadership and commitment to ERM are embodied in the framework.

Integration. A strong and sustained commitment by the organization's top management is the starting point for risk management. Chief among management's responsibilities are:

- defining the risk management policy
- aligning the strategic objectives of the organization with the risk management objectives
- assigning risk management accountabilities and responsibilities at appropriate levels throughout the organization

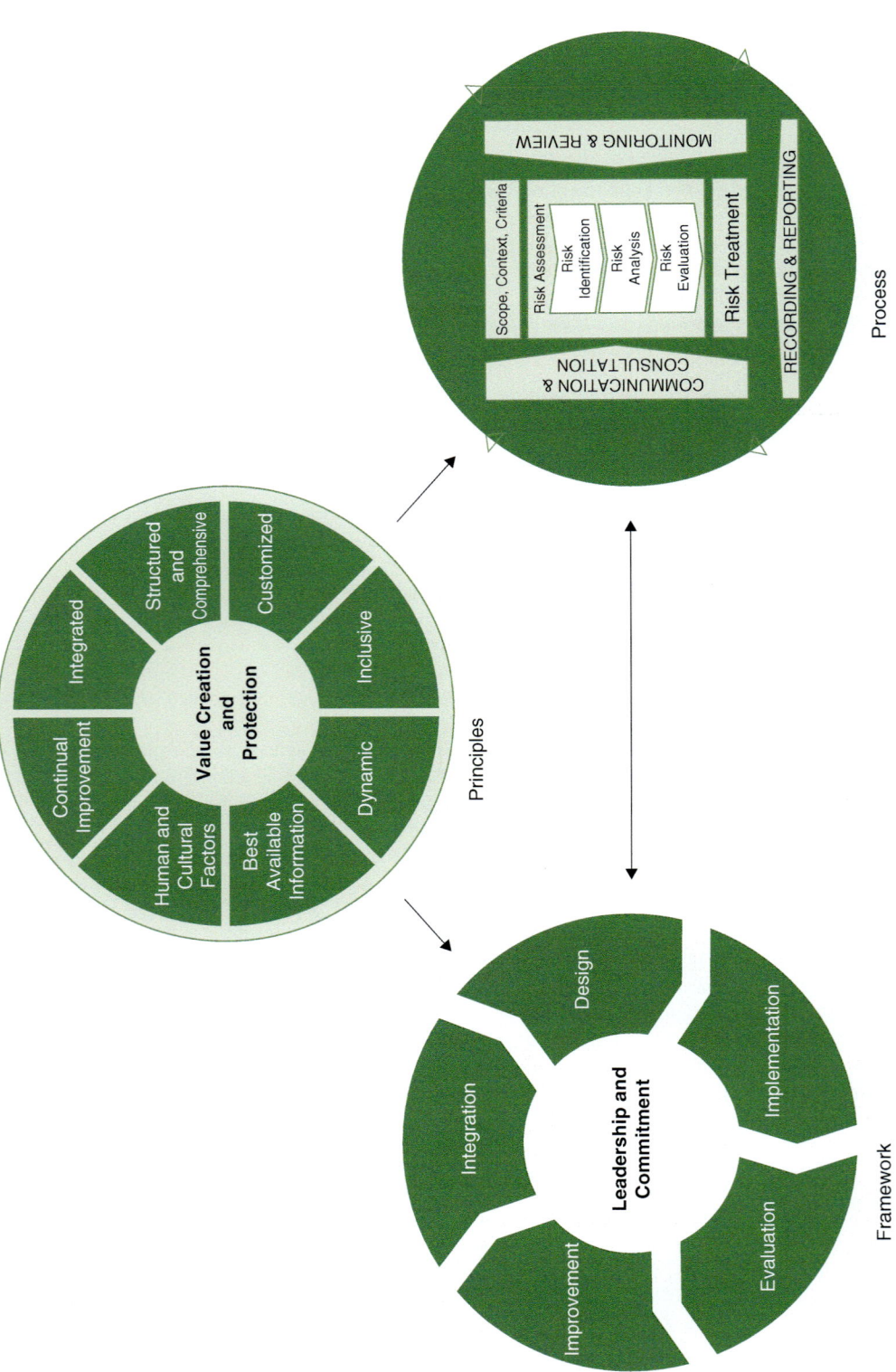

Fig. 17.1. Relationship of the ISO 31000 principles, framework and process. (©ISO. This material is reproduced from ISO 31000:2018 with permission of the American National Standards Institute (ANSI) on behalf of the International Organization for Standardization. (The complete standard can be purchased from ANSI at https://webstore,ansi.org, all rights reserved.)

- allocating sufficient resources to risk management
- ensuring the risk management framework remains appropriate.

Design. Management designs the organization's framework for managing risk. When designing the framework for managing risk the organization should:

- understand the organization and its context
- articulate its risk management commitment
- assign organization roles, responsibilities, authorities and accountabilities
- allocate resources
- establish lines of communication and consultation.

Implementation. Implementing the risk management framework assures that risk management policy and process are embedded in all organizational processes. Personnel must receive training in the use of the organization's risk management process. Successful implementation ensures that the risk management process is a part of all decision-making activities throughout the organization.

Evaluation. Monitoring and review of the framework are necessary and ongoing processes. This ensures that risk management is effectively supporting the organization's performance. The effectiveness of the framework needs to be regularly reviewed and modified as warranted to meet the organization's changing needs.

Improvement. The organization's external and internal contexts will change. Monitoring and adapting the risk management framework to these changes ensures the process can be continually improved. This happens when the framework's suitability, adequacy and effectiveness are regularly improved.

Risk management process

The risk management process describes the manner in which risks are managed. This process can be applied at the strategic, program, project, or operational levels, it should become an integral part of the organization's horizontal and vertical decision-making. The right part of Fig. 17.1 shows the five stages and three processes that comprise the risk management process. The

ISO 31000 risk terminology differs from the International Plant Protection Convention (IPPC) terminology and some of the terminology used elsewhere in this handbook.

The risk management process is applied to a specific risk management activity. It begins by establishing the scope, context and criteria that define the risk management activity. This includes identifying the relevant objectives of the activity and criteria that reflect the entity's risk appetite and tolerance. This means considering how to:

- measure the nature and types of causes and consequences that can occur
- define likelihood
- determine the timeframe(s) of the likelihood and/or consequence(s)
- determine the level of risk
- ascertain the views of stakeholders
- establish the level at which risk becomes acceptable or tolerable, and
- consider combinations of multiple risks when applicable.

Risk assessment comprises the next three stages. The purpose of the risk identification stage is to generate a comprehensive list of risks based on those events that might create, enhance, prevent, degrade, accelerate, or delay the achievement of objectives. This includes identifying the risks of *not* pursuing an opportunity. Risks not identified will not be analyzed further.

The third stage is called risk analysis. This is where an understanding of the risk is developed. It is what the IPPC calls risk assessment. It considers such things as:

- the likelihood of events and consequences
- the nature and magnitude of consequences
- complexity and connectivity
- time-related factors and volatility
- the effectiveness of existing controls
- sensitivity and confidence levels.

Risk evaluation is ISO's fourth stage. Decision-makers compare the level of risk determined during the risk analysis stage with risk criteria established during the context step to determine the need for risk treatment. Deciding which risks need treatment and determining the priority for treatment implementation begins here. Risk evaluation can lead to a decision to:

- do nothing further
- consider risk treatment options
- undertake further analysis to better understand the risk
- maintain existing controls
- reconsider objectives.

The final stage is risk treatment, an iterative process of formulating, selecting risk treatment options and implementing one or more options to modify risks. Risk treatment options include:

- avoiding the risk by deciding not to start or continue with the activity that gives rise to the risk
- taking or increasing the risk in order to pursue an opportunity
- removing the risk source
- changing the likelihood
- changing the consequence
- sharing the risk with another party or parties (including contracts and risk financing)
- retaining the risk by informed decision.

Risk treatment must be planned and implemented. The information provided in the treatment plan should include:

- the selection rationale, including the expected benefits
- identification of those accountable and responsible for approving and implementing the plan
- the proposed actions
- the required resources
- appropriate performance measures
- relevant constraints
- reporting and monitoring requirements
- a schedule for taking and completing actions.

Communication and consultation with internal and external stakeholders are the first process. Communication addresses the risk, its causes, its consequences, and the measures being taken to treat it. Consultation suggests that a team approach to risk management will be employed.

Monitoring and review are the second ongoing process. These activities ensure that risk controls are effective and efficient in both design and operation. Recording and reporting is the third and final process. It is intended to:

- communicate risk management outcomes and activities across the organization
- support decision-making with information
- improve risk management activities
- assist interaction with stakeholders.

17.3.2. COSO ERM Framework

The primary references for this framework are *Enterprise Risk Management: Integrated Framework* (2004) and *Enterprise Risk Management: Aligning Risk with Strategy and Performance* (COSO, 2017).

COSO defines ERM as a process that is executed strategically by the people of the enterprise. It is applied horizontally and vertically throughout the organization and it is geared toward the achievement of objectives. The basic framework is illustrated in Fig. 17.2. The right face of the cube illustrates that risk management occurs across the enterprise. The top face illustrates the enterprise objectives that are served by ERM. The presenting face summarizes the steps in the process. Each face is discussed in turn below.

Across the enterprise

ERM helps a firm's management select strategy consistent with the company's risk appetite. Risk appetite is the amount of risk a firm will accept in pursuit of value. Different strategies expose the entity to different risks. Pursuit of foreign markets for plant commodities while ignorant of the phytosanitary issues that attend that trade is a high-risk position in which to put oneself. Conversely, failure to pursue foreign markets may severely limit a firm's profits and growth. A firm's risk appetite guides its resource allocation. Management considers its risk appetite as it aligns its organization, people and processes, and designs infrastructure necessary to effectively respond to and monitor risks.

Every division, business unit, and subsidiary (see right face of Fig. 17.2) engages in risk management. Thus, ERM encompasses the entire scope of the firm's activities. This includes enterprise-level activities like strategic planning and resource allocation, business unit activities like marketing, production and new customer development,

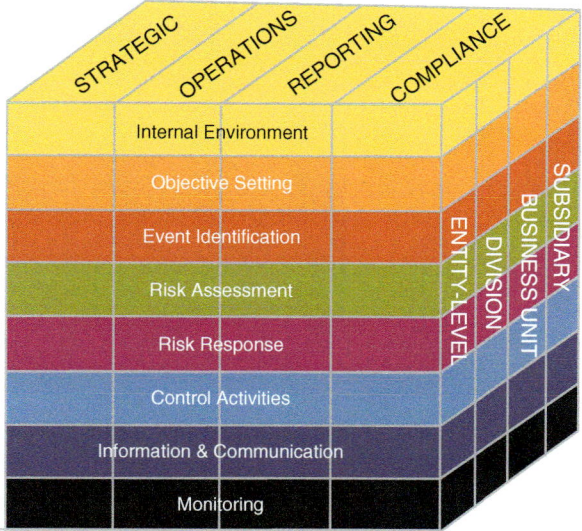

Fig. 17.2. COSO enterprise risk management framework, reproduced with permission.

> **Risk appetite**
>
> An organization with a low appetite for phytosanitary risk and a high appetite for new products in new markets might well see some self-interest in actively collaborating with NPPOs on the choice and use of risk treatments. Such a risk appetite could well provide a competitive advantage to a firm that actively embraces low phytosanitary risk as an enterprise strategy over firms that are phytosanitary naïve.

and special projects and new initiatives that might not yet have found a home in the firm's structure.

COSO's ERM framework requires a portfolio view of risk. Each business function in the entity (right face) has objectives (top face) and must follow the ERM process (presenting face), assessing the risk for each function or unit. It is management's responsibility to manage risk across this portfolio to assure it is commensurate with its risk appetite. Risks for an individual trade decision, for example, may be within the units' risk tolerances, but taken together they may exceed the risk appetite of the firm as a whole. In a portfolio view, management considers potential events to understand how they can affect the enterprise. For example, costly phytosanitary treatments may render trade unprofitable, rejected

consignments due to faulty pest risk management could result in a reversal of fortunes from trade. Careful attention paid to phytosanitary treatments and measures may initially have a negative effect on the firm's cost of capital, but it could positively impact revenue streams for years to come.

Achievement of objectives

ERM is designed to achieve an entity's strategic, operational, reporting, and compliance objectives. Strategic objectives include the firm's high-level goals that are aligned with and support its mission. Operations objectives assure the effective and efficient use of the firm's resources. Objectives relating to reliability of reporting and compliance with laws and regulations are within the firm's control so ERM should be reasonably assured of meeting those objectives. Attention to objectives is what makes ERM enterprise wide.

Components of COSO's ERM framework

Fig. 17.3 provides an effective summary of the nuts and bolts of the COSO ERM framework. There are eight interrelated ERM components that define COSO's risk management process. The ERM process begins by understanding the

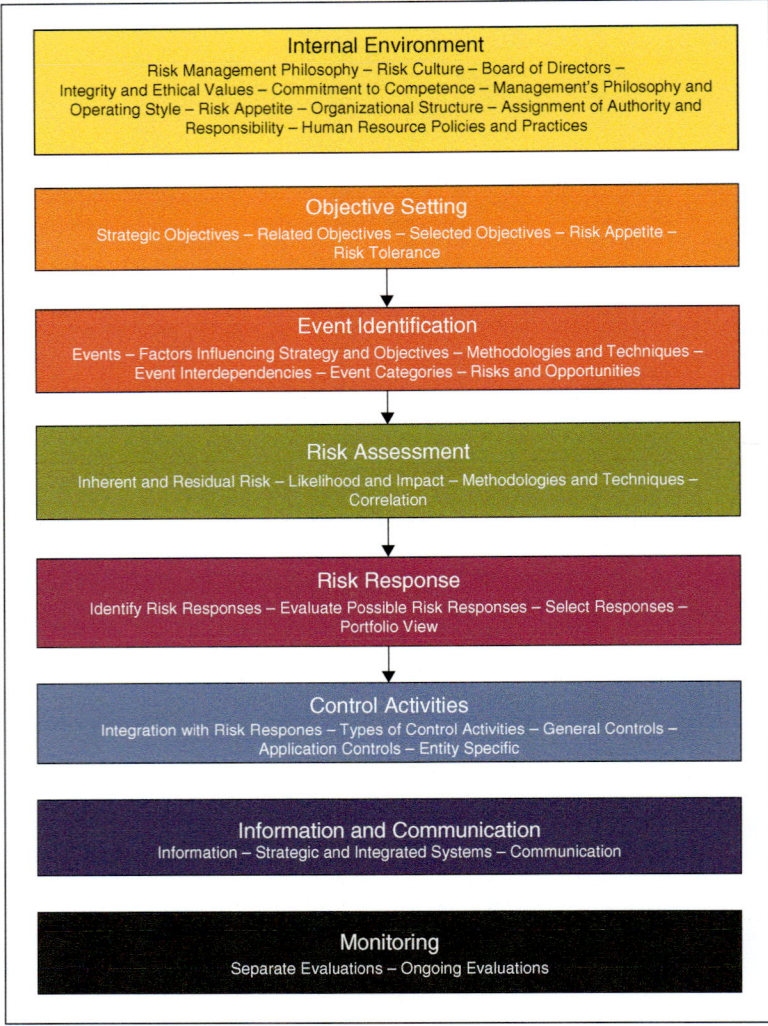

Fig. 17.3. COSO ERM components, reproduced with permission.

firm's internal environment. The internal environment comprises the firm's risk management philosophy, its risk appetite, and risk culture, among the other factors listed in the figure. This internal environment shapes and influences strategies and objectives, how business activities are structured, and how risks are identified, assessed, and managed.

Objectives must exist before they can be achieved. Firms face risks from a variety of external and internal sources. Which of these risks will be assessed and responded to depends critically on the firm's objectives. Internal and external events that can affect achievement of the firm's objectives must be identified. Then management identifies the risks of loss and opportunities for gain among them. New trade markets, pest risk assessments, and phytosanitary risk management decisions are three categories of events of special interest to private firms engaged in international plant commodity trade. These and other events with potential negative or positive impacts on the firm's objectives will require management's assessment and response. It is important that management consider the full scope of the organization when identifying events that could influence achievement of company objectives. Likewise, management must recognize the uncertainties that exist. For example, a company may not know if or how many pests will be identified with its product. Neither might it know if, how, or when phytosanitary measures will be required when it initiates a request for market access.

Firms undertake risk assessment to anticipate the extent to which potential events might impact the achievement of objectives. Both the likelihood and consequence of the risk need to be assessed using either qualitative or quantitative methods. The likelihood and consequence of risks that emerge from the identified events are analyzed to form a basis for determining how they should be managed.

The firm chooses whether or not to respond to the assessed risks. Management can accept, avoid, reduce, prevent, share, or internalize the assessed risk, according to the firm's risk tolerance and risk appetite. Costs and benefits can be expected to weigh heavily in the firm's response to risk. Unlike regulatory risk managers, the enterprise's risk managers will bring the expected likelihood and consequence within the firm's risk tolerances. A risk tolerance identifies the point at which a tolerable level of risk becomes unacceptable.

Control activities are the policies and procedures enterprise risk managers put into place to ensure the risk responses are carried out. With ERM, control activities occur at all levels and in all functions of the organization. They can include changes in business processes, production process changes, compliance changes, and so on. A single control activity could help achieve firm objectives in several categories. For example, a strategic objective of proactive management of phytosanitary pests, an operational objective to reduce pests to acceptable levels, and a compliance objective to meet phytosanitary regulatory requirements are all served by control activities that proactively seek to meet phytosanitary requirements.

Information is needed to identify, assess, and respond to risks. Businesses use their internally generated data and information about external events to manage enterprise risks. Effective risk communication requires that all personnel receive a clear message from top management that ERM responsibilities must be taken seriously. Each person must understand their own role in ERM, as well as how their role affects or is affected by the work of others. Firms need an effective means of communicating information upstream and they must establish effective communication with external parties. All of this information must flow in a form and timeframe that enables everyone to carry out their ERM responsibilities.

ERM decisions must be monitored to assure that the desired effects on objectives are being achieved. This is ordinarily done through ongoing monitoring activities or in separate evaluations. A firm's ERM approach will change over time. Risk responses that were once effective may become irrelevant. Control activities may become less effective, they may be replaced by more effective options, or no longer be performed. The firm's objectives may change. Management needs to determine whether existing ERM measures continue to be effective in the face of such change and to respond accordingly.

17.4. Summary and Look Forward

Here are five things to remember from this chapter.

1. Private enterprises engaged in international plant trade are wise to practice ERM.
2. The underlying premise of ERM is that every entity exists to provide value for its stakeholders and every entity faces uncertainty.
3. ERM encompasses the entire scope of the firm's activities, including enterprise-level activities, business unit activities, and new initiatives that have not yet found a home in the firm's structure.
4. The ISO 31000 and COSO ERM frameworks are most commonly adopted by private industry.
5. ERM decisions must be monitored to assure that the desired effects on objectives are being achieved.

Risk management strategies are the focus of the next chapter, which begins with a detailed consideration of the appropriate level of protection (ALOP) principle before it turns to the practical consideration of strategies that can be used to construct risk management options.

17.5. References

Allianz Global Corporate & Specialty (2019) *Allianz Risk Barometer Results Appendix 2019*. Munich, Germany, https://www.agcs.allianz.com/content/dam/onemarketing/agcs/agcs/reports/Allianz-Risk-Barometer-2019-Appendix.pdf.

Beasley, M.S., Branson, B.C. and Hancock, B.V. (2019) *The State of Risk Oversight: An Overview of Enterprise Risk Management Practices 10th anniversary edition*. North Carolina State University, Poole College of Management, ERM Initiative, https://erm.ncsu.edu/library/research-report/2019-the-state-of-risk-oversight-an-overview-of-erm-practices.

Breukers, A., Bremmer, J., Dijkxhoorn, Y. and Janssens, B. (2009) *Phytosanitary Risk Perception and Management Development of a Conceptual Framework*. LEI Wageningen UR, The Hague, http://edepot.wur.nl/14635.

CGMA (Chartered Global Management Accountant) (2019) *Enterprise Risk Management*. Available at: https://www.cgma.org/resources/tools/essential-tools/enterpise-risk-management.html (accessed June 14, 2019).

COSO (Committee of Sponsoring Organizations of the Treadway Commission) (2004) *Enterprise Risk Management: Integrated Framework*. COSO. Available at: coso.org (accessed February 3, 2017).

COSO (Committee of Sponsoring Organizations of the Treadway Commission) (2017) *Enterprise Risk Management: Aligning Risk with Strategy and Performance*. COSO. Available at: coso.org (accessed February 3, 2017).

ISO (International Organization for Standardization) (2009a) *ISO 31000:2009 Risk Management: Principles and Guidelines*. ISO. Available at: http://www.iso.org/iso/catalogue_detail?csnumber=43170 (accessed February, 2017).

ISO (International Organization for Standardization) (2009b) *ISO Guide 73:2009 Risk Management: Vocabulary*. Available at: http://www.iso.org/iso/catalogue_detail?csnumber=44651 (accessed February, 2017).

ISO (International Organization for Standardization) (2009c) *ISO/IEC 31010:2009, Risk Management: Risk Assessment Techniques*. Available at: http://www.iso.org/iso/catalogue_detail?csnumber=51073 (accessed February, 2017).

WEF (World Economic Forum) (2019) *The Global Risks Report 2019*, 14th edn. Geneva. Available at: http://www3.weforum.org/docs/WEF_Global_Risks_Report_2019.pdf (accessed April 17, 2020).

Part Three:

Risk Management Controls

Risk control is the method by which risk managers evaluate potential losses and take action to reduce or eliminate them. Phytosanitary risk managers are seeking to achieve the appropriate level of protection. Conversely, risk control is the method by which private sector risk managers evaluate potential gains in order to increase or create them. In either case, it is a technique that uses risk assessment findings to identify potential risk factors and measures that can be used to address them. These measures or risk management controls can be combined into risk management options.

The language of risk analysis is far from settled and the actions taken to manage or treat risks are sometimes called *risk responses*. The risk responses or risk management controls are often what people think of when they think of risk management. Thus, risk management is often synonymous with risk management controls in the minds of those unfamiliar with the breadth of the risk management task. Let us be clear: risk management is far, far more than simply identifying measures to respond to a risk. Nonetheless, identifying appropriate risk management measures in response to an assessed risk is an essential part of the risk management process. This is the tactical part of risk management.

This handbook's treatment of this tactical task begins with strategy. Chapter 18 offers a detailed discussion of the appropriate level of protection (ALOP) as a conceptual risk management strategy. Once this concept is developed,

the chapter reviews the systems approach before turning to some practical methodologies for formulating risk management options from control measures.

The tactical discussions begin in earnest in Chapter 19 with certification, an essential part of phytosanitary risk management, that codifies the import requirements for a commodity. Inspection is one of the more commonly used mitigation measures but when it is conducted haphazardly it has limited value at best. Chapter 20 explores the use of inspection then argues for a transition to risk-based sampling to increase the value added of inspections both domestically and globally.

Phytosanitary treatments, pest-free concepts, and irradiation are the topics of Chapters 21, 22, and 23 respectively. The chapter on treatments argues that the strength of risk management measures can and should be tailored to the available scientific information. High risk situations justify strong measures while low risk situations warrant commensurately light measures. A pest-free area is a least trade restrictive measure for managing risk to an acceptable level. Both importing and exporting country national plant protection organizations (NPPOs) stand to benefit from implementing pest-free areas. Before deciding to implement a pest-free area, NPPOs should determine the costs and benefits of establishing a program, and ensure they work with trading partners and their stakeholders. Irradiation may be one of the more poorly understood mitigation measures. It is different from other phytosanitary treatments for a variety of reasons that are considered in Chapter 23.

Post-harvest processing and handling measures are examined in Chapter 24. The primary objective of post-harvest processing and handling is to maintain the plant product's quality. The International Standards for Phytosanitary Measures (ISPMs) have had a great deal to say on this topic and that guidance is reviewed. Post-entry measures are the topic of Chapter 25. These are under-appreciated for their risk management value because risk managers are reluctant to allow potentially infested articles to enter, even when risk analysis supports the use of such measures. Even so, post-entry quarantine is a well-established and widely practiced post-entry measure used for propagative material. The role of these measures in research, analysis, and exhibition as well as in biocontrol is explored in the chapter.

Prohibition, the most trade restrictive measure and, unfortunately, too often a go-to measure for decision-makers less inclined to follow the risk approaches of the Sanitary and Phytosanitary Agreement (SPS Agreement) is the subject of Chapter 26. A dramatically different approach to risk management is the systems approach (Chapter 27). Systems approaches can comprise measures that reduce risk, measures intended to safeguard the commodity, and/or measures intended to reinforce the efficacy of other measures. A significant advantage of a system is that it can include redundant measures as a means to manage uncertainty and ensure risk is adequately managed. Producers are an integral part of implementing systems approaches and should be included in the process of developing any systems approaches that affect them.

18

Risk Management Strategies

Strategy without tactics is the slowest route to victory, tactics without strategy is the noise before defeat.

Sun Tsu, Ancient Chinese military strategist

18.1. Introduction

This chapter on risk management strategies has three themes. The ALOP is the first theme. The discussion of risk management strategies begins with a detailed consideration of the ALOP, a unifying concept conceived in the Sanitary and Phytosanitary (SPS) Agreement that links risk management solutions to assessed risks. The discussion begins with the SPS Agreement definition of the ALOP, which is then used to further develop the concept of the ALOP in order that alternative means of determining an ALOP can be considered. This discussion concludes that a consistent approach to the ALOP is more practical, if not more desirable, than a mythical monolithic expression of an ALOP. The risk management model presented in Chapters 13 and 14 is offered as an example of such a consistent approach.

The second theme is a more practical view of risk management strategy. It focuses on the systems approach and the most basic strategies for developing risk management solutions.

Theme three is tactical: how does a team of risk assessors and managers formulate specific risk management options? Here we borrow heavily from the work of other stewardship organizations with responsibilities for a public trust.

18.2. The ALOP

The World Trade Organizatioin (WTO) has proposed the ALOP as a means for competent authorities to translate sovereign governmental phytosanitary protection into risk-based targets for the plant industry. Annex A of the SPS Agreement defines the appropriate level of sanitary or phytosanitary protection as "The level of protection deemed appropriate by the member establishing a sanitary or phytosanitary measure

---

The World Trade Organization *Agreement* of 1994 refers to the appropriate level of protection (ALOP) and notes that some WTO members refer to this concept as the *acceptable level of risk*. In the phytosanitary risk management community of practice other terms are sometimes used. They include:

- negligible pest risk
- quarantine security
- insignificant risk
- no significant risk
- de minimus risk
- safe.

(Devorshak, 2012)

---

to protect human, animal or plant life or health within its territory."

There is very little guidance currently available on the nature of an ALOP or how to establish a phytosanitary ALOP. A scientific basis for an ALOP is not required, nor must it be expressed in quantitative terms. The SPS Agreement grants each member of the WTO the sovereign right to determine its ALOP. Wilson (n.d.) points out this is not an unfettered right; nations must take into account the objectives of minimizing negative trade effects (Article 5.4 of the SPS Agreement) and consistency in application (Article 5.5). Wilson says the goal of a government ALOP is often expressed in broad terms such as "conservative." He cites the Australia salmon case as an example, where Australia stated before the WTO dispute panel that its ALOP is a high or "very conservative" level of protection aimed at reducing risk to "very low levels ... while not based on a zero-risk approach."

The ALOP is critical to the SPS review and appeal process. It provides a unifying concept for the principles and practices embodied in the SPS Agreement (Wilson, n.d.). A phytosanitary ALOP expresses government policy, which is presumed to be reflective of community expectations with regard to protecting plants, agricultural industries, and the environment while reaping the benefits of international trade. The ALOP concept strives to strike a delicate balance. On the one hand it allows members to adopt levels of protection in accordance with their appetite for and views toward risk. On the other hand, it protects members against trade discrimination, especially through the inconsistent application of import protection requirements.

Wilson points out how the ALOP forces members to decide the balance they want to strike between the risk of pest or disease incursions and the benefits from all trade. Conservative quarantine policy favors reducing pest risk at the cost of trade benefits. Less conservative quarantine policy favors access to products of other countries for both consumption and investment in improved production at the cost of an increase in the likelihood of pest and disease incursions, with their associated costs to industry, the community, and the environment.

The fact of the matter is the ALOP is a critically important concept that is rarely if ever operationalized and articulated. A cynic might suggest the ALOP is a psychological fiction, an idea that all can believe in but which no one can articulate. A pragmatist might suggest the ALOP is, while difficult to define, being clarified in practical terms in cases before the WTO. Wilson has suggested that plant health objectives, the corollary to food safety objectives, should lead to the development of practical links between ALOP and measures facilitating trade negotiations between members.

18.3. The ALOP Concept

Risk was conceptually represented in Chapter 2 by the simple equation, "*Risk = Consequence × Probability.*" Continuing with this conceptual approach, imagine, for convenience of exposition, that phytosanitary risk can be represented by a single numerical value, R_i. In fact, risk estimates are usually far more complex than this. For a given level of risk, R_i, consequence and probability are inversely related as shown in the upper left diagram of Figure 18.1. This diagram shows four iso-risk lines. The level of risk is the same at any combination of consequence and probability along the iso-risk curve. Risk increases as the curves move up and to the right. Every point in the consequence–probability space lies on an iso-risk line, so a limited number are shown to represent the concept.

A risk level of R_2 was arbitrarily chosen to represent a conceptual ALOP. Another nation might choose a higher level of risk, e.g., R_3 or R_4 for its ALOP if they are willing to accept greater phytosanitary risk in exchange for greater benefits from international trade or could choose a more stringent ALOP at R_1 if less phytosanitary risk is preferred. This choice will embody the NPPO's decision on how to minimize negative trade effects. If the ALOP represents an acceptable level of risk, then any area in the consequence–probability space above and to the right of the ALOP iso-risk represents an unacceptable level of risk. Any area on or below the ALOP iso-risk represents an acceptable level of risk. For now, let us simply posit that the NPPO identifies an ALOP that fairly represents government policy preferences for balancing plant protection and the benefits of international trade and it is represented by R_2. The shaded area of the upper right

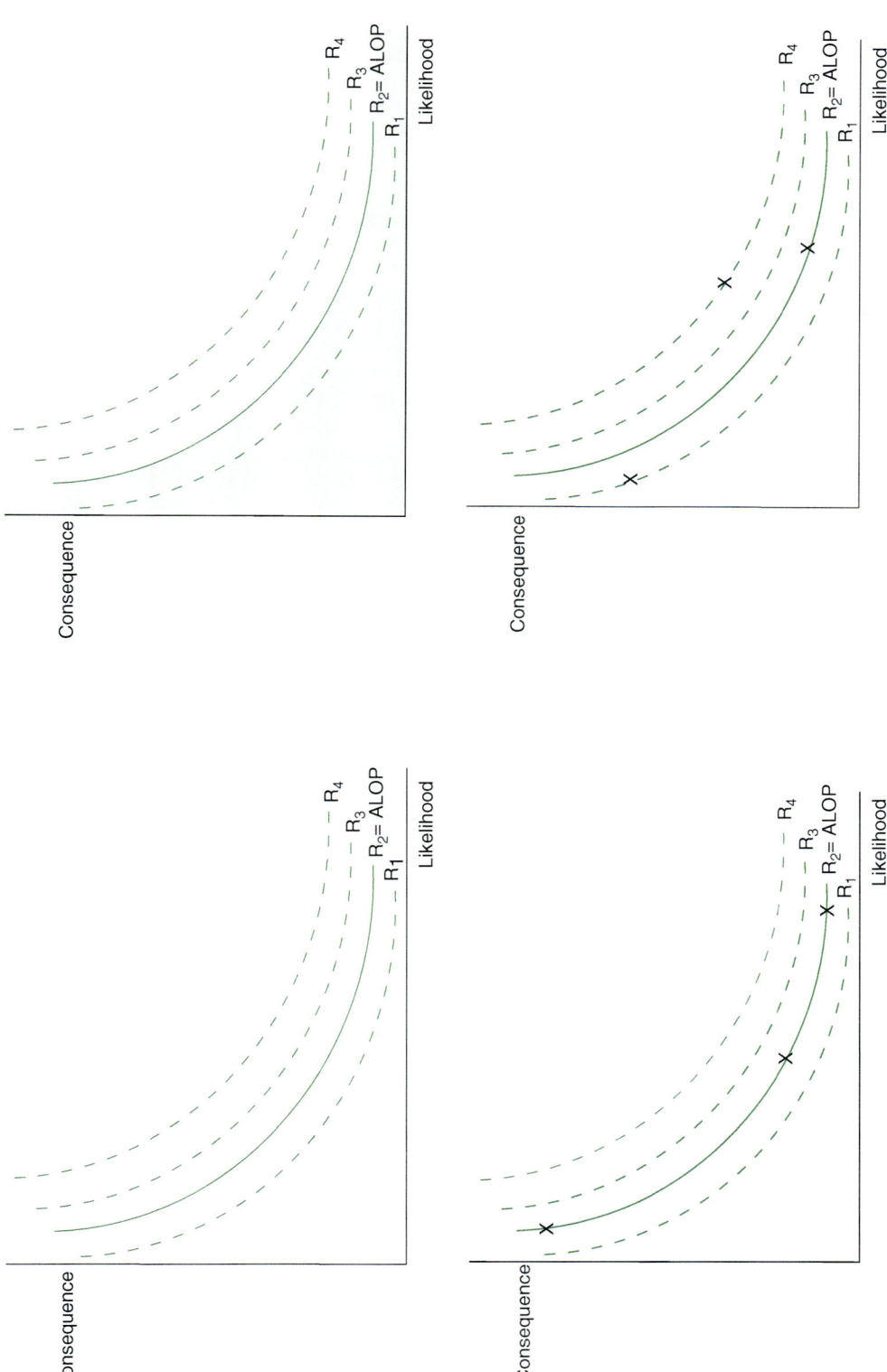

Fig. 18.1. Conceptual explanation of the ALOP.

diagram of the figure shows all those levels of phytosanitary risk that could result from a risk management decision that would be acceptable to the NPPO. Risk management options that resulted in risk levels found in the unshaded portion of this panel would result in an unacceptable level of risk, because they exceed the ALOP.

The lower left diagram of the figure shows three risk management options, represented by Xs, that each achieve the same R_2 level of risk. Each of the risk management options achieves an acceptable level of risk. Conceptually, the SPS Agreement encourages members to identify an ALOP and then to manage all phytosanitary risks to that level. This diagram represents the consistency principle. The lower right diagram depicts a situation which the establishment of an ALOP is intended to avoid. Notice the risk management options, depicted by Xs, all result in different levels of risk. This is a situation that would ideally be avoided. The SPS Agreement says, "With the objective of achieving consistency in the application of the concept of appropriate level of sanitary or phytosanitary protection against risks to human life or health, or to animal and plant life or health, each member shall avoid arbitrary or unjustifiable distinctions in the levels it considers to be appropriate in different situations, if such distinctions result in discrimination or a disguised restriction on international trade."

In a perfect world, risk management options should be formulated so that they bring risks lying outside the line into alignment with the ALOP, by reducing the likelihood of entry, establishment and spread of the pest or disease or by reducing the expected consequences.

> Chapter 2 draws a distinction between an acceptable level of risk and a tolerable level of risk. It is the authors' belief that it might be more accurate to equate the ALOP with a tolerable level of risk. We have endeavored to avoid such language to avoid a potentially confusing contradiction with the SPS Agreement. Nonetheless, the distinction drawn in Chapter 2 suggests a tolerable level of risk is achieved and ALOPs vary among nations because they have different tolerance levels.

A risk management option that would bring the residual risk well below and to the left of the ALOP is likely to be more trade restrictive than necessary.

It is very difficult for any member to explicitly specify its ALOP. As a practical matter, a member's ALOP is often left implicit, e.g. a reasonable certainty of no harm (Havelaar *et al.*, 2004). It is easy to discuss an ALOP in a conceptual framework, it is quite a different matter to be explicit about it. Nonetheless, the ALOP remains a crucial concept in phytosanitary risk management and in trade disputes as well. Effective implementation of an ALOP requires a more explicit articulation of the goal of risk management. The next section considers some different ways to identify an ALOP.

Fig. 18.2 presents a more realistic depiction of a functional ALOP. Levels of risk are less precisely defined in qualitative terms. For the purposes of exposition, imagine that the ALOP is a somewhat flexible concept that aims at reducing pest risks to the area between a low level and a negligible level of risk. Instead of a precise level of risk, the ALOP is defined as a cloud of points that lies between two roughly defined levels of risk as shown in the figure.

The Xs in the figure represent risk management options that achieve the ALOP. Point B represents a situation where the ALOP is exceeded. This is an unacceptable risk. Point A represents a risk management option that is more stringent than the ALOP requires. Such a phytosanitary measure may well be challenged on the basis of discrimination.

Fig. 18.2 suggests that the ALOP is not a precise bright line so much as an imprecise and variable range of acceptable outcomes. The ALOP cloud allows for some reasonable distinctions in risk management solutions that reflect similar but distinct levels of judgment in decision-making. The ALOP is not a static target.

18.4. Identifying an ALOP

This section assumes the somewhat awkward position of describing the current situation while advocating for change to something not yet implemented or well-understood.

18.4.1. Ideal practice

The WTO is quite clear in its intent that the ALOP not be the ad hoc result of a formulated

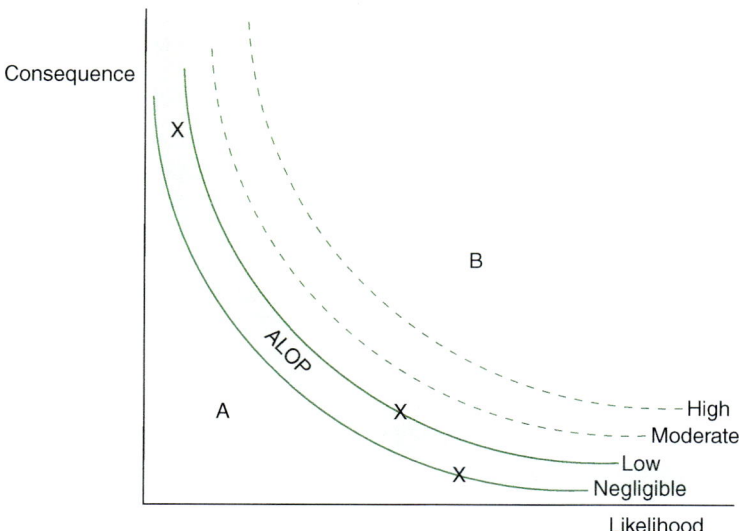

Fig. 18.2. Conceptual depiction of a less precise representation of an ALOP.

risk management option. Conceptually, the ALOP drives the identification of risk management options. Given an assessed risk, the risk management objectives, and a policy-determined ALOP, risk management measures are combined into risk management options intended to achieve the desired level of risk reduction, i.e., the ALOP. This stands in contrast to backing into an ALOP by formulating risk management options, selecting one and letting the resulting level of protection be defined as the ALOP. WTO (2000) says, "The determination of the ALOP is an element in the decision-making process which logically precedes the selection and use of one or more sanitary or phytosanitary measures."

In order to be found to have acted inconsistently under Article 5.5, the Appellate Body (Australia–salmon) considered that a member must be found to have met all of the following three elements:

• the member adopts different ALOPs in several different situations
• those levels of protection exhibit "arbitrary or unjustifiable" differences in their treatment of different situations
• the measures embodying or implementing those differences result in "discrimination or a disguised restriction on international trade."

(WT/DS18/AB/R, para. 140)

The WTO issued its *Guidelines to further the practical implementation of Article 5.5* of the SPS Agreement in 2000 to support the practical implementation of the ALOP. These guidelines focus on two elements of Article 5.5: (1) the objective of achieving consistency in the application of the concept of the ALOP; and (2) the obligation to avoid arbitrary or unjustifiable distinctions in the levels considered appropriate if such distinctions result in discrimination or a disguised restriction on international trade. Several suggestions for members applying the ALOP concept are summarized below.

1. Indicate the level of protection considered appropriate with respect to risks to plant life or health in a sufficiently clear manner.
2. Consider whether there is a difference between the level of protection under consideration and levels already determined by the Member in different situations.
    a. Are these differences arbitrary or unjustifiable?
        i. If so, do they result in discrimination or a disguised restriction on international trade?
3. If the people who decide an ALOP and the people who make a risk management decision are two different groups, clear and effective communication and information flows between these groups is essential.

4. Compare a proposed decision on the level of protection in a particular situation with the level previously used in similar situations with regard to plant life or health.

5. Examine relevant international standards, guidelines or recommendations, or decisions taken by other members facing similar risks and situations.

6. Seek expert advice to determine a new or modified ALOP.

Suggestions for implementing an ALOP are summarized below.

7. Establish clear and effective communication and information flows within and between the authorities responsible for determining an ALOP, and those responsible for selecting and implementing phytosanitary measures designed to achieve the ALOP.

8. Establish a common approach to assessing risks and evaluating risk management options that might be applied to achieve the desired level of protection.

9. Compare any proposed risk management option intended to achieve the ALOP in a particular situation to other phytosanitary measures it has taken with regard to comparable plant life or health risks.

10. Routinely review its ALOP.

11. Consider whether an existing relevant international standard guideline would achieve the ALOP.

12. Examine measures applied by other members facing similar risks and situations.

13. Consider seeking expert advice on the selection and implementation of phytosanitary measures to achieve its ALOP.

These concepts overlap and are more useful for understanding how we arrive at the ALOP than they are for defining the ALOP in advance. Item 1 above advises the member to publish the level of protection and to let interested parties know what it is. Subsequent items suggest some different methods for arriving at an ALOP in a specific situation. In neither the sum total of all these items nor in any one individual one will you find an example of what an ALOP looks like or how to express it. Everyone agrees on its centrality to phytosanitary risk management but no one in authority appears to know exactly what this phytosanitary ALOP is. This has led to some ambiguity in how it is used.

18.4.2. Typical practice

The guiding principle for risk management's ALOP should be to manage identified risk to an acceptable level that can be justified, that is not more trade restrictive than necessary and that follows the principle of minimal impact. Phytosanitary treatments have a long history of development around the assumptions of a worst-case scenario and one-size-fits-all design for a single measure with high-efficacy based on mortality that results in maximum protection. Maximum protection can be and often is an over-reaction to the pest risk identified by PRA. The strength of risk management measures can and should be tailored to the available scientific information; high risk situations justify strong measures while low risk situations warrant commensurately light measures.

The phytosanitary risk management community is still transitioning from the historical paradigm of striving for maximum protection to the measured concepts of the SPS Agreement. In actual practice, the ALOP is often driven by the risk management measures that comprise the risk management option. In other words, the ALOP is what results when risk managers arrive at a recommended option for a given situation. This is quite the opposite of what was intended by the WTO, but for good reason.

The International Standards for Phytosanitary Measures (ISPMs) published by the International Plant Protection Convention (IPPC) do not address the level of protection provided by the various standards that have been developed. Consequently, it is impossible to say whether a member that follows all the international standards has a consistent identification of an ALOP or not. Given the cumulative vagueness surrounding the ALOP, it comes as no surprise that many ALOPs are ad hoc arguments for the desired level of protection in a given situation rather than a coherent universal statement of member government policy.

The range of feasible phytosanitary measures is not extensive for many situations. This means risk managers usually have limited choices for managing the risk. Even if there was an articulated "bright line" ALOP, many risk management options would be unable to achieve it. In such a circumstance, the focus on defining an ALOP is a distraction. Risk assessors and managers alike struggle with understanding how to compare and combine risk management measures with different metrics, e.g., treatment mortality and inspection efficacy. When combined with the multitude of issues associated with qualitative descriptions of risk levels this encourages the slide toward ad hoc and implicit definitions of an ALOP. The pervasiveness of multiple metrics also contributes to a natural tendency to rely on pest prevalence or the reduction in pest prevalence as the common denominator that should be guiding the language and conceptual understanding of risk reduction and levels of protection, but it has neither been seen nor generally adopted as the community of practice clings to the various metrics of the past.

Probit 9 is a measure of mortality that continues to enjoy some popularity as a surrogate measure of an ALOP. Risk managers are concerned about survivors among a pest population. The only way Probit 9 is truly meaningful to risk managers is if the initial infestation rate is known. Using Probit 9 as a "pretty sure we killed them all" standard is inconsistent with member's SPS obligations.

In short, ALOP is typically revealed by what a member state does. In the absence of a detailed statement from a member describing its ALOP in precise terms Wilson (n.d.) says a guide to a member's ALOP may be found in the protection afforded by the body of quarantine policies and practices developed and adopted over the years.

18.4.3. Options

The first obvious choice for selecting an ALOP is to simply adopt international standards. These would include the risk management standards found in the ISPMs of the IPPC as well as regional standards for phytosanitary measures for regional NPPO organizations. There are at least two serious limitations to such an approach. First, the international standards do not specify an ALOP and there is no known way to get from adoption of international standards to a clearly articulated ALOP. Second, only a limited number of international standards exist and phytosanitary risk managers repeatedly find themselves making decisions on issues for which international standards do not exist.

A second obvious choice is to do what other members do by adopting standards already in use by one or more member state. This method has at least the same two serious limitations mentioned above. First, a member that adopts the standards of other members cedes the right to adopt its own ALOP and it ends up with a mongrel ALOP that defies precise definition. Some mimicked members may adopt a stringent approach while others may be more generous. An amalgam of measures from the various members is unlikely to cohere into a unique or identifiable ALOP. Second, only a limited number of pest risks have been addressed by the various member states and it is likely that unique situations will arise that have not yet been addressed by another member.

> In EC-Hormones the Appellate Body found that with respect to the ALOP the goal of consistency does not establish a legal obligation of consistency. The goal of consistency "…is not absolute or perfect consistency, since governments establish their appropriate levels of protection frequently on an ad hoc basis and over time, as different risks present themselves at different times. It is only arbitrary or unjustifiable inconsistencies that are to be avoided."
>
> (WT/DS26/AB/R, para 213)

Cost–benefit analysis can be used to estimate the net benefits of various phytosanitary risk management responses. Choosing the alternative that maximizes relevant net benefits provides a clear and consistent decision rule for judging the economic efficiency of various risk options.

Cost–benefit analysis and other economic tools have been almost absent in the past and are only just being recognized as legitimate and useful. There would appear to be a clear role for cost–benefit analysis in identifying an ALOP but this is not the current reality. Nor is it likely that maximizing net benefits would produce a consistent ALOP.

Advice may be sought from recognized, qualified experts. This advice will solicit comments on any unjustifiable distinctions in levels of protection, potential discriminatory trade effects, or other aspects related to consistency in the application of an ALOP. International standard-setting organizations may be helpful in identifying appropriate experts.

The ALOP, the Food Safety Objective (FSO) and their associated metrics have been proposed by the WTO and Codex Alimentarius as a means for competent authorities to ultimately translate governmental public health policy regarding food safety into risk-based targets for the food industry (Gkogka *et al.*, 2013). Havelaar *et al.* (2004) have argued that while an ALOP is considered an important target for public health policy, it is not useful for the actual implementation of safety controls throughout the food chain. The FSO has been introduced by the International Commission on Microbiological Specifications for Foods (ICMSF) as a methodology for bridging this gap. An FSO is defined by ICMSF as "The maximum frequency and/or concentration of a (microbiological) hazard in a food at the time of consumption that provides the ALOP."

Wilson (n.d.) has argued that plant protection also needs a mechanism to link a measure with the ALOP as the FSO does and he reports that work has commenced on attempts to provide that link through the use of plant health objectives (PHOs). A PHO provides a reason for a phytosanitary measure to address a hazard and describes the expected level of protection from the hazard which would result from application of the nominated measure. Such an objective can facilitate consistent decision-making across commodities if it is practical and technically feasible. Ideally, it will be expressed quantitatively if possible. At the current time it is expected that the objective can only describe the level of protection against a hazard the measure is expected to achieve qualitatively. For example, the standard might be that there would be a very low likelihood of exposure of a plant to the hazard. Though promising, this idea remains in the conceptual stage and is not yet a practical solution.

18.4.4. A way forward

The ALOP is a long way from becoming an explicit policy proclamation. Pest risk is not a monolithic thing. It is part probability of infestation, establishment, and introduction. Its consequences are economic, environmental, and other. The risk–benefit tradeoffs associated with a particular plant–pest pair are extremely variable. A pest risk embodies all of these aspects of

a risk and more. The iso-risk curves above are concepts not realities. An ALOP depends very directly on the decision context and as decision contexts change so will the ALOP.

As the community of practice moves away from maximum protection to a more balanced and reasoned ALOP there is a much stronger case to be made for consistency in approach rather than consistency in outcome. A consistent approach to determining an ALOP should lead to roughly equivalent outcomes for most plant–pest pairs. The WTO (2000) suggests a member should establish common approaches or consistent procedures for use by the authorities assessing risks and evaluating the measures which might be applied to achieve the desired levels of protection. A common approach for risks to plant life or health can be depended on to produce a reasonable and rational ALOP for any member.

The most promising ALOP practice is to focus on the common approach as advocated by the WTO. Focusing on the criteria used for deciding on the ALOP in any given situation and consistency in applying those criteria in other situations is needed. Conceptually, the result will be a history of measures within a range that defines the ALOP as a band with outliers that would be considered unacceptably high or low, as seen in Fig. 18.2. Clearly identified, comprehensive and consistent procedures for assessing risks and for evaluating measures to reduce risk to acceptable levels will assist a member to be more consistent in the application of its ALOP. That is what this handbook offers. Part Two offered a detailed description of a risk management model. In particular Chapters 13 and 14 offer a detailed description of a common and consistent approach to risk management.

A consistent approach can be demonstrated with a simple risk matrix, as seen in Fig. 18.3. The rows of the matrix identify qualitative levels of likelihood for a pest risk. The columns identify qualitative consequence categories. Like levels of risk take the same level of shading. This matrix lends itself to identifying an ALOP consistent with the concepts presented around Fig. 18.2. A member can decide to manage all risks to a certain level. For example, a member might decide to manage all pest risks to the level of the third darkest shade of green. The combinations of row and column headings that result in such a

Likelihood	Consequence						
	None	Negligible	Very Low	Low	Moderate	High	Extreme
High							
Moderate							
Low							
Negligible							
None							

Fig. 18.3. Sample risk matrix that can be used to identify an ALOP.

cell vary, allowing for some differential but consistent treatment of risks. If such a member pushed a risk into the lightest shade of green band this might be deemed an unjustifiable inconsistency if it requires extreme measures or effort to achieve this level of risk management.

18.5. Basic Risk Management Strategy

Effective risk management, both phytosanitary and enterprise, requires an efficacious risk management strategy. That is to say, risk management is more than a specific control measure or set of control measures. Strategy is used to assemble the available measures into a risk management option.

Integrated pest management (IPM) is an example of a grand risk management strategy, as opposed to a specific strategy. FAO says IPM means the careful consideration of all available pest control techniques and subsequent integration of appropriate measures that discourage the development of pest populations and keep pesticides and other interventions to levels that are economically justified and reduce or minimize risks to human health and the environment. IPM emphasizes the growth of a healthy crop with the least possible disruption to agro-ecosystems and encourages natural pest control mechanisms.

(FAO (2020)*"NSP - Integrated Pest Management"*, accessed November 20, 2019)

The most basic strategy for a risk manager is to avoid or minimize the risk of undesirable outcomes. Recall the simple risk equation, $Risk = Consequence \times Probability$ and the two risk management roles introduced in Chapter 2. Risk-taking risk

managers are trying to intensify the consequence and increase the probability of desirable outcomes. Risk-reducing risk managers are trying to reduce the consequences and decrease the probability of undesirable outcomes. These basic strategies can be further parsed as shown in Table 18.1

Enterprise risk managers for firms promoting international plant product trade face an opportunity for an uncertain gain and the risk manager must decide whether or not to assume the risk of pursuing that trade. This is a risk-taking strategy that focuses on increasing the magnitude and likelihood of an uncertain gain. Risk management strategies can be refined by identifying different focal points for the strategy. Risk-takers can cause an opportunity for uncertain gain to come into being by pursuing new trade opportunities, through a risk creation strategy. This could mean new plant products sold to new customers, new plant products sold to existing customers or existing plant products sold to new customers.

A risk enhancement strategy means taking measures to increase the probability that new international plant trade will produce positive outcomes for the firms engaged in trade. Risk

Table 18.1. Risk management strategies for risk-taking and risk reducing along with the strategy targets.

Risk management strategy		
Strategy focal point	Risk-taking	Risk-reducing
Existence of risk	Creation	Avoidance
Consequence of risk	Enhancement	Prevention
Probability of risk	Exploitation	Mitigation
Burden of risk	Sharing	Transfer
Inevitability of risk	Ignoring	Retention

exploitation is a strategy that focuses on the magnitude of the consequences of the proposed trade by maximizing the beneficial consequences from trade. Risk sharing is a strategy that focuses on the burden of the risk or the distribution of its benefits. Risk sharing seeks a partner to manage the trade opportunity. A final risk-taking strategy would be to choose to ignore and not pursue plant product trade opportunities that exist.

Phytosanitary risk managers may also function as risk-takers when they decide to allow trade without any additional risk management measures. Their primary role in international plant trade is most likely to be as risk reducing risk managers. There is a corollary risk reducing strategy for each focal point.

A risk avoidance strategy is most preferred by risk-reducing risk managers because it eliminates the risk. Prohibiting a product's entry into the importing country is an example of risk avoidance. Risk prevention strategies reduce the likelihood of adverse consequences, for example through treatment, risk-based sampling, inspection, and examination. Risk mitigation strategies reduce the magnitude of a risk by reducing the impact of the consequences. Containment and eradication measures are examples of mitigation strategies. Risk transfer is a burden of risk strategy for identifying stakeholders better able to manage the risk or finding a way to share a risk among many stakeholders. Insurance against pest infestation would be an example. Risk retention generally refers to the situation where stakeholders are forced to live with an unacceptable and intolerable level of risk. In such cases, monitoring the status of the risk may be the only viable response to the risk.

18.6. A Systems Approach

A common phytosanitary risk management strategy in the past has been to identify a single measure to meet the appropriate level of phytosanitary protection. A new general strategy pursued with increasing frequency by phytosanitary risk managers is the systems approach to risk management (see Chapters 8 and 27, this volume, for details). A good systems strategy is robust, resilient, and redundant. Robustness is the ability of a system to continue to function across

> Oyarzabal (2015) reports that more and more studies support the concept that the use of many different barriers or "hurdles" may be a more efficient way to control biological hazards in foods. The latest regulations by U.S. FDA recognize that more than one control may be necessary to ensure safety.

a wide range of operational conditions, with minimal damage, alteration or loss of functionality, and to fail gracefully outside of that range. The wider the range of conditions, the more robust the system should be. Resiliency is the ability to avoid, minimize, withstand, and recover from the effects of adversity, whether natural or manmade, under all circumstances of use. Redundancy is the duplication of critical components of a system with the intention of increasing reliability of any system, usually in the case of a backup or fail-safe.

A systems approach recognizes that it is rare for a single risk management measure to meet these conditions or to be perfectly effective in reducing a risk. Consequently, it requires looking at the whole of a complex system and considering opportunities for effective risk management wherever they occur in the system. Thus, a systems approach adopts a production-to-disposition approach to risk management. If there are effective risk management measures that can be adopted in the country of origin during production as well as measures that can be effective at the point of entry or end use, all of these measures may be combined to achieve the desired level of phytosanitary protection.

The real advantage of this strategy is that it has the potential to reduce the risk to the importing country, resulting in increased trade. A systems approach has the potential to unite the risk management interests of risk-taking firms and risk-reducing NPPOs. In addition to reducing

> APHIS suggests that the complexity of a systems approach will depend on the:
> - level of risk involved
> - cost, feasibility, and efficacy of measures
> - suitability of any given management option for managing that risk
> - available information for the pest(s) and associated commodity
> - level of uncertainty
> - ALOP.
>
> (USDA/APHIS/PPQ, 2012)

risks related to targeted pest species, the multiple measures employed with this strategy can mitigate the risk of hitchhikers or other contaminating pests not normally associated with the commodity as well as pests that may not have been accounted for in the pest risk assessment process.

System approaches vary by the number and types of measures that comprise them. These can vary in complexity from simple combinations of independent measures, like low pest prevalence and fumigation to complex control point systems.

18.7. Phytosanitary Risk Management Option Formulation Strategies

Risk management strategy gets tactical in the formulation task. Formulation begins where it begins. Sometimes risk management options appear before the problem is even identified much less agreed upon, especially when risk managers have their "go-to" strategies. Options can and do appear at many points in the risk management process. All of these ideas should be preserved and considered, but every risk management process needs at least one and likely more periods of intensive activity focused on devising measures and building risk management options that meet risk manager's objectives. That is what is meant by formulation.

Only a fool would insist on trying to impose order on such a creative process. Only a bigger fool would attempt such a process without structuring it in some way. A formulation strategy is a disciplined way to produce one or more specific risk management options. Its discipline derives from the structure provided by a more or less orderly sequence of activities. A strategy usually consists of a set of tactics or conditional decisions that shape and guide the development of options. The one golden rule of risk management option formulation is that people need to spend time together doing it. At a minimum, this will include both risk assessors and risk managers. Ideally, input or feedback opportunities would be extended to stakeholders in this part of the risk management process.

A logical first step in formulating risk management options is to identify risk management measures that can be useful for solving the problem at hand. This should include control and critical control points. There are several very practical ways to approach this process. Brainstorming by the risk analysis team may be the most logical way to begin to identify measures that could be used to build a risk management option.

Has anyone else worked on this pest and plant product before? Did they produce a report or document their effort in some other way? If so, begin there. That is your reading assignment—do your homework. If no one has studied your plant–pest pair before, has anyone considered this pest for another plant product? Have they considered other pests for your plant product? Read these reports to see how they managed the risks. What measures were considered? Include them in your initial considerations.

Ask people. Risk management options come from people. Ask people how to manage your risks. Start by asking those closest to the investigation, the risk analysis team. If these people are the insiders, everyone else is an outsider and it can be helpful to get outsiders' viewpoints. Involve stakeholders directly. Ask them how they would manage the risks. Ask the people who are affected by the potential losses and those interested in the potential gains. What measures would they like to see? Which ones do they oppose?

Perhaps someone has already thought systematically about the risk and how to manage it. Look for checklists of measures, like those in Tables 14.1 and 14.2. They can also be found in the documents of the IPPC and of some NPPOs. The APHIS *Guidelines for plant pest risk assessment of imported fruit & vegetable commodities* (USDA/APHIS/PPQ 2012) provides an excellent example of checklists of measures.

Lists of measures are not sufficient for identifying risk management options, especially options developed using a systems approach. There are a number of tried and true methods for building risk management options from measures. Several of them are summarized below.

18.7.1. Control and critical control points in a systems approach

Identify control points in the production chain and select measures that can be applied at these points. Fig. 2.2 (see p. 24) depicts a simple production chain. Control and critical control points are points where risk management measures can be applied in the production chain. It can sometimes be effective to analyze control and critical control points in a systems approach to estimate the efficacy of a risk management option in a with condition scenario. These controls are expected to have a reducing effect on the risk in a system. Critical control points are well defined and their level of efficacy can often be measured, quantified, monitored, and verified. Critical control points are frequently points where controls must be applied in order to reduce risk sufficiently.

By contrast, control points are measures with perhaps an undefined level of efficacy. Field sanitation is an example of a control point. Removing fallen fruit from orchards to manage fruit fly risk would reduce risk, but the exact level of efficacy would be difficult to measure. A critical control point system is one in which most or all of the specific points in the production chain can be well defined, the hazards and mitigations can be measured, each point can be controlled, and the efficacy of each mitigation step can be verified and documented (USDA/APHIS/PPQ, 2012).

18.7.2. Objectives, measures and plans

This may be the most effective and efficient risk management option formulation strategy. It is simple, logical and it produces risk management options. Risk management *objectives* are the cornerstone of a good risk management option formulation process. Begin this method by identifying candidate measures for each objective identified in the risk management process. This simultaneously assures that proper attention is given to achieving objectives in the risk management process and that the risk analysis team thinks carefully about how to manage the risks present. Consider the conceptual example introduced in Table 18.2. Imagine there are three risk management objectives, A, B, and C, and ten different measures that could be used to meet these objectives.

Notice there are multiple measures for each objective and some measures contribute

Table 18.2. Measures identified to achieve specific objectives.

Measure	Objective		
	A	B	C
1	X	X	X
2	X		
3	X		X
4		X	X
5		X	
6		X	
7		X	
8			X
9	X		X
10			X

to multiple objectives. The team's job, in a system approach, is to mix and match the measures, combining them into as many viable risk management options as possible, almost in a "tinker toy" approach. There will be some rules imposed by the physical universe. Some measures may be dependent upon each other. Maybe measure 2 requires measure 9 to be effective while measure 9 may or may not need 2 to function effectively. Measures 2 and 3 may be mutually exclusive because they need to occupy the same physical space and obviously cannot do so simultaneously. Measures 6 and 7 may be redundant, accomplishing exactly the same thing to the same or varying extents in different ways.

With this matrix and these kinds of logical rules in mind the team can begin to build risk management options in a system approach. Measure 1 looks attractive; it contributes to all three objectives. The team may construct a risk management option comprising measures 1, 2, 4, 7, 10. It is easy to imagine how a great many such risk management options can be constructed.

It is not uncommon for objectives to contradict one another at times. Thus, it may be necessary to develop a wide variety of risk management options. It is not essential that each option address every objective, but every objective needs to be addressed by at least one option. Any strategy that has a significant organic element like this does should take care to document the formulation process and rationale. It is important to capture the logic for the options that are formulated as well as the logic for not building others.

18.7.3. All possible combinations of measures

The most comprehensive way to develop risk management options is to make every possible combination of measures identified a separate option. If the team has been comprehensive in its efforts to identify measures then it can take comfort in knowing the best option is somewhere among the combinations generated. Assuming the team applies the rules of dependence and mutual exclusiveness, all the logically viable options will be generated by this mechanistic approach.

When there is a small number of measures, this approach is straightforward and can be completed with pencil and paper. For example, with only two measures, 1 and 2, there are only three possible combinations, or risk management options:

- option 1 = measure 1
- option 2 = measure 2
- option 3 = measures 1 + 2.

Add a third measure, 3, and the total number of options increases from three to seven. The number of possible combinations increases exponentially as the number of measures increases. The governing formula is $N = 2^M - 1$, where M is the number of measures and N is the number of combinations possible. The ten measures shown in Table 18.2 yield 1,023 possible risk management options. (The actual number of viable options will be reduced depending on the nature of the dependencies and

IWR-Plan Decision Support Software was developed specifically to assist natural resources planners in formulating and evaluating all possible combinations of ecosystem restoration plans. Its input data include the measures identified, an estimate of the output of each measure, and an estimate of the cost of each measure. This software tool relies principally on the use of cost-effectiveness and incremental cost analysis rules to screen the results of a formulation effort from a set of potentially thousands down to a much smaller final array of plans that are called "best buy" plans. It is conceptually adaptable to phytosanitary risk management and it is available at http://crbweb01.cdm.com/IWRPlan/default.htm (accessed 20 November 2019).

mutual exclusivities. It is common practice to consider each variation of a measure, based on attributes, a separate measure when using this approach.) Twenty measures can be combined into over a million plans. Computer technology can provide risk managers with the ability to identify and even screen these large numbers of potential options.

18.7.4. Just do it (organic formulation)

Building plans tends to be a mix of mechanistic and organic activities. Mechanistic plan building follows an algorithm or a set of logic rules that automatically leads to the generation of plans. The all possible combinations strategy above is the best example of this. Organic formulation is old school, you make it up as you go. Basically, the team enters a room and starts combining measures into plans. There is no algorithm, no structure, and no software. It only takes knowledgeable people gathered in one place and a few ideas to formulate some risk manamement options.

Organic formulation draws on the experience, knowledge, wisdom, insight, inventiveness, and synergy of the risk analysis team. Risk management options emerge through a collaborative work process that is difficult to typify. Judgment is exercised throughout an organic process, as opposed to codified in a mechanistic process. Organic formulation is a fluid, growing, and changing process. The only real requirement is a genuine effort to identify substantially different combinations of measures that contribute to the achievement of the risk management objectives. If there is any such thing as a traditional option formulation approach, this may be it. People show up and muddle through it.

18.7.5. Jagger–Richards approach

The Rolling Stones were a cover band until their manager confined Mick Jagger and Keith Richards to a room, telling them not to come out until they had written a song. They did. The rest is musical history. This is not a bad strategy to emulate. Pair each member of your risk analysis team with another person. Give them a space and as much of a part of the day as they need to come up with a plan. If you have eight team members, you'll have four risk management options by the end of the day. Change partners and repeat the process until you have a sufficient variety of options to choose from.

18.7.6. Cornerstone strategy

This strategy is useful in those cases where there is a single most important measure in a risk management option that everyone agrees is essential to a successful solution; one critical control point for example. That measure is the cornerstone. Alternative options are then built upon this cornerstone by adding various combinations of measures to the cornerstone measure.

18.7.7. The ideal scenario

This strategy begins by envisioning an ideal future outcome for the proposed trade issue. What does complete success look like? If a consensus vision of this success can be articulated, then the team's formulation task is clearly focused. What has to happen to make this outcome a reality? What are the different ways one could make this future a reality?

It is not easy to arrive at a consensus vision of an ideal solution. Nor is it likely that there will be any one clear and unambiguous path to this future. Most scenarios are heavily influenced by factors that are beyond our control and our knowing, so uncertainty is another hurdle to leap in devising an idealized vision of the future. Alternative risk management options are developed as planners approach this idealized solution from different directions or as they begin to confront the harsh reality that the ideal solution is not feasible.

18.7.8. Something for everybody strategy

This is a pragmatic strategy that recognizes the importance of satisfying stakeholders. The risk analysis team sets out to formulate options that provide outcomes that will satisfy all of the known stakeholders. There are at least two ways to address stakeholder interests. First, an option can be developed for each stakeholder group. Thus, there might be an importer option and a farmer option. Alternatively, options could be developed to ensure that each stakeholder group finds some element of interest to them in each of the formulated options.

18.8. Summary and Look Forward

Here are five things to remember from this chapter.

1. The ALOP is an idea all can believe in, which no one can explicitly specify.
2. An ALOP is not so much a "bright line" policy as a variable range of acceptable outcomes.
3. There is a much stronger case to be made for consistency in approach to an ALOP than there is for consistency in outcomes.
4. A systems approach has the potential to unite the risk management interests of risk-taking firms and risk-reducing NPPOs.
5. Only a fool would attempt to formulate risk management options without structuring the approach in some way.

The next chapter addresses certification as a risk management option.

18.9. References

Devorshak, C. (ed.) (2012) *Plant Pest Risk Analysis, Concepts and Application.* CABI International, Wallingford, UK.

FAO (2020) NSP - Integrated Pest Management. Food and Agriculture Organization of the United Nations. Available at: http://www.fao.org/agriculture/crops/thematic-sitemap/theme/pests/ipm/en/ (accessed June 19, 2020).

Gkogka, E., Reij, M.W., Gorris, L.G.M. and Zwietering, M.H. (2013) Risk assessment strategies as a tool in the application of the ALOP (ALOP) and Food Safety Objective (FSO) by risk managers. *International Journal of Food Microbiology* 167, 8–28.

Havelaar, A.H. *et al.* (2004) Fine-tuning food safety objectives and risk assessment. *International Journal of Food Microbiology* 93, 11–29.

Oyarzabal, O.A. (2015) Understanding the differences between hazard analysis and risk assessment. *Food Safety Magazine eDigest*, November 2015. Available at: http://www.foodsafetymagazine.com/enewsletter/understanding-the-differences-between-hazard-analysis-and-risk-assessment/ (accessed February 3, 2017).

Wilson, D. (n.d.) *The ALOP*. Available at: https://www.ippc.int/static/media/files/publications/en/2013/06/05/1156321210600_11_Wilsonsess5_HOOD.pdf (accessed February 3, 2017).

World Trade Organization (1998) EC Measures Concerning Meat and Meat Products, AB-1997-4. WT/DS26/AB/R, para. 213. Available at: https://www.wto.org/english/tratop_e/dispu_e/hormab.pdf (accessed August 11, 2020).

World Trade Organization (WTO) Committee on Sanitary and Phytosanitary Measures (2000) *Guidelines to Further the Practical Implementation of Article 5.5*. G/SPS/15.

19

Certification

_____

My sincerity is my credential

Malcom X

19.1. Introduction

"Certification" is defined by the online _Oxford English Dictionary_ as "The action or process of providing someone with an official document attesting to a status or level of achievement." In the area of plant health, certification is tied to international and domestic trade in plant commodities that can introduce and/or spread plant pests into new areas. In this context and generally speaking, the "someone" referred to in the definition above is the plant health regulatory authority of each country, i.e., the national plant protection organizations (NPPOs); the "action or process of attesting to a status or level" is termed phytosanitary certification; and the "official document" verifying that a particular status or level of achievement has been reached is called the phytosanitary certificate (PC), which can be in paper form or, recently, have an electronic format.

This chapter examines the role of certification in phytosanitary risk management. It begins by considering why a PC is needed and where one finds the necessary PC information. After offering formal definitions, it outlines the specific requirements for a phytosanitary certification system provided in Article V of the International Plant Protection Convention (IPPC) and in several international and regional plant health standards that deal with phytosanitary certification. These are:

- ISPM 7: Phytosanitary Certification System
- ISPM 12: Phytosanitary Certificates
- ISPM 20: Guidelines for a phytosanitary import regulatory system
- RSPM 8: Authorization of Individuals to Issue Phytosanitary Certificates (PCs).

This chapter explains instances where the certification system is managed through a partnership between two NPPOs or between NPPOs and relevant plant industries. For example, several countries have international bilateral as well as domestic certification systems for plants and plant products and a brief discussion of these is provided. Finally, this chapter discusses the advancements made to date towards an electronic certification system (or ePhyto) to allow trading partners to move away from paper certificates and thus make trade faster, cheaper, and more secure.

19.2. Why is a PC Needed?

As indicated above, trade in plant commodities carries the risk of introducing or spreading plant pests into new areas or countries. As such, when a grower or producer, a company or a broker wishes to engage in trade and, for example, export a type of plant like a poinsettia, plant product like fresh apples or other regulated commodity from their country to another (importing) country, they must first find information on the rules and regulations that govern the importation of regulated plant commodities into the importing country. These rules and regulations are called the phytosanitary import requirements for, say, fresh apples into the importing country.

The phytosanitary import requirements identify the conditions that must be met by the exporting country for that commodity to be allowed entry into the importing country. If an importer in any country is interested in acquiring a commodity from any other country, they must also be aware of the phytosanitary import requirements that exist for that commodity to be imported into their own country.

As can be seen in Fig. 19.1, the PC is issued by the NPPO of the exporting country or its authorized designee and is sent to the NPPO of the importing country. Even though an authorized designee can issue the PC, the document can only be signed by someone working for the NPPO of the exporting country as it is an official document. A PC has three main sections. The first provides information on the consignment including data on the exporter and the types of plant product(s) being exported. The second allows an "Additional Declaration" to be included, for example, addressing pests of concern mentioned in the phytosanitary import requirements of the importing country. The third section indicates what measures, if any, have been applied to the consignment. The completed and signed PC officially attests that a consignment (of plants, plant products or other articles) meets the phytosanitary import requirements of the importing country.

19.3. Where Does One Find a Country's Phytosanitary Import Requirements?

As indicated above, phytosanitary certification, formally defined in Section 19.4, is an important function of an NPPO, as described in Article IV of the IPPC. The phytosanitary procedures in phytosanitary certification may include inspections,

tests, surveillance, or treatments that are focused on pests that are regulated by the importing country. As such, the NPPO of the importing country identifies the pests that are of concern to their country and the import requirements that must be met before effecting trade. The NPPO of the importing country should supply this information to the NPPO of the exporting country.

However, most NPPOs have their own databases containing information on the phytosanitary import requirements of potential trading partners. For example, in the United States, the United States Department of Agriculture, Animal and Plant Health Inspection Service, Plant Protection and Quarantine (USDA-APHIS-PPQ), the NPPO of the United States, maintains the PC Issuance and Tracking System or PCIT which tracks the inspection of plant products and certifies their compliance with the phytosanitary import requirements of trading partners. Inside PCIT, an interested stakeholder can access the Phytosanitary Export Database or PExD to find information on the import requirements to export a plant commodity to a particular country.

Any grower or producer, company or a broker that wishes to export a plant commodity can find PCIT at https://pcit.aphis.usda.gov/pcit/faces/signIn.jsf (accessed January 7, 2020). The first step is to obtain an electronic authentication (eAuthentication) to use the system. Once the requester's credentials are verified by USDA-APHIS-PPQ the stakeholder will have access to PExD. After becoming aware of the phytosanitary import requirements of their potential trading partner, PCIT allows stakeholders to schedule the appropriate phytosanitary procedure(s) for their plant commodity (usually an inspection).

If an importer in one country is interested in acquiring a fresh fruit or vegetable commodity from another country, they must also be aware of the phytosanitary import requirements that exist for that commodity to be imported into their own country. To assist with this information, the USDA-APHIS-PPQ maintains the Fruits and Vegetables Import Requirements database (FAVIR) at https://www.aphis.usda.gov/aphis/ourfocus/planthealth/sa_import/sa_permits/sa_plant_plant_products/sa_fruits_vegetables/ct_favir/ (accessed January 7, 2020). FAVIR has

International Plant Protection Convention   IPPC

ANNEX

Model Phytosanitary Certificate

No. _____

Plant Protection Organization of _____

TO: Plant Protection Organization(s) of _____

I. Description of Consignment

Name and address of exporter: _____

Declared name and address of consignee: _____

Number and description of packages: _____

Distinguishing marks: _____

Place of origin: _____

Declared means of conveyance: _____

Declared point of entry: _____

Name of produce and quantity declared: _____

Botanical name of plants: _____

This is to certify that the plants, plant products or other regulated articles described herein have been inspected and/or tested according to appropriate official procedures and are considered to be free from the quarantine pests specified by the importing contracting party and to conform with the current phytosanitary requirements of the importing contracting party, including those for regulated non- quarantine pests.

They are deemed to be practically free from other pests.*

II. Additional Declaration

[Enter text here]

III. Disinfestation and/or Disinfection Treatment

Date _____ Treatment _____ Chemical (active ingredient) _____

Duration and temperature _____

Concentration _____

Additional information _____

_____

                                    Place of issue _____

(Stamp of Organization)   Name of authorized officer _____

                                    Date _____   _____

                                                                        (Signature)

_____

No financial liability with respect to this certificate shall attach to _____ (name of Plant Protection Organization) or to any of its officers or representatives.*

* Optional clause

Fig. 19.1. Model PC from Annex 1 of the IPPC. Published by FAO on behalf of the Secretariat of the International Plant Protection Convention (IPPC).

easy access to regulations and information pertaining to the importation of fruits and vegetables into the United States, its territories and possessions from other countries. One can search by country, for example Argentina, or by commodity, for example papaya, to acquire "real-time" information on requirements to import fruits and vegetables into the United States.

A stakeholder may be interested in importing other plant commodities, like plants for planting or plants for direct sale. Additional information relevant to plant imports and exports for the United States can be found at https://www.aphis.usda.gov/aphis/ourfocus/importexport (accessed January 7, 2020).

As mentioned above, only the NPPO or its authorized designee can issue PCs as it is the competent plant health regulatory authority for each country. When it comes to signing the PC, only employees of the NPPO have authorization to do this as a PC is an official document. Any grower or producer, company or broker that wishes to engage in trade must work closely with their NPPO to effect lawful certification of plant commodities moving in trade.

19.4. Relevant Definitions from ISPM 5: Glossary of Phytosanitary Terms and the IPPC

Definitions of some relevant terms are offered below.

- *Phytosanitary certification*—Use of phytosanitary procedures leading to the issue of a PC.
- *PC*—An official paper document or its official electronic equivalent, consistent with the model certificates of the IPPC, attesting that a consignment meets phytosanitary import requirements.
- *Additional declaration*—A statement that is required by an importing country to be entered on a PC and which provides specific additional information on a consignment in relation to regulated pests or regulated articles.
- *Consignment*—A quantity of plants, plant products or other articles being moved from one country to another and covered, when required, by a single PC (a consignment may be composed of one or more commodities or lots).

- *Phytosanitary import requirements*—Specific phytosanitary measures established by an importing country concerning consignments moving into that country.
- *Phytosanitary measures*—Any legislation, regulation or official procedure having the purpose to prevent the introduction or spread of quarantine pests or limit the economic impact of regulated non-quarantine pests.
- *Phytosanitary procedure*—Any official method for implementing phytosanitary measures including the performance of inspections, tests, surveillance, or treatments in connection with regulated pests.
- *Regulated article*—Any plant, plant product, storage place, packaging, conveyance, container, soil and any other organism, object or material capable of harbouring or spreading pests, deemed to require phytosanitary measures, particularly where international transportation is involved.
- *Regulated pest*—A quarantine pest or a regulated non-quarantine pest.

Article V of the IPPC states that:

1. Each contracting party, i.e., country that is a signatory to the IPPC, shall make arrangements for phytosanitary certification, with the objective of ensuring that exported plants, plant products and other regulated articles and consignments thereof are in conformity with the certifying statement to be made pursuant to paragraph 2(b) of the Article.

2. Each contracting party shall make arrangements for the issuance of phytosanitary certificates in conformity with the following provisions:

    a. Inspection and other related activities leading to the issuance of phytosanitary certificates shall be carried out only by or under the authority of the official National Plant Protection Organization. The issuance of phytosanitary certificates shall be carried out by public officers who are technically qualified and duly authorized by the official NPPO to act on its behalf and under its control with such knowledge and information available to those officers that the authorities of importing countries may accept the phytosanitary certificates with confidence as dependable documents.

b. Phytosanitary certificates, or their electronic equivalent where accepted by the importing contracting party concerned, shall be worded in the models set out in the Annex to this Convention, shown in Fig. 19.1 above. These certificates should be completed and issued considering relevant international standards.

c. Uncertified alterations or erasures shall invalidate the certificates.

This guidance above can be summarized, in plain language, as:

- PCs facilitate the safe trade of plants and plant products
- they should adhere to the internationally accepted and, thus, harmonized model
- they should only be issued by adequately trained personnel that work for or are authorized by the competent authority (NPPO)
- they should be verifiable and should take into account guidance from adopted international plant health standards.

19.5. Guidance from International Standards on a Phytosanitary Certification System

NPPOs should develop and maintain a phytosanitary certification system to facilitate safe international trade in plants, plant products, or other articles capable of spreading plant pests. The system should certify compliance with the phytosanitary requirements of importing countries as well as freedom from regulated pests. The system should include:

- the legal authority
- administrative responsibilities
- operational responsibilities
- necessary resources and infrastructure
- documentation, communication and review requirements.

19.5.1. Legal authority

The NPPO should have the legal authority to support its phytosanitary certification system, which includes bearing legal responsibility for its actions in using this authority (also see Article IV.2(a)

of the IPPC) and includes preventing the export of consignments that fail to meet phytosanitary import requirements of the importing country.

In the United States, the legal authority for the phytosanitary certification system administered by the USDA-APHIS-PPQ is contained in the Plant Protection Act of 2000. In Canada, the legal authority is provided to the Canadian Food Inspection Agency (the NPPO of Canada) by the Plant Protection Act. S.C. 1990. c.22.

19.5.2. Administrative responsibilities

Each NPPO should ensure that it:

- has adequate personnel and resources to support the certification system
- has clearly outlined the duties and communication channels for personnel involved in phytosanitary certification
- employs or authorizes personnel with appropriate qualifications and skills, and
- provides ongoing training to personnel.

In the United States, the USDA-APHIS-PPQ personnel qualifications include having a bachelor's degree in one of the biological sciences and at least one year of experience identifying pests important in trade. There are also provisions for a combination of education and experience. Furthermore, the Accredited Certifying Officials (ACOs) acting on behalf of the USDA-APHIS-PPQ must go through certification training and require recertification every three years, passing an exam with a minimum of 80% correct answers.

19.5.3. Operational responsibilities of the NPPO

ISPM 7 indicates that the NPPOs of both the importing and exporting country, as appropriate, should ensure that they are capable of performing the following functions in order to meet all responsibilities related to phytosanitary certification:

- document and maintain information regarding the phytosanitary import requirements of its trading partners

- provide appropriate work instructions to personnel consulting the phytosanitary import requirements when issuing PCs
- identify plants, plant products and other regulated articles
- perform inspection, sampling and testing of plants, plant products, and other regulated articles for purposes of certification
- detect and identify pests
- perform, supervise or audit the required phytosanitary treatments
- perform surveys, monitoring and control activities to confirm compliance with the phytosanitary import requirements stipulated in PCs
- complete and issue PCs
- verify that phytosanitary procedures have been established and correctly applied
- investigate and take corrective actions, if appropriate, on any notification of non-compliance with PCs
- develop operational instructions to ensure that phytosanitary import requirements are met
- archive copies of issued PCs and other relevant documents
- review the effectiveness of phytosanitary certification systems
- implement, to the extent possible, safeguards against potential problems such as conflicts of interest and fraudulent issuance and use of PCs
- conduct training for personnel and verify the competency of authorized personnel
- ensure the phytosanitary security of consignments after certification prior to export.

19.5.4. Resources and infrastructure

The NPPO of the exporting country should employ personnel with the technical skills and qualifications necessary to conduct phytosanitary certification activities. All personnel should have adequate training and experience to perform the functions outlined above. In addition, personnel should have no conflict of interest in the outcome of the phytosanitary certification activities. The NPPO of the exporting country may contract personnel outside of the NPPO to undertake many certification activities except for the issuance of the PC itself.

As indicated in the Introduction, phytosanitary certification should be based on the phyto-

sanitary import requirements of the importing country. This official information should be available to the NPPO of the exporting country as stated in several Articles of the IPPC including VII.2(b), VII.2(d) and VII.2(i), and in accordance with ISPM 20. In turn, the NPPO of the exporting country should make the information easily accessible to the grower or producer, company or broker that wishes to engage in trade.

The information on the regulated pests of concern to the importing country should include pest presence and distribution in the exporting country, biology, surveillance, detection, and identification of these pests and means to control these pests, including treatment information where appropriate.

The NPPO of both the importing and exporting country, as appropriate, should ensure that adequate equipment, materials, and facilities are available to carry out sampling, inspection, testing, treatment, consignment verification, and other procedures having to do with phytosanitary certification.

19.5.5. Documentation

The NPPO of both the importing and exporting country, as appropriate, should have a system for documenting the relevant procedures applied and for maintaining records. The system should include efficient ways to store and retrieve the documentation. The system should allow the traceability of PCs and the related consignments and their parts and should allow verification of compliance with the phytosanitary import requirements.

As indicated earlier, in the USDA-APHIS-PPQ the documentation system is called PCIT. The PCIT system tracks the inspection of agricultural commodities and certifies their compliance with phytosanitary import requirements of its trading partners. PCIT provides the USDA-APHIS-PPQ with better security, reporting functions, and monitoring capabilities for exported commodities.

19.6. PCs and ePhyto

As stated in the ISPM 5 definition, a phytosanitary certificate is an official paper document or its electronic equivalent attesting that a consignment meets phytosanitary import requirements. The model PC as described in Annex 1 (see Fig. 19.1) to the IPPC should be used.

Even though many countries still use paper PCs, there is a push for most countries to implement electronic certification or "ePhyto." An ePhyto is the electronic version of the information contained in a paper PC. The information in an ePhyto is in Extensible Markup Language (XML) format, an internationally recognized language used to produce documents in a format that is human-readable and machine-readable. XML is widely used and standardized to allow communication between different computer systems and transmissible over the internet. ePhytos can be exchanged electronically between countries or the data can be printed out into a paper-based PC.

Electronic phytosanitary certification provides several benefits to both exporting and importing countries. ePhyto:

- reduces the possibility for fraudulent documentation
- reduces data entry and validation functions by NPPO staff
- improves security in transmission
- improves planning for arrival and clearance of consignments of plants and plant products at customs
- reduces delays in receiving replacement PCs
- has ability to link into the World Customs Organization "One (single) Window" initiative and harmonize codes and processes.

The development of electronic certification was discussed by the IPPC contracting parties for many years. The first meeting on electronic certification was in 2006 and in 2008 an expert working group (EWG) revised ISPMs 7 Export certification system and 12 Guidelines for phytosanitary certificates and recognized that specific guidance would be needed to deal with electronic certificates. As such, Appendix 1 Electronic certification, information on standard XML schemes and exchange mechanisms was included in the revision of ISPM 12.

An open-ended EWG on eCertification was established by the IPPC contracting parties in 2011 to work on:

- establishing elements and requirements for security and authenticity

- providing advice and recommendations on policy issues
- exploring which entries should be standardized, e.g. names/codes for plants and pests, codes for units, codes for intended uses
- the mode of transmission and standardized language used in electronic certificates
- identifying problems during transition (e.g. concerning transit and re-export).

The term "ePhyto" was coined in 2011 and additional working groups were formed to address other issues such as developing the ePhyto XML map, clarifying the mandatory and optional data elements in relation to the status of issuance of the PC, how to handle re-export PCs, the codes to be used in the electronic certificate and the level of security needed for websites handling the ePhytos. The Appendix to ISPM 12 is focused on the explanation of the elements and codes for ePhyto.

In the last five years, there have been significant and positive advancements towards implementing what is now called the ePhyto solution (https://www.ippc.int/en/ephyto/, accessed January 7, 2020). The IPPC ePhyto solution consists of three main elements aimed at supporting the exchange of ePhytos between NPPOs.

- A central server (hub): to facilitate the transfer of electronic phytosanitary certificates between NPPOs, either from and to their own national electronic system or by using GeNS, described below.
- A Generic ePhyto National System (GeNS): a web-based system that can produce and receive ePhytos, to allow countries that do not have a national electronic system to produce, send and receive ePhytos.
- Harmonization: the structure and transmission of ePhytos will follow a harmonized format through the use of standardized mapping, codes, and lists.

Several countries are currently exchanging ePhytos through the hub and a pilot of the GeNS system is currently underway. Additional information on ePhyto can be accessed here:

- ePhyto video—https://www.youtube.com/watch?v=gjDz7aOv-Ys (accessed January 7, 2020)

- beginners guide—
  https://www.ippc.int/static/media/files/
  publication/en/2019/03/2018-02-20_
  ePhyto_for_beginners_finalized.pdf (accessed
  January 7, 2020)
- ePhyto newsletters— https://www.ippc.int/
  en/ephyto/ephyto-newsletters/ (accessed
  January 7, 2020).

As countries continue to adopt ePhyto, there is still a question concerning a long-term funding mechanism to support it. Additional work will also be required as countries implement the provisions of the Trade Facilitation Agreement as it will affect how electronic data for trade are handled.

19.7. Phytosanitary Certification Procedures

The NPPO should develop and maintain specific work instructions covering all procedures in a phytosanitary certification system, including:

- activities relating to PCs such as inspection, sampling, testing, treatment, and verification of the identity and integrity of consignments
- maintaining security over official seals and marks
- ensuring consignment traceability, including their phytosanitary security through all stages of production, handling, and transport prior to export
- investigation of notifications of non-compliance from the NPPO of an importing country, including, if requested, a report of the outcome of such an investigation. This procedure should be in line with ISPM 13 guidelines for the notification of non-compliance and emergency action
- investigation of invalid or fraudulent PCs.

NPPOs may also have documented procedures related to phytosanitary certification for cooperation with stakeholders (e.g. producers, brokers, traders).

19.7.1. Record-keeping

Copies of the PCs should be kept for purposes of validation and traceability for a period of time, at least one year. For each consignment for which a PC is issued, there should be records concerning: inspection, testing, treatment or other verification carried out; whether samples were taken; personnel responsible; date of activity and results obtained.

Records should be easy to retrieve; a secure electronic storage and retrieval system is recommended. It may also be useful to keep records for non-compliant consignments for which PCs were not issued.

19.7.2. Communication within the exporting country and between NPPOs

Methods and channels should be available for timely communication with relevant departments and agencies, authorized personnel and industry stakeholders (such as producers, brokers and exporters) concerning import requirements of trading partners, geographic distribution and pest status and data on operational procedures having to do with certification. Article VIII.2 of the IPPC indicates that the NPPO of each country should designate a contact point for the exchange of information connected with implementing the IPPC. Official communications should occur between the NPPO official contact points (https://www.ippc.int/en/countries/nppodirectory/, accessed January 7, 2020). However, for specific activities related to certification, e.g. a notification of non-compliance, an NPPO may designate alternative points for contact, e.g. a certification expert, within their own NPPO.

Clear and accurate information should be provided on phytosanitary import requirements by the importing country. Information can be made available through regional plant protection organizations (RPPOs), on the International Phytosanitary Portal (IPP) or, some other mechanism. For example, as discussed earlier in the chapter the NPPO of the United States maintains several databases with real-time information concerning the import requirements of its trading partners.

If after phytosanitary certification the NPPO of the exporting country becomes aware that an exported consignment may not have complied with phytosanitary import requirements, it should inform the importing country as soon as possible. In cases where non-compliance has been identified at import, ISPM 13 Guidelines for the notification of non-compliance and emergency action apply.

19.7.3. Phytosanitary certification system review

The NPPO should periodically review the effectiveness of all aspects of its export phytosanitary certification system and implement changes to the system if required.

19.8. Certification Agreements Between Two Countries—North American Example

Canada is the largest export market for United States' ornamental plant material. In 2011, companies in the United States exported $233.8 million worth of ornamental plants to Canada which represented 52.2% of all ornamental horticulture exports between the two countries (Wocial, 2012). To facilitate trade in plants between their nations, Canada and the United States have a bilateral export certification program for greenhouse-grown plants that uses much of the guidance from ISPM 7, but it plays out through documents other than a PC.

The United States–Canada Greenhouse-grown Plant Certification Program (GCP) is a bilateral export certification program for greenhouse-grown plants shipped between Canada and the continental United States. The GCP is the successor to the greenhouse certification program that was established via a Memorandum of Understanding (MOU) between APHIS and the Canadian Food Inspection Agency (CFIA) in 1996.

The GCP is based on a systems approach (ISPM 14) that integrates different pest risk management measures to achieve the appropriate level of protection against regulated pests. Facilities that enter into a Compliance Agreement with either APHIS-PPQ or CFIA are authorized to export certified plants to the United States or Canada with an Export Certification Label in lieu of a PC. These facilities are known as "Authorized Facilities."

Certified plants are greenhouse-grown plants that meet all phytosanitary import requirements of both Canada and the United States and have completed all the requirements of the GCP. The NPPOs conduct audits at the facilities in their respective countries to authorize facilities and verify compliance with the GCP.

The GCP takes place in three distinct phases:

- determining eligibility of plants
- activities that take place at an authorized facility
- shipment of certified plants.

19.8.1. Phase 1—determining eligibility of plants

Greenhouse-grown plants must originate in either Canada or the United States, or if imported from a third country, they must be allowed into the United States and Canada as per both countries' phytosanitary regulations. Plants that meet these requirements are termed "eligible plants" and may enter into the GCP.

Plants produced under the GCP must be free from regulated pests of concern to both Canada and the United States and must meet the phytosanitary import requirements of both Canada and the United States.

To verify freedom from regulated pests, authorized facilities will:

- inspect plants entering the facility
- implement a scouting program for plants in production
- inspect certified plants when they are shipped
- identify unknown pests
- report new pests and regulated pests found in an area where they have not previously been known to exist to the NPPO where the authorized facility is located.

To be eligible plants, plants must:

- not be prohibited
- not be listed as not authorized pending pest risk analysis (NAPPRA) into either the United States or Canada directly from the country of origin
- meet regulatory size/age requirements, if entering from a third country, and

- meet any other regulatory requirements that may be applicable.

Authorized facilities are required to maintain a current list of plants in production at their facility, indicating which plants are eligible or ineligible for the GCP. The description should include the growth stage and form of incoming plants, e.g. seed, in-vitro/tissue culture plantlets, cuttings, plugs, bare-root plants, or pre-finished plants. The list must be accepted by the NPPO. The NPPO verifies which plants are eligible for inclusion in the GCP, based on taxa and source. The NPPO may be consulted for guidance on how to assess which plants are eligible or ineligible.

The following articles are not eligible plants and may not be certified through the GCP:

- seeds
- grain
- seed potatoes
- regulated invasive plants
- regulated noxious weeds
- fresh fruits or vegetables.

Plants may be grown and shipped in any growing media that is acceptable for trade in plants between the United States and Canada.

19.8.2. Phase 2—activities at an authorized facility

Eligible plants that are produced at an authorized production facility in accordance with the procedures described in their written Pest Management Plan become certified plants. Modules that describe specific pest mitigation or production measures may be required to be included in the Pest Management Plan. Authorized facilities must be maintained practically free from injurious pests.

19.8.3. Phase 3—shipment of certified plants

Certified plants are ready for export and may be shipped between the United States and Canada using an Export Certification Label in lieu of a PC or between authorized facilities in the same country using an "Interfacility Stamp."

For more information on this program see: http://www.inspection.gc.ca/plants/horticul-ture/exports/gcp-technical-requirements/eng/1474666508713/1474666604536 (accessed January 25, 2019).

19.9. Example of Domestic Certification Agreements—A Systems Approach to Nursery Certification (SANC)

The purpose of the SANC initiative is to develop, promote and implement a risk-based nursery certification system utilizing existing state authorities and programs to enhance uniformity, increase efficiency, and reduce pest threats during inter- or intra-state trade of nursery stock in the United States. The SANC program is a voluntary, audit-based program designed to reduce pest risks associated with the domestic movement of nursery stock. Fig. 19.2 illustrates the steps required to become SANC certified.

The SANC standard describes the elements of a harmonized state-level systems approach to nursery certification within the United States. It outlines the responsibilities of state certification agencies and of nursery and greenhouse facilities participating in a SANC program. SANC is designed to address nursery certification of plants moving inter- or intra-state. All nursery and greenhouse operations within states and territories of the United States, in good regulatory standing, are eligible to participate.

As with other international and bilateral certification systems, the SANC, through its standard, clearly outlines the general requirements for participation, including eligible plant material; regulated pests; facility requirements including facility description, staff training, internal audits, records and documents; the certifying authority; what to do when non-compliance occurs; what type of corrective measures can be considered; and the suspension, cancellation or reinstatement of participants.

19.10. The Role of Certification in Pest Risk Management

PCs are one of the best known and most ubiquitous documents in the plant health community. They are important tools for risk management in trade because they codify agreements between trading partners.

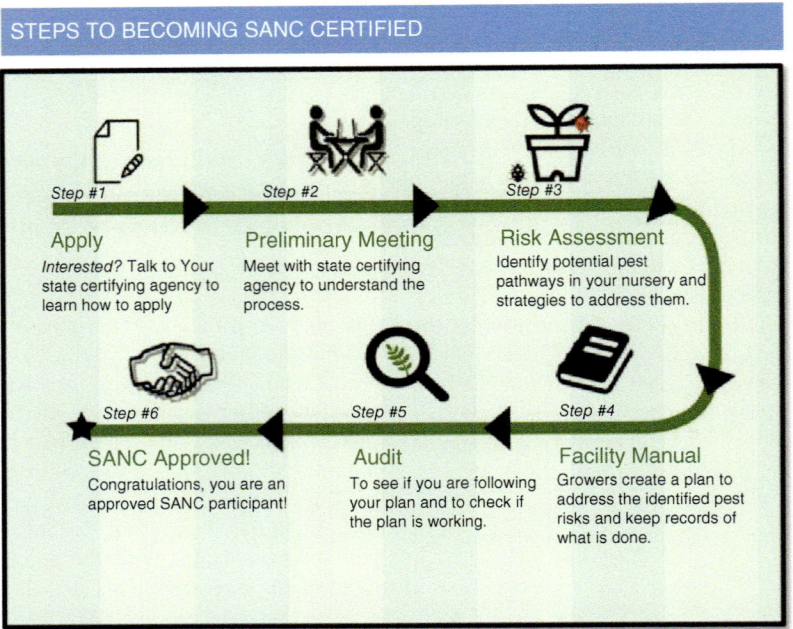

Fig. 19.2. Steps required to become SANC certified.

The exporting countries accept responsibility for meeting the phytosanitary requirements of importing countries by officially "certifying" compliance using a standardized process and format. This single point of international harmonization later provided the basis for the negotiators of the SPS Agreement to adopt the IPPC as the phytosanitary standard-setting organization identified in the Agreement.

19.11. Summary and Look Forward

Here are five points to take away from this chapter.

1. PCs are necessary to facilitate trade in plant commodities.

2. A PC provides information on the consignment, allows an Additional Declaration to be included when needed, and indicates what measures, if any, have been applied to the consignment.

3. The PC document can only be signed by someone working for the NPPO of the exporting country as it is an official document.

4. A PC system requires the legal authority, administrative responsibilities, operational responsibilities, necessary resources and infrastructure, and documentation, communication and review requirements.

5. PCs are important tools for risk management in trade because they codify agreements between trading partners.

Inspection, the subject of the next chapter, is a key element of pest risk management and by far the most frequently applied phytosanitary procedure for both import and export. The pests we can stop with inspection are not a problem, it is the pests that we do not find that are the risk. Risk-based sampling is introduced as a rigorous form of discovery sampling that promises to strengthen inspection.

19.12. References

NAPPO (2013) Regional Standards for Phytosanitary Measures, Publication No. 08: *Authorization of Individuals to Issue Phytosanitary Certificates*. Secretariat of the North American Plant Protection Organization (NAPPO), Raleigh, North Carolina, United States of America.

IPPC (1997) *International Plant Protection Convention.* Rome, IPPC, FAO.

IPPC (2016a) International Standards for Phytosanitary Measures, Publication No. 7: *Export Certification System.* Secretariat of the International Plant Protection Convention (IPPC), Food and Agriculture Organization of the United Nations, Rome, Italy.

IPPC (2016b) International Standards for Phytosanitary Measures, Publication No. 13: *Guidelines for the Notification of Non-compliance and Emergency Action.* Secretariat of the International Plant Protection Convention (IPPC), Food and Agriculture Organization of the United Nations, Rome.

IPPC (2017a) International Standards for Phytosanitary Measures, Publication No. 12: *Phytosanitary Certificates.* Secretariat of the International Plant Protection Convention (IPPC), Food and Agriculture Organization of the United Nations, Rome, Italy.

IPPC (2017b) International Standards for Phytosanitary Measures, Publication No. 20: *Guidelines for a Phytosanitary Import Regulatory System.* Secretariat of the International Plant Protection Convention (IPPC), Food and Agriculture Organization of the United Nations, Rome, Italy.

IPPC (2019a) International Standards for Phytosanitary Measures, Publication No. 5: *Glossary of Phytosanitary Terms.* Secretariat of the International Plant Protection Convention (IPPC), Food and Agriculture Organization of the United Nations, Rome.

IPPC (2019b) International Standards for Phytosanitary Measures, Publication No. 14: *The Use of Integrated Measures in a Systems Approach for Pest Risk Management.* Secretariat of the International Plant Protection Convention (IPPC), Food and Agriculture Organization of the United Nations, Rome, Italy.

Wocial, M. (2012) *U.S. Exports/Imports of Ornamental Plants Between 1992 & 2011.* Growertalks, https://www.growertalks.com/Article/?articleid=19701.

20

Inspection and Risk-based Sampling

In natural science the principles of truth ought to be confirmed by observation.

Carl Linnaeus

> Inspect: Perform an official visual examination of plants, plant products, or regulated articles to determine if pests are present and/or to determine compliance with phytosanitary regulations.
> (ISPM 5, *Glossary of Phytosanitary Terms*)

20.1. Introduction

Inspection is a key element of pest risk management and by far the most frequently applied phytosanitary procedure for both import and export. This is primarily because a high percentage of pest organisms are visually detectable or their signs or symptoms are distinguishable. Accordingly, the results of inspection have traditionally provided critical information forming the basis for decision-making in phytosanitary affairs.

In its broadest context, inspection and examination may imply a wide range of activities, processes, and methods used for various reasons. For instance, the verification of proper documentation is an activity commonly associated with phytosanitary inspection. Likewise, the examination of facilities for compliance or suitability under phytosanitary requirements may fall within the broad interpretation of inspection. Inspection may also be used to gather information or to monitor activities. It can also include laboratory testing. All these possibilities are arguably linked to risk management and extend from the same conceptual foundation reflected in the International Plant Protection Convention (IPPC) definition.

20.2. Three Categories of Inspection

The mainstream understanding of inspection in the phytosanitary community is concerned with the activities performed at ports for the visual detection of pests and regulated articles. This process can be divided into three distinct categories:

1. Examination of documents—ensuring that necessary documents (or electronically submitted information) are complete and correct (customs entry, phytosanitary certificate, import permit).
2. Physical examination—ensuring that the articles presented are consistent with the documentation (100 boxes of fresh blueberries from Guatemala are 100 boxes of fresh blueberries from Guatemala).
3. Phytosanitary inspection—sampling for inspection to detect pests or regulated articles.

The phytosanitary aspect of inspection is central to the idea of using inspection as a phytosanitary measure for risk management and is integral to both the certification of export consignments

---

and the clearance of import consignments. The examination of passenger baggage, called a pre-departure inspection when it is done at the port of embarkation, is focused primarily on the detection of regulated items rather than pests. The "and/or" in the IPPC definition above is significant in this regard. The detection of pests "and" compliance means ensuring compliance with phytosanitary requirements as well as inspection for the detection of pests. The detection of pests "or" compliance means that the inspection may be an examination primarily for compliance, such as with passenger baggage. That is not to say that prohibited items would not be found in an inspection for pests, or pests would not be found in an examination for regulated articles. The concept of inspection is validly applied for both the detection of pests and the detection of regulated articles. The difference is the guiding objective.

20.3. Inspection as a Deterrent

For over a century, national plant protection organizations (NPPOs) have reinforced the importance of inspection as a primary strategy for preventing the introduction of harmful pests. Whether or not anything is inspected, the fact that the international movement of people and goods is *subject* to inspection is a motivation for compliance. Uncertainty is created when traders do not know whether an inspection will be done, the intensity of the inspection, and the outcome. One response to this uncertainty can be efforts to reduce the probability of a negative experience by striving for high compliance. On the other hand, the threat of inspection, or rather the fear of negative repercussions from the results of inspection, can also be a motivation for smuggling or other illicit actions to avoid detection. Some of the positive effect of inspection as a deterrent is therefore lost because of either ignorance or a strong desire to circumvent requirements.

Knowing and accepting the reality that inspection is a deterrent but not a foolproof safeguard against pest entry is an important starting point for risk managers considering inspection as a phytosanitary measure. The risk management discussion begins by asking whether inspection is being done as a general procedure with no specific expectation, or whether risk managers have identified specific targets. Questions then arise regarding the desired effectiveness of inspection, the tolerance for slippage, targeting for higher or lower risk, maximizing the detection value of available resources, and the consistent, fair, and justified use of inspection as a tool for risk management and a phytosanitary measure under the obligations of the World Trade Organization Agreement on the Application of Sanitary and Phytosanitary Measures (the SPS Agreement). The answers to these questions rest in analyses that go far beyond the possibilities of data collected from haphazard inspections. Thus, a central point for risk managers is understanding that there is always some value for inspection as a deterrent—even so, data from consistent, well-designed inspections is essential to provide the insights needed for risk management.

20.4. Inspection as a Procedure

Inspection itself does nothing to reduce pest risk. Although it might be argued that destructive sampling reduces risk by destroying potentially infested material, it is not designed to be a mitigation strategy. Inspection is rather a phytosanitary procedure. It is the actions taken as a result of inspection that will ultimately determine how risk is changed. At an operational level, these decisions will usually be:

● acceptance, e.g., no action
● rejection, e.g., seizure, destruction, re-export or
● application of other measures, e.g., treatment, reconditioning.

In cases where the situation in question has been anticipated, the SPS Agreement tells us that the actions should be based on either a pest risk analysis (PRA) or international standards. In unanticipated circumstances, emergency measures are taken, especially where there is not enough time or information for risk analysis. This is an important point because the majority of actions at ports of entry fall into the category of emergency measures (see Chapter 25, this volume).

Risk managers need to understand which of these actions is most appropriate and what level of detection should trigger action based on the characteristics of the pest(s) or the article of concern. The aim should be to link inspection to the level of pest risk deemed to be acceptable and to the operational feasibility of using inspection to manage risk. The emphasis should be on the use of procedures that are transparent, defensible, practical, and applied in a way that consistently and predictably manages risk.

The primary assumption behind the use of inspection is that the pests of concern are detectable. The organism or its signs/symptoms must be visually discernible and distinct enough to minimize the potential for confusion with non-pest organisms or other conditions. Inspection should not be considered the sole basis for risk management if:

- the pests of concern or their signs/symptoms are not detectable
- the potentially infested article is difficult or unsafe to inspect, or
- the pest of concern is high risk and establishes easily.

Inspection may be a good option for risk management when pests:

- are large, external, and easily recognized
- cause visible damage or have distinct signs/symptoms
- have limited mobility or lack mobility
- do not easily establish
- are highly unlikely to be associated with the article in question, or
- are generally eliminated by harvesting and packing procedures.

A less well-understood assumption associated with inspection is that a certain amount of risk and uncertainty must be accepted. Under normal circumstances, an inspection is not done on 100% of regulated articles, and inspection is not 100% efficient. Since inspection is usually based on a sample and always involves a degree of uncertainty and variability, there will always be some probability that pests will escape detection. Associated with this is a certain degree of confidence in the level of detection achieved using a prescribed level of inspection. The level of possible pest prevalence that is unlikely to be detected may be described as a detection level, threshold prevalence, allowable prevalence, or more often a tolerance.

The acceptance of a tolerance is inherent in the adoption of inspection as a phytosanitary procedure. For this reason, it is not appropriate to use inspection as the basis for phytosanitary decision-making if the objective is true pest freedom. Further, it must be recognized that the role of inspection cannot be properly appreciated for risk management purposes without an understanding of the level of tolerance and variability that is associated with the procedure.

20.5. Inspection as a Data Source

Inspection as a phytosanitary procedure is used primarily for the detection of pests or regulated articles. For practical reasons, inspection is neither a 100% census nor 100% effective. This means that there is always some probability of a non-compliance being detected and a corresponding probability of it escaping undetected. The probabilities for either situation are highly variable when inspections are haphazard, but inspections can be designed to detect a particular frequency of noncompliance based on the statistical parameters of probability. In the absence of such a design, both the importer and the inspector are simply gambling on the outcome of each inspection based on the variables associated with sampling for detection.

The aim of sampling for detection is to bias the inspection as much as possible in favor of detection. Profiling and similar subjective techniques drawing from prior knowledge and the inspector's expertise are important for increasing the probability of finding a regulated pest or article. The detection of a pest or regulated article is interpreted to represent a successful interdiction. The detection of noncompliance usually marks the end of the inspection process and the beginning of a mitigation process. Failure to detect is interpreted as compliance and results in clearance.

The results of sampling for detection can provide information about the relationship of a pest or regulated article to a pathway or situation, and many observations over a period of time may show some trends. A key limitation of sampling for detection is the inability to infer more than the association of the pathway or situation to noncompliance. True action rates for commodities, approach rates for pests, and infestation rates for shipments cannot be usefully calculated because the sampling information is not consistent and includes a high degree of variability associated with the background assumptions and biases. Moreover, the variability in inspection results causes variable levels of action depending on the sampling design, sample size, shipment size, and rigor of the inspection.

Recalling that a central purpose of the post-World War II GATT Agreements was predictable trade, and the WTO (World Trade Organization) *Agreement on the Application of Sanitary and Phytosanitary Measures* (the SPS Agreement), which came into force for most countries in 1994, obliges governments to use the least restrictive measures to achieve their appropriate level of protection, it is difficult to reconcile the arbitrariness of inspection as it is currently practiced every day for thousands of import decisions around the world. That is not to question the value of inspection as a deterrent, but rather to ask whether it is being applied fairly, consistently, and in a defendable way based on risk as envisioned by the SPS Agreement.

Assuming that all inspection agencies are also striving for more efficient and effective pest exclusion, there are additional questions about whether sampling for detection is the best strategy and if the information it provides is helpful for targeting, prioritization, trend analysis and other processes that support the role of pest exclusion as part of risk management.

Inspection designs that properly incorporate statistical conventions of probability do not compromise the deterrence effect of inspection but can substantially increase the possibilities for analysis and the ability to better measure, adjust, and defend the inspection effort. These designs begin with designating the desired level of detection and statistical parameters, such as the confidence level and estimated efficiency which are used to calculate the sample size based on the shipment size. Sample selection is randomized and the entire sample is inspected.

The results of such statistically designed inspections provide a demonstrable level of efficacy for the effort and can be used to calculate true action rates, approach rates, and infestation rates. These calculations in turn can be used to support ranking, targeting, and defendable policy frameworks for strengthening the role of exclusion in managing pest risk: using inspection to improve inspection. What is more, the data offer potent possibilities for trend and pathway analysis as well as a fair, consistent, transparent, and predictable approach to the application of inspection as a phytosanitary measure.

20.6. Inspection as Sampling

The discipline that is most critical to understanding inspection as sampling is acceptance sampling, also called discovery sampling. The application of this statistical concept in risk management allows us to determine whether inspection is the most appropriate phytosanitary procedure to use for managing pest risk and the characteristics of a proper inspection design, recognizing the concepts of tolerance associated with the probability of detection and consideration of the limitations of confidence in acceptance sampling.

For example, finding that two boxes of fruit from a total of ten are free of pests does not provide absolute assurance that all ten boxes are free of pests. There is some probability that pests occur in the remaining boxes and there is a degree of uncertainty, both variability and error, associated with the two boxes that were inspected. The issues that must be considered are the level of tolerance and confidence, which are considered acceptable, and the level of consistency, or the range of variability, in inspection. Note here that the concept of tolerance applies to the entire population, i.e., the whole shipment, not only the sample. The level of presence in a sample is properly known as the acceptance level. The concept of tolerance is often misapplied to samples, as when a "zero tolerance" refers to rejection based on a single detection in a sample. The correct designation is a zero-acceptance level, which translates to some tolerance in the population based on the size of the population, size of the sample, and the confidence level.

A risk-based inspection is one that has as its objective a defined level of possible pest prevalence and a specific level of desired confidence. This contrasts with an inspection that is based on non-transparent criteria that may be arbitrary or intuitive, or one that is designed only for operational convenience that could be haphazard. The development and adoption of risk-based inspection programs enhances the ability of officials to establish priorities for their inspection resources and to design inspection programs that are transparent for trading partners and the private sector. By establishing reference points such as risk-based inspection objectives and a means to measure the results, it becomes possible to identify, in an analytical context and transparent manner, the areas where inspection resources are most needed, and the level of resources required. These determinations then correspond with the acceptable level of risk and the strength of measures to be applied.

20.7. Risk-based Sampling

The pests we can stop with inspection are not a problem. The pests that we do not find present the risk. As discussed above, inspection as it is traditionally practiced is a form of discovery sampling, which means that there is a threshold for pest detection based on the probability of discovery in a sample. The threshold leaves a certain space for leakage which is the inherent tolerance based on whatever sampling design is used. Our ability to adjust inspection to detect different levels of pest infestation, and conversely accept tolerances for leakage, is the key to understanding the value of inspection as a risk management tool.

Each inspection scenario has a shipment size and sample size which can be used to calculate the level of pest infestation that can be detected with a specific sample, level of confidence, and efficiency. Because of the statistical relationship of the probability of detection to sample size and shipment size, the level of detection will vary as the size of the shipment changes unless the sample size is adjusted according to the lot size to maintain a constant level of detection. Since NPPOs generally have no control over shipment size, they can expect to have different detection levels for different size shipments when the sampling is done at a flat rate, resulting in inconsistent levels of risk management. This is a fundamental point to understand if inspection is being used as a risk management tool.

Similar calculations can be used to determine how the sample size can be adjusted to detect a specified level of infestation or contamination in a certain size shipment. These simple

statistical techniques offer risk managers the opportunity to create inspection designs that not only consistently manage risk, but also provide more and better data for risk management. The availability of these data greatly increases the analytical opportunities for risk and resource management. Simple analyses are possible to demonstrate the effectiveness of inspection programs, perform trend analysis, and adjust inspection efforts to better target risks. Targeting is crucial for maximizing the effectiveness of resources, but it can also serve to incentivize safe trade by reducing the inspection effort with "rewards" for demonstrated low-risk trade.

Traditional inspections also frequently stop when a pest is found, whether or not the entire sample has been inspected. The rationale for this is that pest presence represents non-compliance which usually changes the status of the consignment. As noted above, inspection is not absolute. The detection of one pest does not mean it is the only pest present, and the failure to detect a pest does not mean that a shipment is pest-free. By inspecting the entire sample, it may be possible to detect more of the same pest or other pests that can be used to better understand and manage pest risk.

Full inspection of a statistically derived sample size not only provides a more complete picture of non-compliance, but the results support much more robust analysis of approach rates for pests, action rates for the pathway, entity, or country, and infestation rates for the consignment. A data stream based on a history of consistent sampling allows for the analysis of trends and supports ranking and prioritization for risk analysis as well as resource allocation.

In addition to adjusting the sample size to correspond with the shipment size, and inspecting the full sample, it is also crucial that the sampling be truly random. This is important from the standpoint of statistical validity. It is also one of the most difficult aspects of sampling for inspectors to embrace because their tendency is to bias the selection of samples for the detection of pests based on their experience and expertise. Asking an inspector to inspect a sample that he/she does not believe will have a pest, while also ignoring part of the consignment where they feel more confident about detecting a pest, is counterintuitive and may be demoralizing to inspectors accustomed to demonstrating competence by their selection of appropriate samples.

Inspectors need to understand that they still need to be good inspectors and that sampling differently increases the likelihood of finding new pests. Detecting a pest that is commonly intercepted is not as interesting for risk management as finding a new pest. Thus, the bias of inspectors should not be toward the detection of pests that are well-known but rather pests that have not been previously detected.

Based on the discussions above, the best inspection designs for risk management have the following sampling characteristics:

- the sample size corresponds to a fixed detection level for a specific shipment size
- the samples are randomly selected
- the full sample is inspected.

Inspections with these design elements provide more and better data to support risk and resource management decisions. When fairly and consistently applied, such designs are also technically defendable and greatly expand opportunities for a range of useful analyses, including adjustments in inspection intensity and/or frequency to focus more effort on higher risk goods and away from lower risk goods, thereby creating incentives for trade to reduce risk. This is consistent with the obligations of governments under the IPPC, the WTO *Agreement*, and the risk management provisions of the WTO *Trade facilitation agreement*.

20.7.1. Inspection reset

In an era when both the volume and frequency of trade are greatly outpacing the resources devoted to inspection, there is a practical need to prioritize resources by redistributing the inspection effort devoted to low risk trade in order to focus on higher risk trade. The focus may be on business entities such as producers, exporters, shippers, brokers, or importers. It may also be on commodity groups such as cut flowers or specific types of commodities such as cut roses. It could be on countries of origin or on specific ports of entry. Regardless of the focus, the shift from one-size-fits-all inspection to targeted designs requires appropriate data and analysis to identify the concern, the magnitude of the concern, and changes in its status over time. This requires metrics that come from the analysis of data that has not been available or used in the same way previously.

One starting point for this shift is to analyze existing inspection processes in order to calculate the level of detection that is currently achieved and identify weaknesses. This approach can provide insight into the degree of variability in inspection results and issues that limit the use of inspection results for analysis and targeting. Another starting point is to select a desired level of detection and design a pilot inspection process that achieves the specified objective with statistically valid results. This approach is especially useful to understand the resource commitment required to achieve different levels of detection. In either scenario, the objective is to be able to distinguish or rank commodities, entities, countries, or whatever is being targeted using pest detections as a proxy for risk and then adjusting the design to redistribute the inspection effort for better management of the higher risk goods.

Once a design is in place to consistently detect a specific level of infestation and valid data are available to rank results, the risk-basis for actions may be added to the calculus by evaluating biological and economic aspects of pest interceptions that inform risk beyond simply the presence or absence of a pest in the pathway.

Combining statistically-designed inspection results with risk analysis provides a complete and dynamic view of inspection as a phytosanitary measure and opens multiple doors for additional analysis. Phytosanitary actions can be correlated to numerous different trade variables and targeting systems developed for pests, pathways, ports, or any other trade variable that an NPPO may want to correlate with the risk.

Infestation rates can be calculated for individual consignments, true approach rates can be calculated and tracked for pests, and the same can be done for action rates on commodities/pathways. Leakage can be estimated because the effectiveness of inspection can be measured and adjusted according to risk and balanced with the availability of resources.

Perhaps the most important points to make in support of the shift to risk-based sampling is that it is fair and predictable to trade, defendable to stakeholders and trading partners, and provides all involved with a meaningful basis for using inspection as a phytosanitary measure.

In sum, the historical role of inspection as a deterrent to noncompliance based on case-by-case sampling for detection has limited value as a risk management strategy and questionable status as a phytosanitary measure. A contemporary design that acknowledges the relevant statistical concepts of probability and confidence provides a substantially more consistent and effective inspection effort as well as more and better information to support a range of analytical possibilities for improving risk management while also offering fair treatment to trade.

"Inspection for information" builds on the deterrence effect of "inspection for detection" and greatly amplifies the analytical potential of inspection results for improving and defending the role of inspection in risk management.

20.7.2. An example

Consider a numerical example to illustrate the procedure described above. Look at Table 20.1. Risk managers begin by determining an appropriate level of protection for a pest. This can be translated into a level of risk we are willing to accept through the expression of a tolerance. If we can tolerate the risk that results with the presence of a pest in 10% of all product, we would want a detection level of 0.1. That means we expect our inspection to identify all levels or prevalence greater than our tolerance. Because there is some chance in the lots chosen for the sample and inspectors differ in inspection expertise and experience and a host of other factors, we will never be able to be 100% confident in the results of our inspection, so we must also determine the risk manager's desired level of confidence in the inspection results. Confidence levels are given in the first row of numbers in the table.

Thus, if risk managers have determined that a detection level of 0.1 at 0.95 confidence is appropriate, we have determined two important risk-based sampling parameters. Next, we need to know the consignment or lot size, N. Table 20.1 presents two options, N=5000 and N=100. Let us begin with N=5000. A hypergeometric probability distribution has been used to calculate the values in the table. Finding the intersection of row r=0.1 and column P=0.95 for

N=5000 we see the value 29. This means if we inspect 29 randomly selected units of product, we can be 95% confident that we have detected any infestation that exceeds our tolerable detection level of 0.1.

Compare that with a traditional percentage sampling method where the NPPO always samples 2% of the lot. Two percent of 5,000 yields a sample size 100. A traditional approach over-samples the shipment, inspecting 71 more boxes than would be necessary to obtain the desired confidence that we have detected any infestation greater than 10% (0.1). The traditional approach wastes precious inspection resources for this size lot.

Now, consider the same r and P values for the smaller lot, N=100. Locating the intersection of row 0.1 and column 0.95 for N=100 we see a sample size of n=25 is required. The traditional fixed percentage sample of 2% would only inspect two boxes. One could have very little confidence in the results of such a sample. This wastes inspection resources in a different way. It expends time and effort to yield specious results.

What else can we learn from this example? Look at the different confidence levels for the N=100 lot size for r=0.1. To have a confidence level of 0.95 you need a sample of 25. Suppose you sample less than that? Then your confidence begins to decrease. With n=15 you save inspection resources but you have less confidence in your results, 80% instead of the original 95%. Confidence costs you. If a pest can have minor adverse impacts with a prevalence greater than 10% (0.1) an 80% (0.8) confidence level and n=15 may be sufficient. If the pest could have devastating impacts with a prevalence greater than 10% risk managers might insist on 99.9% (0.999) confidence and a more costly inspection of n=48.

Shift your attention to the N=5000 lot size and P=0.95. If risk managers are concerned only with infestations greater than 10% (r=0.1) you need to sample 29 boxes. If concerned with infestations greater than 1% (r=0.01) the sample size increases to 290 boxes. Likewise, more stringent detection limits of 0.1% and 0.01% increase sample sizes to n=2253 and n= 4988, respectively. Consider that last example of 0.01% detection, that means as few as 1 box in 5,000 could be infested. To be 95% sure you have provided that degree of protection you would have to inspect all but 12 of the 5000 boxes. Notice for

Table 20.1. Risk-based sampling example showing tradeoffs among detection limits, confidence levels, and sample sizes.

Detection Level (r)	Confidence level (P)											
	0.8	0.85	0.9	0.95	0.99	0.999	0.8	0.85	0.9	0.95	0.99	0.999
0.0001	4801	4888	4951	4988	5000	5000	100	100	100	100	100	100
0.001	1376	1579	1845	2253	3009	3743	100	100	100	100	100	100
0.01	158	186	22	290	438	643	80	85	90	95	99	100
0.1	16	18	22	29	44	66	15	17	20	25	36	48
Lot size	N=5000						N=100					

low detection levels of a small shipment you would have to open all the boxes. Lower detection levels cost more inspection resources.

There are three parameters in Table 20.1; detection level, confidence level, and sample size. Risk management can choose any two of these three and then the third one is determined for you by the mathematics of the universe. When an NPPO chooses the sample size only, the confidence level and detection level may both be unknown. Risk-based sampling empowers risk managers to define detection levels and confidence levels and it then identifies the sample size, based on the lot size.

20.7.3. A technical note

Tables like that above can be constructed easily using Microsoft Excel or another spreadsheet program. Fig. 20.1 demonstrates one way to calculate these values. In the upper left of the figure you find the basic model. The lower left shows how the model would be used to estimate the confidence levels for a traditional 2% fixed sample assuming a 10% detection level. Sampling for the large lot at 2% produces complete confidence, which comes at the cost of wasted resources as noted above. Sampling 2 boxes out of 100 provides 19% confidence that you have detected any infestation greater than 10%. That level of confidence means you wasted the resources used for inspection.

The right side of the figure shows how to calculate any sample size for a given detection limit, assumed to be 0.1 for the example. Enter any detection level, lot size and confidence limit. To get the sample size use Excel's Goal Seek feature. To access that feature, go to the Data Tab and the Forecast cluster. From it select What-If Analysis, then Goal Seek. The Goal Seek window lower left of right side shows how to fill out the window. Put the desired confidence limit in the "To value" box. Let Goal Seek change the sample size until you get the desired parameters. The Goal Seek Status to the right of the Goal Seek window shows the message obtained when the calculation is completed.

20.8. Summary and Look Forward

Here are five things to remember from this chapter.

1. Inspection has a deterrent effect simply because it exists.
2. Risk managers cannot expect inspection to be 100% effective.

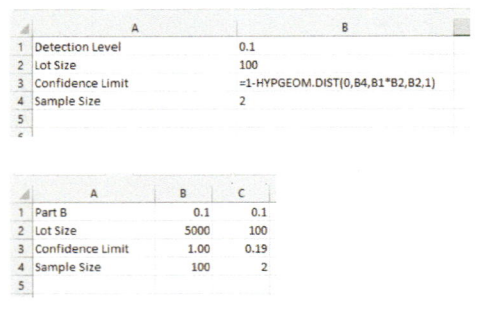

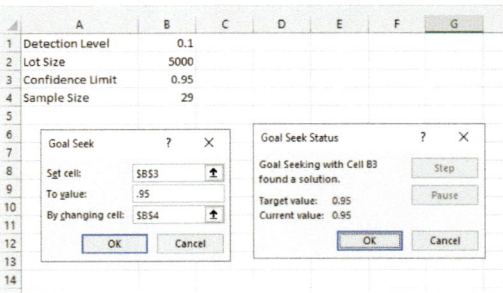

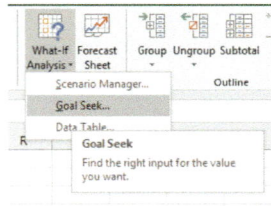

Fig. 20.1. Hypergeometric model used to calculate sample size for a given lot size, detection level, and confidence limit.

3. Haphazard inspections have limited value for risk management.

4. Inspection is sampling and requires a basic understanding of the relevant statistical concepts.

5. Statistically designed sampling that is consistently applied provides useful data for monitoring and adjusting inspection for risk management.

The next chapter explores risk mitigation using phytosanitary treatments.

21

Treatments

_____

Build a better mousetrap and the world will beat a path to your door.

Ralph Waldo Emerson

21.1. Introduction

Treatments are applied to plants, plant products or other regulated articles to reduce the risk of introduction and spread of pests. In this chapter we refer to treatments used in plant health as phytosanitary treatments. Even though treatments, primarily the application of pesticides, are used by growers, producers and others involved in growing/managing/selling plants at several points during the plant production pathway, this chapter will stay within the boundaries of the ISPM 5 definition of treatment, in other words, it will only discuss treatments that are officially established, used or performed by an NPPO.

> Treatment: an official procedure for the killing, inactivation or removal of *pests*, or for rendering pests infertile or for devitalization.
> (ISPM No. 5, IPPC, 2019a)

21.2. The SPS-IPPC Framework

The World Trade Organization (WTO) Agreement on the Application of Sanitary and Phytosanitary Measures (SPS Agreement) sets out the basic principles and obligations to guide the development, adoption, and enforcement of SPS measures applied in international trade. For plant health, the SPS Agreement obliges members to ensure that phytosanitary measures are applied only to the extent necessary to protect plant life or health and that measures are based on scientific principles, i.e., a pest risk analysis, and are not maintained without sufficient scientific evidence. The SPS Agreement is not explicit about domestically applied phytosanitary measures. However, for transparency and ease of harmonization, countries should aim for alignment of their domestic and international risk management procedures.

Article 5.6 of the SPS Agreement indicates that when establishing or maintaining phytosanitary measures to achieve the appropriate level of phytosanitary protection, members shall ensure that such measures are not more trade-restrictive than required. Furthermore, Article 5.8 says that when a member has reason to believe that a specific phytosanitary measure is constraining, or has the potential to constrain its exports and the measure is not based on the relevant international standards, guidelines or recommendations, an explanation of the reasons for such phytosanitary measure may be requested and shall be provided by the member maintaining the measure.

The purpose of the IPPC is "to prevent the spread and introduction of pests of plants and plant products, and to promote appropriate measures for their control" (Article I.1 of the IPPC). The requirement for or application of phytosanitary treatments to plants, plant products, and regulated articles is a phytosanitary measure used by countries to prevent the introduction and spread of regulated pests. Article VII.1 reminds us that countries have sovereign authority to regulate, in accordance with applicable international agreements, the entry of plants and plant products and other regulated articles and, to this end, may prescribe and adopt phytosanitary measures, including, for example, inspection, prohibition on importation, and treatment.

Phytosanitary measures required by a country must also be technically justified (Article VII.2(a) of the IPPC). In the SPS and IPPC plant health framework, high efficacy treatments are valuable measures to deal with new and unexpected, i.e., emergency, phytosanitary situations. It is legitimate for a risk manager to use a high efficacy treatment when no information and time are available to determine if a less rigorous measure is more appropriate for the situation. However, risk managers should remember that in order to maintain this strong measure in place they have the obligation to provide justification for maintaining it.

Emergency Measure: a phytosanitary measure established as a matter of urgency in a new or unexpected phytosanitary situation. An emergency measure may or may not be a provisional measure.

(ISPM No. 5, IPPC, 2019a)

Furthermore, regulatory officials including risk managers should clearly distinguish this situation from one where scientific information is available, i.e., a pest risk assessment has been completed, and where the situation is not an emergency requiring immediate action, such as routinely takes place when managing pest risks in international trade. In this instance, risk managers should choose phytosanitary measures or treatments that are closely aligned with the identified pest risk to ensure that the management measure has a rational relationship with the identified pest risk.

The pest risk assessment provides risk managers with:

- information on specific pests in that pathway
- the pest's ability or inability to enter and establish in the importing country
- the magnitude of the consequences of pest introduction
- production processes that might affect risk.

As such, risk management measures can and should be tailored to the available information. If the identified risk is high, there is technical justification for a proportionately strong measure, e.g., a phytosanitary treatment with high efficacy. Equally, if the identified risk is low, the measure should be commensurately light. In some cases, there will be no need to have risk management measures in place, for example, if the risk is negligible. Unfortunately, this practice is seldom used by NPPOs around the world.

The lack of alignment of the strength of measures with identified risk:

- is troubling from the standpoint of consistent risk management
- can divert plant health resources to where they are not needed
- can result in an unnecessary burden to international trade
- might lead to trade tension and legitimate disputes between trading partners.

Case study: Varietal testing requirements for the importation of fresh apple fruit to Japan from the United States

Under Japan's plant protection law, the importation of certain fruits was prohibited, i.e., the strongest phytosanitary measure, because of the possibility of being hosts for codling moth (*Cydia pomonella*). Eight types of fruit: apricot, cherry, plum, pear, quince, peach (including nectarine) apple and walnut, required testing for methyl bromide fumigation effectiveness against this pest. Japan also insisted that all varieties of each type of fruit be tested despite research showing that there was no difference in efficacy among varieties of the same fruit. The WTO dispute settlement panel and appellate body found that Japan's requirement was being maintained without sufficient scientific evidence and therefore violated SPS Article 2.2. Japan's measure also violated SPS Article 5.1 for four types of fruit, i.e., apricot, pear, plum, and quince, as the measure was not based on a proper risk assessment.

21.3. Relevant Standards

Phytosanitary treatments have a long history of development and many that are still in use today were adopted before pest risk analysis (PRA) was an established practice. Today, the IPPC has adopted several standards that provide guidance on the conduct of PRA, among them ISPM 2 (IPPC, 2016) adopted in 2007, ISPM 11 (IPPC, 2013) adopted in 2013, ISPM 21 (IPPC, 2004) adopted in 2004 and ISPM 32 adopted in 2016.

ISPM 2 reminds us that pest risk management is the third stage of PRA and involves identification of phytosanitary measures that, alone or in combination, reduce the identified risk to an acceptable level. ISPM 11 indicates that the guiding principle for risk management should be to manage identified risk to an acceptable level that can be justified and highlights that phytosanitary measures should not be more trade restrictive than necessary and follow the principle of minimal impact from ISPM 1—*Phytosanitary principles for the protection of plants and the application of phytosanitary measures in international trade.*

Even though pest risk management is mentioned in ISPMs 2, 11 and 21, there is not yet a specific concept standard on the topic. Approved regional concept standards on pest risk management are available that can inform this effort, for example Regional Standard for Phytosanitary Measures—RSPM 40—*Principles of pest risk management for the Import of Commodities* from the North American Plant Protection Organization (NAPPO, 2014).

Several ISPMs also provide guidance on the development and application of phytosanitary treatments, including:

ISPM 18—*Guidelines for the use of irradiation as a phytosanitary measure* (IPPC, 2003)
ISPM 28—*Phytosanitary treatments for regulated pests* (IPPC, 2007)
ISPM 42—*Requirements for the use of temperature treatments as phytosanitary measures* (IPPC, 2018) and
ISPM 43—*Requirements for the use of fumigation as a phytosanitary measure* (IPPC, 2019b)

Each of these is discussed below with additional detail regarding corresponding treatments.

21.3.1. ISPM 18—*Guidelines for the use of irradiation as a phytosanitary measure*

ISPM 18 provides technical guidance on the application of ionizing radiation to prevent the introduction and spread of regulated pests infesting plant commodities. The objective is realized by achieving specific responses in the targeted pests which may include pest mortality, pest inactivation, inability for pests to complete development, or inability for pests to reproduce, i.e., insect sterility. Ionizing radiation treatments may be delivered using radioactive isotopes, electrons generated by machine sources, or X-rays. The unit of measurement for the absorbed dose is the gray (Gy).

ISPM 18 was adopted in 2003. Annex 1 was reserved for listing approved irradiation treatments. However, adopted irradiation treatments are now included as annexes to ISPM 28 and they target specific commodities infested with fruit flies, moths, weevils, and mealybugs, either with specific or generic radiation doses. Phytosanitary uses of irradiation also include the devitalization of plants, for example, treated tubers, bulbs, or cuttings are prevented from sprouting and treated seeds may germinate but seedlings fail to grow, which is a risk management measure for weed seed contaminants.

Several variables need to be considered when applying ionizing radiation as a phytosanitary treatment including dose rate, treatment time, temperature, humidity, ventilation, and modified atmospheres. All variables should be compatible with treatment effectiveness. Another key variable is the tolerance of the commodity. Some commodities such as papaya (*Carica papaya*) and mango (*Mangifera indica*) respond well to irradiation treatment and are radiophilic. They suffer no damage and have improved shelf life. Other commodities such as avocados *(Persea americana)* do not tolerate even very low doses of irradiation and are radiophobic. Most commodities are radiosensitive and may or may not suffer damage depending on the dose and factors such as the cultivar, maturity, and temperature. Although commodity tolerance is a quality concern rather than a phytosanitary concern, it is an important factor in determining the feasibility of a treatment for a specific commodity and therefore merits attention by risk managers (See Chapter 23, this volume).

Irradiation for plant commodities can be conducted in the exporting country or upon entry into the importing country. The United States accepts irradiated commodities from countries where their NPPO has an established preclearance program. Preclearance programs are developed jointly with the host country's NPPO and can include measures such as inspection and treatments conducted under the supervision of the U.S. NPPO. The United States has preclearance programs for irradiation with Mexico, Vietnam, Thailand, India, and Australia. Irradiation for plant commodities can also be conducted upon entry to the importing country. The United States NPPO certifies irradiation treatment facilities for this purpose and their website currently lists five certified facilities (https://www.aphis.usda.gov/import_export/ plants/manuals/ports/downloads/ir_facility_ list.pdf, accessed January 8, 2020.) Examples of irradiated commodities entering the United States include fresh guava fruit (*Psidium guajava*) from Mexico irradiated at origin at 400 Gy.

The International Atomic Energy Agency maintains a publicly accessible International Database on Insect Disinfection and Sterilization (IDIDAS - https://nucleus.iaea.org/sites/naipc/ ididas/SitePages/International%20Database %20on%20Insect%20Disinfestation%20 and%20Sterilization%20(IDIDAS).aspx). It includes data on doses required for phytosanitary irradiation of plant commodities infested with specific pests as well as doses of radiation used to induce sterility in target pests through the application of the sterile insect technique. IDIDAS is complemented by another database, the International Database on Commodity Tolerance (IDCT - https://nucleus.iaea.org/sites/naipc/ IDCT/Pages/default.aspx) which is useful to risk managers for understanding research on the irradiation doses tolerated by most commodities.

21.3.2. ISPM 28—*Phytosanitary treatments for regulated pests*

The purpose of ISPM 28 is to support the application of phytosanitary measures in a wide range of circumstances and enhance recognition of treatment efficacy by NPPOs which may expedite international trade. ISPM 28 lists phytosanitary treatments as annexes for specific pests or for specific pests on specific commodities that have

been evaluated/approved by the IPPC's Technical Panel on Phytosanitary Treatments (TPPT). The phytosanitary treatments provide the minimum requirements necessary to control the specified pest at a stated efficacy.

The body of ISPM 28 also describes the minimum requirements for submission of new phytosanitary treatments by NPPOs. Information should include a detailed description of the treatment including efficacy data, preferably including data obtained under controlled as well as operational conditions. Information on treatment feasibility and applicability should include costs, commercial relevance, versatility, and level of expertise required for application. Submitted treatments may include chemical, physical, mechanical, or controlled atmosphere treatments.

Even though the ISPM 5 definition of pest includes any species, strain or biotype of plant, animal, or pathogenic agent injurious to plants or plant products, most pests for which treatments have been developed and are included in ISPM 28 are insects, with only one treatment affecting nematodes in debarked wood. A list of phytosanitary treatments from ISPM 28 is provided in Table 21.1. Thirty phytosanitary treatments provide efficacy information for physical treatments, 9 using cold, 5 using (vapor) heat and 16 using irradiation, against generic or specific pests or pest/commodity combinations. Except for phytosanitary treatment 19 (irradiation against three species of mealybugs) all other phytosanitary treatments target internal feeding pests. The remaining two phytosanitary treatments deal with sulfuryl fumigation, a chemical treatment for debarked wood. ISPM 28 does not provide specific guidance for phytosanitary treatments for seeds or for plants for planting where the most important pests are usually pathogenic agents. The generalities of physical and chemical treatments are discussed later in this chapter.

> Treatment schedule: the critical parameters of a treatment which need to be met to achieve the intended outcome at a stated efficiency.
> (ISPM No. 5, IPPC, 2019a)

21.3.3. ISPM 42—*Requirements for the use of temperature treatments as phytosanitary measures*

ISPM 42 provides general guidance on the application of temperature treatments, i.e., cold and

Table 21.1. List of phytosanitary treatments from ISPM 28.

#	Treatment	Pest	Pest type	Commodity
PT 1	Irradiation	*Anastrepha ludens*	Fruit fly	Any
PT 2	Irradiation	*Anastrepha obliqua*	Fruit fly	Any
PT 3	Irradiation	*Anastrepha serpentina*	Fruit fly	Any
PT 4	Irradiation	*Bactrocera jarvisi*	Fruit fly	Any
PT 5	Irradiation	*Bactrocera tryoni*	Fruit fly	Any
PT 6	Irradiation	*Cydia pomonella*	Moth	Any
PT 7	Irradiation	Tephritidae generic	Fruit fly	Any
PT 8	Irradiation	*Rhagoletis pomonella*	Fruit fly	Any
PT 9	Irradiation	*Conotrachelus nenuphar*	Weevil	Any
PT 10	Irradiation	*Grapholita molesta*	Moth	Any
PT 11	Irradiation	*Grapholita molesta* hypoxia	Moth	Any
PT 12	Irradiation	*Cylas formicarius elegantulus*	Weevil	Any
PT 13	Irradiation	*Euscepes postfasciatus*	Weevil	Any
PT 14	Irradiation	*Ceratitis capitata*	Fruit fly	Any
PT 15	Vapor heat	*Bactrocera cucurbitae*	Fruit fly	*Cucumis melo* var. *reticulatus*
PT 16	Cold	*Bactrocera tryoni*	Fruit fly	*Citrus sinensis*
PT 17	Cold	*Bactrocera tryoni*	Fruit fly	*Citrus reticulata* x *C. sinensis*
PT 18	Cold	*Bactrocera tryoni*	Fruit fly	*Citrus limon*
PT 19	Irradiation	*Dysmiococcus neobrevipes, Planococcus lilacinus* and *Planococcus minor*	Mealybugs	Any
PT 20	Irradiation	*Ostrinia nubilalis*	Moth	Any
PT 21	Vapor heat	*Bactrocera melanotus* and *Bactrocera xanthodes*	Fruit fly	*Carica papaya*
PT 22	Sulfuryl fumigation	Insects		Debarked wood
PT 23	Sulfuryl fumigation	Nematodes and insects		Debarked wood
PT 24	Cold	*Ceratitis capitata*	Fruit fly	*Citrus sinensis*
PT 25	Cold	*Ceratitis capitata*	Fruit fly	*Citrus reticulata* x *C. sinensis*
PT 26	Cold	*Ceratitis capitata*	Fruit fly	*Citrus limon*
PT 27	Cold	*Ceratitis capitata*	Fruit fly	*Citrus paradise*
PT 28	Cold	*Ceratitis capitata*	Fruit fly	*Citrus reticulata*
PT 29	Cold	*Ceratitis capitata*	Fruit fly	*Citrus clementina*
PT 30	Vapor heat	*Ceratitis capitata*	Fruit fly	*Mangifera indica*
PT 31	Vapor heat	*Bactrocera tryoni*	Fruit fly	*Mangifera indica*
PT 32	Vapor heat	*Bactrocera dorsalis*	Fruit fly	*Carica papaya*

heat, to plant commodities. It contains general information on the operational requirements for each treatment. It complements the 14 temperature treatments that are annexes to ISPM 28. Two chapters in the book by Heather and Hallman (2008) and the chapter on "Phytosanitary treatments for biosecurity" by Hennessey and colleagues in the book edited by Gordh and McKirdy (2014) provide details on the specifications and mechanics of temperature treatments.

In general, cold treatments use refrigerated air to lower the commodity temperature to or below a specific temperature for a specific period of time. Temperature recording generally begins after the commodity has reached the desired temperature, i.e., precooling. Cold treatment is primarily used on fresh fruit commodities that are hosts of internal feeding pests. For example, the (*Phytosanitary*) *Treatment manual for the United States* NPPO lists cold treatment schedules for fresh apple (*Malus domestica*), apricot (*Prunus armeniaca*), avocado, blueberry (*Vaccinium corymbosum*), cape gooseberry (*Physalis peruviana*), carambola (*Averrhoa carambola*), cherry (*Prunus avium*), citrus (*Citrus* spp.), date (*Phoenix dactylifera*), grape (*Vitis vinifera*), guava, kiwi (*Actinidia deliciosa*), loquat (*Eriobotrya japonica*), lychee (*Litchi chinensis*), nectarine and peach (*Prunus persica*), pear (*Pyrus communis*), persimmon (*Diospyros kaki*), plum (*Prunus domestica*), pomegranate (*Punica granatum*), quince (*Cydonia oblonga*), and for pecan (*Carya illinoinensis*) and hickory nuts (*Carya ovata*), among others, against mainly immature stages (eggs and larvae) of Tephritid fruit flies and beetle and moth fruit boring pests. Cold treatments may be used in combination with chemical treatments and are usually applied sequentially. Cold treatments can be applied to commodities at origin, on arrival, or in transit where there is sufficient time and where approved equipment is available.

Heat treatments raise the commodity temperature to the minimum or a higher temperature for a specific period of time. In some instances, rapid cooling is recommended after completion of the heat treatment to preserve commodity quality. Heat treatments may be used in combination with chemical treatments and application of the two is usually sequential. Heat treatments can be administered using hot water, i.e., hydrothermal treatment. The commodity is either submerged for a specific period of time to ensure the required temperature is achieved at the core, or it can involve a sequence of dips at different temperatures.

Hot water immersion is used for certain fruits and vegetables attacked by Tephritid fruit flies and can be used on bulbs for planting and on some seeds. Selected rooted or unrooted plant cuttings can also be treated with hot water, e.g., *Chrysanthemum* spp., to eliminate pathogens. Imported fresh mango fruit from different countries require hydrothermal treatment as a condition of entry into the United States. As with any treatment, there are a number of specific requirements for heat treatments including commodity size, electrical and electronic components, controls and recording devices including those that measure temperature, water quality, and circulation logistics, and options for pre-warming the commodity, for post-treatment cooling and for protecting or safeguarding the treated commodity.

Vapor heat and forced hot air treatments use heated air to increase the temperature inside mostly fruit commodities that are hosts for Tephritid fruit flies. The treatments differ in the amount of relative humidity present in the air. In the United States, vapor heat treatment is used as a condition of entry for fresh bell pepper (*Capsicum annuum*), eggplant (*Solanum melongena*), squash (*Cucurbita* spp.), tomato (*Solanum lycopersicum*) and zucchini (*Cucurbita pepo*) to kill Mediterranean (*Ceratitis capitata*), Oriental (*Bactrocera dorsalis*), and Melon fruit fly (*Bactrocera cucurbitae*) eggs and larvae. Vapor heat is also used to import fruit and vegetable commodities from Hawaii into the U.S. mainland, such as sweet potato (*Ipomoea batatas*), against several beetle species, rambutan (*Nephelium lappaceum*), lychee (*Litchi chinensis*), and longan (*Dimocarpus longan*) against Mediterranean and Oriental fruit fly eggs and larvae.

Dry heat has a long history of use dating back to the late 1700s and employs low humidity hot air to control pests in stored products such as grain, nuts, and dried fruit. Bird seed, like Niger seed (*Guizotia abyssinica*), imported into the United States, is frequently contaminated with weed seeds and a dry heat treatment at 120°C for 15 minutes is effective at devitalizing the weed seeds.

21.3.4. ISPM 43—Requirements for the use of fumigation as a phytosanitary measure

ISPM 43 provides generic requirements for fumigation and is meant to complement the fumigation treatments that are annexes to ISPM 28. Fumigants are chemicals that exist in a gaseous state at normal temperatures and pressures. Fumigants penetrate commodities such as wood and fruits and vegetables to effect internal disinfestation. ISPM 43 indicates that fumigation may be used at any point along a plant/plant product pathway, including during production, packaging or storage at origin, during transport and upon arrival in the importing country. ISPM 43 includes four appendices. Appendix 1 provides a list of the most common single use fumigants. Appendices 2–4 provide formulae to calculate the amount of fumigant required by weight or by volume, the volume for different geometrical shapes of fumigation enclosures, and to calculate gas concentration-time product (CT).

ISPM 43 does not provide details on specific treatments with specific fumigants. Rather, it outlines requirements to ensure that the treatment critical parameters, like concentration/dose, temperature and duration, are achieved so that the stated level of treatment efficacy is reached. It also describes the types of enclosures and equipment needed for fumigation, including dosing, gas vaporizing, heating and circulation equipment, and parameters that should be measured including pressure, temperature and gas concentration, and how to authorize and determine adequacy of third-party treatment providers. It discusses monitoring and auditing activities including documentation and record-keeping.

Fumigants historically used for phytosanitary purposes have included carbon bisulfide, carbonyl sulfide, cyanide, ethylene dibromide, ethylene oxide, methyl bromide, phosphine, and sulfuryl fluoride (Heather and Hallman, 2008). However, health and environmental concerns have reduced the current options to methyl bromide, phosphine and sulfuryl fluoride, with methyl bromide named in the Montreal Protocol as an ozone-depleting substance.

Fumigation involves releasing/dispersing a toxic chemical so that it reaches the target organism in a gaseous state. Chemicals that are applied as aerosols, smokes, mists, and fogs are suspensions of particulate matter in air and are thus not fumigants. Because fumigants are gaseous, their molecules can penetrate and then exit the commodity after fumigation. Fumigation can be used as a condition of entry or applied as an emergency measure. The toxicity of a fumigant depends on the respiration rate of the target pest. In general, pests are poikilotherms and their rate of respiration is lower at lower temperatures. Therefore, fumigation at lower temperatures may require a higher dosage and a longer exposure period. Two chapters in the book by Heather and Hallman (2008) and the chapter by Hennessey and colleagues in the book edited by Gordh and McKirdy (2014) provide details on the specifications and mechanics of fumigation treatments.

Fumigation with methyl bromide is effective against a wide variety of plant pests associated with many commodities and can also be used to devitalize plant material. At the appropriate dosage rate, temperature and time, it is effective against all life stages of insects, mites, nematodes including cysts, snails, slugs, and some fungi such as oak wilt fungus (*Bretziella fagacearum*—formerly *Ceratocystis fagacearum*). Some treatment schedules are as short as two

hours. Pest mortality may not be immediate. In the United States methyl bromide fumigation is used as a condition of entry for numerous fresh fruits and vegetables like apple, pear, apricot, peach, plum, nectarine, asparagus (*Asparagus officinalis*), avocado, banana and plantain (*Musa* spp.), bean (*Phaseolus vulgaris*), beet (*Beta vulgaris*), blackberry (*Rubus* spp.), blueberry, broccoli, brussels sprouts, cabbage and cauliflower (*Brassica oleracea*), rapini (*Brassica ruvo*), bok choi (*Brassica rapa*), cantaloupe (*Cucumis melo*), carrot (*Daucus carota*), cassava (*Manihot esculenta*), celery (*Apium graveolens*), chayote (*Sechium edule*), chicory and endive (*Cichorium* spp.), citrus, cucumber (*Cucumis sativus*), corn (*Zea mays*), garlic (*Allium sativum*), shallot (*Allium cepa*), and strawberry (*Fragaria* spp.), to name a few.

Methyl bromide is also used to fumigate nursery stock, other plants, bulbs, corms, tubers, rhizomes, and roots and some seeds. Fumigation can specify the commodity, e.g., broadleaf and coniferous genera fumigated against external feeders, specify a particular pest, e.g., host plants of citrus blackfly (*Aleurocanthus woglumi*), identify the plant commodity as well as the pest, e.g., *Gladiolus* spp. for gladiolus thrips (*Taeniothrips simplex*), or specify a commodity, its destination, and the pest of concern, e.g., *Pinus* spp. from Canada destined to CA, ID, OR, UT for European pine shoot moth (*Rhyacionia buoliana*). A few examples of seed commodities fumigated with methyl bromide include conifer, chestnut, and cotton seed in bulk, packaged or bagged.

21.4. Technical Attributes

Phytosanitary treatments aim to reduce the prevalence or viability of pests by exposing them to conditions and agents which have a detrimental effect. As indicated earlier, phytosanitary treatments include chemicals, delivered primarily as fumigants or through dipping or spraying the commodity, and physical treatments which include subjecting the commodity to cold, heat or ionizing irradiation. A good treatment should be efficacious against the target pest while minimizing harmful effects on the articles being treated. Heather and Hallman (2008) includes eight chapters dedicated to the development and application of phytosanitary treatments by NPPOs and is a useful reference for risk managers.

NPPOs need assurances that the efficacy of a phytosanitary treatment has been scientifically demonstrated for the pest(s) of concern and for the expected/required response. Efficacy comprises two distinct elements:

- a precise description of required response, such as mortality, sterility, devitalization, and the like
- the statistical level of response including both metric and methodology.

21.4.1. Relationship to pest risk

Phytosanitary treatments have a long history of development around the assumptions of a worst-case scenario and one-size-fits-all design for a single measure with high efficacy based on mortality. As a result, most phytosanitary treatments in use today are an over-reaction to the identified pest risk. Furthermore, in instances of high pest infestation where even a few survivors are problematic, an "uncalibrated" treatment may even be an under-reaction. This lack of alignment with risk is problematic from the standpoint of consistent risk management. It can also result in a waste of resources and an unnecessary burden to international trade.

In emergency situations it may be necessary to use a treatment without regard to its alignment with the risk simply because there is no opportunity to be more precise. However, treatments that are prescribed in advance as conditions for trade are fundamentally different because the strength of measures, i.e., efficacy in the case of treatments, should have a rational relationship with the risk identified in the pest risk assessment.

As indicated earlier, many existing phytosanitary treatments were adopted before PRA was developed and practiced. Now that PRA is practiced routinely when evaluating market access requests, there is opportunity to substantially reduce the strength of many phytosanitary treatments by simply evaluating the "action" policy for the treatment to align it with the identified risk. These risk-based strategies have the added advantage of reducing the use of environmentally harmful pesticides such as methyl bromide.

Methyl bromide fumigation with Probit 9 mortality

Treatment efficacy is correctly described as the desired response and method of measure. In this case, the treatment is a methyl bromide fumigation, the statistical method used for measurement is probit analysis, the level of efficacy to be achieved is Probit 9, and the desired response is mortality.

Risk-based phytosanitary treatments are those that are tailored to the pest risk and situation. These are those prescribed treatments for which their efficacy is "matched" to the risk which has been determined through the conduct of PRA. Such treatments are currently rare because of the reluctance of risk managers to accept such precision in the face of perceived uncertainty. An example of a risk-based treatment would be accepting a less than Probit 9 treatment, the "standard" for fruit fly treatments, for plant/plant product imports from low pest prevalence areas. The idea here is that the threshold for treatment efficacy is lowered because we recognize that pest load in low prevalence areas is greatly reduced.

This is where the SPS Agreement forces us to break from our historical paradigm which is focused on killing/removing pests from the pathway and re-focus instead on a risk-based approach where risk is defined by establishment potential and consequences rather than by presence or absence of a pest in a pathway at import. Adopting a risk-based treatment approach will open new doors of opportunity for treatment research in an area that has stagnated over the last 20 years. Adopting a risk-based treatment approach would also demonstrate that the phytosanitary community takes the SPS Agreement seriously and is making itself contemporary.

21.4.2. Extrapolation

In the case of phytosanitary treatments used as emergency measures, there is a tendency for risk managers to extrapolate treatment efficacy. This means that they may use a treatment in a particular situation where they don't have data or don't have complete data on efficacy for a particular pest situation. Fumigation treatments for "external feeders" and "internal feeders" are good extrapolation examples. Research to demonstrate efficacy for all possible internal or external feeders has not been completed, however, enough research has been done that risk managers feel confident extrapolating the results to similar untested pests. This may be a legitimate approach in emergency situations, but repeated use of extrapolation in non-emergency situations starts to press the legitimacy question. Even though an emergency treatment can be almost anything, it still needs to have a "rational relationship" to the risk; meaning that it is actually known to have some mitigation effect on the pest of concern. For instance, requiring a methyl bromide fumigation for a virus is clearly not legitimate.

> Efficacy (of a treatment): a defined, measurable and reproducible effect by a prescribed treatment.
>   (ISPM No. 5, IPPC 2019a)

There is a wide-open field for discipline and guidance that is a great opportunity for treatment scientists. It is less of an issue with irradiation because that treatment is based on absorbed dose, but across the board, there is a need to understand how far risk managers can extrapolate without specific data. It would seem logical in SPS terms that extrapolation on emergency measures should be more flexible than on prescribed measures, but even that simple point is not yet discussed in any guidance within the SPS-IPPC framework.

21.4.3. Probit 9

Probit 9 mortality was suggested by Baker in 1939 as the required response for fruit fly treatments (Baker, 1939). Probit 9 requires a mortality response of 99.9968%, or 32 or fewer survivors in 1 million individuals treated. No scientific rationale was provided for his choice of mortality as a criterion or his choice of probit 9 as a standard. On this basis, probit 9 has managed to become widely adopted by many countries as the benchmark for quarantine treatments for a wide variety of pests (Roth, 1989; Liquido *et al.*, 1996; Follett and Neven, 2006). Baker reasoned that probit 9 was appropriate because it assured a high level of pest mortality without high levels of damage

to the host. It is a standard based more on tradition than hard science. It was often the best phytosanitary risk management option when few options were available. It is easy to convince an importing trading partner to accept a consignment that has been subjected to such a high degree of quarantine security. It may be unreasonably stringent when an importing trading partner demands it. Many risk managers who would agree that probit 9 treatments achieve a high level of efficacy would be unable to explain why a treatment should achieve 99.9968% efficacy or to define an alternative acceptable limit of efficacy.

Probit 9 treatments are typically favored over other methods when a high infestation rate is expected on a host that can tolerate the treatment and pest organisms can be easily collected, maintained, and treated in large numbers. Under such circumstances, probit 9 level treatments can be more convenient due primarily to the relative speed with which they can be developed, tested, and implemented as compared to the potentially more difficult and complex task of collecting data on pest prevalence, infestation levels, and estimating establishment potential for a customized treatment. The rigorous nature of a probit 9 treatment virtually assures pest freedom for normal commercial shipments and can provide a fast-track to overcoming phytosanitary restrictions. As a result, many important probit 9 treatments have been accepted and are widely practiced based on a design for substantial overkill in order to speed acceptance of the treatment and facilitate trade. Such treatments are difficult to link to a technical justification and are more likely to violate the principles of "equivalency" and "least restrictive measure." The legitimacy of such treatments comes from the fact that they are bilaterally agreed to, not that they are technically justified (Griffin, 2013). Because probit 9 is well-known and has been used for decades, it is often considered to be a universal standard and was once considered the only legitimate technique for analyzing treatment data. Over time however, there has been growing concern about shortcomings in the wholesale use of probit applications for treatment research while at the same time, many different statistical techniques have emerged as alternatives (Landolt *et al.*, 1984; Robertson *et al.* 1994; Chew, 1996; Liquido *et al.*, 1996).

Probit analysis is a statistical methodology used by treatment researchers to derive the dose for a specific level of response. The mathematical derivation of a dose–response relationship assumes that each organism has a tolerance for some stimulus like dose, temperature, time, and the like, that will cause a response if the stimulus exceeds the tolerance threshold of the organism. The desired response for pest treatments is usually mortality, but the response may be different depending on the treatment.

With the advent of phytosanitary risk management, the probit 9 standard has come under considerable criticism in addition to the fact that it is an arbitrary standard that is not based on science. For example, there are other evaluation criteria for risk management that may be more appropriate choices than mortality. Responses such as sterility or preventing maturation, the likelihood of infestation, natural survival, reproductive potential, and establishment potential, packaging and shipping practices and distribution times can also be legitimate options for risk management because although the pests may survive, they are unable to establish. (Landolt *et al.*, 1984; Liquido *et al.*, 1996; Follett and Neven, 2006; Robertson *et al.*, 1994). Probit 9 does not consider the actual risk, i.e., the actual infestation rate, in the consignment before or after the treatment. Requiring probit 9 for commodities rarely infested may be too stringent a measure or difficult to demonstrate. At the other end of the criticism scale is the fact that probit 9 means a 0.0032% survival rate. Commodities shipped in large volumes or with high rates of infestation may still contain substantial numbers of live pests.

Risk levels do not depend on the number of pests that are effectively treated, but rather the number that survive to reproduce and their potential for establishment. Linking the efficacy of the treatment to some level of risk requires, at minimum, a good estimate of the anticipated infestation level in order to estimate survivorship. Beyond this, there are biological, economic, and other important variables to consider for estimating the likelihood and consequences of establishment if the treatment is to have a rational relationship to the risk (see also Chapter 33). From this standpoint, the decision by a risk manager to adopt a treatment should not be based on

an arbitrary treatment efficacy but tailored to account for all the variables and options identified in PRA. Customizing treatments to specific commodities and pest complexes is needed to demonstrate that quarantine treatments are technically justified. The establishment of such a process by risk managers is not only important to ensure compliance with the SPS Agreement, but also allowing the flexibility needed to fairly address different risks for specific pests and pathways and not waste valuable risk management resources on undefendable overkill designs.

Basing treatment decisions on a uniform efficacy requirement also ignores risk-based considerations such as the actual infestation rate, the infestation rate of good hosts vs. poor hosts, natural attrition, different survivor rates, mating and reproduction potential, and colonization potential. For most pests, at least one male and one female need to survive in the same consignment for mating. A founding population usually requires more than one mating pair arriving at the same time in a susceptible area under suitable conditions for establishment to take place.

Pre-harvest pest control and cultural practices, harvest and post-harvest processes, packaging, shipping, and distribution parameters are also missing from the treatment decision when efficacy is the sole criterion. In fact, the focus on achieving a certain level of response is a distraction if it is not technically justified by the risk. This is a key weakness in the probit 9 concept and argues strongly for fully understanding the characteristics of probit 9 requirements in each situation and considering its relationship to the risk.

21.5. Phytosanitary Treatments Used to Reduce/Manage Risk

This section provides examples that illustrate diverse applications of phytosanitary treatments.

21.5.1. ISPM 15 and wood packaging material (WPM) moving in international trade

Wood pallets, crates, boxes, spools/reels, and dunnage made of unprocessed raw wood are internationally recognized as important pathways for the introduction and spread of quarantine pests that pose risks mainly to living trees. Examples of pests introduced into the United States via this pathway include the emerald ash borer (*Agrilus planipennis*) and the Asian long-horned beetle (*Anoplophora glabripennis*). ISPM 15 describes internationally accepted phytosanitary measures that help countries significantly reduce the risk of pest introduction through WPM. Measures described in ISPM 15 include the use of debarked wood, with a specified tolerance for the amount of remaining bark, the application of approved phytosanitary treatments, including fumigation with methyl bromide, heat treatment in a conventional heat chamber using steam or dry heat, or heat treatment using dielectric or microwave heating, and the application of a recognized mark that ensures that treated WPM is identifiable.

NPPOs of both exporting and importing countries have specific responsibilities detailed in ISPM 15. In the United States, the NPPO in association with the wood packaging industry, developed an export program to ensure compliance with WPM import requirements of their trading countries. The quality control program certifies, i.e., authorizes, registers and accredits, WPM treatment facilities, manages/monitors the wood mark, establishes inspection, verification and auditing procedures for facilities and assists in traceability of the WPM. The integrity of the program is dependent upon compliance by industries. The NPPO of the importing country authorizes the entry of the treated WPM or verifies, on import, that ISPM 15 requirements have been met. NPPOs are also responsible for notifications of non-compliance.

Methyl bromide fumigation of WPM must be done in accordance with a treatment schedule specified or approved by the NPPO. The schedule should achieve a minimum concentration-time (CT)

over 24 hours at a specified temperature and have a final residual concentration. Monitoring of gas concentrations must be carried-out periodically and the CT must be achieved throughout the profile of the wood including its core. Examples of the treatment schedule for methyl bromide fumigation of wood from Annex 1 of ISPM 15 are provided in Table 21.2.

For heat treatments using conventional technology a minimum temperature of 56°C for a minimum duration of 30 continuous minutes throughout the entire profile of the wood is the fundamental requirement. For microwave heat treatment, a minimum temperature of 60°C for 1 continuous minute must be achieved throughout the profile of the wood. As new technical information becomes available, existing treatments may be modified or reviewed or new treatments for WPM may be developed.

21.5.2. Cold treatment for fresh fruit commodities

Holding commodities at temperatures near the freezing point of water was one of the first major disinfestation methods against quarantine pests. Cold treatment was used during an incursion of Mediterranean fruit fly into Florida, USA, in 1929 and is still being used today against several pests attacking different fresh fruit commodities.

The advantages of using cold treatment is its tolerance by a wide variety of plant commodities for consumption including subtropical and tropical fruits. Cold treatment systems are usually inexpensive compared to other treatments, are free from chemical residues and are accepted by organic growers. The chief disadvantage is the prolonged treatment time which might adversely affect the quality of the commodity.

Table 21.2. Minimum required CT over 24 hours for wood packaging material fumigated with methyl bromide.

Temperature (°C)	Minimum required CT (g/m³) over 24 hours	Minimum final concentration (g/m³) after 24 hours*
21.0 or above	650	24
16.0–20.9	800	28
10.0–15.9	900	32

*In circumstances when the final concentration is not achieved after 24 hours, a deviation in the concentration of ~5% is permitted provided additional treatment time is added to the end of the treatment to achieve the prescribed CT.

Cold treatment in trade situations is typically applied while the commodity is in transit, especially sea transport aboard ships. Cold treatment equipment must be certified to ensure it is able to produce and uniformly maintain the required temperatures without fluctuation. If cold treatments are interrupted and temperatures are not maintained, the shipment may be rejected on import or require retreatment. Temperature is measured in degrees Celsius, the internationally accepted standard for biological research and technology.

Table 21.3 below provides details of the nine ISPM 28 adopted treatment schedules and their stated level of efficacy for specific citrus commodities infested with different species of Tephritid fruit flies.

In some cases, commodities such as apples are stored post-harvest at temperatures that are lethal to pests and in this case the cold treatment occurs alongside post-harvest storage.

21.5.3. ISPM 38 and seeds moving in international trade

ISPM 38 provides guidance to NPPOs in identifying, assessing and managing pest risks associated with the international movement of seeds. It discusses the establishment of import

Table 21.3. Nine phytosanitary treatment schedules and efficacy levels for selected citrus commodities infested with different species of Tephritid fruit flies.

Annex	Pest species	Commodity	Cold treatment schedule—time begins after fruit has reached target temperature
PT 16	*Bactrocera tryoni*— Queensland fruit fly	*Citrus sinensis*— orange	3°C for 16 continuous days (CD); Navel 95% confidence killing no less than 99.9981% of eggs and larvae (E&L); for Valencia confidence killing no less than 99.9973% of E&L
PT 17	*B. tryoni*	*Citrus reticulata* x *C. sinensis*—tangor	3°C or < for 16 CD; 95% confidence killing no less than 99.9986% of E&L
PT 18	*B. tryoni*	*Citrus limon*—lemon	2°C or < for 14 CD—95% confidence killing no less than 99.99% or 3°C or < for 14 CD—95% confidence killing no less than 99.9872% of E&L
PT 24	*Ceratitis capitata* Mediterranean fruit fly	*Citrus sinensis*	2°C for 16 CD—95% confidence killing no less than 99.9937% or 2°C for 18 CD—95% confidence killing no less than 99.999% of E&L or 3°C or below for 20 CD—95% confidence killing no less than 99.9989% of E&L
PT 25	*C. capitata*	*Citrus reticulata* x *C. sinensis*	2°C or < for 18 CD—95% confidence killing no less than 99.9987% of E&L or 3°C or < for 20 CD—95% confidence killing no less than 99.9987% of E&L
PT 26	*C. capitata*	*Citrus limon*	2°C or < for 18 CD—95% confidence killing no less than 99.9975% of E&L or 3°C or < for 20 CD—95% confidence killing no less than 99.9973% of E&L
PT 27	*C. capitata*	*Citrus paradisi*	2°C or < for 19 CD—95% confidence killing no less than 99.9917% of E&L or 3°C or < for 23 CD—95% confidence killing no less than 99.9916% of E&L
PT 28	*C. capitata*	*Citrus reticulata*	2°C or <for 23 CD—95% confidence killing no less than 99.9918%
PT 29	*C. capitata*	*Citrus clementina*	2°C (max. fruit core temperature) for 16 CD—95% confidence killing no less than 99.9900% of E&L

requirements to facilitate seed movement and discusses inspection, sampling, testing, and phytosanitary certification of seeds.

Seed pests moving in international trade can include weed seeds as contaminants and seed-borne pests. According to definitions contained in ISPM 38, seed-borne pests are carried by seeds externally or internally and may or may not be transmitted to plants growing from these seeds and cause their infestation. Seed-transmitted pests are seed-borne pests that are transmitted via seeds directly to plants growing from these seeds and cause their infestation.

Seed certification, inspection, and cleaning to remove weed seed contaminants and testing to certify that seed lots are free from weed seed contaminants are common management practices. Seed testing certificates from internationally recognized/accredited laboratories or using internationally recognized testing methods can provide evidence of freedom from regulated weed species in seed lots. If NPPOs agree to mutually accept these measures, it would eliminate the need for testing seed lots in both the country of origin and destination.

ISPM 38 mentions that seed treatments may be used as phytosanitary measures, but that their application may be unrelated to pests, for example when seeds are treated with a growth enhancer. In general, seed treatments include application of pesticides including fungicides, insecticides, nematicides and bactericides, or disinfectants as well as physical treatments such as dry heat, steam, hot water, irradiation by ultraviolet light, high pressure, deep-freezing, and biological treatments. Recall that phytosanitary treatments are applied to kill, inactivate or remove pests or for rendering pests infertile or for devitalization.

Seed treatments are often "protective," e.g., intended to prevent a soil borne pest from infesting the growing seedling or used to remove microbial pests from the seed surface. A biological treatment for seed may directly, e.g., by antibiotic production or niche competition, or indirectly, by inducing plant resistance, reduce the risk of pathogen introduction and epidemic development. A seed treatment may be effective against one or more pathogens, or two or more treatments may be combined to target one or more pathogens.

Evaluating the efficacy of phytosanitary treatments for seeds can be challenging. Depending on the diagnostic test used to verify efficacy, seed may still test positive for pest presence even if the treatment has successfully inactivated it, e.g., serology may still detect pathogen proteins or DNA-based tests may still detect residual DNA. In some cases, it may be appropriate to grow out the seed under conditions conducive for disease expression to determine if viable pathogens are still present. Furthermore, seed phytosanitary treatments may not achieve 100% pest devitalization. As such, the risk identified for the pest and the appropriate level of protection of the importing country should determine the necessary level of efficacy for the treatment. A combination of factors may help evaluate the true risk of a particular infected/infested seed lot including pest biology, minimum founder rate, seed treatment efficacy, and acceptable level of risk.

Although seed trade has increased dramatically in the last 40 years, the tools available to regulators to manage seed transmitted pests remain limited and this is an area that needs further research. Many treatments currently available for phytosanitary purposes are not appropriate for dealing with pathogens in or on the seed coat of seed being used for propagative purposes as it dramatically decreases seed germination. Development of additional treatment options for seed borne and seed transmitted pathogens is needed. Several private industries are investigating new technologies to treat seed borne pathogens, but the treatments need to be validated by NPPOs in order to use these treatments for phytosanitary purposes. Additional options for both internal and external pathogens need to be developed by both industry and government scientists and validated for use by NPPOs. In advance of the adoption of ISPM 38, NAPPO developed a discussion document on *Criteria for evaluating phytosanitary seed treatments* (NAPPO, 2017).

As discussed above, ISPM 28 (*Phytosanitary treatments for regulated pests*) provides guidance on treatment feasibility. For seeds, a special consideration is the effect of treatments on viability or germination rate. Treatments that negatively impact these attributes should be carefully considered before use. Information on the following seed attributes should be included in the experimental design for seed treatments:

- vigor and viability
- physical damage and physiological alterations

- promotion of latency or premature germination
- effects on seed storage
- shelf-life of the treatment.

Other aspects of feasibility should include cost, practicality of application, and accessibility of inputs, e.g., equipment and materials. It should be noted that pesticide treatments for seeds may need to be registered in both the exporting and importing countries.

Additionally, seed sampling methodologies is an area for further research. Sampling can influence the detectability of pathogens in seed lots. Required sample sizes could vary depending on the pathogen and seed type. Many lots of seeds that move around the world contain less than 20,000 seeds. These small seed lots pose unique challenges to NPPOs in terms of sampling and testing. Often lots of seeds are too small for testing. Developing effective treatments for quarantine pathogens may facilitate movement of small lots of seeds.

21.5.4. ISPM 36 and plants for planting moving in international trade

Plants for planting are considered to pose a higher pest risk than other regulated articles and therefore additional specific guidance on pest risk management is needed to help address this elevated pest risk. The initial plant-related pest risk factors to be considered are plant species, cultivar, and area of origin. Within any given plant species, there is a range of pest risks associated with the type of plant material. Ranked from lowest to highest pest risk and recognizing that these rankings may vary depending on specific circumstances, are meristem tissue culture, in vitro culture, budwood/graftwood, unrooted cuttings, rooted cuttings, root fragments, cuttings, rootlets or rhizomes, bulbs and tubers, bare root plants, and rooted plants in pots. Pest risk may increase with plant age as older plants have increased exposure to potential pests.

21.6. Domestic Applications

Broadly speaking, the phytosanitary treatment toolbox is used when the NPPO is managing domestic programs for well-established pests or when it is dealing with issues associated with international trade in plants, plant products, or other regulated articles. In both scenarios, a risk manager may be faced with a new and unexpected phytosanitary emergency. Domestic programs with the goal of pest containment, suppression, or eradication generally use approved risk management actions and procedures, including phytosanitary treatments, codified in a country's regulations or may, on occasion, require emergency legislative exemptions for some treatments such as pesticides.

Domestic programs may use phytosanitary treatments to:

- prevent a well-established pest from expanding its range, e.g., the Japanese beetle (*Popillia japonica*) program in the Eastern United States (regulations can be found in 7 CFR 301.48) or the European cherry fruit fly (*Rhagoletis cerasi*) program in the province of Ontario, Canada
- slow the seasonal spread of a well-established pest, e.g., the gypsy moth (*Lymantria dispar*) program in the United States, or
- maintain/enforce a pest-free area (PFA) or area of low pest prevalence (ALPP) to enable export of a specific commodity from the area, e.g., maintenance of a Caribbean fruit fly (*Anastrepha suspensa*) PFA for white grapefruit in Florida, USA, destined for export to Japan.

In the latter example, risk management efforts begin to blur between domestic issues and international trade.

The historical model for a phytosanitary treatment is a single, high-mortality treatment prior to export or immediately upon entry of the traded commodity. The possibilities for greater flexibility and creativity for phytosanitary treatments have increased substantially as countries continue to explore alternatives to fumigation with methyl bromide and translate the SPS Agreement principles into practice. Combination treatments, low-dose treatments, non-mortality treatments, and treatments as part of systems approaches are becoming more common, dramatically increasing the options which can be considered for risk management.

Pest risk management associated with international trade in plants, plant products, or regulated articles requires bilateral or multilateral and harmonized actions. For example, in bilateral market access negotiations, the result of the PRA will identify the pest risks and indicate whether a specific commodity might require a phytosanitary treatment before export, during shipment or on arrival in the importing country. Once negotiations are concluded and market access is granted, exporting and importing countries will maintain documentation/records on negotiated and agreed-upon import requirements. For example, the importation of fresh apple fruit from Argentina into the United States has, as a condition of entry, the requirement for a cold treatment, usually, but not always, completed while the commodity is in transit. However, cold treatment is not required for Argentinian apples grown in areas that are free of quarantine significant fruit fly pests for the United States, e.g., medfly and South American fruit fly. Multilateral harmonized actions are necessary when trade of plants and plant products and movement of regulated articles do not adhere to the traditional "one country one commodity" model. Examples include international trade of seeds, movement of wood packaging material, and movement of sea containers.

> **Provisional measure:** A phytosanitary regulation or procedure established without full technical justification owing to current lack of adequate information. A provisional measure is subjected to periodic review and full technical justification as soon as possible.
>
> (ISPM 5, IPPC, 2019a)

As mentioned earlier, risk managers may be faced with an unanticipated plant health emergency such as a new pest incursion. In this situation, the risk manager must act quickly, and risk management decisions must be made under uncertainty. For example, no data may be available on the level of pest infestation/infection, the extent of the pest incursion, or the commodities affected by the pest. Furthermore, there may not be a PRA on that specific pest available anywhere. In extreme cases, the complete identity of the pest may be unknown. In these situations, risk managers apply an emergency measure. In a plant health emergency, the risk manager might choose a high efficacy treatment to deal with the new pest detection/incursion. In some instances, emergency exemptions might be needed for phytosanitary treatments, especially pesticides, that are not approved for use for that specific pest, commodity or combination. With no pest risk assessment and little or no additional information, the risk manager assumes the worst-case scenario. However, they should have some indication that the treatment chosen is effective against that pest type.

The ISPM 5 definition of emergency measure indicates that it may or may not be provisional. A provisional measure is one that is established without full technical justification. However, as indicated in the definition, a provisional measure is subjected to provision of full technical justification as soon as possible. Failure to provide justification for maintaining a provisional measure has been the subject of a WTO-SPS phytosanitary dispute (see case study in the Chapter). Japan did not seek information to evaluate the provisional measures it put in place for requirements for methyl bromide fumigation of fruit varieties susceptible to codling moth (*Cydia pomonella*).

21.7. Summary and Look Forward

Here are five things to remember from this chapter.

1. The strength of risk management measures can and should be tailored to the available scientific information; high risk situations justify strong measures while low risk situations warrant commensurately light measures.

2. The guiding principle for risk management should be to manage identified risk to an acceptable level that can be justified, that is not more trade restrictive than necessary and that follows the principle of minimal impact.

3. Phytosanitary treatments have a long history of development around the assumptions of a worst-case scenario and one-size-fits-all design for a single measure with high efficacy based on mortality that can be on an over-reaction to the pest risk identified by PRA.

4. Probit 9 level treatments, as a standard for phytosanitary security, are not based on any specific scientific data and can actually provide an insufficient level of security in some instances and too much in others.

5. The possibilities for greater flexibility and creativity for phytosanitary treatments have increased substantially as countries continue to explore alternatives and translate the SPS Agreement principles into practice.

The next chapter covers the concept of pest freedom as it applies to consignments and areas.

21.8. References

Baker, A.C. (1939) *The Basis for Treatment of Products Where Fruit Flies are Involved as a Condition of Entry into the United States*. Vol 551. Circular. United States Department of Agriculture, Washington, D.C.

Chew, V. (1996) Probit analysis and Probit 9 as a standard for quarantine security, in (P.W. Bartlett, C.R. Chaplin, and R.J. van Velsen, eds) *Plant Quarantine Statistics, a Review*. Proceedings of an International Workshop. Horticulture Research and Development Corp. Sydney, Australia.

Follett, P.A. and Neven, L.G. (2006) Current trends in quarantine entomology. *Annual Review of Entomology* 51, 359–385.

Gordh, G. and McKirdy, S. (eds) (2014) *The Handbook of Plant Biosecurity — Principles and Practices for the Identification, Containment and Control of Organisms that Threaten Agriculture and the Environment Globally*. Springer, Dordrecht.

Griffin, R.L. (2013) *Review of Phytosanitary Security Based on Probit 9 Treatment Standard*. Eighth Session of the Commission on Phytosanitary Measures: Agenda item 14. Secretariat of the International Plant Protection Convention, FAO, Rome Italy.

Heather, N.W. and Hallman, G.J. (2008) *Pest Management and Phytosanitary Trade Barriers*. CAB International, Wallingford, UK.

IPPC (2003) International Standards for Phytosanitary Measures, Publication No. 18: *Guidelines for the Use of Irradiation as a Phytosanitary Measure*. Secretariat of the International Plant Protection Convention (IPPC), Food and Agriculture Organization of the United Nations, Rome.

IPPC (2004) International Standards for Phytosanitary Measures, Publication No. 21: *Pest Risk Analysis for Regulated Non-quarantine Pests*. Secretariat of the International Plant Protection Convention (IPPC), Food and Agriculture Organization of the United Nations, Rome.

IPPC (2007) International Standards for Phytosanitary Measures, Publication No. 28: *Phytosanitary Treatments for Regulated Pests*. Secretariat of the International Plant Protection Convention (IPPC), Food and Agriculture Organization of the United Nations, Rome.

IPPC (2012) International Standards for Phytosanitary Measures, Publication No. 36: *Integrated Measures for Plants for Planting*. Secretariat of the International Plant Protection Convention (IPPC), Food and Agriculture Organization of the United Nations, Rome.

IPPC (2013) International Standards for Phytosanitary Measures, Publication No. 11: *Pest Risk Analysis for Quarantine Pests*. Secretariat of the International Plant Protection Convention (IPPC), Food and Agriculture Organization of the United Nations, Rome.

IPPC (2016) International Standards for Phytosanitary Measures, Publication No. 2: *Framework for Pest Risk Analysis*. Secretariat of the International Plant Protection Convention (IPPC), Food and Agriculture Organization of the United Nations, Rome.

IPPC (2016) International Standards for Phytosanitary Measures, Publication No. 32:*Categorization of Commodities According to their Pest Risk.* Secretariat of the International PlantProtection Convention (IPPC), Food and Agriculture Organization of the United Nations, Rome.

IPPC (2017) International Standards for Phytosanitary Measures, Publication No. 38: *International Movement of Seeds*. Secretariat of the International Plant Protection Convention (IPPC), Food and Agriculture Organization of the United Nations, Rome.

IPPC (2018a) International Standards for Phytosanitary Measures, Publication No. 15: *Regulation of Wood Packaging Material in International Trade*. Secretariat of the International Plant Protection Convention (IPPC), Food and Agriculture Organization of the United Nations, Rome.

IPPC (2018b) International Standards for Phytosanitary Measures, Publication No. 42: *Requirements for the Use of Temperature Treatments as Phytosanitary Measures*. Secretariat of the International Plant Protection Convention (IPPC), Food and Agriculture Organization of the United Nations, Rome.

IPPC (2019a) International Standards for Phytosanitary Measures, Publication No. 5: *Glossary of Phytosanitary Terms*. Secretariat of the International Plant Protection Convention (IPPC), Food and Agriculture Organization of the United Nations, Rome.

IPPC (2019b) International Standards for Phytosanitary Measures, Publication No. 43: *Requirements for the Use of Fumigation as a Phytosanitary Measure*. Secretariat of the International Plant Protection Convention (IPPC), Food and Agriculture Organization of the United Nations, Rome.

Landolt, P.J., Chamber, D.L. and Chew, V. (1984) Alternative to the use of probit 9 mortality as a criterion for quarantine treatments of fruit fly (Diptera: Tephritidae) infested fruit. *Journal of Economic Entomology* 77, 285–287.

Liquido, N.J., Griffin, R.L. and Vick, K.W. (1996) *Quarantine security for commodities: current approaches and potential strategies*. Proceedings of Joint Workshops of the Agricultural Research Service and the Animal and Plant Health Inspection Service. U.S. Dept. of Agriculture, Agricultural Research Service publication 1996-04 (Issued August 1997), Beltsville, MD.

NAPPO (2014) *Principles of Pest Risk Management for the Import of Commodities*. Available at: http://www.nappo.org/files/8314/3889/6413/RSPM40-e.pdf (accessed January 17, 2020).

NAPPO (2017) *Criteria for Evaluating Phytosanitary Seed Treatments*. Available at: http://www.nappo.org/files/5415/0970/9624/Seed_Treatment_Criteria_DD09-FINAL-e.pdf (accessed January 17, 2020).

Robertson, J.L., Preisler, H.K. and Frampton, E.R. (1994) Statistical concept and minimum threshold concept. In: Paull, R.E. and Armstrong, J.W. (eds) *Insect Pests and Fresh Horticultural Products: Treatments and Responses*. CAB International, Wallingford, UK.

Roth, H. (1989) Concepts and recent developments in regulatory treatments (R.P. Kahn, ed.) *Plant Protection and Quarantine* 3, 117–144. CRC Press, Boca Raton, Florida.

22

Pest-free Concepts

There are degrees of freedom. Complete freedom isn't always good, nor is the lack of it always bad.

Susan Dennard

22.1. Introduction

Pest freedom is obviously a desirable condition, especially when paired with a reliable mechanism to verify its status. Whether this condition is attributed to a growing area, a consignment, or even a period of time, there is substantial risk management value in having confidence that pests of concern are absent.

While the concept of pest freedom is broad enough for a range of applications, the primary focus for phytosanitary purposes is its geographic application for pest-free areas. The utility of this application is demonstrated by the number and types of standards established by the International Plant Protection Convention (IPPC) for different applications: pest-free areas, pest-free areas for fruit flies, pest-free areas of production, and pest-free production sites. ISPM No. 4 (*Requirements for the establishment of pest free areas*; IPPC, 2017a) was one of the first and most useful standards put in place by the IPPC because it was such a natural and universally appreciated concept. It also had significant positive implications for trade based on the numerous risk management programs successfully

implemented around pest-free area designs (Liquido and Griffin, 1997). The phytosanitary community is thus very familiar with the concept and experienced with its application in trade.

22.2. Degrees of Freedom

The general declaration for the model phytosanitary certificate annexed to the *New revised text* of the IPPC (1997) states:

> This is to certify that the plants, plant products or other regulated articles described herein have been inspected and/or tested according to appropriate official procedures and are considered to be free from quarantine pests specified by the importing contracting party and to conform with the current phytosanitary requirements of the importing contracting party, including those for regulated non-quarantine pests.

This statement was updated from the 1979 version which was slightly revised from the original 1952 version. The new statement includes several technical modifications to be consistent with changes to the Convention, but the conceptual core remains unchanged and fundamentally flawed in two key areas. First, the implied requirement for shipments to be free of quarantine pests is inconsistent with the WTO/SPS (World Trade Organization/Sanitary and Phytosanitary

Measures) which requires measures to be based on risk assessment or international standards. From this standpoint, the presence of a quarantine pest should be accepted if it is unlikely to establish via the pathway in question. The SPS Agreement argues that the requirement for pest freedom should be justified by a high probability of establishment for a pest with potentially serious consequences.

The second conceptual problem is "considered to be free from." The word "considered" provides an ambiguous space for interpretation that requires greater definition or discipline in additional guidance. Clarification is provided in ISPM 12 (*Guidelines for phytosanitary certificates*, revised 2014) (IPPC, 2017b):

> "Considered to be free from quarantine pests" refers to freedom from pests in numbers or quantities that can be detected by the application of phytosanitary procedures. It should not be interpreted to mean absolute freedom in all cases but rather that quarantine pests are believed not to be present based on the procedures used for their detection or elimination. It should be recognized that phytosanitary procedures have inherent uncertainty and variability and involve some probability that pests will not be detected or eliminated. This uncertainty and probability should be taken into account in the specification of appropriate procedures.

The central issue here is the technical reality that neither inspection nor testing or treatment can assure absolute freedom in every instance. The sensitivity and the intensity of the inspection, testing, or treatment, and the variability associated with those processes, determine the degree to which a shipment may be "considered" to be pest free. The key to applying this guidance in the real world is accepting and accounting for the fact that different procedures will have different probabilities of detection for different pests, and all procedures have some level of uncertainty about their efficacy.

The phytosanitary certification statement is an important example of the controversy that can be created when pest freedom is interpreted differently by trading partners as it is applied to a consignment. The same is true when the concept is applied to pest-free areas (ISPM 4), areas of low pest prevalence (ISPM 22), pest-free places of production and pest free production sites (ISPM 10). Likewise, for eradication (ISPM 9),

surveillance (ISPM 6), pest status (ISPM 8), and treatments (ISPM 28). In all cases, the significance of pest presence or absence depends on the ability to detect (sensitivity), the effort (intensity) and the uncertainty (variability and error) associated with the results. Assumptions are made about the conditions associated with the official finding of pest freedom in the absence of explicit information about the conditions.

Making pest freedom a requirement, declaring eradication, or recognizing a pest-free area, requires an understanding of the process and uncertainty associated with the conclusion in order to be complete and fully understandable. The absence of this information leaves the conclusion open to challenge and misinterpretation. Unfortunately, this point is not made frequently and clearly enough to ensure that an official statement of pest freedom associated with phytosanitary measures is always complemented by official recognition of the relevant conditions in order to be meaningful.

Shifting from the unrealistic idea of absolute freedom for a consignment or area to the more realistic concept of freedom in degrees qualified by conditions is helpful for understanding the concept of "strength of measures" and the "rational relationship" of phytosanitary measures to pest prevalence (see Chapter 33, this volume). Confusion around these concepts has been at the root of nearly every WTO dispute in the SPS domain and continues to cause trade tension across the phytosanitary community.

22.3. The WTO-SPS Agreement

Article 5.2 of the WTO-SPS Agreement (1994) refers to taking into account "the existence of pest- or disease-free areas" in the assessment of risk. Article 6 refers entirely to "adapting [measures] to the sanitary or phytosanitary characteristics of an area." One result of these references in the Agreement is establishing the legitimacy of risk management applied at other than a national boundary.

The animal health community refers to this concept as "regionalization" and has struggled to internalize the concept because it breaks with their century-long practice of basing disease status on a country. The phytosanitary community had no such history and has easily adopted and successfully applied the concept but struggles instead with recognizing pest freedom qualified by conditions related to risk. All freedom is not the same and the SPS Agreement provides no guidance on this technical aspect of understanding and applying pest-free concepts.

The SPS Agreement states that "Members *shall* take into account available scientific evidence ... prevalence of specific diseases or pests; existence of pest- or disease-free areas..." (Article 5.2, emphasis added). Article 6 describes key concepts related to pest free areas including:

- areas can include all of a country, part of a country or all or parts of several countries (Article 6.1)
- members shall recognize concepts of pest-free areas and areas of low pest prevalence, taking into account factors such as geography, ecosystems, epidemiological surveillance, and effectiveness of sanitary and phytosanitary controls (Article 6.2)
- exporting countries should furnish necessary evidence supporting claims of pest freedom (Article 6.3)
- the SPS Agreement provides definitions of pest-free areas and areas of low pest prevalence. (Annex A).

It is worth noting that these articles describe obligations for both importing and exporting countries. In the case of importing countries, they are obligated to take into account pest freedom and low pest prevalence in deciding on phytosanitary measures. In the case of exporting countries, they are obligated to provide supporting evidence and make such evidence available to importing countries. This collaborative design with shared burdens is a prominent feature throughout the SPS Agreement.

22.4. The IPPC

Like the SPS Agreement, the IPPC also includes specific language relating to pest-free areas. Article IV.2(e) states that NPPOs should make provision for the protection of endangered areas and the designation, maintenance, and surveillance of pest-free areas and areas of low pest prevalence. Article VII of the IPPC lays out requirements

related to imports. Article VII.7(g) states that "contracting parties shall institute only phytosanitary measures that are technically justified, consistent with the pest risk involved and represent the least restrictive measures available, and result in the minimum impediment to the international movement of people, commodities and conveyances." Article VII.7(j) specifies that "Contracting parties shall, to the best of their ability, conduct surveillance for pests and develop and maintain adequate information on pest status in order to support categorization of pests, and for the development of appropriate phytosanitary measures. This information shall be made available to contracting parties, on request."

Like the SPS Agreement, the IPPC creates clear obligations for both importing and exporting countries to participate in the identification and recognition of pest status and pest-free areas particularly with respect to phytosanitary measures, and to exchange information that supports such designations.

Guidance on the requirements for establishment of pest-free areas and areas of low pest prevalence are described in several International Standards for Phytosanitary Measures (ISPMs):

- ISPM 4: requirements for the establishment of pest-free areas.
- ISPM 10: requirements for the establishment of pest free places of production and pest-free production sites
- ISPM 22: requirements for the establishment of areas of low pest prevalence
- ISPM 26: establishment of pest-free areas for fruit flies
- ISPM 29: recognition of pest free areas and areas of low pest prevalence.

Other standards that play an important role with respect to pest free areas are:

- ISPM 6: surveillance
- ISPM 8: determination of pest status in an area.

22.4.1. Purpose of pest-free areas

NPPOs establish pest-free areas for the purpose of exporting plants, plant products, and other regulated articles to other countries without additional phytosanitary requirements related to pests at the origin. In terms of risk management, pest freedom provides for managing risk to the appropriate level of protection or acceptable level of risk because the regulated pest in question is not present in the exporting area. This is consistent with the principles of "least trade restrictive measures" and "managed risk" in the SPS Agreement and the IPPC.

The purpose of establishing pest-free places of production (PFPP) and pest-free production sites (PFPs) is the same as for establishing pest-free areas. The primary difference is that PFPPs and PFPSs are more limited in geographic scale, where specific establishments (places of production or production sites) are able to establish pest freedom even if a pest occurs in a country.

NPPOs may establish areas of low pest prevalence for the purpose of facilitating exports or to limit the impact of a pest in an area. Areas of low pest prevalence (ALPP) differ from pest-free areas because they may not be sufficient as a stand-alone phytosanitary measure for the purpose of exporting plants, plant products, or other regulated articles. However, any phytosanitary measures that are added to products moving from an ALPP should be consistent with the principle of managed risk.

22.4.2. Role of pest-risk analysis

Whether the decision is to require a pest-free area (for import) or to establish a pest-free area (for export), there are multiple factors to weigh and important information to collect and analyze. Pest risk analysis (PRA) provides a structured mechanism to support this process.

Information and analysis

Establishing pest freedom is linked to PRA in several ways. The information needed to establish pest freedom overlaps with information used to conduct a PRA. NPPOs should have a thorough understanding of the biology of the pest, how likely the pest is to establish in an area, how the pest can reproduce and spread, its host range, and other ecological and climatic factors. This type of information will help determine how to define a pest-free area, and how likely it is that

the pest-free area can be maintained naturally or through the application of official measures.

The importance of timely exchange of information in PRA and in establishing pest-free areas cannot be overstated. Article VIII of the IPPC clearly establishes requirements for information exchange for the purpose of communicating about pest status and for PRA and these requirements apply equally to importing and exporting countries.

Economic assessment

Another aspect of PRA that becomes important is when an NPPO undertakes the decision to establish a pest-free area. Recall that PRA includes analysis to determine the economic impact of a pest. NPPOs should analyze the potential economic impacts of a pest balanced against the cost of establishing a pest-free area. If the area will require significant and costly intervention by the NPPO or stakeholders but result in marginal benefits, e.g. high cost to establish and low benefit overall, then the establishment may not be economically efficient and worth doing. On the other hand, if the overall benefit, i.e., reduced costs to producers or increased opportunities for exports, exceeds the cost of establishing and maintaining the pest-free area, then NPPOs may decide that establishing the area is useful. Furthermore, such analyses can be useful in identifying the resource requirements needed for a program before the program is initiated. Thus, an economic analysis, similar to what would be conducted for a PRA, should be conducted by NPPOs that are considering establishing pest-free areas.

Lichtenberg and Lynch (2005) provide an excellent analysis of factors NPPOs should consider in determining the economic costs and benefits of establishing pest-free areas. They found that establishment of pest-free areas was more beneficial if countries faced higher treatment requirements and for countries that had geographic limits to pest introduction. However, they also found that where costs of eradication of a pest were high, the increased costs of control did not lead to overall benefits that justify establishing a pest-free area. They in addition reported that in some cases there can be an uneven distribution of benefits in that costs fall mainly on producers. This is particularly worth noting in the sense that it is often the industry and producers that bear the costs of a program rather than an NPPO. This should also be taken into consideration in determining whether to establish a pest-free area.

Risk and uncertainty

The decision to undertake establishing a pest-free area for a pest can also be impacted by the understanding of risk and uncertainty related to that pest. For pests that are known by the NPPO to be absent and it is reasonable to project that the pest will remain absent for the foreseeable future, there is a very low level of uncertainty concerning the resources needed and likely success of establishing a pest-free area. In such cases, the decision to establish such an area may be simple and straightforward.

Lesser known pests introduce greater uncertainty. New pests are usually detected as a result of some kind of surveillance. In many cases, there are "program pests," i.e., pests for which the NPPO is conducting regular surveillance based on pre-existing knowledge and an understanding of the actions that should be taken if those pests are detected. Thus, there is a distinction between new exotic pests for which we have little experience, and pests that are occasionally introduced, but which are well understood and for which responses have already been designed prior to introduction.

For example, fruit flies in the family Tephritidae are a taxonomic group of pests which are considered to be economically important by many, if not most NPPOs. Any time a new species of fruit fly is introduced into an area from which it was previously absent, we already understand the pest is important, i.e., an analysis has already been done, or we have past experience with the pest, and there is usually a response plan already in place to eradicate any new infestations and re-establish a pest-free area. That situation is considerably different than dealing with, for instance, a new species of wood boring beetle where we don't know the full host range or other important biological information about the pest.

There are key differences here. For pests that are well understood, the level of uncertainty is usually much lower. We understand potential impacts, management strategies, costs and benefits of taking different types of actions

and we often do not need to spend efforts engaging stakeholder support. On the other hand, for pests that are poorly understood or completely new, we have a much higher level of uncertainty related to likely success of establishing a pest-free area and the resources required. The decision-making becomes much more difficult. We may be uncertain about the pest's ability to spread in the new area, if it will have significant impacts, i.e., what kind of impact and what sector(s) of agriculture or the environment will be impacted, or if management options will be feasible and effective.

Furthermore, the level of uncertainty can affect how a pest and a pest-free area for a pest are perceived by decision-makers and stakeholders. Risks that are poorly understood, or with high levels of uncertainty, are often perceived to be much higher than they actually are. Thus, if a program is proposed that has an uncertain outcome, e.g., due to lack of certainty about the pest, or lack of certainty about management options, support for the program from decision-makers or stakeholders may be lacking.

Uncertainty in a pest-free program is best addressed by early, clear communication with all stakeholders internal and external to the NPPO. Identifying areas of uncertainty that can be reduced by obtainable information can help to eventually address information gaps. In addition, clear communication of uncertainty can help identify where a program may need to be strengthened in order to compensate for uncertainty.

Decision-making for pest-free areas involving eradication of a pest

Whether the pest is already well understood or the pest is a new pest that is poorly understood will have major effects on the types of programs that are implemented. The big difference is that we are dealing with risks that are well understood and with low uncertainty (in the first case) versus risks that are unknown and have a high level of uncertainty in the latter case. This affects decision-making and any logistical decisions that subsequently arise.

Depending on the pest, the NPPO may decide to:

- do nothing if the pest is not predicted to spread, for instance, or if the pest will spread naturally

- do surveillance, i.e. "wait and see what it does"
- suppress or contain the pest
- manage the pest in an "area-wide pest management program"
- conduct an eradication program
- establish a pest-free area.

Fig. 22.1 provides a graphic representation of the continuum of options available to NPPOs in responding to a new pest introduction and the resources required for responding. Eradication programs are at one end of the spectrum for controlling pests as they typically require the greatest commitment of resources like time, money, staff, and so on. Not all pests are suitable for eradication (Myers *et al.*, 1998). The biology of the pest and the characteristics of the environment will be the greatest determining factors as to how likely an eradication program is to succeed, but other factors such as costs and benefits and stakeholder support are also important.

Within the range of options from "do nothing" to "eradicate" there are a number of factors for the risk manager to consider. Factors favoring eradication include:

- cost–benefit analysis shows significant economic loss to industry or the community if the organism establishes
- physical barriers and/or discontinuity of hosts between production districts
- cost-effective control is difficult to achieve, e.g., limited availability of protectant or curative treatments
- the generation time, population dynamics, and dispersal of the organism favor more restricted spread and distribution
- pest biocontrol agents are not known or recorded
- vectors are discontinuous and can be effectively controlled
- outbreak(s) are few and confined

- trace-back information indicates few opportunities for secondary spread
- weather records show unfavorable conditions for pest development
- ease of access to outbreak site and location of alternate hosts.

Factors favoring alternative strategies include:

- cost–benefit analysis shows relatively low economic impact if the organism establishes
- major areas of continuous production of host plants
- cost-effective control strategies are available
- short generation times, potential for rapid population growth, and long-distance dispersal lead to rapid establishment and spread
- widespread populations of known pest biocontrol agents are present
- vectors are unknown, continuous or difficult to control
- outbreaks are numerous and widely dispersed
- trace-back information indicates extensive opportunities for secondary spread
- weather records show optimum conditions for pest development
- difficult terrain and/or problems accessing and locating host plants.

All these factors are weighed while taking account of political and stakeholder concerns regarding possible options to understand the best strategy and alternatives, including identifying the point where a program is no longer justified and should be abandoned or shifted to a different strategy.

Pest-free areas as risk management

As mentioned above, the establishment of a pest-free area is in itself a phytosanitary measure

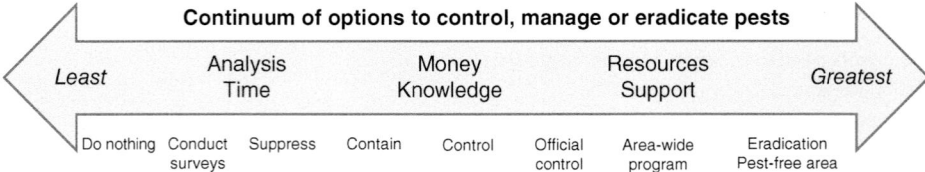

Fig. 22.1. Continuum of options for responding to a new pest introduction and resources that affect decision-making.

that is used to manage risk by both importing and exporting countries. This establishes the linkage of pest-free areas to pest risk management. While the SPS Agreement does not specifically refer to pest risk management, the entire agreement is itself about risk management and the SPS Agreement includes significant language around the concept of pest-free areas as measures to mitigate risk. In cases where it is economically beneficial, pest-free areas can offer highly cost-effective risk mitigation for both importing and exporting country NPPOs and their industries (Lichtenberg and Lynch, 2005).

22.5. Establishing Pest-free Areas

The definition of a pest-free area is "An area in which a specific pest is absent as demonstrated by scientific evidence and in which, where appropriate, this condition is being officially maintained." The first step for an NPPO to establish that an area is free of a pest is to define the area that is being considered for pest freedom. The area may be the entire country, parts of a country or a group of countries. In many, if not most cases, pest-free areas may be delimited by natural barriers such as bodies of water, mountains, deserts, or other geographic features that prevent pest movement from one area to another. The pest-free area should be defined specifically enough that it can be readily identified and communicated to importing country NPPOs as part of the evidence provided to support the claim that an area is free of a pest. This underscores again the importance of exchanging information between exporting and importing country NPPOs.

Once the NPPO has defined the area that is under consideration for a pest-free area, the NPPO should take steps to determine if the pest is present or absent. The second step for an NPPO to establish that an area is free of a pest is to conduct surveillance consistent with requirements in both ISPM 6 (surveillance) and ISPM 8 (determination of pest status in an area). These standards identify how surveillance can be conducted and how to establish the evidence necessary to determine the status of the pest in an area. Surveillance helps to establish the scientific

evidence necessary to demonstrate that a pest is absent from an area.

In order to declare an area as "pest free," an NPPO must be able to demonstrate through surveillance that the pest is absent (see ISPM 6). An NPPO can conduct general surveillance or specific surveillance for a pest, as necessary to make a determination that the pest is absent. General surveillance typically includes reviewing relevant scientific literature, pest records, or other information that can demonstrate a history of pest absence in an area. Specific surveillance includes detection and delimiting surveys that are conducted specifically for a pest in an area to determine if it is present or absent. Depending on the biology of the pest, negative specific survey results may need to be provided for a few or several growing seasons. ISPM 8 describes what criteria should be met to declare a pest is absent from an area.

22.5.1. Maintaining pest freedom

Pest-free areas may be maintained naturally or may require specific phytosanitary measures to maintain pest-free status. Even with areas that have large geographic barriers, phytosanitary measures may be needed to assure pests do not spread into an area.

Examples of measures that are used to help maintain pest freedom in an area include:

- determining the pest is a quarantine pest
- establishing import requirements on material that could introduce the pest
- preventing the movement of regulated articles that could introduce the pest
- regular monitoring surveys to demonstrate the pest continues to be absent.

More stringent measures, including increased monitoring surveys, may need to be applied depending on the pest distribution. In cases where a part of a country has been declared a pest-free area, but the pest occurs in other parts of the country, the NPPO should apply official measures within its territory to ensure the pest does not spread to the pest-free area. If a pest has a restricted distribution within an area that is otherwise free of the pest, the NPPO may need to put in place official measures to contain the movement of the pest out of the infested area to an area where the pest is absent.

It is essential that NPPOs keep accurate survey records because these records help fulfill the requirement for providing scientific evidence that establishes pest freedom. These records are used to monitor programs and should be furnished to importing country NPPOs on request.

22.6. Operational Considerations for Pest-free Areas

There are a number of operational factors that should be taken into consideration when establishing a pest-free area. As mentioned above, an analysis to determine economic impacts and to identify necessary resources is useful before an NPPO undertakes the work necessary to establish a pest-free area. Resources for NPPOs to establish pest-free areas can include administrative, analytical, and operational resources. Such resources include, for example:

- Administrative resources to establish regulations and requirements related to establishing and maintaining the pest-free area.
- Analytical resources to analyze pests, costs, and benefits of establishing the pest-free area.
- Operational resources to conduct surveys, enforce regulations on domestic and imported products, and implement control or other phytosanitary measures as needed to establish and maintain the pest-free area. Operational resources include supplies, e.g. survey supplies, as well as staff resources to conduct the work. Table 22.1 provides a comparison of the types of resources needed depending on whether a pest is unmanaged versus establishing a pest-free area.

22.6.1. NPPO tools to support pest-free areas

NPPOs should plan well ahead of implementing a program for a pest-free area, taking into account all of the projected resources and regulatory tools that will be needed to establish and maintain the area. The most important general tools the NPPO has for managing pests post-border are:

- phytosanitary measures such as laws and regulations, e.g., domestic quarantines; granting authority to the NPPO to take actions to eradicate pests or limit their spread
- surveillance aimed at early detecting, delimiting and monitoring pest populations
- analytical tools including risk analysis, economic analysis, climatological tools, and prioritization methods to aid decision-making for pest management programs
- pest management tools such as biological control, chemical control, cultural control measures, quarantines, and the like
- risk communication and involving stakeholders, including getting support for programs and garnering stakeholder involvement with programs to increase chance of success.

22.7. Stakeholder Considerations

In addition to NPPO resources, additional resources may be necessary to establish and maintain the pest-free area. Depending on the type of pest, various groups such as producers or homeowners may have to participate in a pest-free program or contribute financial and other resources to ensure its success.

Effective communication with stakeholders is critical for the success of a program. Stakeholders are those individuals, groups, or organizations that can affect, or be affected by, the actions of the NPPO. Examples of stakeholders include:

- growers
- property owners
- naturalists
- environmental groups
- industry groups
- researchers and academia
- state and local governments
- NPPOs and regional plant protection organizations (RPPOs) (USDA/APHIS, 2010).

There are numerous reasons to include stakeholders in the planning and execution of pest-free

Table 22.1. Comparison of resources for pest-free areas versus unmanaged pests.

Resources	Management options		
	Unmanaged	Managed but not pest free	Pest-free area
Pest status	Not regulated	Potentially a quarantine pest	Quarantine pest
Administrative	No regulations	Potentially some regulations to limit movement of regulated articles	Usually regulations are required, including designation of pest as quarantine pest, and regulations on host material or regulated articles
Surveillance	Local or state surveillance may occur; NPPO surveillance may be done if a pest is regulated by other countries	Detection, delimiting, and monitoring surveys may be conducted if the pest is subject to a control program or is a quarantine pest for other countries	Surveys are required, including detection, delimiting, and monitoring surveys to provide scientific evidence to support designation of the pest-free area
Analysis(PRA and economic)	Analysis is likely unnecessary	Analysis may inform whether a program should be continued, expanded, or discontinued.	Analysis is required to determine costs, benefits, and likelihood of success of pest-free area, as well as to identify what measures may be needed to maintain the area
Pest management operations	No NPPO pest management; may be managed by industry or local government	NPPO may provide pest management or official control, including measures to limit spread	NPPO must provide measures to prevent introduction and spread of pest into pest-free areas
Stakeholder involvement	Minimal—stakeholders may be informed of pest status	Stakeholders may be consulted or involved in pest management. This may include industry as well as local governments	Extensive stakeholder involvement may be required, including to limit the movement of regulated articles to prevent pest spread

areas at all stages of a program. While the NPPO is ultimately responsible for coordinating measures related to establishing and maintaining a pest-free area, NPPOs usually cooperate with stakeholders in executing a pest-free area program. Depending on the nature of the pest, stakeholders may be affected by control programs if the NPPO must remove or control regulated material from private properties.

More often, stakeholders have a central, active role to play in pest management programs. Control programs often require certain actions on the part of growers and other industry groups that include everything from field pest management practices to observing specific quarantines such as restricting the movement of host material. In these cases, the NPPO must communicate and work with stakeholders to ensure that the purpose of the program, the objective of specific actions and the respective roles of the NPPO and stakeholders are clearly identified.

22.8. Other Pest-free Concepts

The bulk of this chapter has focused on the geographic application of the concept of pest freedom known as a pest-free area and its variations. Pest-free areas are a well developed and commonly practiced phytosanitary measure with great value for risk managers but the concept of pest freedom is not limited to this single application. As noted in the introduction, pest freedom can also be designed as a measure for other phytosanitary applications.

Best known is perhaps the application to a consignment which is confirmed by issuance of a phytosanitary certificate. As noted in 22.2, there are technical problems with simply certifying "freedom from pests" unless relevant conditions accompany the statement. It is quite a different thing to state that a consignment is found free of quarantine pests based on visual examination of samples taken randomly to detect a 5% infestation rate with 95% confidence (see Chapter 20, this volume). This specification makes it possible for the importing country to understand exactly the basis for certification of freedom and decide if it is appropriate. Such a specification puts consignment freedom on a par with pest-free areas in terms of its technical rigor. Without such a specification, the exporting country uses one design and the importing country uses another for the same consignment. Of course, the results are also likely to be different which leads to concerns about the phytosanitary status of consignments and unfair treatment of the trade. Unfortunately, this has been the practice of NPPOs for decades and this poor risk management practice continues today.

Another example of applying the concept of pest freedom in a different way involves identifying a period of time when commodities may be harvested and shipped without risk of contamination by the pests of concern. This measure requires knowledge of pest biology and a means to monitor for pest presence. This strategy is often used for contaminating pests that are attracted to conveyances. Gypsy moth, Japanese beetle, and Brown marmorated stink bug are examples of pest for which the United States has designed programs around pest-free periods.

22.9. Summary and Look Forward

Here are five things to remember from this chapter.

1. Pest-free areas are an important option for pest risk management.
2. When a pest-free area is feasible, it represents a least trade restrictive measure for managing risk to an acceptable level.
3. Both importing and exporting country NPPOs stand to benefit from implementing pest-free areas; at the same time, both importing and exporting country NPPOs have obligations related to information exchange and the recognition of pest-free areas.
4. Before deciding to implement a pest-free area, NPPOs should determine the costs and benefits of establishing a program, and ensure they work with trading partners and their stakeholders.
5. Economic and biological analysis can provide the necessary analytic support for NPPOs to make appropriate decisions in establishing and maintaining pest-free areas.

The next chapter presents irradiation as a promising phytosanitary measure that faces significant challenges. It has emerged as a viable quarantine treatment in recent years but some of its unique characteristics have slowed the adoption of this technology.

22.10. References

FAO (1997) *New Revised Text of the International Plant Protection Convention*. Food and Agriculture Organization of the United Nations, Rome.

IPPC (2017a) *ISPM 4: Requirements for the Establishment of Pest Free Areas*. Secretariat of the International Plant Protection Convention (IPPC), Food and Agriculture Organization of the United Nations, Rome.

IPPC (2017b) *ISPM 12: Guidelines for Phytosanitary Certificates*. Secretariat of the International Plant Protection Convention (IPPC), Food and Agriculture Organization of the United Nations, Rome.

Lichtenberg, E. and Lynch, L. (2005) Exotic pests and trade: when is pest-free status certification worthwhile? *Agricultural Resource and Economics Review* 35(1), 52–62.

Liquido, N. and Griffin, R.L. (1997) *Quarantine Security for Commodities: Current Approaches and Potential Strategies*. US Agricultural Research Service publication 1996-04. Beltsville, MD.

Myers, J. H., Savoie, A. and van Randen, E. (1998) Eradication and pest management. *Annual Review of Entomology* 43, 471–491.

USDA/APHIS (2010) *Risk Analysis 101. Training Manual*. On file with USDA APHIS PERAL, Raleigh, NC.

WTO (1994) *The WTO Agreement on the Application of Sanitary and Phytosanitary Measures*. World Trade Organization, Geneva.

23

Irradiation

_____

The electric light did not come from continuous improvement of candles.

Oren Harari

23.1. Introduction

Phytosanitary irradiation (PI) has emerged in recent years as a viable quarantine treatment but some of its unique characteristics have retarded adoption of the technology. The primary challenge from a risk management standpoint is not efficacy, it is virtually 100% effective when the right dose is used, but rather the mindset that dead insects are needed to verify that treatments are performed correctly. Although the issue is sometimes cast as a question of confidence that the national plant protection organizations (NPPOs) can provide adequate assurances, the solutions are often confounded by a background of misconceptions about irradiation and assumptions about the relationship of irradiation to other phytosanitary measures, especially inspection. Irradiation is unique in many respects and needs to be properly understood by risk managers to be appropriately applied as a phytosanitary measure.

23.2. Why Is Irradiation Different?

It is important to understand at the outset that irradiation is based on absorbed dose as opposed to the delivered dose typical of treatments such as fumigation. This characteristic allows PI to be pest specific without being commodity specific and creates the opportunity for generic treatments as well as generic regulatory approaches. Doses can be applied to entire groups of pests, e.g., 150 grays (Gy) for all fruit flies, irrespective of the commodity. This makes commodity-by-commodity evaluation unnecessary. All commodities sharing the same pest concern can have the same treatment and authorization. The only limitation is the tolerance of the commodity to the dose, a practical issue related to product quality.

Another aspect of irradiation that distinguishes it from other phytosanitary treatments is the range of possible responses, which can include mortality, inactivation (for microorganisms), devitalization (for propagules), sprout inhibition, arrested development, or sterilization. The response is a function of the absorbed dose. A very low dose (ca. 50 Gy) will stop potatoes from sprouting, making it possible to regulate potatoes as vegetables rather than potential propagules and overcoming a significant global regulatory challenge. Doses above 10 kGy are used to inactivate microorganisms such as _Salmonella_ in food. The wide range of possible responses is an advantage that makes irradiation unique among phytosanitary treatments which typically aim only for mortality as a response.

Where mortality is not the response, there is always the potential to encounter live pests—so called "wigglers." The idea of accepting wigglers can be counterintuitive to inspectors but it represents an important principle of risk: the presence of a quarantine pest in a pathway is not justification for action if there is no probability of establishment. The application of this concept is not limited to PI; many other situations exist where quarantine action is not justified for a quarantine pest because the pest cannot establish via the pathway in question, e.g., armored scales on fruit for consumption. The key issue is shifting the focus of concern from the presence of the pest in the pathway to the likelihood of establishment via the pathway in question. There is no technical justification for phytosanitary measures against live pests in a pathway when the likelihood of establishment is negligible. This is a fundamental principle for risk managers to adopt.

In the absence of dead pests to validate a treatment, the phytosanitary official must have confidence that:

1. Research has established the proper dose.
2. Dosimetry confirms the product has been appropriately treated.

This is a significant departure from historical import program designs but central to accepting PI as well as other current and future risk management concepts such as systems approaches.

23.3. International and National Regulatory Frameworks

It is first crucial to recognize that nuclear technologies in general are highly regulated in every part of the world. In most countries there are at least three and usually more agencies responsible for regulating irradiation facilities and multiple layers of international and national regulatory frameworks that apply. The regulatory requirements of the national nuclear regulatory agency are foremost and strictly follow International Atomic Energy Agency (IAEA) guidelines. There are also detailed ISO standards (formerly American Society for Testing and Materials (ASTM)) for all types of irradiation treatments, facilities, and dosimetry. In addition, the irradiation of food products is covered by Codex Alimentarius standards which have been jointly developed by the Food and Agriculture

Organization of the United Nations (FAO), and the World Health Organization (WHO) (Codex, 1983). Phytosanitary treatments are specifically covered by the International Plant Protection Convention (IPPC) in International Standards for Phytosanitary Measures (ISPMs), especially ISPM 18, *Guidelines for the use of irradiation as a phytosanitary measure* (IPPC, 2016).

National regulatory structures vary in their design and structure, but every country has at least a nuclear regulatory agency, a food safety agency, and a national plant protection organization who all have the authority to regulate PI treatments and facilities. Many countries have other national and local agencies involved with the oversight of facilities and treatments. The magnitude of the international, national, and local regulatory framework for irradiation far exceeds that of any other phytosanitary measure in the breadth of regulatory authorities, the depth of regulatory detail, and the level of accountability.

23.4. Facilities

Irradiation treatment facilities are more expensive to build and operate than other phytosanitary treatment facilities. High levels of safety, accountability, and quality assurance are essential and strictly practiced in order to meet regulatory requirements and the commercial demands of clients that use the technology. Any commercial benefit from cutting corners is far outweighed by the risk of being shuttered by regulatory officials. Although there is always the possibility for mistakes and perhaps corruption, the probability is substantially lower than for other treatments and the likelihood that mistakes would go unnoticed is very low simply because the technology requires a high level of care and accountability to be commercially viable.

Ionizing radiation occurs naturally in the environment and may also be artificially produced using X-ray, electron beam accelerators, or gamma sources such as cobalt-60. Most PI treatments require extremely low doses, i.e., less than 1 kGy. Low-dose treatments effectively inhibit sprouting, e.g., in potatoes, onions, garlic, and can also increase the shelf life of many fresh commodities by delaying ripening, improving quality factors such as juice yield and hydration, and killing or sterilizing arthropod pests. Treatments done for food safety (up to 10 kGy for spoilage organisms) or for bandages, tampons, medical instruments or hospital and astronaut food (up to 30 kGy for all microbiological hazards), make irradiation an important strategy for controlling foodborne illness and decreasing waste. The human life-threatening consequences of sloppiness associated with a treatment in these latter categories, and the lack of direct oversight programs for these treatments is indicative of the high level of confidence associated with irradiation treatment technology. There is an argument here for justifying any special phytosanitary oversight with evidence of variability, errors, or abuse that raise the risk to levels which would require a higher standard of accountability than what is already in place for other types of irradiation treatment.

23.5. Policy Issues

The long evolution and slow growth of PI as a viable commercial treatment can be traced mainly to worries about the high initial investment to build a facility in the light of uncertainty regarding public acceptance. Despite more than 60 years of research that has consistently shown ionizing radiation to be an effective, safe, and feasible technology for food processing, and multiple statements to that effect by credible national and international institutions and experts, precautionary perceptions and policies have continued to dog the contribution this technology can make to safe trade and food security.

Three unfortunate circumstances that are largely beyond the control of NPPOs have worked against PI:

- irradiation designated as a food additive
- special labeling requirements
- 1 kilogray (I kGy) ceiling for insect disinfestation treatments of food.

Each is addressed in turn below.

23.5.1. Irradiation designated as a food additive

The Joint FAO/IAEA/WHO Expert Committee on Food Irradiation (WHO, 1966) defined irradiation as an additive rather than a process, despite there being substantial and consistent research to show that there is no trace of irradiation or harmful by-products resulting from a phytosanitary treatment. In most countries, regulating additives invokes an entirely different approach than regulating processes which has unnecessarily complicated national regulatory frameworks for PI.

23.5.2. Special labeling requirements

During the 1960s, food products that had been treated with irradiation in the Netherlands were marked with a "treated by irradiation" logo, also known as the "Radura." This logo (see Fig. 23.1) was later included in the Codex Alimentarius Standard on irradiated food as an option to label irradiated food but was made mandatory by the US Food and Drug Administration (FDA) for irradiated food in the United States. This requirement is both discriminatory and misleading. No other phytosanitary treatment requires such labeling, and the label has different meanings depending on the type of treatment. In the case of meat and other products treated with high doses, the label signifies freedom from microbiological hazards. This is not the case with phytosanitary treatments. A consumer purchasing a labeled mango may be confused to find that they

Fig. 23.1. The Radura logo communicates a product has been treated by irradiation.

can become ill from contamination by *Salmonella* and find live insects in fruit that is fully compliant with phytosanitary treatment requirements and carries the Radura label.

> The irradiation process creates no chemical or physical changes in products that affect human health. Irradiation has no residues and the doses used for food are far too low to make any product radioactive. The Radura mark therefore only identifies the type of treatment that was applied without conveying any useful safety or health information.
>
> The same Radura mark is used for both sanitary and phytosanitary treatments. This can be misleading to consumers who may mistakenly believe that a marked product is microbiologically safe when it has received a phytosanitary treatment.

23.5.3. 1 Kilogray (I kGy) ceiling for insect disinfestation treatments of food

A key limitation to the use of PI in the United States is the 1 kGy limit arbitrarily established by the FDA for treatments of food. The FDA acknowledges that there is no technical justification for this limit but has given no priority to changing the regulation. Eliminating the limit or revising it upward would open many possibilities for PI treatments that need to exceed 1 kGy in commercial circumstances to be effective.

Beyond these external issues, one of the main internal barriers for national regulatory frameworks is the tendency for risk managers to require PI authorizations on a commodity-by-commodity basis when only commodity tolerance, a nonphytosanitary factor, distinguishes one commodity from another. Recognizing and using generic applications provides legitimate regulatory shortcuts compared to other treatments but taking advantage of these shortcuts requires a break from the dogma of the usual regulatory designs.

23.6. Programmatic Issues

There is a programmatic tendency to create strong treatment verification mechanisms for PI in the absence of a conventional method to detect treatment failures, e.g., dead pests. This desire for tangible post-treatment verification of a PI treatment is rooted in traditional inspection paradigms based on official procedures implemented at a single point in the risk management continuum, usually at the border. These designs typically focus on the commodity or the pest, or both.

One popular commodity-linked proposal that has been raised and debated numerous times in the past is the use of radiation-sensitive indicators (RSIs). RSIs are adhesive-backed substrates coated or impregnated with radio-sensitive materials which may be affixed to or printed on treatment lots and which undergo a visual change when exposed to ionizing radiation. The visual change does not occur at a specific dose but over some dose range above the background, thereby demonstrating whether a product has been exposed to some level of irradiation.

RSIs are not a substitute for proper dosimetry but may be helpful in other aspects of the treatment process such as locating a specific zone of process interruption or to monitor a multiple-sided irradiation process. However, these are all technical applications for a treatment facility. The limitations of RSIs make them unsuitable for applications outside the treatment facility. RSIs have nonlinear response characteristics and environmental susceptibilities that make them unsuitable for accurate dose measurement. Exposure to environmental conditions such as heat, daylight, ultraviolet radiation, and certain gases may cause undesirable changes to indicator materials, and external environmental influences may make the interpretation of the indicators meaningless outside the irradiation facility without exceptional controls. Finally, some irradiation or storage conditions may result in false positive or negative observations. For these reasons, indicators should not be used as a regulatory criterion for effective treatment or the means for distinguishing treated from untreated products.

The value of RSIs for inspectors at the border is dubious at best. Aside from the technical limitations described above, RSIs are also easily manipulated and therefore cannot substantially reduce the potential for fraud.

Examples of pest-based strategies can be seen in the long history of research and repeated calls for additional research investments into scientific techniques that verify post-treatment efficacy via analysis of the pests, e.g., "squash tests."

These techniques are very important for the Sterile Insect Technique because it is crucial for sterile adult insects that are collected in traps to be reliably distinguished from their wild cousins. Routine port-of-entry analyses of larval stages in treated commodities is much different. Unlike surveillance where pests are taken from traps and sent to a laboratory for careful analysis under controlled conditions, port of entry processes require destructive sampling of the commodities and processing specimens under port conditions. Processing specimens requires expertise, equipment, and laboratory facilities that are not feasibly deployed in a port environment. Alternatively, specimens collected at the port may be sent to laboratories for processing. Either scenario adds substantially more time, cost, and complexity to the clearance process and is unnecessary for routine operations in the absence of evidence that indicates potential problems.

23.6.1. Oversight and inspection

A central question for PI program design is the role and value of oversight and inspection. Some form of oversight, often including inspection as a cornerstone, is generally a common element of regulatory program designs, but it deserves special attention in PI programs because of heightened concerns about verifying that the prescribed treatment has been effectively applied. The idea, however, should not be to add phytosanitary measures to compensate for uncertainty, but to apply the right measures at the right place and time for the right reason so they complement an effective program design. This approach has the advantage of maximizing risk management with technically defendable measures that are also resource sensitive.

Oversight is used to prevent and detect non-compliance in the PI process. The use of RSIs discussed above would be an element of oversight because it confirms exposure to radiation. Inspection is used to detect pests and determine their regulatory significance. The laboratory analysis of pests following treatment could be part of a post-treatment inspection process. Inspection is often an important element of the oversight process. It is important to be clear about these concepts, their relationship to each other, and their role in risk management to avoid simply piling measures on other measures to create the illusion of greater security without a true sense for the level of risk management being achieved.

Oversight

Oversight designs can include one or a combination of three approaches. First, is direct oversight by the NPPO of the importing country. This requires officials from the importing country to be onsite for the verification of the treatment and safeguarding consignments. The second approach is oversight by the NPPO of the exporting country. In this case, phytosanitary officials in the country of origin verify compliance. The last option is third-party verification by the treatment facility or other commercial entity accredited to certify compliance. Combinations of these approaches can also be constructed with emphasis on one or another element. Adding to this complexity is the possibility to provide full direct oversight for every treatment or only audit (monitor) via spot-checks of critical aspects with scheduled or unscheduled visits.

The most rigorous and expensive approach is full, direct oversight by the importing country. At the other end of the spectrum are third-party audits, including self-audits by the treatment facility. Ultimately, a risk management decision regarding appropriate design elements and the necessary rigor of oversight involves questions of resources, trust, policies, and understanding which aspects of the treatment program are critical for ensuring an effective treatment. The last point is simple for PI.

An effective PI treatment requires the application of the appropriate dose. The appropriate dose is established by research and then calculated for a specific facility based on the type and arrangement of equipment and commodity. As noted above, the actual dose may be two or more times higher than the dose indicated by research in order to ensure that every part of the treatment load receives at least the minimum absorbed dose.

If we accept that research establishing the appropriate dose is correct, then dosimetry is the key to verifying its application in practice. Dosimetry is used to measure the strength of the radiation source, to calibrate the exposure for specific load designs, and to validate each treatment. This process and its records are therefore vital to verifying the efficacy of any irradiation treatment, phytosanitary or otherwise. Since NPPOs do not typically employ dosimetry experts but rely upon the dosimetry expertise in facilities, there is no need for phytosanitary officials to do more than simply check dosimetry records.

There is nothing for a phytosanitary official to see during the treatment except products entering and exiting the secure area where exposure occurs. The value of being onsite is primarily associated with checking dosimetry records and observing safeguarding operations, e.g., ensuring the segregation and security of treated and untreated products. A critical analysis of all possible aspects of PI that can be objectives for oversight will show that dosimetry and security are the vital points for validating PI. Similar procedures are used to verify other treatments.

The design for oversight will depend on available resources, the policy framework for oversight (typically modeling policies after other treatments), and the degree of confidence and trust the NPPOs have for each other and the PI facilities in the implementation of treatment programs. The WTO-SPS Agreement would argue for the least restrictive design in the absence of evidence that a more rigorous design is justified. Unfortunately, treatment program designs generally, and especially PI programs, typically begin with strong measures which are rarely revisited, even when a solid record of successful treatment is established. Oversight designs also tend to include unnecessary oversight requirements that add substantial cost and effort without increasing confidence or give a false sense of confidence. Some applications of inspection fit this category.

Inspection

Inspection has a long history of service to the phytosanitary community. It is by far the most widely used and common phytosanitary measure for both import and export. Inspection is a comfortable measure because it provides tangible results linked to universally practiced policies. The reality of inspection as a risk management measure is however highly variable, poorly understood, and inconsistently practiced. Two

important points to understand in the context of PI are that inspection is far from being a highly effective measure, and targeting PI treated pests with inspection is the equivalent of searching for non-quarantine organisms. Both points lead to important questions about the role of inspection in oversight designs for PI.

Inspection may be required before treatment, after treatment, or both. Inspection can also be repeated. For instance, an inspection may be done in the country of export and another inspection done at the border in the destination country. Furthermore, the inspection may be a simple visual examination or involve destructive sampling (cutting) and laboratory analyses. In all these cases, inspection does nothing to directly mitigate pest risk but rather provides information that can result in other actions being taken which change the phytosanitary status of a consignment. Current inspection designs vary considerably and are often not transparent regarding the risk management strategy and objectives. The role and value of inspection in PI oversight designs is especially interesting because the unique features of PI challenge traditional assumptions about inspection.

Recall from the discussion above that inspection is for the detection of pests and the determination of their regulatory significance. The first question to ask in the context of PI is the value of this process for a treatment that is virtually 100% effective for target pests if the dose and application are correct. The obvious response is to detect non-target pests for which there may be no approved dose.

The detection of non-target pests however should not automatically result in the rejection of a consignment before evaluating the possible efficacy of the prescribed treatment for the non-target pest of concern. Because every part of the treatment lot receives at least the minimum absorbed dose, much of the treatment lot will receive a higher dose which can be over two times the minimum, depending on the irradiation source and configuration of the treatment lot. From this standpoint, the effectiveness of PI for non-target pests is likely to be significant and easily within the range of efficacy that APHIS accepts for other treatments that are extrapolated when efficacy data are incomplete or lacking, e.g., fumigation for external feeders. A plethora of PI data exists to support the extrapolation of approved PI treatments for many non-target pests. A policy and criteria need to be in place to make this possibility available for PI in the same way that it is for other treatments.

Inspecting for the detection of a target pest after treatment is equivalent to looking for non-quarantine pests unless the objective is to detect target pests for the verification of treatment effectiveness via post-treatment laboratory analysis of the pests. The shortcomings of post-treatment laboratory analysis have been discussed above as a programmatic issue because of the operational challenges, but it is important to understand that incorporating this as part of the oversight design also creates a quarantine enigma where target pests that are not wanted in a commodity in the first instance, must be present at some level for a poor measure (inspection) to verify the effectiveness of a measure that fully mitigates the risk of those same target pests. If no target pests are found, a likely and desirable scenario, it is not possible to validate the effectiveness of the treatment with this nonsensical design. Nevertheless, an occasional analysis to verify the effectiveness of treatment can be useful to bolster confidence in the PI program and reinforce why routine pest analysis is unnecessary as a required measure.

In cases where PI is not required in advance as a condition of entry, an inspection may result in the detection of pests for which treatment is required and PI is a viable option as an emergency action. This is currently a rare scenario because so few facilities exist nearby ports of entry and the logistics for ad hoc emergency treatment make it more expensive and complex. Regulatory policy designs nevertheless need to account for this possibility because of the important potential PI has as an alternative to methyl bromide fumigation, especially as more facilities become available near ports.

An argument can be made that the presence of any pests should raise concern about the effectiveness of pest management processes prior to treatment. Routine inspection to detect unanticipated pests, or spikes in the prevalence of anticipated pests, provides a barometer on the overall phytosanitary state of the commodity and allows for potentially important changes to be observed and adjustments made in the program before failures occur. It is important to understand that in this case, inspection is not being used as a phytosanitary measure, i.e., a requirement for trade, but rather to monitor the baseline effectiveness of measures applied by the country of origin to manage pest risk. These inspections provide no technical justification for pre-treatment action based on the detection of target pests, especially when an action decision results from inspection, a measure with a high level of inherent slippage.

Post-treatment inspection for pests at the border only makes sense if the integrity of the shipment was violated or is in question. As explained above, the post-treatment detection of target pests is not a legitimate basis for action, and the post-treatment detection of non-target pests is also likely to be inconsequential depending on the possibilities for extrapolation. Oversight designs that include both a pre-treatment and post-treatment inspection grossly exaggerate the role and value of inspection in PI programs. But in the instance where the oversight design includes both a pre-treatment and post-treatment inspection, the inspection designs should at least be thoughtfully aligned to provide a specific level of detection for meaningful results and to account for natural variability. Over-emphasis on the role of inspection in PI programs demonstrates a lack of understanding regarding the high degree of variability associated with inspection versus the extremely low level of variability associated with PI.

When a shipment is inspected, either before or after treatment, or both, and no pests are found, the shipment is considered compliant whether or not the treatment was performed or was effective. This is the century-old inspection paradigm that continues to misguide phytosanitary programs. The reality is that there is substantial uncertainty and variability around that inspection result, and unless the inspection was designed for a specific level of detection and confidence, the results do very little to accurately characterize the phytosanitary status of the commodity. Additional haphazard inspections do not substantially improve the situation from the statistical standpoint of the probability of presence or absence of pests.

In the case of PI, where there is an approved dose that has been demonstrated by research to be effective and an approved treatment that is proven by dosimetry to have been properly applied, there is essentially no uncertainty about the phytosanitary status of the commodity for target pests and even many non-target pests. Questions about maintaining the security and integrity of the treated shipment, i.e., safeguarding, are all that remain for a PI program to be highly effective.

Based on the discussions above, a thoughtful analysis of the role and value of inspection in PI oversight designs results in little justification for a strong application of this measure in PI programs.

23.6.2. Mistakes and fraud

Every phytosanitary program has the potential for mistakes and fraud, and every good program design includes mechanisms to prevent or detect and correct these events. Mistakes are unintentional failures in the program that may result from human error or poor design. Fraud is a deliberate action taken to circumvent requirements. Both are important factors in the question of confidence between trading partners.

A certain amount of natural variability surrounds every event. Any event that is repeated continuously by humans has an increasing likelihood of failure in one or more of the processes. By virtue of creating a phytosanitary program of any kind, we accept some natural background failure rate. The more complex the process, the greater the likelihood of mistakes. Fortunately, PI is not complex. But because it is a nuclear technology, it is surrounded by fear, misinformation, and malevolent perceptions that drive the need for greater precaution.

Fraud is typically driven by the need to gain a financial or other advantage. In the case of PI there is much more to lose from deceptive practices simply because of the relatively high investment required to establish suitable treatment facilities and to procure all the relevant regulatory approvals. It is much more likely that fraud would be associated with processes such as phytosanitary certificates and other aspects of the PI program not directly related to the treatment itself and treatment facility processes.

Thinking about PI program design as a simple Hazard Analysis Critical Control Point (HACCP) process helps to provide an objective,

analytical focus. Three critical elements and their controls emerge:

- establish the dose through credible research
- ensure that the dose is applied correctly through dosimetry
- safeguard by maintaining consignment integrity.

Each specific program has its unique strengths and vulnerabilities, and the degree of confidence between trading partners also varies so that there can be no generic formula for program design but establishing regulatory mechanisms to avoid, detect and correct mistakes and fraud in the three core elements form a central strategy for determining the right type and level of oversight for PI.

23.7. Closure

The discussion above crosses a range of issues associated with establishing oversight designs for phytosanitary irradiation programs. One thread that flows through all the discussion is the importance of being thoughtful about program objectives and the designs that achieve those objectives in a sound and defendable way. PI does not fit nicely into traditional policy frameworks for treatment, nor does it align well with historical risk management concepts and operational paradigms. It does however offer the opportunity to add an important new tool to a desperately small risk management toolbox if the technology can be embraced with a strong analytical focus and objective, albeit different, policy designs.

Although unconventional in many ways and strongly over-disciplined compared to other phytosanitary treatment technologies, PI has slowly grown to take a small but central role among risk management strategies for several NPPOs. A thoughtful approach that incorporates contemporary risk management concepts, sound analysis, and progressive regulatory designs is important to continue demonstrating the value of the technology as a phytosanitary measure without propagating misconceptions, distrust, unjustified fear, and trade tension.

23.8. Summary and Look Forward

Here are five things to remember from this chapter.

1. PI needs to be understood as different from other treatments.
2. Irradiation is highly regulated in all countries.
3. The role of inspection with irradiation should be questioned.
4. Policy and program frameworks for irradiation are overly restrictive compared to other treatments.
5. Phytosanitary measures should be evaluated in terms of their effect on the risk of pest establishment rather than presence or absence of a live pest in a pathway.

The next chapter will explore post-harvest processing and handling.

23.9. References

Codex (Alimentarius Commission) (1983) *Codex General Standard for Irradiated Food and the Recommended International Code of Practice for the operation of irradiation facilities used for treatment of foods*. Vol. 15 FA/WHO, Rome, Rev 1-2003.

IPPC (2016) International Standards for Phytosanitary Measures, Publication No. 18: *Guidelines for the Use of Irradiation as a Phytosanitary Measure*. Secretariat of the International Plant Protection Convention (IPPC), Food and Agriculture Organization of the United Nations, Rome.

WHO (World Health Organization) Joint FAO/IAEA/WHO Expert Committee on the Technical Basis for Legislation on Irradiated Food & World Health Organization (1966) *The Technical Basis for Legislation on Irradiated Food: Report of a Joint FAO/IAEA/WHO Expert Committee* (April 1964). WHO, Rome.

24

Post-harvest Processing and Handling

_____

Little by little, a little becomes a lot.

Tanzanian proverb

24.1. Introduction

Traditionally, risk-reducing measures applied to commodities moving in trade have relied on phytosanitary treatments aimed at reducing the presence of a pest on a given commodity. However, a wide array of additional measures may be applied to manage risk, either singly or in combination. These include measures (IPPC, 2013a, 2013b; NAPPO, 2014) that are:

- applied to the area or place of production, e.g., pest-free areas or areas of low pest prevalence
- applied to the commodity in the field before harvest, e.g., field treatments, integrated pest management (IPM) programs, field sanitation, fruit bagging, and timing of harvest
- applied to the commodity during post-harvest processing and handling, e.g., safeguarding in the field, washing, cleaning, drying, waxing, treatments, safeguarding during packing, and storage.

This chapter discusses post-harvest processing and handling practices and their effects on plant commodities/products moving in trade. Because they typically reduce the risk of introduction or spread of pests, they are considered phytosanitary measures, see definition below.

It is not easy to find a common-language definition for post-harvest. However, harvest is defined by the online Dictionary.com (accessed January 10, 2020) as "the gathering of a ripened crop." Following logically then, post-harvest, includes processes that the plant commodity/product is subjected to once it is harvested or gathered. "Processing" is defined as "to put through the steps of a prescribed procedure." Finally, handling—for commerce—is defined by the same source as "the process by which a commodity is packaged, transported, etc."

Post-harvest processing and handling measures are part of the biosecurity continuum at pre-border, border, and post-border stages. They are an important part of modern agricultural production.

24.2. Post-harvest processing and handling measures

Post-harvest processing and handling measures may include cleaning, washing and/or disinfecting, grading or culling, sanitation, drying, treatment, protecting/safeguarding (from, for example, reinfestation), cooling, storage type, and packing or re-packing. These measures, alone or in combination, are applied after harvest or during the shipping, distribution and end use parts of the commodity pathway (Fig. 24.1). Visual examination or inspection can also take place after harvest and can include packinghouse inspections

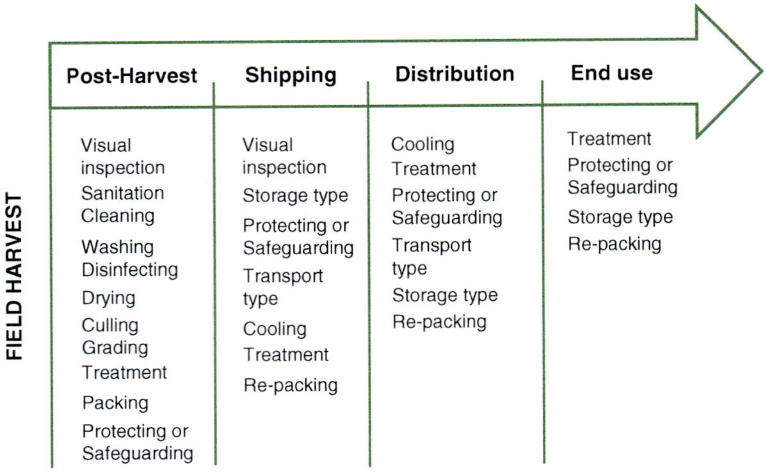

Fig. 24.1. Some examples of post-harvest processing and handling measures applied to plants and plant products.

> Visual examination: Examination using the unaided eye, lens, stereoscope or other optical microscope (ISPM No. 5).
> Inspection: Official visual examination of plants, plant products or other regulated articles to determine if pests are present or to determine compliance with phytosanitary regulations (ISPM No. 5, IPPC, 2019a).

and reinspection of the commodity before shipment. Post-harvest fruit cutting, a form of visual examination, should also be mentioned here. Inspection is discussed in detail in Chapter 20. In some instances, these measures form part of a systems approach. Systems approaches are discussed in detail in Chapter 27.

The plant commodities/products subjected to post-harvest processing and handling measures can be intended for consumption, e.g., fresh fruits and vegetables for direct sale to consumers or for additional processing into juice, frozen or canned products; cut flowers for direct sale to consumers; logs or grain for further processing; or for propagation, e.g., seeds, unrooted and rooted cuttings, plants in vitro or whole plants. The trade can be domestic or international.

Post-harvest processing and handling measures will vary depending on the plant commodity/product. They are applied after the commodities/products leave their place of production, e.g., planted field, orchard, greenhouse, unmanaged or managed forest, orchard, and the like, and before they reach the consumer either as fresh or processed plant commodities/products.

24.3. Objectives of Post-harvest Processing and Handling Measures

According to Watkins and Nock (2012) and others, the primary objective of post-harvest processing and handling is to maintain plant commodity/product quality by:

- reducing metabolic rates that result in undesirable changes in color, composition, texture, flavor and nutritional status, and undesirable growth such as sprouting or rooting
- reducing water loss that results in wilting, softening, and loss of salable weight and crispness
- minimizing bruising, friction damage and other mechanical injuries
- reducing spoilage caused by decay, especially of damaged or wounded tissue and preventing contamination by pests or by human pathogens
- preventing development of freezing injury or physiological disorders, such as chilling injury or senescent disorders.

It is also useful when grading, categorizing, and labeling the commodity. However, from the phytosanitary risk management perspective, the focus is on the risk-reducing effects.

24.4. Relevant ISPM 5 Definitions

Additional definitions that are useful in the context of this chapter are provided below.

Commodity—A type of plant, plant product, or other article being moved for trade or other purpose.

Consignment–A quantity of plants, plant products or other articles being moved from one country to another and covered, when required, by a single phytosanitary certificate. A consignment may be composed of one or more commodities or lots.

Contaminating pest–A pest that is carried by a commodity, packaging, conveyance or container, or present in a storage place and that, in the case of plants and plant products, does not infest them.

Packaging–Material used in supporting, protecting or carrying a commodity.

Pathway–Any means that allows the entry or spread of a pest.

Pest–Any species, strain or biotype of plant, animal or pathogenic agent injurious to plants or plant products. Note: In the IPPC, "plant pest" is sometimes used for the term "pest."

Plant products–Unmanufactured material of plant origin, including grain, and those manufactured products that, by their nature or that of their processing, may create a risk for the introduction and spread of pests.

Phytosanitary measure–Any legislation, regulation or official procedure having the purpose to prevent the introduction or spread of quarantine pests, or to limit the economic impact of regulated non-quarantine pests.

24.5. General Types of Post-harvest Processing and Handling Measures

Post-harvest processing and handling measures are specific to the commodity type. For example, washing is appropriate for some commodities, e.g., papaya (*Carica papaya*) and apple (*Malus domestica*) fruit, while not indicated for more perishable fresh commodities such as berries, lettuce or cut flowers. In general, post-harvest processing and handling procedures include the following.

Protecting or safeguarding the commodity during and immediately after harvest and while the commodity is being transported to the packing-house or processing facility. These practices can preserve the commodity quality by avoiding bruises or other physical damage, which can, in turn, facilitate infestation by pests and prevent pests present in the place of production from infesting the harvested commodity. Specific examples include placing harvested fruit inside baskets lined with foam, using screen-protected baskets, field boxes, or bins when harvesting pome or stone fruit, celery or peppers, or harvesting berries directly into plastic clamshells in the field (Fig. 24.2). In the case of wood commodities,

Fig. 24.2. Harvesting berries directly into clamshells in the field.

protecting the harvested commodity might include moving the felled timber away from the forest and into an environment where the wood is safeguarded, such as a processing yard, where logs are stored on top of a cement pad and covered by a tarp or by mesh.

Ensuring harvesting and packing house/shed equipment cleanliness. Examples include cleaning/disinfecting the packinghouse/shed prior to harvest; disinfecting equipment prior to use and in between uses in different processing facilities; having screen protected entrances to the packing and storage sheds; using only clean or new packaging materials (Fig. 24.3); and storing packaging materials in a clean place.

General sanitation can include removal of contaminants from the packing line, such as leaves, stems, soil, and removing residues and trash that may attract regulated pests to the packinghouse.

Ensuring enforcement of packinghouse/shed processes/plans that move the commodity from dirtier to increasingly cleaner areas during commodity post-harvest processing and handling. In some cases, this may also affect how labor is allocated to processing and handling tasks. An example would be packinghouse employees starting their workday by showering-in, changing into clean coveralls, and being assigned to work on packing the washed/disinfected commodity for the first part of their workday. They may then be reassigned to either harvesting the commodity or receiving the commodity from the field, a "dirtier" part of post-harvest, for the rest of the day.

Monitoring for target pest(s) in and around the packinghouses/sheds. Depending on the pest, monitoring can be done using active or passive traps or, in the case of pathogens, Petri dishes with appropriate growing media. In general terms, active traps are those that use an attractant to lure the pests to the trap, e.g., light, food baits, or synthetic pheromones that are involved in mating and attract one or the other gender, depending on the pest species. Passive traps are those that capture flying or crawling pests that encounter the trap by chance during flight or movement. Growing media would be selected based on expected pests such as fungi or bacteria.

Scheduling/conducting post-harvest processing and handling activities when pests are not active, e.g., flying in the early morning.

Avoiding the use of bright lights that might attract flying insects into packinghouses/sheds while commodity is being processed and when loading transport trucks.

Fig. 24.3. New boxes being fed into the packing line for papaya fruit.

Cleaning/washing/disinfecting the commodity. Cleaning/washing may use only water or add a mild detergent or disinfectant. Water should be of potable quality. Some commodities undergo sequential washing. For some commodities drying after washing is recommended. Figs 24.4–24.6 show post-harvest processing for papaya fruit for export. After fruit is harvested, it is taken to the packinghouse and emptied into large freshwater baths containing a mild bleach solution. The fruit spends a prescribed amount of time in the bleach solution and is manually moved to a second freshwater pool containing a mild detergent solution. Packinghouse workers ensure, either by rubbing the fruit or agitating the water bath, that the bath is effective at removing soil, dust and/or other contaminants on the fruit surface.

Brushing instead of washing. Some commodities are placed on conveyor belts that move

Fig. 24.4. Papaya, variety Maradol, for export in Chiapas, Mexico, fruit receive first wash/scrub in tank with potable water and mild detergent; workers use soft mits to remove dust and external pests from the harvested fruit.

Fig. 24.5. Second tank where papayas are subjected to an antifungal bath before being dried and packed.

the commodity under a series of soft (horsehair) brushes to remove dust and other contaminants. A specific example is the brushing of nopalitos (cladodes of *Opuntia* cactus) after harvest to remove the glochid spines and externally attached pests such as mealybugs.

Specific processes for removing known contaminants. Some post-harvest processing and handling measures are commodity-specific and tailored to pests commonly associated with the commodity.

Culling and/or sorting which involves removing/discarding harvested commodity that is unacceptable, e.g., small size, discoloration, or shows signs of damage such as malformed fruit, broken skin, evidence of pest infestation, and the like. For some commodities, e.g., pears and radishes, this is done after washing, but in some cases, it is done before the commodity is washed (Fig. 24.7).

Safeguarding the commodity after packing. This can be done, for example, by covering packing boxes or crates with insect-proof screen, wrapping the commodity boxes with plastic, wrapping each commodity unit in paper.

Fig. 24.6. Previously washed and dried papaya Maradol are subjected to air from a compressor to remove external pests that hide near the fruit peduncle. Note the storage of clean packing materials inside the packinghouse.

Fig. 24.7. Previously washed oranges are visually examined and culled and packaged under ultraviolet lights (photo by Jeannette Warnert, USDA image gallery).

Fig. 24.8. Box of packed papaya fruit showing the scannable codes for traceability. The code in each box allows traceback to field of harvest, date of harvest and other details for each box of export quality papayas.

Traceability of consignments can be ensured by using appropriate identification of the packaged commodity (Fig. 24.8).

Storage and transportation under commodity/ plant product appropriate conditions which might include prechilling the commodity, cold storage, modified atmosphere, and the like.

Timing shipping/transportation activities can be managed to take advantage of periods when pests are absent/inactive.

24.6. Who Applies Post-harvest Processing and Handling Measures?

In general terms, each commodity/plant product has one or more production models that are followed. Designated individuals are usually responsible for each step of the production pathway, from pre-planting activities to pre-harvest commodity production through harvest, post-harvest, transport and distribution. Post-harvest processing and handling measures are usually industry practices performed by individuals hired to do specific jobs. These measures can be validated, accredited, inspected, audited and certified by or under the supervision of the national plant protection organization (NPPO) of the exporting country.

24.7. Evaluating the Efficacy of Post-harvest and Handling Measures

Efficacy evaluates the extent to which a given measure reduces pest risk (see Chapter 21, this volume). When a quantitative calculation of efficacy is not possible, efficacy may be expressed in qualitative terms such as high, medium, and low. Post-harvest and handling measures do not directly affect the pests so other indirect measures of efficacy may be more appropriate. For example, measuring an activity like the number of culled commodity units in the packinghouse, negative trap counts, results of inspections and audits (NAPPO, 2014).

24.8. International Plant Health Standards and Post-harvest Processing and Handling Measures

This section summarizes what the International Standards for Phytosanitary Measures (ISPMs) have to say that is germane to post-harvest and handling measures.

24.8.1. ISPM 1: *Phytosanitary principles for the protection of plants and the application of phytosanitary measures in international trade*

In Section 2, "Operational principles of the IPPC," ISPM 1 mentions the use of a systems approach (2.5) where integrated measures for pest risk management may provide an alternative to single measures in order to meet the appropriate level of phytosanitary protection of an importing country. Section 2 also indicates that countries shall ensure, through appropriate measures, that the phytosanitary integrity of the composition, substitution, and reinfestation of consignments (2.9) after certification is maintained prior to export. Post-harvest processing and handling measures can be part of a systems approach and can also maintain the phytosanitary integrity and security of consignments.

24.8.2. ISPM 2: *Framework for pest risk analysis*

In Section 3, "Aspects common to all PRA stages," Section 3.3.2 indicates that for pathway-initiated risk analysis a complete commodity description is needed. This description should include commodity post-harvest and handling practices used for that commodity, as they will most likely reduce the risk of the commodity serving as a pathway for pests.

24.8.3. ISPM 6: *Surveillance*

Section 2.2.8, "Biosecurity and sanitation," mentions that when developing surveillance protocols, the plant health regulatory authorities should consider procedures to ensure that spread of pests is not facilitated during a survey. Post-harvest processing and handling measures can maintain the phytosanitary integrity and security of consignments thereby reducing the risk of pest spread.

24.8.4. ISPM 7: *Phytosanitary certification system*

Section 3.3, "Technical information on regulated pests," indicates that personnel involved in phytosanitary certification should be provided with adequate technical information concerning regulated pests for the importing countries including the means to control such pests. Many post-harvest processing and handling measures are important contributors to controlling pests.

24.8.5. ISPM 8: *Determination of pest status in an area*

Section 3.3 indicates that the determination of pest status in an area is based on supporting information that includes the phytosanitary measures used to prevent introduction or spread of the pest in question. Post-harvest processing and handling measures on the host commodity contribute to maintaining the phytosanitary integrity and security of consignments.

24.8.6. ISPM 10: *Requirements for the establishment of pest free places of production and pest free production sites*

Section 2.1.1 highlights that the availability of effective and practical measures for control and management of the pest is an advantage in establishing and maintaining a pest-free place of production or pest-free production site. Section 2.2 mentions the four main components necessary to establish and maintain pest-free places of production or pest-free production sites. These include verification that pest freedom has been attained or maintained and product identity and phytosanitary security of the consignment. Also, Section 3.1 mentions documentation of procedures to ensure the phytosanitary security of a consignment. Post-harvest processing and handling measures contribute to maintaining pest freedom and maintain the phytosanitary integrity and security of consignments.

24.8.7. ISPM 11: *Pest risk analysis for quarantine pests*

Section 2.2.1.2 discusses the probability that a pest will be associated with the pathway at origin and suggests that factors to consider when making this determination should include cultural and commercial procedures applied at the place of origin, e.g., application of plant protection products, handling, culling, roguing, and grading. Section 2.2.1.3 discusses the probability of pest survival during transport or storage and indicates that factors to consider when assessing this probability should include commercial procedure, e.g., refrigeration, applied to consignments in the country of origin, in transport or storage or in the country of destination; it suggests that the vulnerability of the life stages present during transport or storage of the commodity also be considered. Section 2.2.1.4 discusses the probability of the pest surviving existing pest management procedures applied to consignments from origin to end use and indicates that these pest management procedures should be evaluated for effectiveness against the pest in question. Many post-harvest processing and handling measures are important contributors to controlling pests.

24.8.8. ISPM 14: *The use of integrated measures in a systems approach for pest risk management*

Systems approaches provide, where appropriate, an equivalent alternative to procedures such as treatments or replace more restrictive measures like prohibition. This is achieved by considering the combined effect of different conditions and procedures. Systems approaches provide the opportunity to consider both pre- and post-harvest procedures that may contribute to the effective management of pest risk. Systems Approaches are more fully discussed in Chapter 27.

24.8.9. ISPM 18: *Guidelines for the use of irradiation as a phytosanitary measure*

Section 7.1 of ISPM 18 indicates that consignment handling procedures before, during and after receiving treatments should be properly documented.

24.8.10. ISPM 32: *Categorization of commodities according to their pest risk*

The standard provides guidance to importing countries on how they can categorize the risk of plant and plant products enabling the introduction and spread of quarantine pests based on the methods and degrees of processing they have been subjected to before being exported. The standard does not consider recontamination of the products after processing. However, it does consider the intended use of the products in the country of import. This categorization helps importing counties decide on possible establishment of import requirements.

The standard identifies four categories of pest risks. Category 1 includes products that have been processed to the point where they do not remain capable of being infested with quarantine pests. Category 1 products should not require phytosanitary measures or need phytosanitary certification. Processing methods that would make a plant product eligible for Category 1 include carbonization, cooking, pasteurization, fermentation, roasting, pureeing, sterilization, and the like. Examples of products that are considered Category 1 are in Appendix 2 of ISPM 32.

24.8.11. ISPM 36: *Integrated measures for plants for planting*

ISPM 36 outlines the main criteria for identification and application of integrated measures at the place of production for plants for planting moving in international trade. Integrated measures are applied throughout the production and distribution processes for these commodities. Measures may include visual examination of plants and sanitation, as well as specific packing and transportation requirements. Section 2.2.1.2 describes the pest management program for the place of production and includes many handling practices described earlier in this chapter. For example, disinfection of tools and equipment, treatment of water, personal hygiene of facility workers, routines for use of packaging materials and packing facilities; pest monitoring procedures and physical barriers such as screens and double doors. Section 2.2.1.6, "Packaging and transportation," indicates that plant material should be packed in a manner that prevents infestation by regulated pests, packing material should be clean, conveyances used to move plant material should be examined and cleaned prior

to loading, and each lot in the consignment should be identified in a way that can be traced back to the place of production.

24.8.12. ISPM 40: *International movement of growing media in association with plants for planting*

ISPM 40 indicates that origin and production methods of the components of growing media may affect the pest risk of the growing media associated with plants for planting. They suggest that growing media should be produced, stored, and maintained under conditions that will prevent its infestation or contamination. The pest risk may also depend on factors related to the production of the plants (addressed by ISPM 36) as well as the interaction between the two. Measures that reduce the risk of growing media contamination include storing and maintaining it under conditions that keep it free from quarantine pests, using clean tools, containers and equipment when working with the growing media and the plants, and using physical isolation, e.g., protected conditions, prevention of pest transmission by wind, production on benches separated from contact with soil. Section 3.3. suggests that places of production for growing media may be inspected. Depending on the specific type of growing media, disinfection through sterilization or autoclaving can be used. The measures outlined in the standard are handling measures that reduce the risk of moving pests in trade, even though growing media is not a harvested commodity.

24.8.13. ISPM 41: *International movement of used vehicles, machinery and equipment*

ISPM 41 identifies and categorizes the risk associated with used vehicles, machinery, and equipment moving in trade and identifies appropriate phytosanitary measures to reduce this risk. The measures include cleaning, treatments, prevention from contamination, requirements for facilities and waste disposal, and verification procedures. Used vehicles may have become contaminated by regulated pests or regulated articles, like soil, in the country of origin or of original use, and when moved internationally present a risk to the country of destination. Used vehicles may have been used in farms, crop fields, forests, or in close proximity to other vegetation, like forests. They may have been stored outdoors close to vegetation or under lights that attract insects.

Cleaning for these regulated articles may include emptying water reservoirs, removing filters and debris, abrasive blasting, pressure washing, steam cleaning, sweeping and vacuuming, and compressed air cleaning. Partial dismantling of the used vehicles may be necessary. Prevention of contamination/recontamination after cleaning may include storage in appropriate areas and on surfaces that prevent contact with soil, and managing the vegetation around storage areas. Even though used vehicles are not commodities that are harvested, the measures outlined in the standard are handling measures that reduce the risk of moving pests in trade.

24.8.14. ISPM 42: *Requirements for the use of temperature treatments as phytosanitary measures* and ISPM 43: *Requirements for the use of fumigation as a phytosanitary measure*

Both standards have sections that refer to prevention of infestation after treatment. They indicate that measures should be implemented to prevent possible infestation or contamination of the commodity after temperature or fumigation treatments and suggest that the following measures may be applied for that purpose:

- keeping the commodity in a pest-free enclosure
- packing the commodity immediately in pest-proof packaging

- dispatching the commodity as soon as possible.

The measures suggested are safeguarding/protection post-harvest processing and handling measures. Furthermore, Section 6.1 in ISPM 42 and 43, "Documentation of procedures," indicates that commodity handling procedures before, during and after fumigation should be specified.

24.9. Summary and Look Forward

Here are five things to remember from this chapter.

1. The primary objective of post-harvest processing and handling is to maintain plant commodity/product quality.

2. Risk-reducing measures applied to commodities moving in trade include measures applied to the commodity during post-harvest processing and handling, e.g., safeguarding in the field, washing, cleaning, drying, waxing, treatments, safeguarding during packing, and storage.

3. Post-harvest processing and handling phytosanitary measures, applied after the commodities/products leave their place of production, will vary depending on the plant commodity/product.

4. Post-harvest and handling measures do not directly affect pests so indirect measures of efficacy like the number of culled commodity units in the packinghouse, negative trap counts, or the results of inspections and audits may be more appropriate measures.

5. There are at least 14 ISPMs with information germane to the use of post-harvest and handling measures for risk management.

The next chapter introduces a relatively new innovation in phytosanitary risk management, post-entry measures. These are an important, but traditionally overlooked, part of the risk management continuum that are underutilized.

24.10. References

IPPC (1998) International Standards for Phytosanitary Measures, Publication No. 8: *Determination of Pest Status in an Area*. Secretariat of the International Plant Protection Convention (IPPC), Food and Agriculture Organization of the United Nations, Rome.

IPPC (1999) International Standards for Phytosanitary Measures, Publication No. 10: *Requirements for the Establishment of Pest Free Places of Production and Pest Free Production Sites*. Secretariat of the International Plant Protection Convention (IPPC), Food and Agriculture Organization of the United Nations, Rome.

IPPC (2002) International Standards for Phytosanitary Measures, Publication No. 14: *The Use of Integrated Measures in a Systems Approach for Pest Risk Management*. Secretariat of the International Plant Protection Convention (IPPC), Food and Agriculture Organization of the United Nations, Rome.

IPPC (2003) International Standards for Phytosanitary Measures, Publication No. 18: *Guidelines for the Use of Irradiation as a Phytosanitary Measure*. Secretariat of the International Plant Protection Convention (IPPC), Food and Agriculture Organization of the United Nations, Rome.

IPPC (2009) International Standards for Phytosanitary Measures, Publication No. 32: *Categorization of Commodities According to Their Pest Risk*. Secretariat of the International Plant Protection Convention (IPPC), Food and Agriculture Organization of the United Nations, Rome.

IPPC (2011) International Standards for Phytosanitary Measures, Publication No. 7: *Phytosanitary Certification System*. Secretariat of the International Plant Protection Convention (IPPC), Food and Agriculture Organization of the United Nations, Rome.

IPPC (2012) International Standards for Phytosanitary Measures, Publication No. 36: *Integrated Measures for Plants for Planting*. Secretariat of the International Plant Protection Convention (IPPC), Food and Agriculture Organization of the United Nations, Rome.

IPPC (2013a) International Standards for Phytosanitary Measures, Publication No. 1: *Phytosanitary Principles for the Protection of Plants and the Application of Phytosanitary Measures in International Trade*. Secretariat of the International Plant Protection Convention (IPPC), Food and Agriculture Organization of the United Nations, Rome.

IPPC (2013b) International Standards for Phytosanitary Measures, Publication No. 11: *Pest Risk Analysis for Quarantine Pests*. Secretariat of the International Plant Protection Convention (IPPC), Food and Agriculture Organization of the United Nations, Rome.

IPPC (2016) International Standards for Phytosanitary Measures, Publication No. 2: *Framework for Pest Risk Analysis*. Secretariat of the International Plant Protection Convention (IPPC), Food and Agriculture Organization of the United Nations, Rome.

IPPC (2017a) International Standards for Phytosanitary Measures, Publication No. 40: *International Movement of Growing Media in Association with Plants for Planting*. Secretariat of the International Plant Protection Convention (IPPC), Food and Agriculture Organization of the United Nations, Rome.

IPPC (2017b) International Standards for Phytosanitary Measures, Publication No. 41: *International Movement of Used Vehicles, Machinery and Equipment*. Secretariat of the International Plant Protection Convention (IPPC), Food and Agriculture Organization of the United Nations, Rome.

IPPC (2018a) International Standards for Phytosanitary Measures, Publication No. 6: *Surveillance*. Secretariat of the International Plant Protection Convention (IPPC), Food and Agriculture Organization of the United Nations, Rome.

IPPC (2018b) International Standards for Phytosanitary Measures, Publication No. 42: *Requirements for the Use of Temperature Treatments as Phytosanitary Measures*. Secretariat of the International Plant Protection Convention (IPPC), Food and Agriculture Organization of the United Nations, Rome.

IPPC (2019a) International Standards for Phytosanitary Measures, Publication No. 5: *Glossary of Phytosanitary Terms*. Secretariat of the International Plant Protection Convention (IPPC), Food and Agriculture Organization of the United Nations, Rome.

IPPC (2019b) International Standards for Phytosanitary Measures, Publication No. 43: *Requirements for the Use of Fumigation as a Phytosanitary Measure*. Secretariat of the International Plant Protection Convention (IPPC), Food and Agriculture Organization of the United Nations, Rome.

NAPPO (2014) *Principles of Pest Risk Management for the Import of Commodities*. Available at: http://www.nappo.org/files/8314/3889/6413/RSPM40-e.pdf (accessed January 17, 2020).

Watkins, C.B. and Nock, J. (2012) *Production Guide for Storage of Organic Fruits and Vegetables*. New York State Dept. of Agriculture and Markets. Publication # 10. Available at: https://ecommons.cornell.edu/bitstream/handle/1813/42885/organic-stored-fruit-veg-NYSIPM.pdf?sequence=1&isAllowed=y (accessed January 17, 2020).

25

Post-entry Measures

_____

A man who does not plan long ahead will find trouble at his door.

Confucius

25.1. Introduction

Effective mitigation of pest risk at origin is traditionally preferred over measures that may be applied after import, but thinking of risk management in terms of a continuum rather than a one-sided process opens other opportunities for safe trade. This chapter explores risk management measures that may be applied following import.

Foremost among post-entry measures is post-entry quarantine (PEQ), a well-established practice that has been widely used for decades to facilitate the exchange of plants and especially new germplasm. Other post-entry measures are conceptually, legally, and technically similar but operationally different enough to warrant separate discussions. Potential post-entry measures that may be legitimately considered but have not yet been used are also worthy of discussion within the broad scope of risk management options after import.

25.2. Why Post-entry?

Conventional wisdom holds that risk management should favor measures applied prior to import or, as a last resort, at the border. A big part of the underlying philosophy is that pest risk managers become more uncomfortable as the pest gets closer to the area being protected; the further the mitigations are from the destination, the better. This mindset however comes from an outdated view of quarantine and discounts the value of pest risk analysis as an objective, science-based method to understand the risk, the uncertainty, the efficacy, feasibility, and impact of options for risk management.

Consider for instance that the pest risk associated with weed seeds contaminating grain for processing will be fully mitigated when the grain is ground into flour. There is no need to mitigate for the weeds at origin or the border if safeguards are in place to maintain the integrity of the consignment and prevent weed seeds from escaping before the grain reaches the mill. In this example, processing becomes a post-entry measure and the best option for effective risk management.

ISPM 11 (*Pest risk analysis for quarantine pests*) tells us that post-entry measures may be necessary for "pests not detectable on entry." ISPM 20 (*Guidelines for a phytosanitary import regulatory system*) provides additional scope and detail.

Notice the subtle distinction between "measures that may be required after entry" and "quarantine." Quarantine is a specific type of post-entry measure. Not all post-entry measures are quarantine.

Phytosanitary measures that may be required after entry include the following.

- Detention in quarantine (such as in a post-entry quarantine station) for inspection, testing or treatment.
- Detention at a designated place pending specified measures.
- Restrictions on the distribution or use of the consignment (e.g. for specified processing).

(ISPM 20, 4.2.1)

In some cases, NPPOs may decide that a period of quarantine is necessary for a specific type of consignment. This is usually because of the difficulty verifying the absence of quarantine pests upon entry. Post-entry quarantine provides controlled conditions for testing for the presence of pests, time for the expression of signs or symptoms, and appropriate treatment if necessary. The key elements of quarantine are the same as in animal or human health: secure confinement, observation, testing, or research. Treatment may also be included if it involves confinement *after entry*, which is different than treatment that is prescribed as a condition of entry or treatment done as an emergency measure to gain entry. These latter situations may require confinement for some period until the treatment is completed, but the consignment is not released until afterward.

Another point we can extract from the guidance in ISPM 20 is "restrictions on distribution." This means that limited distribution, geographically and/or by time, is another post-entry option to the extent national plant protection organizations (NPPOs) can identify appropriate areas or times for consignments to enter and be distributed. The United States has a long history of import requirements that limit the entry of consignments to certain ports, distribution to specific states, or entry only during specified times of the year, especially for fruit fly host material. In many cases, these requirements are put in place as provisional measures that are later evaluated and modified based on experience. From this standpoint, post-entry measures can also become a useful tool for addressing uncertainty.

The asynchrony of seasons between northern and southern hemispheres creates many opportunities to move commodities with pests that would be unable to establish in the opposite hemisphere because they arrive during the wrong season for their life stage. The ecological difference between tropical and temperate climes offers similar possibilities, but these are rarely explored. Many more prospects exist for risk management designs that are based on or include seasonal or climatological asynchrony. The problem is overcoming the traditional mindset that live pests should not be in the pathway even if they cannot establish.

25.3. Legal Frameworks

Post-entry measures can include any risk management action applied or required by phytosanitary authorities following release of a consignment by Customs. This means that as a first requirement, the NPPO must have legal authority to apply regulatory controls and take regulatory actions following release at the border. There are three primary legal mechanisms that are used:

1. Direct statutory authority.
2. Indirect authority or extension of authority under compliance arrangements.
3. Collaboration with sub-national authorities.

25.3.1. Direct statutory authority

The legal authority of NPPOs to regulate consignments and take phytosanitary actions is traditionally limited to imports but it may not stop at the border. Depending on the design of the statutory authority, i.e., acts and laws, NPPOs may be able to directly exercise their authority after a consignment has entered domestic commerce. This is the most desirable and least complicated situation. It allows the NPPO to take advantage of the entire risk management continuum and create holistic risk management designs for the application of measures where they are most effective, including post-entry. The main disadvantage of this design is the resource requirements for supporting domestic operations which are typically much more dispersed than border operations, especially in geographically large countries.

25.3.2. Indirect authority

The most common situation is a mix of full authority for imports with limited or indirect authority for consignments post-border. In this case, the NPPO will only have direct responsibility for regulations applying to imports but partial or no authority over consignments after entry. When legal systems allow, NPPOs can extend their border authority by putting in place legal arrangements that make the release of the consignment at the border contingent on cooperation by domestic entities for follow-up requirements.

In the example above where wheat grain is authorized for movement to a mill for processing, the NPPO may have an agreement with the entity responsible for transportation and another agreement with the processing facility. Violations of these agreements could result in the NPPO revoking the authorization for wheat to enter if associated with these entities. The primary drawback to indirect authority is the complicated legal nature of arrangements and enforcement. Indirect controls also tend to be less rigorously managed than border operations, leading to weaker compliance.

25.3.3. Collaboration with sub-national phytosanitary authorities

In cases where the NPPOs authority ends at the border and the legal system makes no provision for the extension of authority through post-border arrangements, NPPOs may collaborate with provincial or local authorities to implement post-entry measures. Such arrangements depend on the legal force of sub-national authorities and their political will to consistently collaborate. Again, using the example of grain imported for milling, provincial officials could provide oversight and enforcement for domestic entities based on the understanding that they

make it possible for national authorities to authorize consignments that would otherwise require stronger measures or be impossible to import. The key issue with this approach is enforcement and the ability to defend local actions based on national requirements.

25.4. Research, Analysis and Exhibition

Most NPPOs have special provisions in their statutory framework to authorize the import of pests, plants and plant products and other regulated articles that are needed for research, analysis, or exhibition but might be restricted or prohibited as commodities. These imports are a low volume, unique and highly variable portion of commerce that represent a critical area for the application of post-entry measures.

There is frequently no precedent or consistent guidance for these imports, so many decisions are highly dependent on one-time judgments that demand a high level of risk management experience and insight. Import permits with very specific requirements tailored to the situation are usually required. Such authorizations may be one-off or ongoing and can involve any or all the legal designs described above. Arrangements will typically be with a research organization such as a university, private laboratory, or government agency, but could also be as simple as allowing an item to be imported once in limited quantities and controlled conditions for demonstration purposes and then destroyed.

History has shown that some of the worst pest introductions have been as a result of escapes or releases from research. A well-known example is the introduction of Gypsy moth (*Lymantria dispar*) into North America. The pest became a major forest pest in the United States after escaping from research in 1868. This raises one of the more difficult aspects of risk management associated with consignments for research: the relationship of the risk manager with the researcher or research institution. The importer is typically an expert on the organism(s), whether they are plants, pests, or both, and may resent the intervention of regulatory officials who complicate their work. Communication is very important for developing an effective working relationship, so both the researchers and regulators understand the position and motives of the other.

25.5. Biocontrol and Beneficial Organisms

ISPM 3 (2017) is a useful reference for both regulatory and technical guidance, including post-entry measures used for the handling and movement of biological control agents and beneficial organisms.

The standard emphasizes the important role of pest risk assessment (PRA) in supporting the understanding of risk and especially decisions on appropriate measures. Another key aspect of the standard is the detailed guidance on the shared roles of NPPOs in importing and exporting countries.

25.6. Plants for Planting

Live plants, consisting of entire plants, plant parts, and seed intended for planting, comprise a crucial subsector of agriculture trade. The live plant pathway also poses significant phytosanitary threats for two reasons. First, living hosts are ideal for carrying regulated pests which may not be obvious due to the plant life stage, pest life stage, or time of year. Second, phytosanitary treatments for the live plant pathway can be logistically difficult, expensive, potentially damaging, and, if administered improperly, ineffective.

PEQ was designed as a special mechanism to facilitate the importation of small quantities of new germplasm that could be grown and tested under biosecure conditions before being released to domestic growers. Facilities used for PEQ range from fully enclosed, highly secure facilities on par with animal and human quarantine designs, to simple screen houses or even open areas specially designated and isolated from other hosts of concern. Such facilities are either operated by the NPPO or other government agency, a research university, or a commercial research organization under NPPO oversight.

The specialized nature of the containment structures, equipment, and personnel make PEQ an expensive risk management option that must be undertaken with greater care than routine measures. Cost recovery designs are common for offsetting the investment by the NPPO and co-operators. High costs however tend to limit the number of users and volume of consignments to the minimum amount, highest priority and highest value plant materials.

25.6.1. Intermediate PEQ

Propagative material is sometimes "cleaned" in a country other than its destination. This is often the case with new germplasm that is sent from its natural origin to international research centers including but not limited to Consultative Group for International Agricultural Research (CGIAR) Centers (see CGIAR, 2019). These research facilities are typically not in the destination country and are often far removed from the areas where wild or cultivated hosts may be endangered by pest escape.

Intermediate facilities operate under the authority of their host country. These arrangements offer both advantages and disadvantages for the destination country. It is obviously an advantage to have PEQ offshore from a risk management standpoint, however, the destination country has limited control over policies, operations, and priorities of an organization in a third country. Even if the ultimate importing

---

ISPM 3 addresses biological control agents capable of self-replication (including parasitoids, predators, parasites, nematodes, phytophagous organisms, and pathogens such as fungi, bacteria and viruses), as well as sterile insects and other beneficial organisms (such as mycorrhizae and pollinators), and includes those packaged or formulated as commercial products. Provisions are also included for import for research in quarantine stations of non-indigenous biological control agents and other beneficial organisms.

The scope of this standard does not include living modified organisms, issues related to registration of biopesticides, or microbial agents intended for vertebrate pest control.

(Scope, ISPM 3, *Guidelines for the export, shipment, import and release of biological control agents and other beneficial organisms*)

country is not fully satisfied with intermediate PEQ, it can substantially reduce the number of pests, time and effort required to complete PEQ at the destination by taking advantage of intermediate facilities.

25.6.2. Open PEQ

Certain types of PEQ may not require confinement. For instance, a virus may not be able to spread without its vector. Potentially infected plants grown in an area where the vector is absent are not a threat to other hosts and therefore require no structural separation. Although rarely used, there are clearly situations where open PEQ is a cost-effective option as a post-entry measure.

25.6.3. Commercial nursery PEQ

The importation of commercial lots of plants for planting poses special challenges for risk managers. The material is highly perishable, difficult to handle and inspect without damage, and it increases in value over time. Consignments of nursery plants are not typically one-off authorizations, but rather routine commerce supported by a PRA and import requirements elaborated in regulations. These requirements often include post-entry arrangements such as a specified growing period in a greenhouse or other controlled environment favorable for pest detection and control before release for commercial distribution.

25.6.4. Seeds and tissue culture

The NPPO of the importing country may also require post-entry quarantine for seeds, including confinement in a quarantine station, in cases where a quarantine pest is difficult to detect, where symptom expression takes time, or where testing or treatment is required and no alternative phytosanitary measures are available. Guidance on post-entry quarantine stations is provided in ISPM 34 (IPPC, 2016).

As part of PEQ, a representative sample of the seed lot may be sown and the plants growing from these seeds tested; this may be an option for small seed lots used for research. The NPPO of the importing country may consider, based on the findings of a PRA, that the pest risk can be adequately managed by requiring the imported seeds to be planted in a designated planting area isolated from other host plants.

25.7. Sampling for Inspection and Testing

Many of the post-entry measures described here and practiced around the world have sampling for inspection or testing as a central feature. It is important for risk managers to recall that, as with the inspection of other consignments, there are statistical parameters to consider, including the desired level of detection and confidence interval which will determine the appropriate sample size for a given lot size.

When sampling for testing, the overall confidence level is compounded by the sensitivity of the test method. This means that if the sampling has 95% confidence, we are accepting the possibility of being wrong 5% of the time. If that sampling is designed to detect a 5% infestation level, then we are accepting 4% or less infestation escaping detection. Finally, if the testing method is 90% accurate, we are also adding the possibility that 10% of the tests will be incorrect. The important point is not the mathematics but for risk managers to recognize that while testing may be a more sophisticated method for detecting a pest, it does not necessarily provide greater precision. Statistical designs for testing are needed to understand the true sensitivity of this measure in practice, just as they are needed for inspection (see Chapter 20, this volume).

25.8. Summary and Look Forward

Here are five things to remember from this chapter.

1. Post-entry measures are under-appreciated for their risk management value because risk managers are reluctant to allow potentially infested articles to enter even when risk analysis supports the use of such measures.

2. PEQ is a well-established and widely practiced post-entry measure used for propagative material.

3. Post-entry measures are essential for biocontrol and beneficial organisms.

4. Imports of restricted plants or pests for research require unique designs and careful risk management.

5. Sampling for inspection and testing in post-entry programs requires the same attention to statistical parameters as routine commodity inspection.

The next chapter discusses one of the most commonly used, most over-used and most misused of all phytosanitary risk management measures, prohibition.

25.9. References

CGIAR (2019) Research Centers. Available at: https://www.cgiar.org/research/research-centers/ (accessed September 20, 2019).

IPPC (2013) International Standards for Phytosanitary Measures, Publication No. 11: *Pest Risk Analysis for Quarantine Pests.* Secretariat of the International Plant Protection Convention (IPPC), Food and Agriculture Organization of the United Nations, Rome.

IPPC (2016) International Standards for Phytosanitary Measures, Publication No. 34: *Design and Operation of Post-entry Quarantine Stations for Plants.* Secretariat of the International Plant Protection Convention (IPPC), Food and Agriculture Organization of the United Nations, Rome.

IPPC (2017) International Standards for Phytosanitary Measures, Publication No. 3: *Guidelines for the Export, Shipment, Import and Release of Biological Control Agents and Other Beneficial Organisms.* Secretariat of the International Plant Protection Convention (IPPC), Food and Agriculture Organization of the United Nations, Rome.

IPPC (2019) International Standards for Phytosanitary Measures, Publication No. 20: *Guidelines for a Phytosanitary Import Regulatory System.* Secretariat of the International Plant Protection Convention (IPPC), Food and Agriculture Organization of the United Nations, Rome.

26

Prohibition

26.1. Introduction

The spectrum of measures available to the phytosanitary community for risk management can be viewed on a scale of strength ranging from the least trade restrictive measure to the most trade restrictive measure. The lowest end of this scale is the "do nothing" option which implies the risk in question is acceptable under the circumstances or perhaps nothing can be done so the risk must be accepted. The latter situation might occur with a pest that is spreading naturally and has no practical control measures. In such cases, the risk management strategy shifts to planning for the impacts and managing the consequences. This chapter is concerned with the other end of the scale for the strength of measures which is typically characterized by an unacceptably high level of risk demanding the most trade restrictive measure.

The most trade restrictive measure is one that results in no trade—typically a prohibition. From a trade facilitation standpoint, prohibition should be the final option when all other options that might allow trade have been exhausted. From a protection standpoint, prohibition can seem to be a simple solution to unacceptable risk but it may not provide the expected results.

The objective of prohibition in a phytosanitary context is to prevent an activity from occurring or an article from moving in trade. The principle of transparency would argue that the legal authority for prohibition exists, the prohibited article or action is publicly identified, and the rationale for its status is publicly available. A declared prohibition is a phytosanitary measure and therefore the basis for the prohibition should be either a risk assessment or an international standard.

> To harmonize sanitary and phytosanitary measures on as wide a basis as possible, Members shall base their sanitary or phytosanitary measures on international standards, guidelines or recommendations, where they exist, except as otherwise provided for in this Agreement.
>
> (SPS Art 3.1)
>
> Members shall ensure that their sanitary or phytosanitary measures are based on an assessment, as appropriate to the circumstances, of the risks to human, animal or plant life or health, taking into account risk assessment techniques developed by the relevant international organizations.
>
> (SPS Art 5.1)

Other factors may enter into the final risk management decision (see Chapter 14, this volume), and operational policies or regulatory designs can also lead to the effect of prohibition without the benefit of a risk analysis to justify the result. These situations also need to be understood by risk managers as forms of prohibition to the extent they result in the same outcome—no trade.

This chapter contrasts prohibition with operational policies and regulatory designs which also create "no trade" scenarios. It looks at the circumstances under which prohibition is justified as a risk-management measure and the cultural, trade, political, and enforcement challenges for risk managers using it.

26.2. Declared Prohibition

The idea of using regulatory authority to stop a harmful article or activity is easily seen as a first resort for intractable risks. Human nature draws us toward prohibition as an intuitive solution to a problem—simply stop it. The same human nature however can greatly complicate the reality of implementation, as demonstrated many times through history. One famous example is the 18th Amendment of the US Constitution which put in place a nationwide prohibition on the production, importation, transportation, and sale of alcoholic beverages from 1920 to 1933. This controversial regulatory action had the intended effect of reducing alcoholism and some other health concerns, but it created new health problems associated with the consumption of tainted homemade and bootleg products. Prohibition also promoted organized crime and substantially increased violence. A very large and rich underground economy developed with the exchange of substantial sums that were invisible to the government and therefore unavailable to tax at a time when the economy was struggling. The law was ultimately repealed because of the extensive damage to society and government from unintended effects (PBS, 2011).

The US experience with the 18th Amendment was a hard lesson to learn and remains a pivotal case-study for regulators and legislators. The fundamental lesson from the experience is to take full account of tradeoffs, repercussions, and the potential for risk to shift in a way that exacerbates the problem or creates new problems that may be as bad or worse than the issue being addressed. Similar situations occur frequently with

similar results in other regulatory disciplines when risk managers do not carefully consider the entire panorama of lateral and downstream impacts when prohibition is raised as a viable option.

A common misunderstanding associated with prohibition is that it is a highly effective measure. Phytosanitary prohibitions are expected to close a pathway in the same way the 18th Amendment was supposed to stop the consumption of alcohol. There are at least two situations where this is not the case. One is where natural spread is a viable pathway. Any regulatory strategy should be weighed against the likelihood of the hazard occurring naturally or being uncontrollable. For instance, what would be the value of prohibiting the movement of plant material from affected areas to adjacent areas if the pest is a fungus that releases spores into the wind?

Promoting a shift toward illicit movement is the other situation where the effectiveness of prohibition is compromised. Although prohibition clearly represents a worst-case situation for legitimate trade, it doesn't always represent the most effective measure for risk management. National plant protection organization (NPPO) and Customs officials constantly monitor trade for prohibited articles that are smuggled in baggage, cargo, mail, and conveyances. In some instances, a lack of awareness may be to blame, but many cases are also deliberate. The greater the interest in the prohibited article, the less effective we can expect prohibition to be. In fact, prohibition can increase the risk of illegitimate trade everywhere there is a strong motivation for a pathway to exist. Consider for instance a prohibition on the importation of a popular fruit or vegetable or a valuable new plant variety. The high demand for these commodities in the face of prohibition will increase the probability for smuggling and the likelihood of pest introduction. A better strategy may be to accept a somewhat less comfortable level of risk by authorizing the articles or activity under less restrictive measures that also reduce the motivation for smuggling while increasing the ability of regulators to monitor imports.

The key point for risk managers to understand regarding the use of prohibition is distinguishing its strength as a phytosanitary measure from its efficacy as a risk mitigation measure in the context of unintended effects. Trade restrictiveness does not correspond directly to risk mitigation effectiveness, and the difference can be strikingly apparent when prohibition fails. This is an excellent example of a factor that may change a risk management strategy from what the risk assessment would argue to what a risk manager would consider more appropriate.

The economic costs versus the expected benefits of implementing any measure, but especially prohibition, is another factor that should be part of the calculus prior to prescribing prohibition. More protection is not always better when viewed from a cost–benefit standpoint. If the investment required for enforcement, outreach, and mitigating unintended effects is greater than the value of the protection afforded, then the prudence of prohibition again needs to be questioned.

Where prohibition is an unavoidable result, efforts need to be directed not only to strong enforcement, but also to raising awareness and providing information in a way that promotes compliance. The objective should be to balance enforcement with outreach to reduce both deliberate and unintentional actions that undermine the risk management value of prohibition. This social science aspect of risk management is often overlooked in the application of prohibition as a phytosanitary measure and discussions of risk communication. This is partly due to the lack of specific guidance on methods but also because of the tendency for risk managers to focus on technical aspects of the specific and immediate risk challenge. A risk management calculus that opens the boundaries to social science aspects of the risk and broader implications of a strategy can open new avenues for risk management.

Imagine for instance that a beautiful flowering plant popular in exotic locations is prohibited because of its propensity for carrying a harmful plant virus with a broad host range. The plant is frequently intercepted in the baggage of air travelers and cruise passengers, but some plants certainly escape detection despite the best efforts of inspectors. Now imagine that the plants can enter if purchased from merchants belonging to a special certification program that tests plants for virus freedom. The temptation to simply prohibit the plants continues to be great because of the risk, but prohibition is more likely to increase the risk than a program that makes it possible to import legitimately. In this case, an uncomfortable measure that allows the risk to be monitored and managed will be more effective than a comfortable measure which does not have the intended effect.

26.3. Not Authorized

Because prohibition is a phytosanitary measure, it must be based on an international standard or a risk assessment according to the Sanitary and Phytosanitary Agreement (SPS Agreement). However, there are many articles unable to move in trade which are not prohibited but simply have not been authorized. The result is the same—no trade, but the rationale and authority behind the condition is extremely important for trading partners to determine the actual status.

Different types of regulatory designs can have a significant effect on the processes that provide the authorizations for regulated articles to be imported. One design may identify restricted or prohibited articles. Another may identify authorized products. Some are based on the type of product, e.g., fruit, vegetable, cut flower, others may be based on the intended use such as consumption, propagation, or processing. Still other regulatory designs may be for specific articles such as machinery, roses, cotton, and the like. Finally, there may be combinations of these designs, e.g., cut roses versus roses for planting. In each case, the regulatory design requires different processes to arrive at a decision on the entry status and conditions for import. In the meantime, the article is "not authorized."

The "not authorized" category is for those articles that need to be evaluated for their measures to be decided; in other words, for a pest risk assessment (PRA) to be completed. Typically, the process is initiated by a request for market access from a trading partner, although NPPOs may initiate the process for any number of internal reasons. Once a trading partner makes an official request, it is incumbent on both the importing NPPO and the exporting NPPO to begin collaborating on the exchange of information necessary to properly evaluate the request, including the completion of a risk assessment and the development of viable risk management options. Failing to follow-through in good faith is abusing the approval process to create an unjustified barrier to trade. Creating an unnecessarily cumbersome approval process results in the same problem.

Unfortunately, the term "prohibited" is used too loosely and erroneously implies that a justification exists for a measure when in fact a PRA has not been done to establish the official status of the article or activity in question. Risk managers should be especially careful to recognize this distinction and avoid confusing "prohibition" with "not authorized" to ensure trading partners understand the basis for "no trade." The former is a phytosanitary measure which should be supported by a risk assessment; the latter is a question of process and requires clarification on the actions needed to evaluate the import and arrive at a decision on its status. In either case, the central question will be the risk and appropriate measures for risk mitigation.

26.4. Emergency Actions

An operational intervention in trade that can lead to a "no trade" scenario is the rejection of consignments at the port of entry based on inspection. Although the imported articles may be authorized for import, they must also meet entry requirements which are verified with inspection. The rejection of a single consignment based on inspection is a situation-specific prohibition based on a single event, typically the detection of a pest. This represents an unanticipated situation that requires the risk to be addressed urgently. Consignments are safeguarded to ensure their phytosanitary security and emergency actions are prescribed based on the nature of the problem, local conditions, and available options. The most common risk management options for emergency actions include:

- rejection—re-exportation or destruction
- treatment —applying a treatment that may or may not be specific for the pest in question or
- reconditioning—removing contaminated materials, e.g., wood packing.

All NPPOs have a sovereign and fundamental right to implement emergency actions when faced with new or unanticipated risks. The International Plant Protection Convention (IPPC) identifies this as a principle (ISPM 1). The SPS Agreement is less clear in this regard, referring instead to "urgent concerns" in Annex B Notification (SPS). The concept is the same: the ability to take emergency measures is necessary and practical but it must also be carefully practiced to avoid becoming an unjustified barrier to trade. This is especially true when emergency actions are not followed by risk analysis to determine whether the initial reaction was correct and if adjustments are needed in risk management for more appropriate measures to be applied based on the risk.

In an emergency situation, the reaction of the NPPO can be whatever it considers necessary to safeguard against the immediate risk and possible future risks. The response need not be supported by a PRA or other analysis but it should be transparently tied to a policy framework that ensures consistent actions to avoid being completely arbitrary. The first response in most port of entry situations is to put in place the strongest possible measures—usually rejection or treatment if it is feasible.

In principle, such strong reactions should only be necessary until the situation can be evaluated and the risks addressed in a systematic way. NPPOs, however, do not have the resources to follow every emergency action with a risk analysis and it would not be prudent to perform risk analyses for events that may never occur again. This means that judgments must be made by risk managers to determine which emergency actions require follow-up. Some may receive immediate attention because they represent a very high risk or have broader implications than just the consignment in question. For instance, an unanticipated pest may be found that significantly changes the risk for the import in question and perhaps affects the phytosanitary status of related products and the exporting country generally. Examples of this have occurred as a result of new fruit fly outbreaks.

Most emergency actions are viewed by risk managers as unique events, and even though the same situation and reaction may be repeated dozens of times over long periods, the idea that these are all emergency measures is likely to continue without a technical justification until challenged. This is where risk managers need to be careful. Serial rejections become a de facto prohibition that fits with the "no trade" pattern. At some stage, it can be argued that repeated emergency actions no longer represent emergencies because they are not anticipated and have been repeated enough times and over a long enough period to be analyzed for their technical justification. Risk managers may overlook this aspect of policy under the ruse of providing protection from contaminated imports; i.e., precautionary policies are justified because there should be no pests, irrespective of the risk. While this approach appears to take the high road from a protection standpoint, it is inconsistent with the principle of managed risk and represents an overly simplistic and unfair approach to phytosanitary policy that is not defendable if challenged in the SPS framework.

There is no clear point where a series of similar emergency actions can no longer be justified. It is, however, clear that the more such actions are repeated for a pattern of similar scenarios and the longer the emergency actions continue without review, the more vulnerable the NPPO becomes to an SPS challenge.

26.5. Summary and Look Forward

Here are five things to remember from this chapter.

1. Prohibition is likely to have unintended effects which need to be considered.
2. Risk can increase or shift as a result of using prohibition.
3. Prohibition and "not authorized" have different bases but the same effect on trade.
4. Regulatory designs and approval processes affect how measures like prohibition are put in place.
5. Emergency actions can become de facto prohibitions.

The next chapter will move beyond traditional risk management options to explore alternative approaches.

26.6. References

IPPC (2016) International Standards for Phytosanitary Measures, Publication No. 1: *Phytosanitary Principles for the Protection of Plants and the Application of Phytosanitary Measures in International Trade*. Secretariat of the International Plant Protection Convention (IPPC), Food and Agriculture Organization of the United Nations, Rome.

PBS (2011) *Prohibition: Unintended Consequences|PBS*. Pbs.org Available at: http://www.pbs.org/kenburns/prohibition/about/ (accessed January 2, 2020).

27

Systems Approaches

Let our advance worrying become advance
thinking and planning.

Winston Churchill

27.1. Introduction

The benefits of international trade in plants and
plant products are many, evident and significant
(see Chapter 1, this volume) while the tools that
have been used to manage pest risk in trade are
relatively limited. Systems approaches are increasingly accepted and implemented as an important addition to the toolbox of options to
manage phytosanitary risk because they offer a
multitude of alternatives to standalone phytosanitary measures (e.g., EPPO, 2013; USDA/
APHIS/BAPHIQ, 2013; Jang *et al.*, 2015;
NWHC, 2017). This chapter considers the role
of systems approaches in current and future
phytosanitary risk management decisions.

Several key points about systems approaches provide a starting point for this discussion. These are:

- systems approaches add possibilities for
  trade that single point mitigation cannot,
  while providing an equivalent or better
  level of protection
- systems approaches provide flexibility and
  precision as alternatives to traditional single point mitigations

- systems approaches can provide an effective
  approach to the uncertainty that is identified
  in a risk assessment or the evaluation of the
  efficacy of risk management measures
- systems approaches allow risk managers to
  formally include practices and conditions
  that are not traditionally prescribed as
  measures
- systems approaches create opportunities
  for new risk management designs, taking
  us far beyond the current toolbox
- systems approaches may provide an equivalent or better level of protection at a lower
  total cost than using a single measure
- the analysis for a systems approach is useful
  for aligning our thinking with the risk management continuum
- systems approaches encourage collaboration between national plant protection organizations (NPPOs) and industry
- systems approaches can range in complexity from a collection of qualitative measures
  to a sophisticated quantitative control-point
  design like Hazard Analysis Critical Control
  Point (HACCP) systems.

27.2. What is a Systems Approach?

The International Plant Protection Convention
(IPPC) defines a systems approach as "The integration of different risk management measures,

at least two of which act independently, and which cumulatively achieve the appropriate level of protection against regulated pests" (ISPM 5). Note that systems approaches may also be referred to as "integrated measures" in some standards or other documents. Aluja and Mangan (2008) define a systems approach as "the integration of pre- and postharvest practices, from the production of a commodity to its distribution and end use that cumulatively meet predetermined requirements for quarantine security."

Systems approaches are addressed generally in ISPM 14 (2019a), *The use of integrated measures in a systems approach for pest risk management*. In addition, ISPM 35 (2019b), *Systems approach for pest risk management of fruit flies (Tephritidae)* and ISPM 36 (2019c), *Integrated measures for plants for planting* provide specific guidance on their respective specific topics. The North American Plant Protection Organization (NAPPO) Regional Standard for Phytosanitary Measures (RSPM 40), *Principles of pest risk management for the import of commodities* provides a good discussion on principles of risk management and systems approaches.

27.3. When are Systems Approaches Appropriate?

Systems approaches are most appropriate when single measures cannot reduce risk to an acceptable level, single measures are highly uncertain, or a single measure is not available. In addition, systems approaches may be useful when production systems incorporate mitigation practices as part of the normal production process. For this reason, systems approaches are often the least trade restrictive measures available and also the most feasible, but systems approaches may not be feasible for some commodities if the costs to apply multiple measures are excessive, or the production processes or pests are poorly understood.

One of the main advantages of using systems approaches is that even if one of the measures fails, the system includes additional measures that still mitigate risk. A second advantage of using systems approaches is that they can provide a mechanism for addressing uncertainty by varying the types and strengths of measure. Lastly, an important advantage of systems approaches is

that measures can be applied throughout a pathway, from the point of origin of a commodity through to the end use of that commodity in an importing country. This provides maximum flexibility to the advantage of both importing and exporting country NPPOs.

NAPPO RSPM 40 notes several important concepts regarding systems approaches:

- by definition, systems approaches require that two or more measures act independently
- systems approaches may include "safeguards"—measures that do not kill pests or reduce their prevalence but reduce the potential for their entry and establishment
- systems approaches consider the combined effect of different conditions and procedures
- systems approaches address uncertainty by varying the number and strength of measures
- systems approaches consider the "chain" from field to distribution in the importing country
- systems approaches provide an alternative to single and other less practical measures.

Griffin (2012) describes conditions that favor the success of a systems approach:

- the pest and pest/host relationship are well known
- practical systems exist for pest detection in the field and in consignments
- growing, harvesting, packing, transportation, and distribution practices are well-known and standardized
- pests of concern are typically absent or rare
- the volume/value of the commodity offsets increased program costs
- pest mitigation and safeguard measures can be identified, monitored, and corrected
- phytosanitary security is apparent through either qualitative or quantitative assessment.

27.4. ISPM 14

This international guidance says systems approaches can provide an alternative to single measures to meet the appropriate level of phytosanitary protection of an importing country. They are also useful in situations where no single phytosanitary measure is available. To qualify as a systems approach, the approach must combine different measures, at least two of which act independently of one another. The independent measures may be single measures or they may be a combination of dependent measures. The cumulative risk managing effects of a systems approach are derived from the measures that comprise the system.

Fig. 27.1 depicts a generic pathway or supply chain that shows how a systems approach might work. The plus signs at each stage represent the relative risk level at that stage of the pathway—more pluses equal more risk. Measures may be applied at any point in the pathway. This simple graphic shows that a specific measure applied at a stage of the pathway reduces the risk incrementally. The cumulative effect of this conceptual system is attrition of the risk to an acceptable level.

The process that leads to a systems approach begins with the conclusions from the pest risk assessment, Stage 2 of pest risk assessment. These are used to decide whether pest risk management is required and the strength of measures to be used. In Stage 3 of PRA, pest risk managers identify ways to respond to the assessed risk. The efficacy of the considered risk management options is evaluated and a most appropriate option is recommended.

ISPM 14 identifies potential circumstances when a systems approach may be used. The first set of circumstances occurs when an individual measure:

- is not available or is likely to become unavailable
- is detrimental to the commodity, human health, or the environment
- is not adequate to meet phytosanitary import requirements
- is not cost-effective
- is overly trade restrictive
- is not feasible for any other reason.

Other circumstances that favor the development of a systems approach include when the pest and pest–host relationship are well known and where multiple intervention points are clearly identified. Similarly, a systems approach may be appropriate if the relevant growing, harvesting, packing, transportation, and distribution practices are well-known or if the prevalence of the pest(s) is known and can be monitored. When a systems approach has been demonstrated to be effective for a similar pest/commodity situation it is worth considering it again. When data are sufficient to assess the effectiveness of individual measures either qualitatively or quantitatively or when individual measures can be monitored and corrected, a systems approach may be appropriate. Anytime a systems approach provides cost-effective risk management that is equivalent to a single measure it should be considered. Thus, there is a wide range of circumstances where a systems approach for risk management can be used. These systems can vary greatly in their complexity from a simple aggregation of measures to a sophisticated quantitative control-point design, as discussed in the following section.

A systems approach may be championed by the importing country, the exporting country, or ideally, collaboratively, by both countries. It will be the importing country that ultimately decides on the suitability of the systems approach in meeting its phytosanitary import requirements. ISPM 14 identifies the need to evaluate the efficacy of a systems approach quantitatively, qualitatively, or through a combination of both. It provides guidance on how to evaluate the effectiveness of a systems approach.

27.5. Types of Systems Approach

Systems approaches can range from relatively simple combinations of two or more measures to

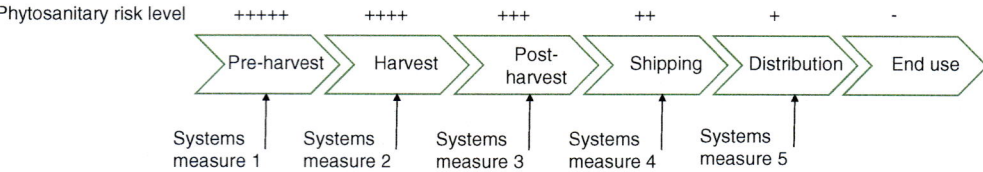

Fig. 27.1. Simple plant product supply chain.

highly complex systems that manage risks throughout the entire pathway. The complexity of a systems approach is tied to the commodity in question and the risk presented by the quarantine pest(s) associated with that commodity. At their simplest, systems approaches can be two independent measures, for example, a phytosanitary treatment and an import inspection. From there, systems approaches can increase in complexity both in the types of measures applied, and in terms of where in the pathway the measures are applied.

It is also useful to consider in general terms the purpose of required measures in systems approaches. Generally speaking, measures fall into three categories:

1. *Measures to reduce risk.* This includes measures such as phytosanitary treatments that kill pests but may also include other measures such as culling or limiting imports to off-season periods.
2. *Safeguarding measures.* Safeguarding measures prevent pests from entering a pathway. This may include field sanitation or packaging that prevents pest infestation in harvested commodities.
3. *Verification measures.* Verification measures can include practices such as system audits or inspections designed to verify that the system approach is efficacious and has not been compromised or otherwise failed.

Devorshak (2012) suggested that systems approaches fall into three broad categories: mitigation systems, quality systems, and control point systems.

27.5.1. Mitigation systems

These systems consist of combining phytosanitary procedures. They do not consider the quality of the commodity, the benefits of trade, the costs of mitigation, or anything but the measures prescribed for achieving phytosanitary security and their effectiveness.

Consider an example using papayas and Tephritid fruit flies, adapted from Devorshak (2012). A single point mitigation response might consist of harvesting only quarter-ripe fruit. This becomes the keystone element in a

systems approach. Because it is not infallible, a mitigation system might also include cultivating a papaya variety known to be a poor host for the fruit flies of concern. The system can be further strengthened by adding surveillance in and around the growing area to document fruit fly population pressure. These additional measures function independently of the harvesting at a specific degree of maturity measure.

The bulk of the phytosanitary protection is provided by the ripeness of the fruit at harvest. The additional measures of variety selection and surveillance provide additional phytosanitary security by reducing the likelihood of fruit fly introduction and establishment. If the primary measure fails, the additional measures provide redundancy in risk management. Note that even more measures could have been added to this system. Additional field practices or packinghouse culling could have been added to the system, but the harvest, variety selection, and surveillance measures may be judged to cumulatively provide the required level of protection to the importing country.

27.5.2. Quality systems

Quality systems consist of phytosanitary and other procedures. They typically include a range of processes that are designed to ensure the quality of the commodities as well as the phytosanitary security provided by the system of measures.

Consider an example using cut flowers, leaf miners and other pests, adapted from Devorshak (2012). Imagine that visual inspection is the only officially prescribed phytosanitary procedure in this hypothetical quality system. Why might this even be called a systems approach? Other risk attrition effects come from field and packinghouse processes the grower/producer has put in place to assure and maintain suitable commodity quality. Examples of these measures may include a program of specific cultivation practices, pest surveillance, harvesting practices, and field or glasshouse treatments. The harvest and packing operations may have strict culling and sanitation guidelines, including washing and cooling. Furthermore, a risk-based designed inspection prior to export provides a high degree of confidence in pest freedom (see Chapter 20, this volume). When a grower/producer is accredited by the NPPO and has a record of conformity, confidence in that grower/producer extends to phytosanitary security, especially when all the pests associated with the commodity are visibly detectable.

In this example, demand for the commodity requires a high level of quality. The commercially necessary quality processes also reduce the pest load and therefore can be considered part of the phytosanitary systems approach without being officially prescribed. Recognition of these commercial processes is based on a high degree of confidence that they will occur and have consistent mitigation effects in addition to the prescribed phytosanitary measures.

27.5.3. Control point systems

A control point is a practice, procedure, process, or location at or by which a measure can be applied, or conditions occur that will contribute to reducing risk. A critical control point is a control point where controls must be applied in order to reduce risk sufficiently. Acceptance criteria at any control point require the systematic observation of measurable characteristics and the ability to exercise action when criteria are not met (Griffin, 2012). A control point system is the plant health conceptual equivalent of the food safety community's well-established and widely applied HACCP (Chapter 28, this volume). HACCP is a process that identifies the point(s) at which potential contamination can occur, i.e., the critical control points or CCPs, and it strictly manages and monitors these points as a way of ensuring the process is "in control" and that the safest product possible is produced. Control point systems for phytosanitary purposes follow a similar conceptual model (ISPM 14) and include the following elements:

- determine the hazards and the objectives for measures within a defined system
- identify independent processes or actions that can be monitored and controlled
- establish criteria for acceptance/failure to assure control
- monitor control points

- take corrective action when monitoring results indicate that criteria are not met
- review or test to validate system efficacy and confidence.

HACCP is a relatively advanced application of systems approaches and represents the most rigorous and rigid type of systems approach that can be used for phytosanitary risk management. HACCP is probably most appropriate for industries producing high value commodities using sophisticated and well-controlled production systems.

A control point system typically requires a well-defined plant production/processing system such as a packing facility where measures and controls can be established and managed to obtain the required stringent level of phytosanitary protection. Because of the extra demands such a system places on risk managers, the systems are most often characterized by high-volume, high-value commodities combined with concerns for high-risk pests. Control point systems also typically involve the highest technology inputs.

Consider an example using bulbs and soil, adapted from Devorshak (2012). In this example, there are three independent processes that provide control points for the production and phytosanitary risk management of bulbs for export. All three processes occur within the closed system of a bulb sorting, culling, and packing process. The goal is to provide bulbs that are free from pests and contamination by soil.

The first control point involves a cleaning machine whose output is routinely sampled to ensure 95% efficacy. The second control point is a combination of machine sorting and hand-culling to eliminate misshapen, miscolored and otherwise unacceptable bulbs. The outputs from this process are continuously sampled to verify 95% efficacy. The last control point is a mechanical packing process that includes the application of a light coating of a pesticidal preservative that neutralizes most microorganisms associated with small bits of soil. This process is also continuously monitored for efficacy. Each of these control points has an associated acceptance criterion. Bulb lots that enter the system are not released until all sampling and monitoring results indicate an acceptable level of efficacy. These measures, in total, provide the required level of phytosanitary security.

27.6. Systems Approaches and Pathways

As noted above, an advantage of systems approaches is that measures can be applied anywhere and in a range of combinations throughout a pathway. In order to identify points in a pathway where mitigations can be applied, it is useful to "map" the pathway or describe the different parts of a pathway. Fig. 27.2 demonstrates a simple method for illustrating a commodity pathway that will help inform the decision on where mitigations might be applied.

Once a pathway has been described, the risk manager can determine what and where measures might be applied in the pathway to reduce risk to an acceptable level. Table 27.1 describes general types of measures that can be applied throughout a pathway to manage risk (Griffin, 2012). Measures can be applied to commodities at the very earliest stages, including pre-planting, e.g. requiring the use of certified pest-free planting material. Likewise, measures can be applied throughout the pathway including through end use, e.g. limiting the geographic distribution of an imported commodity or limiting imports to certain times of year when pests will not establish.

27.7. Independent and Dependent Measures

A systems approach may be composed of independent or dependent measures. By definition, a systems approach must include at least two independent measures. Phytosanitary measures are independent if the performance of one measure has no effect on the performance of the other measure, and vice versa. To illustrate with a simple example, if the mortality rate effected by measure A is unaffected by and has no effect on the mortality rate effected by measure B, then measures A and B are independent of one another. Independent measures provide a measure of redundancy. If, for any reason, one measure fails, this will have no impact on the expected effect of the other measure.

Visual inspection and limiting the area for distribution of a commodity are two independent measures. A suboptimal inspection has no impact on the effectiveness of limiting the movement of the commodity. Likewise, if the commodity is moved improperly that will not affect the inspection.

Dependent measures interact with and affect one another. The interrelations and interdependencies between two or more measures can vary. For example, measure C may not be effective unless measure D is used, or measures C and D may interact in a synergistic way to produce the desired risk-reducing effect. Dependent measures cannot stand alone, they depend on at least one other measure. Consider a pest-free place of production, e.g., a glasshouse, where being pest free depends on all openings having both double doors (C) and screening (D) at every opening. Both must work for the glasshouse to remain pest free.

A systems approach may be composed of independent and dependent measures. An independent measure may be composed of several dependent measures. Fig. 27.3 provides a conceptual example of how a systems approach can be formulated. There are two independent measures in this example, but each is composed of a series of dependent measures. Note that independent measures do not all comprise a series of dependent measures as shown in this example, many of them are stand-alone measures.

By mixing and matching and adding measures into a system, risk managers can achieve higher levels of protection and finer granularity in the strength of measures than is possible with a stand-alone measure. Think of independent and dependent measures as the building blocks of systems. Alternative systems can be formulated and evaluated to obtain the desired level of protection for any commodity–pest combination that arises.

27.8. Evaluating Systems Efficacy

Describing the efficacy of systems approaches presents certain challenges. For instance, phytosanitary treatments can be quantified in terms of pest mortality, e.g., "Probit 9 mortality" and inspection can be described in terms of likelihood

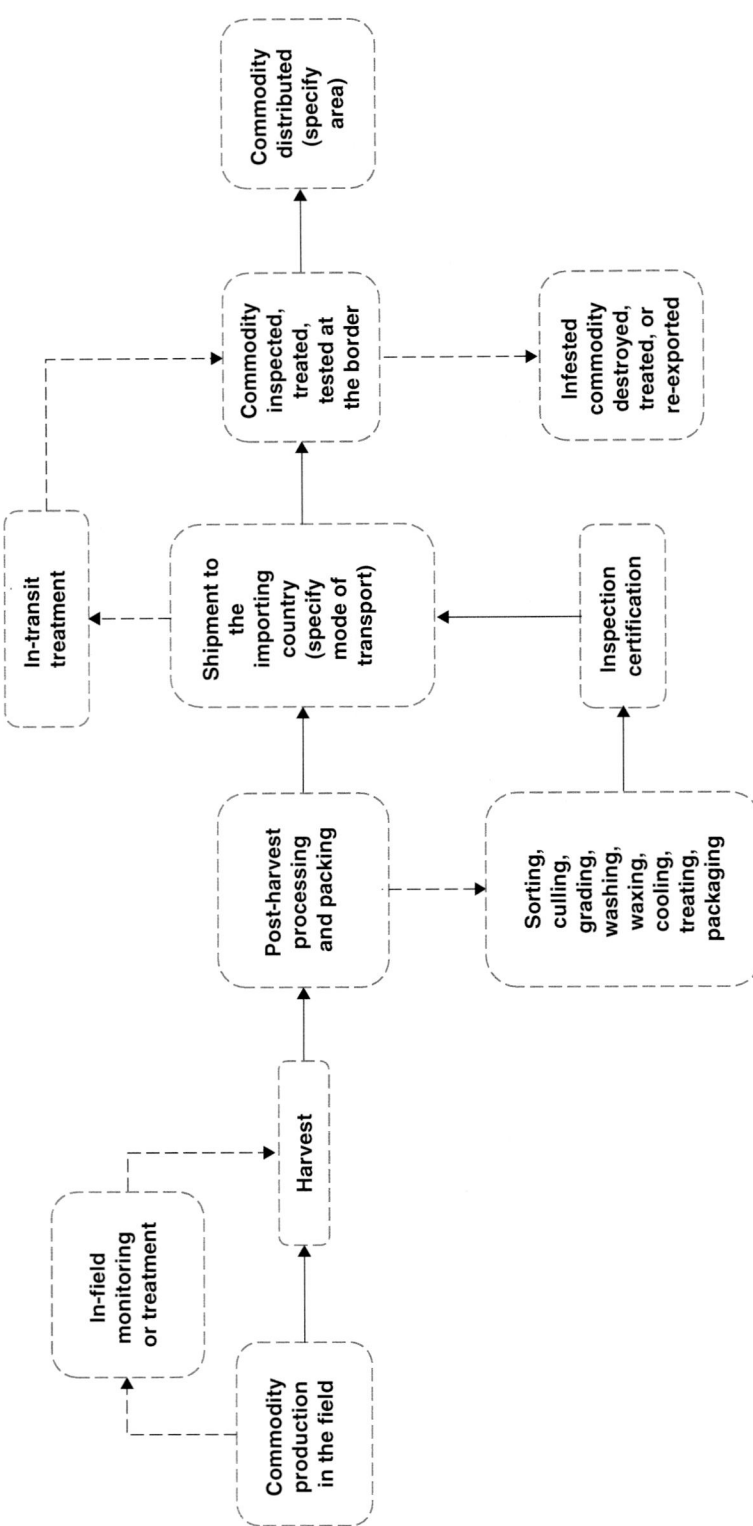

Fig. 27.2. A basic commodity pathway illustration.

Table 27.1. Examples of measures that can be applied to manage pest risk in a systems approach^ (adapted from Jang and Moffit, 1994).

Pre-harvest	Harvest	Post-harvest handling	Shipping	Distribution	End use
Pest-free areas or areas of low pest prevalence	Harvesting at specific times or specific stages of ripeness	Post-harvest treatments (e.g. chemical, heat, waxing, washing brushing, etc.)	Treatment in transit (e.g. cold treatment)	Restrictions on ports of entry	Restrictions on end-use
Resistant cultivars	Culling infested products	Testing	Speed and type of transport	Restrictions on time of year	Post-entry processing
Healthy planting material	Field sanitation	Culling	Pre-shipment inspection	Post-entry quarantine	Packaging*
Pest mating or development disruption	Harvest technique	Packing house inspection	Testing	Post-entry inspection	
Sanitation and cultural controls	In-field chemical treatments	Processing (degree & type)	Sanitation*	Post-entry treatment	
Certification schemes	Field surveillance	Method of packing*	Type of packaging*	Packaging*	
Testing	Tarping*	Screening*			
Protected conditions*	Sanitation*	Sanitation*			

* Indicates a safeguarding measure.
^ This table is not inclusive of *all* potential measures; it lists the most common used on imported commodities.

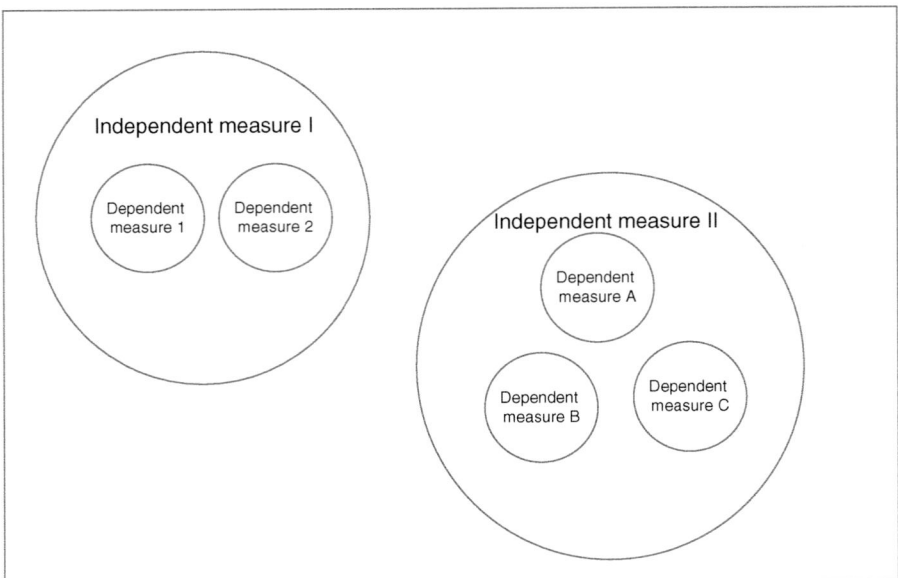

Fig. 27.3. Conceptual construction of a systems approach from dependent and independent measures.

of detection giving a certain inspection rate. Because systems approaches can employ a number of different types of measures, all with different effects or endpoints, the overall efficacy of a systems approach can be difficult to describe. For instance, pesticide applied during field production may result in reduced trap captures or lower pest finds during fruit cutting. Culling at the packing facility may remove almost 100% of the visually detectable pests but only some of the pests that are not detectable. Then, of course, some components of a systems approach are not even mitigation measures. Trapping surveys and risk-based inspection, for example, serve as means to monitor and verify the prevalence of pests but have no mitigation effect.

RSPM 40 lists some options for describing efficacy as well. It describes efficacy as the specification of the desired response or endpoint and a measurement of that response or endpoint, represented as a metric. Endpoints may include:

- mortality
- sterility, including sterility of F1 generation
- inactivation
- altered behavior.

Mortality may not be the most appropriate endpoint for measuring efficacy of a systems approach because many of the measures do not specifically cause mortality. Griffin (2012) suggests that in order to express the overall efficacy of a systems approach, a "common currency," i.e., term of expression, is needed. The chosen endpoint affects how phytosanitary security will be expressed. Examples of endpoints include:

- prevalence of pests in a consignment or proportion of pests removed
- frequency of entry, i.e., number of pests entering per unit time
- probability of entry, e.g., probability of pest entry per unit of commodity imported
- frequency or probability of establishment
- frequency or probability of pest outbreaks.

An approach that is sometimes easier said than done, is to deconstruct the system into its individual components and evaluate the contribution of each component to phytosanitary security in whatever way it is measured. Qualitative, semi-quantitative or quantitative analyses have been used to estimate efficacy endpoints.

Fig. 27.4 provides a hypothetical example adapted from Devorshak (2012) of how one might estimate the efficacy endpoint for a systems approach for a single pest. The pathway is familiar from Fig. 27.1. The plus signs indicate relative risk levels. In this example, zero risk is not an option. Some specific measures have been identified by using letters. The first example considers measures A, B, and C together. The second considers A, B, C, and D.

Beneath the pathway are two tables. The first table provides efficacy information for the endpoint proportion of pests removed/remaining. Effectiveness measures the percentage of pests removed by that measure. Its complement shows the pests remaining. Any one of these measures taken as a stand-alone measure would likely be unacceptable. Systems approaches provide a cumulative impact that is evident in the calculations of the second table.

The calculations begins with 100% of the pests present before measures are applied, assuming that more pests cannot be introduced along the pathway. Choosing a poor host for this pest reduces the prevalence by 75%, thus 25% of the original number of pests remain. Field treatment removes 40% of the remaining 25% that survived the first measure. That leaves 15% of the original number of pests. Cold treatment in transit reduces the remaining 15% of the pests by another 80%, leaving 3% of the original number of pests. The hypothetical system has eliminated 97% of all pests. This level of protection may be deemed appropriate or other measures may be added to the system if risk needs to be reduced further.

If we add commodity washing to the post-harvest activities, we reduce the 3% by half,

eliminating 98.5% of the original number of pests, leaving 1.5%. If desired, additional measures can be added until the appropriate level of protection is met.

A systems approach enables risk managers to consider more than phytosanitary effectiveness. For example, the cost of adding the commodity washing could be weighed against the value of reducing the number of pests another 1.5%. If the risk assessment indicates such a prevalence, 3% of the unmitigated number of pests is a low risk, and risk managers may determine that it is not cost-effective to require a commodity washing.

27.9. Evaluating Cost-effectiveness

As with any other type of risk mitigation, systems approaches should also be evaluated for cost-effectiveness. Efficacy refers to the overall effect of the systems approach in reducing phytosanitary risk while cost-effectiveness refers to the efficacy of the measures compared to the total cost of the measures.

Cost-effectiveness analysis can be a useful tool in selecting measures. This type of analysis determines which risk management option meets the specified desired response or endpoint at the lowest cost. It is appropriate when weighing options with similar outcomes, such as mortality

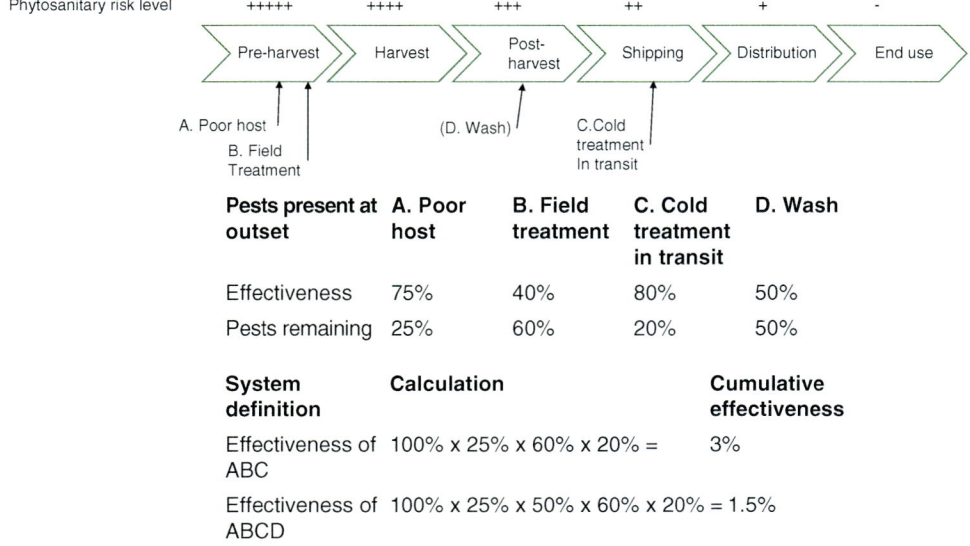

Pests present at outset	A. Poor host	B. Field treatment	C. Cold treatment in transit	D. Wash
Effectiveness	75%	40%	80%	50%
Pests remaining	25%	60%	20%	50%

System definition	Calculation	Cumulative effectiveness
Effectiveness of ABC	100% x 25% x 60% x 20% =	3%
Effectiveness of ABCD	100% x 25% x 50% x 60% x 20% = 1.5%	

Fig. 27.4. Calculation of hypothetical cumulative efficacy endpoint.

rate, but it should not be used to compare programs that have different outcomes.

RSPM 40 notes that to determine the cost-effectiveness of each mitigation option, first identify and place a monetary value on the costs of a program, then divide these costs by the specific metric of the system's effectiveness, e.g. mortality, probability of establishment, packinghouse culls, trapping counts, and so on.

$$Cost\ effectiveness\ ratio = \frac{Total\ cost}{Units\ of\ effectiveness}$$

The result is the cost per unit of effectiveness.

27.10. Uncertainty and Redundancy

It is the risk assessor's responsibility to characterize the uncertainty in a decision process. It is the risk manager's responsibility to address that uncertainty in an appropriate manner in the recommended risk management option. For phytosanitary risk management, one common response to uncertainty is to err on the side of caution. This sometimes means risk managers will rely on conservative assumptions and worst-case data to determine the strength of measures. A worst-case scenario is usually some combination of measures failing or operating at minimal efficacy while pest challenges are maximized.

Adding redundant risk management measures to the risk management system of measures is one way to compensate for the uncertainties that risk managers are trying to address. Lack of experience with a particular commodity–pest pair is a common source of uncertainty that has been addressed through redundancy. In fact, ISPM 14 states, "A systems approach may include measures that are added or strengthened to compensate for uncertainty due to data gaps, variability, or lack of experience in the application of procedures." It is important to recognize that measures added in response to uncertainty are not "based on scientific evidence" as required by the Sanitary and Phytosanitary Agreement (SPS Agreement) and must therefore be provisional measures. Recall that provisional measures are not permanent and must be reviewed at some stage to determine if the redundancy is still necessary or if measures

can be reduced while still managing risk to an acceptable level.

The central role of uncertainty in determining the nature of and response to phytosanitary and other risks associated with international trade of plant commodities has been a major focus of this handbook (see especially Chapters 3 and 15, this volume). Every risk management process is hampered by some degree of uncertainty, including both knowledge uncertainty and natural variability. This will include elements of the risk assessment as well as the efficacy of the recommended risk management measures.

If redundant measures are to be added to a system, it is important to be aware of two significantly different views of the system. One is the purist's view. In this view the system is basically what the risk manager has specified and can control. For example, if the risk manager's systems approach consists of measures A, B, and C but the grower does D and E for quality control and they also have phytosanitary effectiveness, the purist manager says the system is A, B, and C. Imagine A, B, and C reduce a high risk to a low risk and the addition of D and E would reduce the risk to a negligible level. Adding redundancy in the form of additional phytosanitary requirements to A, B, and C in this situation or increasing the strength of A, B, and C may waste the resources associated with the additional measures or strength of measures.

Difficulties can arise when measures are added, or the strength of measures is boosted for additional redundancy to cover undefined contingencies and unknown uncertainties that may not even exist. Such redundancy lacks a clear justification and finds no basis in the standard. This rather routine practice is usually not challenged unless the requirements severely limit trade.

Because of the inherent complexity of evaluating systems efficacy there can be a general perception that uncertainty is higher and less easily controlled. The response to this is often even greater redundancy. In truth, phytosanitary systems themselves are often not complex and systems approaches are often more precisely measured and monitored for efficacy so that the overall efficacy uncertainty may actually be significantly less than single-measure strategies.

A more holistic view of redundancy considers all measures along the pathway that may have phytosanitary effects. This begins with the independent and dependent measures that form the core of the systems approach. It also includes other conditions or procedures that have a pest mitigation effect. In this instance if the core measures are A, B, and C the holistic view would recognize the effects of D and E as well, even though they may not be independent, measured, controlled or prescribed, or formally considered part of the systems approach. In this holistic view the redundancy required to address concerns about remaining uncertainty may be fully or partially met by the unofficial measures.

One complication that arises from these philosophical differences about whether or not to consider the effects of unofficial elements is that a systems approach designed with a holistic view cannot be based on the standard and would require a clear bilateral understanding of what is expected and required by all parties.

27.11. Benefits of Systems Approaches

This chapter began by identifying several key points about systems approaches. We return to each now.

27.11.1. Greater benefits from trade

Alternatives to stand-alone mitigation strategies that rely on prohibition or other trade restricting stringent measures are now available. The systems approach makes it possible for risk managers to achieve their appropriate level of protection while permitting, if not encouraging, more trade. With a systems approach, the risk management choice is no longer between phytosanitary protection and trade. The systems approach makes both possible.

27.11.2. Greater flexibility and precision

Stand-alone mitigation measures provide whatever protection they provide and that is it. If a measure is too stringent, then trade suffers. If that measure is not stringent enough, plant health can suffer. Finer degrees of control can be exercised by adding more and more actions to the system strategy. There can be multiple ways of obtaining the appropriate level of protection. This provides risk managers with flexibility in the tradeoffs their alternative strategies present.

Systems provide flexibility. It is not necessary to treat each pest exactly the same when the commodity or pathway vary. Neither is it necessary to treat each commodity the same way when the pests vary. The layers of a system allow risk managers greater precision in achieving the desired level of phytosanitary protection.

27.11.3. Effective hedge against uncertainty

Are you uncertain about the efficacy of one or more of your phytosanitary measures with no effective options for reducing that uncertainty? Add another measure of protection to backstop the measure with the uncertain performance. Not sure just how serious the risk is? Strengthen the line of protection with additional measures. The advantage of a systems approach is the ability to address variability and uncertainty by modifying the number and strength of measures to meet the appropriate level of phytosanitary protection. This will provide the confidence required by producers and consumers that the product has acceptably low levels of infestation.

27.11.4. Includes nontraditional measures

Quality systems are a type of systems approaches that comes to the fore when a holistic approach to risk management is employed. Good risk management makes effective use of scenario comparisons (see Chapter 14, this volume) using a "without and with" condition comparison of conditions. These conditions should reflect real conditions in the world.

In Chapter 1 the risk management task for international trade in plant commodities was described as production risk management and phytosanitary risk management. In any given plant trade issue, there is going to be the phytosanitary risk management strategy and the rest of the world's response to that strategy. When there is strong evidence to suggest that stakeholders in the exporting and importing countries are going to take steps that reduce phytosanitary risks beyond what the phytosanitary core of measures produce, it is unrealistic to ignore how the world responds to bilateral trade agreements.

Nontraditional risk management measures can be incorporated into the core of the strategy by the NPPO. Or they can be employed by stakeholders in response to the risk management decision or the exigencies of the market. A strong argument can be made for widening one's view of a system to include phytosanitary risk managers and the production risk managers who respond to them in their pursuit of the benefits of international trade. That wider view would realistically include the practices and conditions that are not traditionally prescribed as measures.

27.11.5. New risk management toolbox

The nontraditional methods described above would, alone, significantly expand the risk manager's toolbox. If risk managers could come to rely on the voluntary initiatives of industry, many new systems synergies could be possible. A systems approach makes it possible to consider measures that, alone, would never be adequate, but in combination with other measures might just provide the increment of protection that a situation calls for.

The traditional toolbox was filled primarily with tools to kill pests. Pesticides are, increasingly, being considered the last line of defense rather than the weapon of choice they were for decades. Measures that reduce the prevalence of pests or reduce their potential for entry or establishment can be included in a systems approach. Systems can include tools like monitoring, exclusion of pests and host fruit, community awareness and control of incursions.

27.11.6. Cost considerations

One of the more promising aspects of the systems approach is that finer levels of granularity

of control are possible and there are more tools in the toolbox. In addition to all that this means for the ability to more precisely manage risks while promoting trade, it also offers the possibility of considering cost as a tradeoff in the design of risk management strategies. The economic costs of prohibition can be staggering. Prohibition has been considered a viable option when there are no other suitable measures for managing a risk. For decades a suitable measure was a phytosanitary treatment. The systems approach frees risk managers from this line of thought. Other suitable measures can now be thought of as collections of traditional and nontraditional tools, tools enacted by the NPPO and tools enacted by industry voluntarily or for quality reasons. The end result is more options and options mean varying costs, both of implementation as well as the impact of the phytosanitary measures on the economics of the exporting and importing countries. Those costs can now be a consideration in decision-making.

27.11.7. Aligns thinking with the risk management continuum

The analysis done to formulate a systems approach is useful for aligning our thinking with the risk management continuum. This continuum can be thought of in at least two ways. First, as noted above there is the continuum of production risk management conducted by private industry and phytosanitary risk management practiced by NPPOs. The expanded toolbox above and partnering possibilities described below are two critical ways the systems approach connects us with this continuum.

The other and, perhaps, more obvious continuum is that risk management is no longer a Bernoulli trial with exactly two outcomes, success or failure. It is no longer necessary to think of risk management as succeeding only when all pests have been eradicated. The presence of pests is not tantamount to establishment and damages. There is a continuum between eradication, 0, and do nothing, 1. There may not be as many options now as there are numerical values between 0 and 1, but we do recognize that total eradication is not always or even often necessary to successfully manage risks. You do not need a zero probability of damages for successful risk management.

27.11.8. Systems encourage collaboration

A systems approach can be used to empower private sector interests to become, if not full partners, at least junior partners or collaborators in the phytosanitary risk management process (see Chapter 8, this volume). There are likely many instances where the private sector would be willing to take on voluntary responsibilities in preference to facing regulatory strictures. Conceptually, a systems approach can consider any stakeholder anywhere in the supply chain or pathway as a potential resource in devising a system. As opportunities for collaboration grow, aided by technology, there is hope that systems will be formulated more by collaborative partnerships that rely less on the dictates of government regulators.

27.11.9. Systems complexity

Systems approaches can range in complexity from a collection of familiar and time-tested measures to a sophisticated control-point design like HACCP. Moreover, the analysis of these systems can range from simple qualitative characterizations to complex quantitative or probabilistic risk assessments.

27.12. Operational Considerations

Unlike many types of phytosanitary measures that are implemented almost exclusively by NPPOs or other authorized entities, many elements of systems approaches may be implemented by producers or industry entities. Therefore, when risk managers undertake to develop a systems approach, they should be prepared to actively engage producers in the process. Indeed, this may be an absolutely critical element to developing a systems approach. Where systems approaches involve measures applied pre-planting through harvest, producers may be involved in mitigations such as field sanitation, certification programs, growing season field inspections, and other similar activities that are typical components of a systems approach. After harvest and through shipping, other entities may be involved, including secondary packing houses, shippers, and importers and distributors. This may include everything from packinghouse inspections, culling, packing requirements, and requirements related to distribution of the commodity in an importing country. Note as well that it is often the producers that will bear the costs of a systems approach, which makes it all the more critical that NPPOs involve their producers in the development of the systems approach in order to ensure the producers are able to implement the program.

To illustrate the critical nature of involving producers in systems approaches, two examples are provided. The first examines the requirements for the movement of citrus nursery stock from areas where certain citrus diseases occur in the United States. The second looks at good seed production practices developed by seed companies to ensure tomato seed is produced free of *Clavibacter michiganensis*, a major tomato seed pathogen.

The United States Department of Agriculture (USDA) Protocol for Interstate Movement of Citrus Nursery Stock from Areas Quarantined for Citrus Canker, Citrus Greening, and/or Asian Citrus Psyllid (CNS Protocol) provides a systems approach to allow producers of citrus nursery stock (CNS) to safely move CNS from areas where certain quarantine pests occur. This program is applied at the domestic level only in the United States. The protocol relies on a combination of measures:

- all source material for CNS must originate from a clean stock program
- all CNS must be grown in approved structures that prevent pest infestation
- all CNS must be treated with approved insecticides at specified intervals
- all CNS must be inspected by the NPPO during production and prior to shipment
- CNS from certain areas is allowed only a restricted distribution to non-citrus producing areas.

Note that this systems approach relies on mortality, i.e., phytosanitary treatments, safeguarding, i.e., growing protected structures, and verification, i.e., inspection of CNS. Also note that this systems approach relies on the actions of producers ensuring they select source material from a clean stock program and growing their CNS in approved structures that exclude pests. Producers are also responsible for treating their CNS as required in the CNS protocol and agree to only sell their material to areas where they are allowed to distribute it. In order to ensure that producers adhere to requirements of the systems approach, they must sign a compliance agreement with the NPPO. This demonstrates the need for close coordination between the risk managers and producers.

While the systems approach for CNS represents a certain burden on producers because of additional costs involved with implementing requirements, the producers also benefit by having an assurance that they will be able to sell their product under the program. Likewise, the NPPO benefits by ensuring that infested material will not be shipped to citrus producing areas while managing their own costs by sharing responsibility for implementing the program with producers. And quite simply, this program could not be accomplished without the full participation of the producers enrolled in the program. The larger citrus producing community benefits as well because the overall likelihood of citrus pests spreading through the movement of CNS is managed to an acceptable level.

The second example of industry involvement in systems approaches is the Good Seed Production Practices (GSPP, 2020). This is an industry-led and developed program to prevent *Clavibacter michiganensis* subsp. *michiganensis* occurring in tomato seed production. This program is an important one for NPPOs to be aware of as seed health has become a more prominent concern for NPPOs globally over the past several years. Seed health has become a concern for at least four reasons.

First, production chains for seeds may involve movement of material through multiple locations and countries. During the production process seed may undergo different types of processing and handling that may change the phytosanitary status of seed. Mixing of seeds combines different species, varieties or cultivars into a single lot, e.g., lawn grass or wildflower mixtures. Blending of seeds combines different seed lots of the same variety. Seeds from various origins and different harvest years may be mixed and blended.

Second, we may have limited knowledge of how the seed has been produced, tested and/or treated during the production chain. Third, from a scientific and technical perspective, seeds differ from other commodity classes because of the cryptic nature of seed diseases, uncharacterized or poorly described seed diseases, and how such diseases can be detected both in terms of sampling and diagnostics, as well as difficulties in treating seeds for pathogen eradication or disease management. Finally, seed trade has increased exponentially over the past decade and continues to increase.

The GSPP system developed to manage the risk is aligned with a HACCP type of approach. Specifically, the program includes the following elements:

- isolation of the seed and seedling production location from the environment
- prevention of infection by managing the risk factors
- constant monitoring during the growing season of both seeds and young plants
- checking before delivery: all seed lots must be tested by seed tests approved by GSPP
- independent audits.

The main components of the systems approach in this case involve safeguarding, i.e., preventing infection in the first place, and verification through inspection and audits.

One of the most notable points to understand about the GSPP is that the program is driven entirely by the industry itself and supported by participating companies. The GSPP uses a private standard as opposed to a standard developed by an organization like the IPPC or NAPPO. Lastly, the GSPP is a purely voluntary program in which companies can elect to participate.

Similarly, the American Seed Trade Association (ASTA) has produced quality management manuals (e.g. 2010) for their member companies. While quality management systems (QMS) are not synonymous with the "systems approach," both QMS and systems approaches employ similar tactics to prevent the presence of pests in seed production. QMS in seed production is therefore an important starting point for considering the wider application of systems approaches in managing seed health challenges. As of the writing of this text, the seed industry, NPPOs and the IPPC are working to develop an international standard for systems approaches for seeds and many of the QMS elements used by seed companies are likely to be incorporated into a wider systems approach.

27.13. Summary and Look Forward

Here are five things to remember from this chapter.

1. Systems approaches usually require a good understanding of the system being managed and the ability to manage risk at different control points in the system, e.g. throughout the pathway from point of origin to end use.

2. The types of measures applied in a systems approach include measures that reduce risk (e.g., removing, killing, deactivating, or otherwise reducing the presence of a pest), measures intended to safeguard the commodity (e.g. packaging, preventing infection), or measures intended to verify the efficacy of other measures (e.g., inspection).

3. Systems approaches often include redundant measures as a means to manage uncertainty and ensure risk is adequately managed, but redundant measures should be reassessed as knowledge of the system improves.

4. In designing a systems approach, it is important to consider defining endpoints of measures (e.g. mortality, pest freedom, and cost effectiveness).

5. In many, if not most cases, producers are an integral part of implementing systems approaches and should be included in the process of developing any systems approaches that affect them.

A great deal of what currently passes as risk management in the phytosanitary community might more appropriately be called "hazard management." It is important to re-orient this phytosanitary community of practice away from hazard management in favor of risk management. The differences in these two concepts are explored in the next chapter.

27.14. References

Aluja, M. and Mangan, R.L. (2008) Fruit fly (Diptera: Tephritidae) host status determination: critical conceptual, methodological, and regulatory considerations. *Annual Review of Entomology* 53, 473–502.

ASTA (2010) *Guide to Seed Quality Management Practices*. American Seed Trade Association, Alexandria, Virginia.

Devorshak, C. (ed.) (2012) *Pest Risk Analysis: Concepts and Application*. CAB International, Wallingford, UK.

EPPO (2013) *Pest Risk Analysis for Thaumatotibia leucotreta*. EPPO, Paris, France. Available at: www.eppo.int/QUARANTINE/Pest_Risk_Analysis/PRA_intro.htm (accessed January 29, 2020).

Griffin, R.L, in Devorshak, C. (2012) *Pest Risk Analysis: Concepts and Application*. CAB International, Wallingford, UK

GSPP (2020) *Good Seed Production Practices*. Available at: https://www.gspp.eu/ (accessed January 29, 2020).

IPPC (2019a) *ISPM 14: The use of Integrated Measures in a Systems Approach for Pest Risk Management*. Secretariat of the International Plant Protection Convention (IPPC), Food and Agriculture Organization of the United Nations, Rome.

IPPC (2019b) *ISPM 35: Systems Approach for Pest Risk Management of fruit flies (Tephritidae)*. Secretariat of the International Plant Protection Convention (IPPC), Food and Agriculture Organization of the United Nations, Rome.

IPPC (2019c) *ISPM 36: Integrated Measures for Plants for Planting*. Secretariat of the International Plant Protection Convention (IPPC), Food and Agriculture Organization of the United Nations, Rome.

Jang, E.B. and Moffitt, H.R. (1994) The systems approach to achieving quarantine security, in J.L Sharp and G.J. Hallman (eds) *Quarantine Treatments for Pests of Food Plants*, 225–237. Westview Press, Boulder, CO.

Jang, E. B., Miller, C.E. and Caton, B. (2015) *Systems Approaches for Managing Risk of Citrus Fruit in Texas During a Mexican Fruit Fly Outbreak*. Technical report to USDA-APHIS-PPQ, https://www.aphis.usda.gov/plant_health/.

NAPPO (2014) *RSPM 40: Principles of Pest Risk Management for the Import of Commodities*. Ottawa, Canada. http://nappo.org/files/8314/3889/6413/RSPM40-e.pdf.

NWHC (Northwest Horticultural Council) (2017) *Work Plan for the Exportation of apples from the United States to Mexico*, http://nwhort.org/export-manual/countries-toc/mexico/.

USDA/APHIS/BAPHIQ (2013) *Systems Approach Work Plan for the Exportation of Apples from the United States into Taiwan*. Proposed modifications to systems approach work plan for U.S. Apples to Taiwan by BAPHIQ, April 16, 2013. http://www.calapple.org/media/files/20130416AppleWorkPlanBAPHIQRevised Version.pdf).

Part Four:

Issues in Pest Risk Management

Part 4 is the handbook's attic. This is where you will find things that do not fit anywhere else but which could not be left out. A potpourri of topics is found here.

Phytosanitary risk managers sometimes focus on managing hazards rather than managing risks, and this can often result in costly misallocation of resources. A hazard, like a plant pest, is a thing that has the potential to cause harm. A hazard alone is not sufficient for the existence of a risk. The hazard must cause actual harm or have the potential for causing actual harm to rise to the level of a risk. This chapter explains how a risk management decision based on a hazard assessment may do nothing to reduce risk and can result in a misallocation of scarce risk management resources.

Chapter 29 provides a detailed overview of the most common tools and types of analysis used to assess the economic consequences of international plant trade. It begins by considering the various perspectives that are available and defined by the wide array of stakeholders. Economics is well equipped with a rich tool set, more than sufficient for estimating the economic consequences of international plant trade. The Sanitary and Phytosanitary (SPS) Agreement's impact on economic analysis is reviewed and challenges to that framework are identified. There is room for considerable advance and improvement in the quality of economic analyses done in support of phytosanitary risk management decisions.

Knowledge management is the topic covered in Chapter 30 and it is introduced as a strategic approach to, over time, reducing the substantial uncertainty that plagues risk management decision-making. The phytosanitary risk management community has learned many valuable lessons in its years of global practice. Knowledge management provides the community of practice with a strategy for harnessing that knowledge in service to the global community.

Commodities for consumption include a wide range of items. Some, such as empty shipping containers or the superstructure of vessels, may have no direct relationship to agricultural products but have demonstrated the potential to present a threat for pest introduction. Plants for planting and fresh fruit, on the other hand, are clearly agricultural products that can serve as a pathway for introducing harmful pests. Chapter 31 examines the unique aspects of risk management for those items that are identified as commodities for consumption.

Genetically modified organisms and invasive species are the focus of Chapter 32. This odd couple is combined because these two topics share a common background in risk management through their relationship with the precautionary approach. This chapter examines the background and conceptual foundation that links these topics and aims to clarify their role in risk management from a phytosanitary standpoint.

The risk-informed philosophy of the SPS Agreement shook traditional phytosanitary risk management to its core. Once focused on keeping pests out of a country at any cost, risk management has contributed significantly to the evolutionary transformation of early plant protection programs into the phytosanitary systems we have today. The final chapter looks at central concepts like the strength of measures, a rational relationship, and the continuum of risk management that have caused the break from the past and redefine the future of pest risk management.

28

Hazard Analysis vs. Risk Analysis

The general public has trouble differentiating between the terms hazard and risk.

John A. Singley

28.1. Introduction

There is considerable confusion about the use of the terms hazard and risk. Oyarzabal (2015) has consistently observed that participants of Hazard Analysis and Critical Control Point (HACCP) classes think that "hazard" and "risk" are synonyms that can be used interchangeably. This misconception is understandable given the messy and imprecise manner in which the risk vocabulary is applied. The misunderstanding may be exacerbated by the complexity of people's perception of risk and the fact that, in many cases, perceived risks are not always well correlated to the actual risks associated with hazards, at least those known to harm humans (Slovic, 1987).

> Hazard ≠ risk
>
> - Hazard is the potential to cause harm.
> - Risk is the likelihood of harm under defined circumstances.

A hazard is not a risk. Hazard assessment is not risk assessment. A pest of quarantine concern is a hazard. Not all species introduced to a new area can survive, not all become established, and not all of those that establish become pests. A pest of quarantine concern does not become a risk until it is likely that it will cause some harm. There is no risk without a consequence. Thus, entry of a pest into a new country is not, in and of itself, sufficient for a risk. If the entry of a limited number of individual organisms poses no threat of an undesirable consequence there is no risk. In the unlikely event that a pest would develop a sustainable breeding colony and spread without causing any consequences, there would be no risk.

When it comes to phytosanitary risk management there is a concern that, at times assessors and managers may equate entry with establishment or introduction with impacts, either of which can lead to an inappropriate leap from entry to a presumption of high risk. So, for clarity, let us begin by stating that establishment entails a great deal more than entry and impacts require more than introduction. Only a complete risk assessment can determine that a pest is high risk.

This chapter is devoted to making the distinctions between hazard and risk and between hazard assessment and risk assessment. To do so, it proceeds by defining hazards more completely than was done in Chapter 2. Risk-based approaches to analysis are contrasted with hazard-based approaches. This is followed by a discussion of the need to focus risk management measures on actual harm rather than perceived harm. The chapter then describes a few hazard analysis tools.

28.2. What is a Hazard?

The language of risk analysis is messy and it is still evolving. Different dialects of risk are spoken by different communities of practice and the definition of "hazard" can be confusing because of these dialects and because some dictionaries use the term "risk" in their definitions of hazard. Dictionary.com, for example, defines hazard as "an unavoidable danger or risk." The EPA *Thesaurus of Terms Used in Microbial Risk Assessment* has 14 definitions for hazard. Chapter 2 defines a hazard as anything that is a potential source of harm to a valued asset. Here we refine that definition for a plant hazard to mean any source of potential damage, harm or adverse effects on plants, plant yields, plant health, or plant life.

A common way to classify hazards is by category

- **Biological**: bacteria, viruses, insects, plants, birds, animals, and humans.
- **Chemical**: depending on the physical, chemical, and toxic properties of the chemical.
- **Ergonomic**: repetitive movements, improper set-up of workstation.
- **Physical**: radiation, magnetic fields, pressure extremes (high pressure or vacuum), noise.
- **Psychosocial**: stress, violence.
- **Safety**: slipping/tripping hazards, inappropriate machine guarding, equipment malfunctions or breakdowns.

(Canadian Centre for Occupational Health and Safety, https://www.ccohs.ca/oshanswers/hsprograms/hazard_risk.html, accessed November 21, 2019)

Notice that a hazard, as used here, is not the harm itself but the potential to cause harm. When we specify the harm and measure that potential to cause harm as a probability we transition to a risk, which was earlier defined as "*Risk = Consequence × Probability.*" In the food safety community of practice this equation is sometimes modified to read "*Risk = Hazard × Exposure.*" Hazard stands for the

harm and exposure stands for the probability. There is no wonder the terms are confused. Risk is more than a hazard; hazard is part of a risk but it is less than a risk.

28.3. Risk-based or Hazard-based Approaches?

Hazards can be avoided, prevented, eliminated, or reduced to acceptable or tolerable levels. Hazard analysis and risk management both include avoidance, prevention, and elimination of hazards. A total absence of the phytosanitary hazard should be acceptable to everyone. The food safety community of practice is well acquainted with the notion that the mere existence of a hazard is not sufficient for a risk. Acceptable daily intakes (ADIs) establish limits for the amount of substances of toxicological concern that may be safely consumed daily over a person's lifetime. It is not a specific point at which safety ends and possible health concerns begin. Occasional intake above the ADI is not of concern.

Maximum residue levels (MRLs) have been established for many pesticides. An MRL is the maximum amount of pesticide residue that is expected to remain on food products when a pesticide is used according to label directions, that will not be a concern to human health.

The European Food Safety Authority (EFSA) uses a margin of exposure (MOE) to determine the dangerousness of substances that are both genotoxic and carcinogenic. The MOE is the ratio of a toxin's no-observed-adverse-effect level to its theoretical, predicted, or estimated dose or concentration of human intake. It is another threshold measure that helps to establish the distinction between a hazard and a risk.

Tolerable thresholds for the presence of a microbial hazard may be expressed. Examples include an absence in 25 grams, which is a more stringent requirement than absence in 10 grams. Both measures allow that there may still be low levels of hazard presence. Such risk-based approaches recognize that the presence of a hazard is not tantamount to harm. Many risk-based approaches imply that a non-zero level of a hazard is acceptable or tolerable. ADI,

MRL, and tolerable thresholds help illustrate the point that the mere presence of a hazard need not imply a risk.

There are, of course, also examples where the line between hazard and risk is blurred. Overbosch (2013) provides one example of the comingling of hazard and risk by pointing out the increasing reliance on very different considerations of the concept of acceptability. He points to an emerging tendency to treat contaminants with a scientifically determined MRL on a zero-tolerance basis. The thinking behind this is that detected levels vary and the presence of a contaminant hazard suggests that other samples might exceed established limits. Given this logic, some see the decision to allow such a product into the market as negligent and indefensible.

Europe halted imports of American long-grain rice in 2006, when traces of genetically modified LL601 were detected, even though the protein found in LL601 is approved for use in other products (Environment News Service, 2006). The "Sudan Red recall" in the UK is an example of a zero-tolerance hazard-based approach. In laboratory animals, this dye has been shown to cause cancer. There's no evidence that it is a human carcinogen, although the evidence is not definitive. The UK and Europe banned foodstuffs containing the dye in 2003 to be on the safe side. In 2005, a consignment of Worcestershire sauce was found to contain chili powder contaminated with the dye. Many foods used the sauce as an ingredient. Because the amounts of the dye in these products was so small, and because the link to cancer in humans hadn't been proven, the overall risk to health was small. Nonetheless, it triggered the largest recall in UK history to that point in time. No amount was considered safe and products were recalled based on their traceability rather than on the risk they posed to the public (Overbosch, 2013). These are examples where the mere presence of a hazard was not tolerated; they are also examples of decisions that were not based on risk assessment results.

In the food safety community of practice (CoP), risk assessment consists of hazard identification, hazard characterization, exposure assessment, and risk characterization. Hazard analysis or hazard assessment is, like hazard, an imprecisely defined term whose meaning varies with the context of its usage. In a food safety context, one could define hazard identification

and hazard characterization as the essence of a hazard analysis or hazard assessment. This makes the distinction between a four-step risk assessment and a two-step hazard assessment vivid. Nonetheless, there are definitions of hazard assessment that make it sound indistinguishably close to risk assessment.

Food safety is not the only risk assessment community of practice that focuses on hazards. Occupational safety, industrial hygiene, and environmental health are three more CoPs that tend to follow the logic that if you eliminate hazards you eliminate risks. These CoPs are primarily concerned with accidents and risks in the workplace. These communities have long focused on hazard elimination as a primary risk management strategy. HACCP places a strong emphasis on hazard identification and subsequent control of the hazard.

Some pest risk managers likewise seem to want to manage hazards rather than risks. However, in international trade the benefits of the activity are essential to life on the planet. On a smaller and more immediate scale, trade produces significant amounts of consumer and producer surplus. Trade generates jobs and income and produces a wide range of effects, as detailed in Chapters 1 and 7. As a result, the mere existence of a hazard is not sufficient for preventing trade. There must be more than a source of potential harm, there must be a *probability* of harm that warrants a risk management response.

Codex Alimentarius (the international food safety community) risk assessment has the four steps mentioned above. It can be efficient and effective at times to do a partial risk assessment. If there is no exposure or if there is no hazard or if there is no harm, there is no risk. Establishing what is missing can be an effective way to establish the absence of a risk. A partial risk assessment cannot, however, establish the existence of a risk.

Different communities of practice have different triggers for risk management. Occupational safety and environmental health respond to the existence of hazards. Food safety often responds to exposure. Pest risk needs to respond to harm because there is a significant tradeoff in benefits gained vs. risk to plant life.

It does matter, perhaps a great deal, however, whether decision-making is risk- or hazard-based. Overbosch (2013) offers several arguments for why this is so. A zero-tolerance hazard-based approach sets regulatory agencies on a collision course with reality. As analytical methodologies become ever more sensitive, selected hazards can now be found where traditionally none could be detected. We get ever closer to the point where every possible environmental substance can be detected in every substrate on the chemical side. Consequently, levels of hazard below the sensitivity of previous detection methodologies once, by default, regarded as tolerable are no longer seen as so. Detection limits push us to a zero-tolerance world that may be more stringent than necessary and, ultimately, financially unsustainable.

The tendency, by some, to designate hazards on grounds other than strictly scientific ones increases the number of hazards. So, too, does pushing scientific requirements to allow for fast-track hazard designation. Such aversion-based hazard management tends to end up in the zero-tolerance category either directly, as genetically modified organisms (GMOs) do for some nations, or indirectly, as happened with Sudan Red. Once the hazard warning bell has been rung there is no way to unring it.

As more zero-tolerance hazards are identified, noncompliant test results will become more frequent, leading to more "scandals," recalls, feelings of uncertainty in the general public, and wasted resources as perfectly acceptable food is destroyed as if it was toxic waste. When the public feels uncertain, this typically drives calls to carry out more tests at lower levels of sensitivity and this accelerates the wasteful cycle.

A runaway internal logic can develop with these hazard-based approaches. The more often we adopt a zero-tolerance approach the more difficult it becomes to argue with it. That makes different and more efficient approaches difficult to consider. Moreover, efforts to enforce zero tolerances divert resources from more urgent priorities.

Phytosanitary risk management is hampered by the same limitations when risk assessors and risk managers mistakenly equate a pest's entry into a country with harm. Depending on the life history and life requisites of a pest, allowing limited numbers of a pest to enter a country could be part of a cost-effective risk management strategy. This would certainly be so when the likely cumulative numbers of organisms entering

the country are well below the threshold required to establish a sustainable breeding colony.

Much of the driving force behind zero-tolerance hazards is aversion-based. Overbosch (2013) suggests the best option for opposing this approach might be in stressing the negative consequences of many hazard-based approaches. These include the:

- drive to completely eliminate implicated products from the market in the absence of any significant risk
- food waste associated with this practice
- inherent tendency to "discover" more of these instances
- considerable resources required
- tendency to reconfirm existing feelings of uncertainty among the public
- absence of a contribution toward reducing the harmful effects risk managers intended to address.

Moving forward, phytosanitary risk managers need to become more vocal in challenging hazard-based, zero-tolerance approaches that may effectively undermine the benefits of international trade and the wise use of scarce risk management resources. To summarize the primary distinction between a hazard approach to risk management and a risk approach to risk management, hazard approaches tend to be zero tolerance measures that rely on avoidance, prevention, or elimination. Risk approaches allow that the mere presence of a hazard is not sufficient cause for risk management. The probability of harm resulting from a low-level presence of a hazard may well constitute an acceptable or tolerable level of risk that requires no intervention.

28.4. Focus on Actual Harm

The risk of introduction of a pest of quarantine concern includes the probability of introduction and the harm that would be caused by that introduction. That harm typically includes some mix of undesirable economic, environmental, political, and social consequences. The main goal of phytosanitary risk management must be to target risk management efforts toward the prevention of actual harm. That means reducing the consequences of introduction or, perhaps, reducing the probability of introduction to an acceptable or tolerable level. That also means that the presence of a hazard or, more specifically, a pest in numbers insufficient to cause harm, may be an acceptable or tolerable risk. A pest's entry into a nation without the potential for harm through establishment is not in itself a sufficient trigger for risk management measures.

Sanitary and Phytosanitary Agreement (SPS Agreement) jurisprudence has made an important distinction between the concepts of probable (some nonzero likelihood) and possible (imaginable), which bears directly on this question of hazard vs. risk. A harm that is possible may be of far less risk management concern than a harm that is probable. Many possible scenarios can be described as representing risks without credible evidence. A single pregnant organism can destroy a nation's crop if we are unfettered by common sense and scientific evidence. Expert opinion and assumptions may be sufficient to establish the possibility of an assertion but they are far from sufficient for establishing the probability of an incident. APHIS (2012) asserts that the results of trade disputes on this point clearly and consistently support the position that scenarios used in risk analysis under the SPS Agreement must have a demonstrated probability and cannot simply be possible. Thus, harm must be more than a conceptual possibility. The probability of harm must be unacceptably high.

28.5. Hazard Assessment Tools

The Pillsbury Corporation and NASA developed the HACCP control system in the 1960s to ensure food safety for the first manned space missions. A HACCP plan provides a structure for identifying hazards encountered in a process and putting controls in place at critical control points to protect against the hazards and to maintain the quality, reliability and safety of the system's outputs. HACCP seeks to minimize risks by controlling the hazards in a process rather than by end product inspection.

HACCP plans are used primarily by food companies to ensure food safety. The concept is used anywhere within the food chain to control risks from physical, chemical, or biological contaminants of food. HACCP operates on the principle of identifying things that can influence system output quality, and identifying points in the process where critical parameters can be monitored so that these hazards can be controlled. This is a principle that can be generalized to technical systems other than food.

Hazard operability study (HAZOP) is the structured and systematic examination of a planned or existing product/project, process, procedure, or system. The technique uses guidewords to question how the design intention or operating conditions may not be achieved at each step in the design, process, procedure, or system. A multidisciplinary team usually conducts a HAZOP in a series of meetings. It is especially useful for identifying and dealing with deviations from a design intent, due to deficiencies in the design, component(s), planned procedures, and human actions. It has been widely applied to software design review.

The objective of a preliminary hazard analysis (PHA) is to identify hazards and events that can cause harm for a given activity, facility, or system. PHA focuses only on identifying the hazard, i.e., the thing, event, or circumstances that might cause harm. It makes little effort to estimate the probability or consequences of that harm. A PHA should be updated as design detail is increased as well as during construction, testing, and operation. Predicting the emergence of new hazards and the correction of identified hazards is the goal of these ongoing updates.

The greatest strength of the method is that it can be used when there is limited information about risks. Identifying the hazard is the first essential step toward risk identification. The advantage of this technique is early identification of potential risks. PHA does little to help risk managers know how to manage a risk.

> Zero hazard implies zero risk. Because a hazard implies the potential to cause harm, an undesirable consequence exists in principle. As the hazard level rises above zero the probability of that undesirable consequence may begin to rise. It is entirely possible that a finite level of pests could result in a probability of introduction that is so low as to be regarded as an acceptable or tolerable risk.

28.6. Summary and Look Forward

Here are five things to remember from this chapter.

1. A hazard is anything with the potential to cause harm.

2. A hazard alone is insufficient for a risk to exist, there must be some likelihood of harm from the hazard.

3. Pests of quarantine concern are hazards.

4. Entry of a pest into a country is not sufficient proof of risk.

5. A pest must present a sufficient likelihood of harm to warrant risk management.

The next chapter examines examples of how analysis of economic consequences can contribute to the phytosanitary risk management process.

28.7. References

APHIS (2012) *Plant Epidemiology and Risk Analysis*. Laboratory Center for Plant Health Science and Technology. Raleigh, NC.

Environment News Service (2006) *US: Anapproved Transgenic Rice Found in U.S. Rice Supply*. Available at: http://www.cbgnetwork.org/1628.html (accessed February 3, 2017).

Overbosch, P. (2013) Food safety management: hazard- or risk-based? *Food Safety Magazine eDigest*, February/March 2013. Available at: http://www.foodsafetymagazine.com/magazine-archive1/februarymarch-2013/food-safety-management-hazard-or-risk-based/ (accessed February 3, 2017).

Oyarzabal, O.A. (2015) *Understanding the differences between hazard analysis and risk assessment. Food Safety Magazine eDigest*, November. Available at: http://www.foodsafetymagazine.com/enewsletter/understanding-the-differences-between-hazard-analysis-and-risk-assessment/ (accessed February 3, 2017).

Slovic, P. (1987) Perception of risk. *Science* 236, 280–285.

USDA/APHIS/PPQ (2012) *Guidelines for Plant Pest Risk Assessment of Imported Fruit and Vegetable Commodities*. Plant Epidemiology and Risk Analysis Laboratory, Center for Plant Health Science and Technology.

29

Economic Consequence Assessment

The economy is a wholly owned subsidiary of the environment, not the reverse.

Herman E. Daly

29.1. Introduction

Phytosanitary risk management measures constitute non-tariff barriers to trade (NTBs). Sanitary and phytosanitary (SPS) measures can present significant obstacles to agricultural exporters, especially for small producers in developing countries. SPS measures can be complex and less than transparent. They are sometimes implemented differently and in ad hoc ways across destination markets. Different conditions of access across markets are relatively difficult for small producers to assess, resulting in uncertainty for prospective entrants. Studies have shown there are often opportunities for domestic producer lobbies to influence the regulatory approval process and potentially "game the system" to the detriment of developing country exporters (Jouanjean *et al.*, 2016).

These SPS NTBs need more economic analysis. Pest risk management is mature enough and its economic consequences are important enough to warrant greater consideration. It is time to give more serious attention to the economic consequences of pest risk management. "*Risk = Consequence × Probability*" is a definition presented early in this handbook. There is no

risk unless both factors are present. There is no true risk assessment unless both consequences and probability are considered. When consequences are neglected all you have is an exposure assessment. That means pest risk assessment relies disproportionately on the probability of establishment and spread of a quarantine pest. If you want to improve phytosanitary risk management decisions, consider the consequences of trade and of phytosanitary risk management more explicitly.

> **The risk managers**
>
> Risk management along the international plant and plant product trade continuum comprises production risk management and phytosanitary risk management.

Phytosanitary risk management could do a better job of considering the consequences of allowing or disallowing trade as well as of whether the risk treatments are proportionate to the consequences they are intended to reduce. Economic consequences, especially the potential loss of gains from trade and rising costs of phytosanitary risk management, are often the adverse consequences that *production* risk managers are most concerned about.

Economic consequences, broadly construed, are of less interest to *phytosanitary* risk management decisions. A number of pest risk assessors and managers alike have described the legacy of pest risk management as "find a risk and eliminate it," though historically, little consideration has been explicitly given to the economic rationality of the balance of costs of control vs. damage caused, and no consideration has been given to weighing this tradeoff against the potential economic gains from trade.

This chapter examines the assessment of economic consequences. It proceeds by acknowledging there are winners and losers in terms of the economic impacts of risk management decisions. Who these people are often depends on the accounting stance one assumes. Several different accounting stances are considered in the section on economic viewpoint. At that point the official economic guidance of the SPS Agreement and the International Plant Protection Convention (IPPC) are considered followed by a discussion of four challenges that result from this guidance. Three types of losses that result from phytosanitary measures are defined next, along with four types of market failure. The chapter ends by looking at five kinds of economic tools used in decision-making as well as five types of economic analysis that can be used to assess the economic consequences of risk management.

Risk analysis is science-based. Economics is science. PRA's that ignore economics ignore a vitally important decision criterion.

29.2. Who Wins, Who Loses?

One of the most fundamental questions we can ask about plant and plant product trade and the phytosanitary issues that can arise because of it is: who are the potential winners and losers? This is closely followed by: whose interests should the risk manager consider? To answer these questions, we must begin by recalling there can be many risk managers along the continuum of international trade in plants and plant products. Many are private sector risk managers who can be expected to protect the interests of their organization. The most prominent risk managers are the public sector national plant protection organization (NPPO) risk managers with stewardship responsibility for their nation's human, animal, and plant life and health.

Some of the benefits of international plant trade accrue to the firm(s) that grow the traded plants, some to the shippers of those plants. The port that receives the plants, the workers who unload them, and the firms that sell them benefit, as do the people who buy and consume the products. The costs of damages associated with a potential infestation may be borne by producers of crops, owners of affected forest lands, and others who realize no direct benefits from the trade. Their increased costs of eradication and control will benefit the companies that supply those goods and services. Risk management measures may impose additional costs on producers of plant products that raise the price that consumers must pay. How much of this very complex scene should be of interest to the risk manager?

Is the NPPO of an importing nation obligated to consider the benefits and costs to the exporting nation when making a risk management decision? Is the importing NPPO obligated to consider the benefits to consumers and producers in their own nation? Should the importing NPPO consider only the impacts on plant resources when making a decision?

It may represent some sort of economic ideal to take all the benefits and costs into account "to whomsoever they accrue" (U.S. Interagency Committee on Water Resources, 1950). Then we could be assured that the decision made would be in the world's interests and not simply the interests of the locality where an infestation might occur or of the purveyors of international trade.

There are at least two very practical problems with this approach (Howe, 1971). First, it would be very costly and time consuming to search for benefits and costs to the entire world. Second, the NPPO risk manager's incentive system orients her to his own constituency. But just who is that constituency?

It is not clear that there is a right answer to this question. It is, however, quite clear that each interest in international trade and plant production as well as consumers of the relevant products will have its own natural viewpoint of what benefits and costs are important to consider. It is also clear that it is the phytosanitary risk manager's responsibility to make the viewpoint used to estimate the economic impacts of proposed risk management decisions clear to all who will consider those impacts.

Bottom line up front

Consider a hypothetical situation in which risk managers are considering a market access request, which if granted would produce C100M (M for millions) in net total surplus (see Chapter 7, this volume). Imagine the import would result in infestation that would cause a total loss to domestic producers of C50M and that there is no effective treatment for the infestation. Suppose the risk manager denies market access on the basis of the C50M in damage that could not be mitigated.

Granting market access would produce C100M in benefits at a cost of C50M to producers. From a national perspective (broader than a producer perspective) that is a good deal for the nation. Theoretically, those who gain the C100M could compensate producers with C50M and still have C50M for themselves.

Arthur (2004) provides a compelling example that makes this point in his analysis of Australia's ban on New Zealand apple imports to prevent fire blight disease. He identifies the asymmetric viewpoint risk managers adopt when applying a quarantine measure as the core problem with economic analysis. Arthur notes that economic welfare in the SPS Agreement and subsequently in domestic quarantine policy processes primarily quantifies the impact on domestic producers of the importing country. Theirs is the phytosanitary risk management viewpoint most often adopted. The problem is that risk management options evaluated in this way could be economically inefficient from the viewpoint of a more balanced national perspective (see box).

It is incumbent upon risk managers to make clear the nature and extent of the economic benefits and costs considered in decision-making. Some of the obvious choices for benefit and cost viewpoint include:

1. Direct impacts on the importing country's domestic producers only.
2. Direct impacts on domestic welfare.
3. Direct and indirect domestic economic impacts (economic impacts include domestic welfare plus other impacts like changes in jobs, income, tax revenues, and the like).
4. Direct impacts on foreign and domestic producers only.
5. Direct impacts on foreign and domestic welfare.
6. Direct and indirect global economic impacts.

A domestic perspective appears to be most realistic. Considering the direct impacts on domestic producers, number 1, above, is consistent with the explicit guidance found in the SPS Agreement and appears to be the most common extent of economic analysis, when economic consequences are even considered. Domestic producers have a right to expect to know the potential impacts on them of a public policy decision. To make a well-informed economic decision, however, risk managers should at least have estimates of the direct impacts on the domestic welfare of producers and consumers. This remains an aspirational goal for most PRAs, which tend not to consider economic benefits foregone. Most economists would consider the current viewpoint adopted for economic consequences to be too limited, preferring a more complete consideration of national benefits and costs.

29.3. Why Does Economic Viewpoint Matter?

In a world of limited resources, the world we all live in, it is unwise to waste resources. Let us use a simple definition of waste: if the costs of an action exceed the benefits of an action, that action wastes economic resources. Its corollary is failure to take action when the benefits of that action exceed its costs. When several actions are available, economic efficiency would dictate choosing the alternative that maximizes net benefits. This is by no means a rigorous definition but it will do for now. An incomplete view of the benefits and costs of an action can lead to inefficient or wasteful allocations of resources.

Wilson and Antón (2006) argue that the best phytosanitary risk management measures are the ones that are the least trade distorting, superior in terms of economic welfare, and provide protection of health and safety for all concerned. If measures mitigate the SPS risks equally, the best measures have the least negative effects on trade and domestic economic welfare. Wilson and Antón assert that the economics literature finds bans to be the most trade restrictive SPS measures, followed by tariffs and risk mitigating strategies. Consequently, bans and tariffs affect economic welfare more than other risk mitigating strategies. They argue that SPS measures are simply trade barriers that protect producers.

Most PRAs are qualitative and they do not explicitly estimate economic values associated with the international plant and plant product trade risk management continuum. Typically, economic effects are considered when determining the strength of risk management measures because the strength of measures must be consistent with the magnitude of the risk. Thus, if risk managers want stringent measures in place, they must be able to demonstrate higher consequences in their absence. This is normally done in a qualitative manner.

Economists would consider the lack of rigor in the economic analysis to lead to ill-informed and possibly wasteful decision-making. Including explicit qualitative or quantitative economic information in the decision-making process is one of the single best ways to improve the quality of the decisions produced by phytosanitary risk management. The process is mature enough to accommodate at least qualitative economic rationality as a required decision criterion. All decision-making in the production risk management arena is presumably based on economics-informed risk information. It is time for the phytosanitary risk analysis community to embrace more economics in its decision-making.

This section presents a series of hypothetical examples that illustrate how decisions can be changed by economic information. This helps to establish the practical value of economic analysis as part of phytosanitary risk management decision-making. The values in all examples are contrived to keep the examples and their mathematics simple. The hypothetical currency unit "C" is used once again. All monetary values should be considered annual equivalents, properly discounted and amortized. Quantities can be considered to be millions of C to make the units more compelling. The analytical perspective is assumed to be domestic unless otherwise specified.

Examples in the sections that follow make use of the following terms:

A. "Without condition damages" to domestic crops as a result of trade are the damages that are expected to occur if no risk management measures are required for the imported product.
B. "With condition damages" to domestic crops are the damages that are expected to remain after risk management measures are imposed. These are *residual damages*.

C. "Reduction in damage to domestic crops" is the difference between the without and with condition damages.
D. "Domestic cost of risk management measures to reduce domestic damage" includes all direct costs to domestic producers and consumers plus all indirect impacts, including rising costs of imports borne by the domestic economy.
E. "Total costs with phytosanitary risk management" is the sum of the second and fourth terms.
F. "Net gain to phytosanitary risk management" is the difference between the third and fourth terms or the first and fifth terms.

29.3.1. Minimizing costs to the domestic economy, no benefit estimate

Imagine a situation where allowing market access to an exporting country with no risk management measures in place would result in crop damage losses of 9C sustained by domestic producers of the importing country. The benefits of trade are not known but are implicitly assumed to be significant. Suppose a pest risk management option could reduce this damage to 1C, i.e., the residual crop damage with trade is 1C. The damages reduced by the risk management option are 8C. See Table 29.1 for data and the list of terms above for entry explanations.

Now consider that risk management option costs 6C to implement. The net gain from pest risk management compared to conditions without the additional risk management is 2C. Thus, pest risk management in this case reduces the cost of trade from 9C to 7C, a saving of 2C. Clearly, risk management is more cost-effective than no risk management in this example. The expenditure of 6C annually is justified by being less than the 8C in annual damage that would be reduced by the pest risk management measures. This approach establishes that risk management is a cost-effective way to realize the unquantified benefits of trade. It does not, however, tell us if the risk management is "worth it" because we do not know the benefits from trade.

29.3.2 Maximizing net benefits to the domestic economy

Suppose the sum of all the domestic benefits is 4C, but that fact is not known. In the example above, risk managers have just decided to bear costs of 7C because it is the cost-effective way to realize the unknown benefits. Once we know the benefits are 4C, the decision now is: would you spend 7C to get 4C? This is clearly an irrational decision based on economic criteria. It yields negative net benefits of −3C (Table 29.2). No risk manager would trade 7C for 4C in a world of limited resources.

Unless risk managers make some effort to estimate or understand the magnitude of the economic benefits of trade, there is no objective way to know if any decision being made is economically efficient. It may well be expedient, in the traditions of phytosanitary risk management, but it may not be efficient. For now, think of efficiency as benefits exceeding costs.

Let us now consider another possibility, where benefits are 20C. In such a case, risk management yields net benefits of $20C - 7C = 13C$. That represents a net gain for society and it is a better return than no risk management would yield, i.e., $20C - 9C = 11C$. Allowing trade with phytosanitary risk management is a rational economic decision in this case (Table 29.3).

A fourth example shows trade is most efficient with no risk management measures implemented. This last example of this section is presented in

Table 29.1. Hypothetical data representing a phytosanitary risk management decision without benefit estimates.

Item	Annual monetary value(1,000,000's C)
1. Without condition damage to domestic crops as a result of trade	9C
2. With condition damage to domestic crops (residual damage)	1C
3. Reduction in damage to domestic crops (1 − 2)	8C
4. Domestic cost of risk management measures to reduce domestic damage	6C
5. Total cost with phytosanitary risk management (2 + 4)	7C
6. Net gain (cost saving) from phytosanitary risk management (3 − 4 or 1 − 5)	2C

Table 29.2. Hypothetical data representing a phytosanitary risk management decision that results in negative net benefits.

Item	Annual monetary value(1,000,000's C)
0. Estimated gross domestic benefits associated with trade	4C
1. Without condition damage to domestic crops as a result of trade	9C
2. With condition damage to domestic crops (residual damage)	1C
3. Reduction in damage to domestic crops (1 − 2)	8C
4. Domestic cost of risk management measures to reduce domestic damage	6C
5. Total cost with phytosanitary risk management (2 + 4)	7C
6. Net benefits associated with trade and its phytosanitary risk management (0 − 5)	−3C

Table 29.3. Hypothetical data representing a phytosanitary risk management decision that results in positive net benefits.

Item	Annual monetary value(1,000,000's C)
0. Estimated gross domestic benefits associated with trade	20C
1. Without condition damage to domestic crops as a result of trade	9C
2. With condition damage to domestic crops (residual damage)	1C
3. Reduction in damage to domestic crops (1 − 2)	8C
4. Domestic cost of risk management measures to reduce domestic damage	6C
5. Total cost with phytosanitary risk management (2 + 4)	7C
6. Net benefits associated with trade and its phytosanitary risk management (0 − 5)	13C

Table 29.4. Hypothetical data representing positive net economic value of trade without risk management measures in place from a domestic accounting stance.

Item	Annual monetary value(1,000,000C)
0. Gross economic benefits to importing country (total surplus)	15C
1. Without condition damage to domestic crops as a result of trade	5C
2. Without condition net benefits subtotal (0 − 1)	10C
3. With condition damage to domestic crops (residual damage)	1C
4. Domestic cost of risk management measures to reduce domestic damage	17C
5. With Condition Subtotal (1 − 4 − 5)	−3C

Table 29.5. Hypothetical data representing net economic value of trade from a global accounting stance.

Item	Annual monetary value(1,000,000's C)
1. Gross economic benefits to importing country (total surplus)	4C
2. Gross economic benefits to exporting country (total surplus)	10C
3. Without condition damage to domestic crops	9C
4. Without condition net benefits subtotal (1 + 2 − 3)	5C
5. With condition damage to domestic crops (residual damage)	1C
6. Domestic cost of risk management measures to reduce domestic damage	6C
7. With condition net benefits subtotal (1 + 2 − 5 − 6)	7C

Table 29.4. In this instance trade is efficient, producing net benefits of 10C, without additional pest risk management measures in place. When pest risk management measures are imposed in the second scenario (rows 0, 3–5) trade is no longer economically efficient yielding a net return of −3C. The damage is modest compared to the cost of measures required to reduce the damage.

The examples of this section were contrived to make several points. It may be economically more efficient to deny trade all together, to allow trade with no additional risk management, or to allow trade with new risk management measures. It is impossible to know which of these situations prevails unless risk managers have economic information about the benefits of trade.

29.3.3. A global perspective

Economic analysis provides a rational framework for decision-making in a world of limited resources. Consider the example of Table 29.5. Rows 1 and 2 represent the net economic gains from trade in both the exporting and importing countries. These would be measured by total surplus consistent with the presentations of

Chapter 7. In the scenario depicted in rows 1–4 no new or additional pest risk management is applied. The net benefits of 5C indicate the trade is economically efficient from a global perspective. It takes 9C in global costs to produce 14C in global benefits, a net change of 5C. Notice that the trade is not a good deal for the importing country, which has benefits of 4C and costs of 9C for a net effect of −5C. Thus, the accounting viewpoint is very important to decision-making.

The second scenario (rows 1, 2, 5–7) considers the addition of pest risk management measures in place at a cost of 6C reducing residual damages to 1C, for total costs of 7C. This is 2C cheaper than the no risk management option and it results in net global benefits of 7C as shown in row 7. From a global perspective, trade with pest risk management is the best deal. It produces 7C instead of 5C in net benefits. From a purely domestic economic perspective this trade would not take place. Without risk management the domestic economy loses −5C. With risk management the domestic economy loses 4C − 7C = −3C. None of this information is available unless economic values are considered, of course.

Murina and Nicita (2017) found that agricultural products entering the European Union market must comply with a substantial number of SPS measures. The proliferation and increased stringency of SPS measures favors exporters capable of SPS compliance at the expense of exporters operating in other countries, resulting in a competitive repositioning of international trade in favor of nations that can comply more easily. Regulatory constraints and requirements in importing countries may differ substantially from those in exporting countries. This asymmetry directly impacts the costs of compliance.

29.3.4. Economically efficient pest risk management measures

Table 29.2 presents an example from a domestic perspective where trade should not take place with or without pest risk management measures. For contrast, Table 29.6 presents an example where trade without pest risk management is not economically efficient but where trade with risk management is. The first scenario (rows 1–3) shows the damage to domestic crops exceeds the benefits of trade. The second scenario (rows 1, 4–6) shows that trade with pest risk management measures in place produces a net gain of 5C.

Once again, the importance of economic benefit information for rational decision-making is clear.

29.3.5. Strength of measures

The strengths and costs of risk management measures are presumed to have an increasing relationship, i.e. stronger measures cost more, weaker measures cost less. Justifying the strength of measures is a simple thing when economic

information is available. Consider the example in Table 29.7.

There are three strength of risk management measures levels, each has its corresponding cost, which comprises the cost of pest risk management measures plus residual damages. Total economic benefits vary because low and medium strength measures have more crop losses averted associated with them. Net benefits are total economic benefits minus their associated costs.

The data show that high strength measures are economically efficient in the sense that the value of that response (28C) exceeds the cost (25C) of that response. Economists would optimize the net benefits and identify medium strength measures as the most efficient response. Risk managers, of course, are free to consider values in addition to net benefits when they make their final choice of strength of measures.

Table 29.6. Hypothetical data representing positive net economic value of trade with risk management measures in place from a domestic accounting stance.

Item	Annual monetary value(1,000,000C)
1. Gross economic benefits to importing country (total surplus)	15C
2. Without condition damage to domestic crops	20C
3. Without condition subtotal (1 − 2)	−5C
4. With condition damage to domestic crops (residual damage)	0C
5. Domestic cost of risk management measures to reduce domestic damage	10C
6. With condition subtotal (1 − 4 − 5)	5C

Table 29.7. Hypothetical data representing strength of a phytosanitary risk management measures choice based on economic information.

Risk management option	Costs of risk management option & residual damage	Total economic benefits	Net benefits
Low strength	10C	14C	4C
Medium strength	15C	25C	10C
High strength	25C	28C	3C

29.4. Economics in the SPS Agreement and IPPC Guidance

Article 5.3 of the SPS Agreement provides the first mention of economic consequences in the agreement. It says:

> In assessing the risk to animal or plant life or health and determining the measure to be applied for achieving the appropriate level of sanitary or phytosanitary protection from such risk, Members shall take into account as relevant economic factors: the potential damage in terms of loss of production or sales in the event of the entry, establishment or spread of a pest or disease; the costs of control or eradication in the territory of the importing Member; and the relative cost-effectiveness of alternative approaches to limiting risks.

Wilson and Antón (2006) say this definition of relevant economic factors suggests that producer concerns are the primary economic concerns of the SPS Agreement. This guidance has been interpreted by some NPPOs to preclude consideration of the economic effects that are not specified. Chief among these are changes in total surplus of producers and consumers who stand to benefit from the proposed trade. As illustrated in Section 29.3 of this chapter, that can result in the misallocation of resources. At a minimum, it ignores the interests of international trade stakeholders.

Article 5.6 goes on to say:

> Members shall ensure that such measures are not more trade-restrictive than required to achieve their appropriate level of sanitary or phytosanitary protection, taking into account technical and economic feasibility.

Annex A provides the final reference to economics:

> 4. Risk assessment—The evaluation of the likelihood of entry, establishment or spread of a pest or disease within the territory of an importing Member according to the sanitary or phytosanitary measures which might be applied, and of the associated potential biological and economic consequences; or the evaluation of the potential for adverse effects on human or animal health arising from the presence of additives, contaminants, toxins or disease-causing organisms in food, beverages or feedstuffs.

This guidance does not specify any particular methodology or type of analysis. Damage as

lost sales or production and costs of control are clearly focused on the impacts on domestic producers. The explicit reference to cost-effectiveness at the end of Article 5.3 suggests that a cost–benefit analysis is not required. The subsequent reference to economic feasibility in Article 5.6 is somewhat vague. In general usage, economic feasibility tends to mean the benefits of the activity exceed the costs of the activity. There are legitimately different viewpoints that can be used to define benefits and costs used to establish economic feasibility. The economic consequences of spread of a pest disease are identified as a risk assessment output, but their nature is not further specified beyond what appears in Article 5.3.

There is a popular interpretation of this guidance to narrowly constrict the economic data that can be considered in a pest risk *assessment*. There is, however, nothing in the SPS agreement that precludes the consideration of additional economic criteria in the establishment of a nation's appropriate level of protection (ALOP).

Looking to the ISPMs for further guidance, ISPM 2 suggests a qualitative assessment is sufficient when it says:

> In order to estimate the potential economic importance of the pest, information should be obtained from areas where the pest currently occurs. For each of these areas, note whether the pest causes major, minor or no damage. Note whether the pest causes damage frequently or infrequently. Relate this, if possible, to biotic and abiotic effects, particularly climate.

ISPM 2 goes on to say "expert judgement is then used to assess the potential for economic importance." It offers the following examples of economic factors to consider:

- type of damage
- crop losses
- loss of export markets
- increases in control costs
- effects on ongoing integrated pest management (IPM) programmes
- environmental damage
- capacity to act as a vector for other pests
- perceived social costs such as unemployment.

Significantly, qualitative assessments seem to be acceptable and the impacts retain a focus on domestic producers. Ironically, the third bullet above recognizes the importance of trade benefits, although it does it from the perspective of the domestic producer's loss of access to export markets as a result of reduced crop yields. Access to export markets can also be restricted by phytosanitary risk management measures.

ISPM 5 makes it clear that economic effects can include monetization of environmental and social effects. It also says economic effects should not be interpreted to be only market effects. The strongest statement in favor of cost–benefit analysis is found in ISPM 5 where it says the following:

4.2 Costs and benefits

A general economic test for any policy is to pursue the policy if its benefit is at least as large as its cost. Costs and benefits are broadly understood to include both market and non-market aspects. Costs and benefits can be represented by both quantifiable measurements and qualitative measurements. Non-market goods and services may be difficult to quantify or measure but nevertheless are essential to consider.

Economic analysis for phytosanitary purposes can only provide information with regard to costs and benefits, and does not judge if one distribution is necessarily better than another distribution of costs and benefits of a specific policy. In principle, costs and benefits should be measured regardless to whom they occur. Given that judgments about the preferred distribution of costs and benefits are policy choices, these should have a rational relationship to phytosanitary considerations.

Costs and benefits should be counted whether they occur as a direct or indirect result of a pest introduction or if a chain of causation is required before the costs are incurred or the benefits realized. Costs and benefits associated with indirect consequences of pest introductions may be less certain than costs and benefits associated with direct consequences. Often, there is no monetary information about the cost of any loss that may result from pests introduced into natural environments. Any analysis should identify and explain uncertainties involved in estimating costs and benefits and assumptions should be clearly stated.

This description suggests that cost–benefit analysis is an appropriate tool, although it allows that it may be qualitative or quantitative. Significantly, the guidance suggests a broad point of view be adapted when it says, "costs and benefits should be measured regardless to whom they occur" and then goes on to suggest both direct and indirect effects be considered. This guidance would seem to be compatible with a global accounting stance for estimating benefits and costs. ISPM 5 cautions that economic analysis alone is insufficient to say what distribution of costs and benefits is best. A global perspective seems more aspirational than practical at this point in time.

ISPM 11 introduces economic analysis as part of Stage 2 of PRA, which calls for the assessment of potential economic consequences (including environmental impacts). Later the ISPM expresses a preference for quantitative economic data and suggests it may be useful to discuss these impacts with an economist. It says: "Wherever appropriate, quantitative data that will provide monetary values should be obtained. Qualitative data may also be used. Consultation with an economist may be useful."

When the magnitude of potential negative economic consequences associated with a market access request is generally agreed to be unacceptable, detailed analysis of economic consequences is not required. When such agreement exists the risk assessment will focus more on the probability of introduction and spread. However, when the level of economic consequences is in question, greater detail in the analysis is expected. More economic analysis is also expected to evaluate the strength of measures or when assessing exclusion or control. It is important to establish that the damage prevented exceeds the cost of risk management control. This is especially important when more stringent and costly measures are proposed.

> In the event that the magnitude of potential negative economic consequences associated with a market access request is negative, it is imperative that the evidence for this judgment be produced and documented.

Bear in mind, as shown in the examples of Section 29.3, damages prevented in excess of the cost of risk management are not sufficient to establish economic efficiency. To do this cost–benefit analysis may be warranted.

ISPM 11 cautions that attention be paid to the dispersion of costs and benefits over time and place. Most of the direct and some of the indirect effects of a pest will be commercial or will impact an identified market. These effects that should be identified and quantified may be positive or negative, and include:

- effect of pest-induced changes to producer profits that result from changes in production costs, yields or prices
- effect of pest-induced changes in quantities demanded or prices paid for commodities by domestic and international consumers
- quality changes in products and/or quarantine-related trade restrictions resulting from a pest introduction.

Consideration of these economic impacts would mark a significant improvement in economic analysis and the decisions based on that analysis.

Three analytical techniques identified in the ISPM guidance and described later in this chapter that can be used when economic consequences are quantified are:

1. partial budgeting
2. partial equilibrium
3. general equilibrium.

In summary, then, the available guidance seems to suggest that qualitative assessment of the economic consequences will suffice for a wide variety of circumstances. This includes when it is generally agreed that the economic consequences of infestation are unacceptable and no further economic information is offered. The economic consequences include commercial, environmental, and social effects although as a practical matter the primary consequences tend to focus on losses sustained by domestic producers. Quantitative assessments, broad viewpoints, and cost–benefit analysis are not only not precluded, they are encouraged by ISPM guidance.

29.5. Challenges Encountered in the SPS Economic Framework

Four challenges arise in the practice of assessing economic consequences of pest introduction in the SPS economic framework. These are:

1. Most such assessments are qualitative.
2. Ex-post analyses are easier than ex-ante analyses.

3. Analyses focus on producer impacts rather than a more complete array of economic consequences.
4. There are substantial uncertainties to be addressed in these analyses.

Economic assessments in most PRAs are primarily qualitative and rely on expert judgment (Sansford, 2002; Brunel *et al.*, 2009). Expert judgment has a tremendous cost advantage but it is not, generally, a good substitute for doing the actual analysis. In addition, qualitative assessments based on expert judgment suffer a lack of transparency and they are not as repeatable as other analyses. Worst of all, qualitative approaches may be (ab)used for political or protectionist goals (Soliman *et al.*, 2012).

Sometimes the four challenges all run together. One of the reasons that so many risk assessments rely on qualitative consequence assessments is that ex-ante consequence estimates are very difficult due to the uncertainty that attends them. In addition to the uncertainty about the pest and its disease symptoms there is considerable geographic and temporal uncertainty. What percentage of the vulnerable crops will be affected? In which direction will the pest head? How soon might the pest establish? How fast will it spread? Will there be yield loss, quality degradation, or both? In the face of this uncertainty, qualitative assessments are often easier.

Ex-post estimates of consequences are not easy but by contrast they are far more straightforward than ex ante estimates. Unfortunately, ex ante estimates are required to properly evaluate the economic impacts of risk management proposals. Breukers *et al.* (2008) poins out the need for ex-ante assessments of a control policy in order to have insight to potential impacts of control policies.

Soliman *et al.* (2012) provide an entrée to the third weakness. While noting that plant import regulation is indispensable for protecting against pest invasions, he observed that overly strict import restrictions can unnecessarily limit trade and reduce welfare. This aligns with Arthur's (2004) argument for using comprehensive economic analyses to quantify the effects of risk management measures on market equilibrium, trade, economic efficiency, and net social welfare. That means considering both the costs and benefits of quarantine measures to evaluate the tradeoffs between alternative regulatory and non-regulatory actions, as demonstrated in Section 29.3. Roberts *et al.* (1999), Beghin and Bureau (2001), and Maskus *et al.* (n.d.) join in arguing for economic analysis that demonstrates the economy-wide welfare effect of a quarantine measure, as opposed, for example, to a focus on impacts on domestic producers. There is general recognition within the economics profession that there is substantial room for improving the economic analysis used in support of phytosanitary risk management decision-making.

Arthur argues that the economic analysis framework set out by the SPS Agreement is asymmetric in its consideration of the "relevant economic factors" (WTO, 1995) to consider for implementation of a quarantine measure. The guidance suggests that costs faced by producers in the event of disease entry and establishment are counted, but the forgone benefits of trade in products that would occur without the quarantine measure in place are not. For example, risk management options are implemented to reduce damages to producers that result from pest infestations. These measures may reduce the supply of product imported or they may increase the cost of imports. Both of these phenomena reduce consumer surplus. Thus, any decision that reverses or avoids the need for such risk management measures averts these adverse impacts, restoring the benefits to consumers. These are legitimate benefits to the unregulated trade option. Thus, a decision not to regulate a commodity–pest pair may be a "bad" decision based on its SPS risk reduction capability but it may be a very good decision based on economic consequences.

Consider Arthur's example. New Zealand is more efficient at producing apples than Australia. Therefore, allowing New Zealand apples into the country would increase the number of apples available to consumers as it reduces the price of those apples. More and cheaper apples benefit Australian consumers. Fig. 29.1 shows Supply (S) and Demand (D) for apples in Australia.

Without New Zealand apples Australia would have Q_{AD} apples at the Australian price. Recall from Chapter 7 that value exceeds price and Australian consumers are enjoying "A" in benefits (consumer surplus) and Australian producers enjoy 'B + C' in benefits (producer surplus). If New Zealand apples are introduced to Australian markets, consumers would enjoy $Q_{w/NZ}$ apples at the New Zealand price, i.e., more apples and cheaper apples. Consumers win and now enjoy

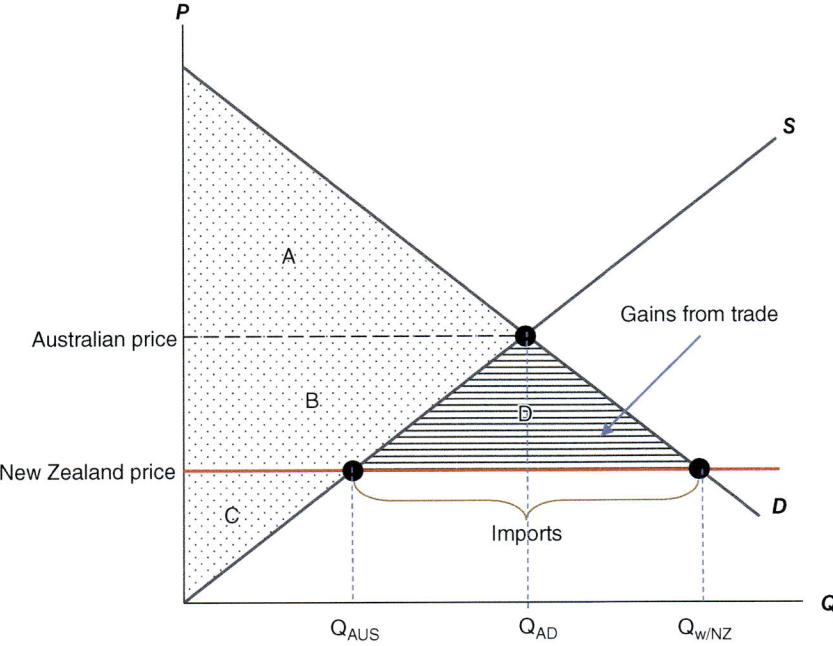

Fig. 29.1. Australian gains from trade as a result of allowing New Zealand apples into Australian markets.

benefits of "A + B + D" while Australian producers are adversely affected and now enjoy benefits of "C" alone. Some of the consumers' gain comes
at the cost of producers' loss, but D is a net gain
to society that results directly from trade.

In Arthur's study the value of D was much
greater than the losses of Australian apple producers to fire blight disease. This, Arthur, claimed,
may be a powerful argument for not regulating
the entry of apples at all. He notes, however, the
"political sensitivity factor" is important. The
distribution of gains and losses is unequal. All
the losses in Australia are borne by the relatively
few (1,500) apple producers while the gains are
enjoyed by millions of individual apple consumers. In Arthur's worst case scenario, trade
would be allowed without risk management
measures and fire blight would enter all Australian production areas. Under this scenario, total
gains to consumers was AC201 million against
producer losses of AC111 million for a net gain
to Australia of AC90 million. Arthur does not
advocate no regulation. He does, however, advocate a symmetry of emphasis for considering
consumer and producer impacts and then making risk management decisions fully informed
about the economic consequences of those
decisions.

Uncertainty, addressed extensively in Chapter 15, presents a significant challenge to the assessment of economic consequences. Soliman
et al. (2012) identify model outcomes and model
parameters as important issues. They go on to
assert that whether the impact assessment is accurate or not, is not as important as whether the
risk management action justified by the assessment was correct. It is more important, in their
view, for the assessment to point toward the
proper action than it is for the economic assessment to be accurate.

They frame the issue in terms of type I and
type II errors. A type I error rejects the null hypothesis of no risk management is required,
when, in fact, no risk management is required.
This would be providing unnecessary measures.
This could occur when the economic assessment
suggests the economic risks justify phytosanitary measures when they do not. Type II error is
believing no risk management action is required,
when in fact, it is. It occurs when the economic
analysis does not correctly detect risks that
would warrant phytosanitary measures. It is the

uncertainty in pest risk assessment that could
lead to over- or underestimation of economic impacts. Soliman et al. argue that overestimation is
more likely when the precautionary principle is
applied. Thus, the precautionary principle increases Type I errors and decreases Type II
errors. These errors can be reduced by the use of
more advanced types of economic analysis, such
as those discussed later in this chapter.

29.6. Structural, Incidental and Export Losses

A great deal has been learned about the economics of infestation by nations that have struggled to prevent or respond to infestations within
their own countries. Breukers et al. (2008), in a
study of Dutch potato rot, identified three kinds
of losses associated with phytosanitary risk
management: structural, incidental, and export
losses. These are costs associated with the exporting country.

The enforcement of preventive measures is
largely independent of the level of disease inci-
dence, so the corresponding enforcement costs
are called structural costs. These are costs that
are built into the structure of the risk management decision. They include the costs of monitoring disease prevalence. This encompasses the
monitoring intensity specified by the control
strategy and the costs of taking and analyzing
samples. It also includes costs resulting from restrictions, the cultivation of susceptible crops or
plants, and other highly individualized details.
Because structural costs do not depend on the
level of infestation, they are relatively constant
over time for a given control strategy.

There is a class of costs incurred when a nation reacts to an outbreak of the disease. These
costs are related to the outbreak incident and are
called incidental costs by Breukers. Incidental
costs arise during the year of detection and in
subsequent years. Using brown rot to illustrate
incidental costs, potato lots with brown rot detected must be destroyed. Infections may still be
missed due to a limited sample size, so a negative
outcome does not guarantee the tested lot is free
of brown rot. Potato lots not found to be infected
that had (indirect) contact with an infected lot
may be classified as "probably infected." There

are limited marketing possibilities for these lots. They cannot be replanted, for example, so their market value is reduced. The total incidental costs in any particular year will depend on the number, size, and category of detected lots, the number of farms involved, and the crop production characteristics of these farms. Incidental costs can vary dramatically from year to year.

Quarantine disease can cause crop losses and export losses. Export losses occur when importing countries lose their confidence in the phytosanitary quality of an import, resulting in reduced exports.

29.7. Market Failures

A market is a group of willing buyers and sellers. Supply, i.e. willing producers and sellers, and demand, i.e. willing consumers and buyers, interact in a market and determine both prices and quantities of goods and services that change hands. Prices function to allocate society's scarce resources. Price simultaneously reflects the good's value to buyers and the supplier's cost of producing the good. Market determined prices guide self-interested consumers and firms to make decisions that, in most cases, maximize society's economic well-being. Markets are used to decide what goods to produce, how to produce them, how much of each to produce, how much to charge for them, and who gets them. Market economies answer these questions through the decentralized decisions of many consumers and firms as they interact in markets.

Prices provide incentives to consumers and producers. Risk managers are well-advised to consider the price/incentive effects of any risk management option they are considering. When markets work, they can work very well in allocating resources. But sometimes markets fail to allocate society's resources efficiently for a variety of reasons. When this happens, government intervention has the potential to alter and improve market outcomes.

Common causes of market failure include:

- asymmetric information
- market power
- nature of certain goods
- externalities.

Asymmetric information occurs when one party knows more about a product than another. If exporters sell a product known to be pest infested, this is asymmetric information. Buyers, unaware of the facts about the quality of the product, may overestimate the value of that product and unwittingly pay a price that is greater than the true value of that product to them.

When a single buyer or seller can exert significant influence over prices or output, we can observe a market power failure of a market. Monopolies are the best example. Here a single firm can set the price for their good at any level they desire. Although monopolies provide the best-known example of market power, any imperfectly competitive situation can distort the market outcome. Risk management options that reduce competition or help consolidate market power can contribute to market failures.

The third cause of market failure is related to the nature of the good itself. There are some things that can only be consumed by everybody or nobody at all. These are called public goods and they have distinct characteristics. They are:

- non-rival—one person's consumption of them does not stop another person from consuming them, i.e., the supply of the good is not reduced when a person consumes it
- non-excludable—if one person can consume them then anyone can consume them, i.e., it is impossible to stop another person consuming them.

'Plant protection' is an example of such a good. A subset of these public goods is also non-rejectable. People can't choose not to consume them even if they want to. National defense, homeland security, and public health are examples of non-rejectable public goods. Plant protection, which may be produced by phytosanitary risk management measures is non-rejectable. Risk management options can increase the supply of public goods.

Some risks threaten common goods. Common goods are goods that are non-excludable but rival. Many environmental resources are common goods. Government regulations, including pest risk management measures, can sometimes effectively reduce the risk to common resources.

An externality is an economic side-effect and it occurs when an economic activity (consumption or production) affects a bystander to that activity. These side-effects are never fully reflected in the prices of goods. Externalities don't enter the cost or benefit decisions of either buyers or sellers, that is why they are called external. They can be harmful (negative externalities) or beneficial (positive externalities), but the harmful ones seem to attract more attention. The rationale for government intervention with quarantine measures is based on the existence of negative externalities. In the case of trade, importing products infested with pests imposes some of the costs of importation on domestic producers in the form of pest and disease risks. Products are imported for consumption but the full or social cost of importation risk is not borne entirely by the importer or final consumer of that product, to the extent that domestic producers are harmed by importation.

Fig. 29.2 illustrates the effect of an infestation on the welfare of importers (curve D) and exporters (curves S and S'). The market, without any quarantine measures, would produce the market price and Q_{MKT}. Importers would enjoy consumer surplus of "A+ B + C + D" while exporters enjoy producer surplus of "E + F + G." The difference between S and S' is caused by the existence of a negative externality, which is damage

to the crops of domestic producers. If the market bore the full cost of this externality it would produce Q_{SOC} at the social optimum price. This would result in total surplus of "A + B + E" instead of "A + B + C + D + E + F + G." The difference, "C+ D + F + G," is the total burden of the externality. From an economic perspective, the ideal government intervention (risk management) would restrict the output of this product to Q_{SOC}.

Arthur (2004) argued that correcting for this externality with quarantine measures could be inefficient. Consider Fig. 29.3, which shows the market for consuming this hypothetical good. Q_D and the price without imports is the market result for this country in the absence of trade. With imports the price falls to the price with imports and the quantity consumed increases to the quantity with imports ($Q_{w/M}$). Take that as the starting point for this example. Now imagine that to halt the damage to domestic producers seen in Fig. 29.2, importation is prohibited. (It is not necessary to prohibit importation but that is the easiest case to illustrate.) Quarantine measures can have the effect of increasing the price with imports that would result in different sized

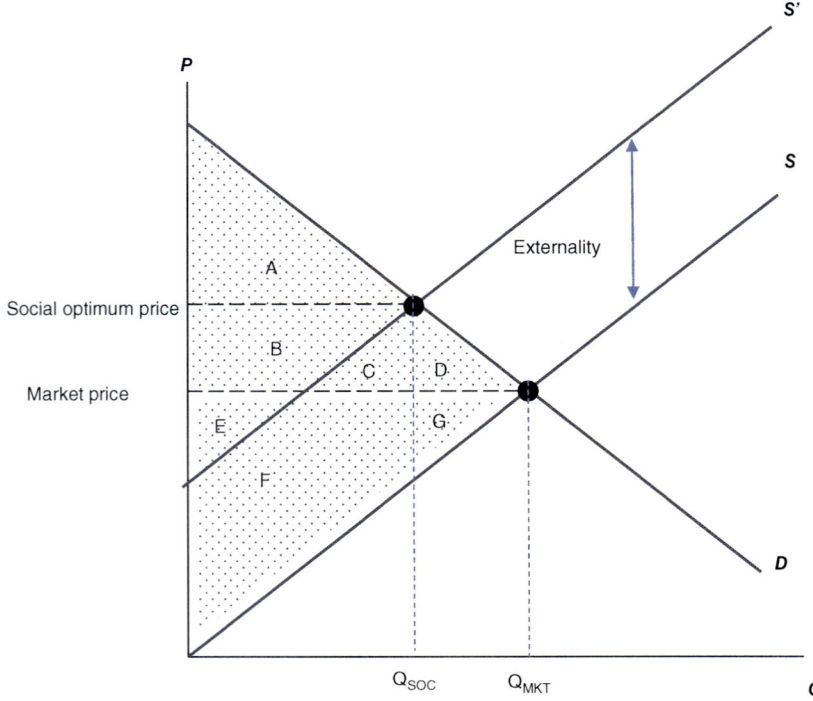

Fig. 29.2. Price, quantity, and welfare effects of negative externalities in the market for imports illustrated.

effects for areas B and D. In Fig. 29.3 we see that means consumers lose benefits equal to "B + D." Arthur's argument is that if "B + D" in Figure 29.3 is greater than "C + D + F + G" then restricting importation is not economically efficient. Measures designed to protect producers from quarantine risks may hurt consumers a great deal more than they help producers. Applying quarantine measures to correct for an externality on producers is not always beneficial at the social level.

> When an importing country imposes a phytosanitary regulation, compliance with that regulation imposes a cost on foreign suppliers, which acts like a trade tax, resulting in a deadweight loss as well as transfers from consumers to producers (Beghin and Bureau, 2001).

Government intervention via quarantine measures is an economically efficient course of action when the level of social welfare in affected markets can be improved. That means the benefits of intervention outweigh the costs of intervention and the intervention is efficient. (The Kaldor and Hicks Criterion identifies a policy action as beneficial when the overall level of welfare is increased, such that the gains are greater than the losses. In such a case, those who benefit could, hypothetically, compensate those who bear the costs and still

have money left over, leaving them better off even after compensating those who lost money.) When costs exceed benefits the measures are not economically efficient. Economic efficiency does not address questions of equity or fairness. When consumer benefits accrue to millions of individuals and damages to producers accrue to hundreds of producers, fairness is a legitimate concern. Such policy decisions might be more practical if they provide means for redress by damaged producers.

29.8. Economic Tools for Use in Risk Management Decision-making

Economic information is useful for decision-making. While good economic information does not guarantee a good risk management decision, the lack of good economic data always leads to a partially informed decision. Analysts have a variety of alternative economic approaches to generate and present economic data. Five approaches for providing economic information are identified:

1. Economic impact analysis
2. Cost-effectiveness analysis
3. Incremental cost analysis
4. Cost–benefit analysis
5. Damage assessment

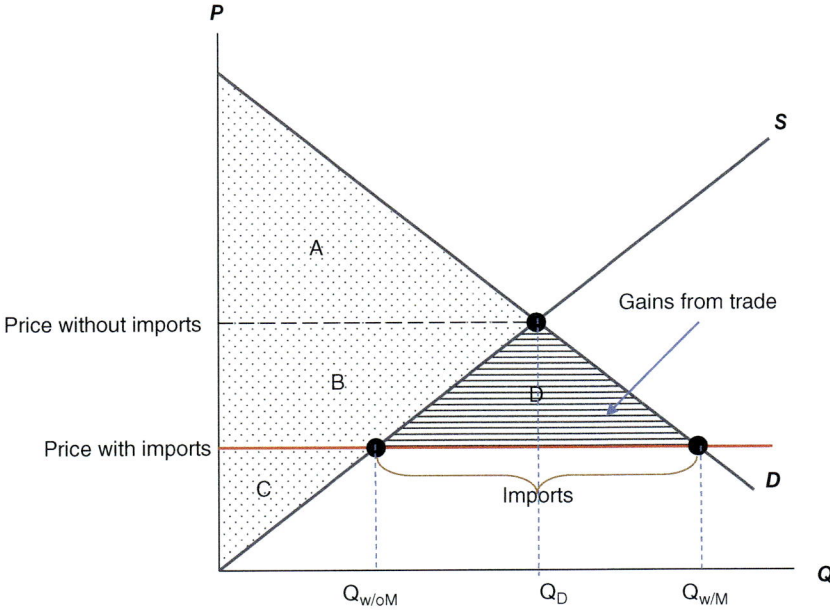

Fig. 29.3. Benefits from trade forgone by consumers as a result of trade restrictions.

29.8.1. Economic impact analysis

Economic impact analysis (EIA) can be used to determine how a change in trade, regulation, or risk management affects the local, regional, national, or global economy. The economic impacts of trade decisions and risk management options usually include effects on jobs, incomes, prices, taxes, and possibly measures of economic welfare like consumer and producer surplus. Thus, economic impact analysis is intended to measure these types of economic effects associated either with the status quo or with particular risk management decisions and options in place. Cost–benefit analysis measures direct benefits and costs of a risk management option. It typically does not convert these direct effects into their indirect effects on the economy, such as changes in employment, wages, business sales, or land use. This is the role of EIA.

The most common forms of EIA trace spending through an economy and measure the cumulative effects of that spending in the impact region. The impacts can, for example, forecast the number of jobs created or lost by an infestation or a risk management option. Many economic impact models also predict impacts on personal income, business production, sales, profits and tax collections. It is not unusual for an EIA to show impacts on dozens of different sectors when it relies on input–output models for its analysis.

29.8.2. Cost-effectiveness analysis

Cost-effectiveness analysis (CEA) is concerned with achieving a fixed objective (the effect) with the least expenditure of resources (the cost). In a risk management context, it might mean finding the least costly way to obtain a given level of protection for domestic producers. For example, what is the least cost way of cutting a risk from high to low? Alternatively, it may be looking for the risk management option that provides the greatest risk reduction (effect) for a given expenditure (cost).

Cost-effectiveness relies on a measurable output or effect like the number of infested hectares reduced, pests intercepted, tons of product received, and the like. It also requires estimates of the costs of obtaining different levels of that output. CEA is most useful when the outcomes of a risk management option can be quantified but not readily or reliably monetized. So, it is a useful methodology for situations where it is unnecessary or impractical to consider the monetary value of benefits. For example, if each risk management measure produced exactly the same non-monetary benefits, it may be sufficient to identify the least costly of the risk management options.

A risk management option is cost-effective if it has the lowest cost of all alternatives that produce a given amount of benefits. It is the cheapest way of obtaining a specific and fixed outcome. When the levels of output vary, as they usually will, cost-effectiveness is usually determined based on the lowest cost per unit of output. Suppose a decision is made to permit trade, subject to achieving a low level of risk and there are several options for achieving that level of risk. The most cost-effective measure might be defined as lowest risk management and crop damage cost per ton of import.

29.8.3. Incremental cost analysis

When risk managers are not considering a fixed level of risk reduction or a fixed expenditure on risk reduction, incremental cost analysis (ICA) is an effective technique. Incremental cost is the cost of a little more (i.e., an increment) of an output. When the increment is one additional unit of output or an arbitrarily small increment of output, incremental cost is often called marginal cost. It can be used to compare different resource allocation options in like terms. The healthcare profession has used CEA and ICA to guide healthcare budget allocation decisions and they are routinely used in economic analyses,

Consider Fig. 29.4. The marginal damage associated with a risk of introduction is shown at the origin and it increases with the risk of introduction. The marginal cost of risk management is zero when the risk of introduction is high. The optimal level of risk management is reached where the marginal risk management cost (MC_{RM}) equals the marginal damage of introduction (MD_E) at Q^*, the cost-effective level of risk of introduction or, conversely, the cost-effective level of risk management.

The area under MC_{RM} from right to left up to Q^* is the total cost of risk management. The area under MD_E from the origin up to Q^* is the total damage cost. The sum of these two quantities is the cost-effective total for optimal risk based on incremental cost. Alternatively, one might speak of the incremental cost of reducing the risk of introduction from high to low or to keep P(Introduction) below some specific level, say 5%. A more practical example of this theory is presented in Table 29.8.

Consider the risk of hectares of production being infested by a pest. Imagine several independently implementable risk management measures have the costs and impacts shown

below. The incremental cost is defined as the change in total cost/change in output. In this instance the change in output is infested hectares prevented. The incremental costs (C15,000,000/ 650,000 hectares = C23 per hectare) show heat treatment is the best buy for C15 million, followed by chemical treatment at a cost of C30 million, now a cumulative cost of C45 million.

Risk managers must make at least a subjective judgment about the value of a hectare of infestation prevented. If its value exceeds C23 per hectare, use heat treatment; if its value exceeds C27 per hectare, use heat treatment, then use chemical treatment. When incremental

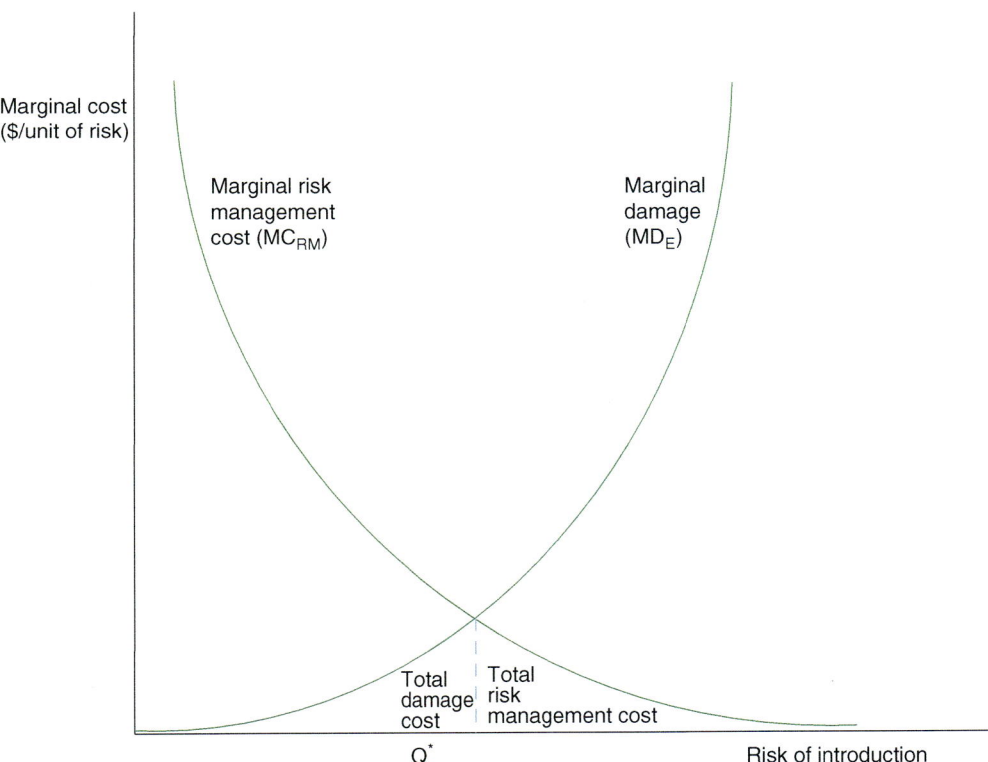

Fig. 29.4. Incremental cost of risk management.

Table 29.8. Hypothetical example of incremental cost analysis.

Risk management measure	Additional cost	Hectares of infestation prevented	Incremental cost of hectares prevented
Inspection	C10 million	250,000	C40
Chemical treatment	C30 million	1,100,000	C27
Heat treatment	C15 million	650,000	C23
Irradiation	C20 million	100,000	C200

costs take an order of magnitude jump, as they do for the irradiation option, this break point often signals when enough has been spent. The analysis here helps risk managers allocate a limited budget to risk management measures. Or it may suggest it is reasonably cost-effective to implement the first three measures to prevent infestation of 2,000,000 hectares at a cost of C55 million but it is not worth spending an extra C20 million for only 100,000 more hectares.

29.8.4. Cost–benefit analysis

Cost–benefit analysis or benefit–cost analysis compares the present value of all social benefits to the present value of all opportunity costs associated with a risk management option. If the net benefits (benefits – costs) of the risk management option are zero or greater the option is considered economically efficient. If there is more than one risk management option, then the greater the net benefits the better off society is, in economic terms, as a result of the option.

Sample CBA	
Costs	
Private compliance	C1,140
Public monitoring and enforcement	96
Total	C1,236
Benefits	
Consumer surplus	C1,896
Producer surplus	382
Total	C2,278
Net Benefits	C1,042
Benefit–cost Ratio	1.8

29.8.5. Damage assessment

Damage assessment is used to determine the monetary value of injury due to plant mortality, reduced yield and quality, as well as mitigation costs that result from the actions of others. This is simply estimating losses quantitatively then monetizing the value of that loss.

29.9. Types of Analysis

Four types of analysis seem to be most common in phytosanitary risk management economic consequence assessment. Three of these are identified in ISPM 11. They are partial budgeting, partial equilibrium analysis, and general equilibrium analysis. The fourth method that shows up in the literature is input–output analysis a type of EIA.

29.9.1. Partial budgeting

Partial budgeting is the most common type of analysis used to evaluate phytosanitary economic consequences. ISPM 11 says this will be adequate when the economic effects of a risk management activity are generally limited to producers and are relatively minor. Soliman *et al.* (2010) describe this as a marginal approach that shows the net increase or decrease in farm income due to a risk management decision, rather than the profit or the loss of a farm as a whole, thus, this is a form of EIA. The appeal of partial budgeting arises from its simplicity and transparency. Holland (2007) includes the limited data, time, and skill required to use it among its advantages. Partial budgeting tends to overlook effects on consumers and contributes to the asymmetric consideration of economic impacts described earlier in the chapter.

In a 2003 study, Macleod *et al.* conducted an ex-ante partial budget economic analysis of the introduction of *Thrips palmi* in England. The present value of ten years of negative economic impact of *T. palmi* is estimated to be between £16.9 and £19.6M, depending upon the rate of pest spread. Damages include yield and quality losses, additional research, plant health certification costs, and loss of exports. If export losses do not materialize, losses would drop to £0.6 and £3.3M over ten years. The benefit–cost ratios for eradicating the outbreak and maintaining an exclusion policy towards *T. palmi* range from 4:1 to 19:1 with no loss of exports, to a range from 95:1 to 110:1 with export losses. Impacts on consumers of the vegetable and ornamental plants that could be affected are not included.

Risk-based partial budgeting, which accounts for the uncertainty in budget inputs, may be a useful refinement. Fig. 29.5 provides a simple example. The number of hectares affected is uncertain and is represented by the pert

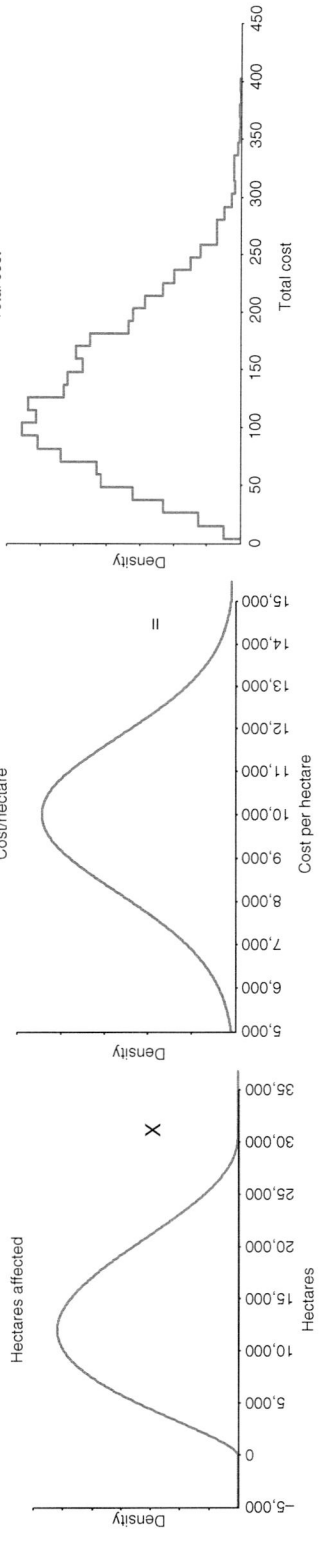

Fig. 29.5. Conceptual illustration of a risk-based partial budget calculation.

distribution on the left. The average amount of damage per hectare is also uncertain and is represented by the normal distribution in the middle. A simple simulation of this multiplication calculation yields an uncertain consequence represented by the distribution on the right. Simulation offers analysts a way to address many of the uncertainties they face in ex-ante partial budgeting estimates of economic consequences. Simulation also provides a step forward from the qualitative estimates that are so common.

29.9.2. Partial equilibrium analysis

Partial equilibrium analysis takes a market-based approach rather than a budget approach. It considers only part of the market, however, usually focusing on the equilibrium in a market or markets that are directly affected by a risk management decision. A partial equilibrium analysis can measure net welfare changes arising from the pest or risk management option impacts on producers and consumers. The examples shown in Figs 29.2 through 29.4 use partial equilibrium analysis.

Partial equilibrium can be used to evaluate the welfare effects on participants in a market that has been affected by a shock like a risk management intervention or the introduction of a pest. The approach requires functional relationships for supply and demand for the commodity of interest. Market equilibria before and after the shock can be analyzed to determine changes in total surplus, i.e., social welfare.

Partial equilibrium analysis is appropriate when pest or risk management intervention impacts are expected to change prices significantly. Extensive data are likely to be required to develop supply and demand functions. Although this approach is good for evaluating the effects on specific commodity markets, it is limited in its ability to account for economy-wide effects.

29.9.3. General equilibrium analysis

When the economic effects of infestation or risk management intervention are significant to the national economy, general equilibrium analysis is the desired analytical framework. General equilibrium analysis takes all markets into account and can consider changes to wages, interest rates, or exchange rates, for example. Sometimes called "computable general equilibrium analysis" (Soliman *et al.*, 2010), this methodology combines the strengths of input–output analyses and partial equilibrium models to answer a wide range of economic questions. General equilibrium models are complex, expensive to develop, and more difficult to run and interpret. These models are, therefore, only appropriate to address situations likely to generate measurable macroeconomic impacts.

29.9.4. Input–output analysis

Input–output (I–O) analysis is ideal for identifying economic impacts on economic flows like employment, income, sales, tax revenues, and the like. I–O analysis requires a model that describes the monetary flows of inputs and outputs among the productive sectors of an economy. A simple example is provided in Table 29.9.

The rows of the table show total output of a specific sector sold to other sectors, i.e. to intermediate demand, or to final demand. Thus, agriculture sells 80 of its 470 total output back to agriculture and it sells 30 to transport. A change in final demand for the products of a sector, as a result of a risk management intervention or infestation, generates direct and indirect effects on the economy as a whole. A change creates a large initial effect by causing a direct change in the purchasing patterns of the affected sector. Then suppliers of the affected sector will alter

Table 29.9. Soliman *et al.*'s hypothetical I–O table indicating monetary flows of an economy.

Selling sector	Agriculture	Industry	Transport	Final demand	Total output
Agriculture	80	300	30	60	470
Industry	120	500	150	350	1120
Transport	40	200	10	10	260
Value-added	25	120	15	100	260
Total Input	265	1120	205	520	2110

their purchasing patterns to meet the demand changes that result from the initial change. This creates a smaller second round, and this pattern repeats itself until the adjustments are exhausted. Multiplier effects generated in this way characterize the impact of the initial sector change on the entire economy.

I–O analysis does a great job of capturing spillover effects among economic sectors. However, I–O models only account for changes in the economy due to shifts in demand. Supply is usually assumed to be perfectly elastic. Soliman *et al.* (2010) argue that real supply constraints can cause I–O models to miss important effects of a pest introduction. Also, I–O models cannot account for price changes or changes in the structure of a sector over time because the model structure assumes fixed prices, no substitution between inputs, and constant returns to scale. This limits the applicability of I–O analysis for pest infestation problems.

29.9.5. Which analysis type should be used?

Soliman *et al.* (2010) consider this question. Partial budgeting is a basic and easily understood technique that is good for assessing direct impacts. Its scope is limited, however, and it does not include the effects of pest damage on market prices, supply, and demand. It also fails to address spillover effects to other sectors of the economy. Partial equilibrium and general equilibrium broaden the scope of the analysis to include those price effects. Partial equilibrium considers the affected commodity market only. General equilibrium considers the whole economy. I–O analysis occupies a position between partial budgeting and general equilibrium.

Soliman *et al.* conclude that despite its limitations, partial budgeting is the default method

of choice for most risk assessments. It provides insight into the most immediate impacts of the pest. It is also easily understood and explained. Perhaps, most practically, the required data can often be obtained with reasonable accuracy and modest effort. It is, however, flawed by its limited focus on domestic producers. Significantly, the results of partial budgeting evaluations are necessary inputs for the other techniques. Partial equilibrium modeling is preferred when the changes in production volumes are large enough to affect prices.

29.10. Summary and Look Forward

Here are five things to remember from this chapter.

1. The SPS Agreement effectively limits economic analysis to the consideration of damages sustained by domestic producers and the costs of risk management in risk assessment.

2. ISPM guidance tends to favor expert judgment, which leads to qualitative economic consequence assessments that are difficult to reproduce.

3. Economic assessments are challenged by the preference for qualitative assessments, the difficulty of ex-ante estimates, ignoring impacts on consumers, and significant amounts of uncertainty.

4. Five useful economic tools are: economic impact analysis, cost-effectiveness analysis, incremental cost analysis, cost–benefit analysis, and damage assessment.

5. Partial budgeting that focuses on the direct impacts on domestic producers is a reasonable level of economic consequence assessment at this point in time.

Knowledge management, an efficient strategy for reducing uncertainty, is the subject of the next chapter.

29.11. References

Arthur, M. (2004) *An Economic Analysis of Quarantine: The Economics of Australia's Ban on New Zealand Apple Imports*. 2006 Conference New Zealand Agricultural and Resource Economics Society, Nelson, New Zealand. 31959. Available at: http://ageconsearch.umn.edu/bitstream/137985/2/2006_arthur.pdf (accessed February 3, 2017).

Beghin, J.C. and Burea., J-C. (2001) *Quantification of Sanitary, Phytosanitary and Technical Barriers to Trade for Trade Policy Analysis*, working paper 01-WP 291, Centre for Agriculture and Rural Development, Iowa State University.

Breukers, A., Mourits, M., van der Werf, W. and Oude Lansink, A. (2008) Costs and benefits of controlling quarantine diseases: a bio-economic modeling approach. *Agricultural Economics* 38, 137–149.

Brunel, S., Petter, F., Fernandez-Galiano, E. and Smith, I. (2009) Approach of the European and Mediterranean plant protection organization to the evaluation and management of risks presented by invasive alien plants, in Inderjit (ed.) *Management of Invasive Weeds*. Springer, Netherlands, 319–343.

Holland, J. (2007) *Tools for Institutional, Political, and Social Analysis of Policy Reform. A Source Book for Development Practitioners*. The World Bank and Oxford University Press, Washington, District of Columbia, USA.

Howe, C. (1971) *Benefit–cost Analysis for Water System Planning*. American Geophysical Union, Washington, District of Columbia, USA.

Jouanjean, M-A., Maur, J-C. and Shepherd, B. (2016) US phytosanitary restrictions: the forgotten non-tariff barrier. *Journal of International Trade Law and Policy*, 15, 1, 2–27.

Macleod, A., Head, J. and Gaunt, A. (2003) The assessment of the potential economic impact of Thrips palmi on horticulture in England and the significance of a successful eradication campaign. *Crop Protection* 23, 601–610.

Maskus, K., Otsuki, J. and Wilson, T. (n.d.) Quantifying the impact of technical barriers to trade: a framework for analysis. Paper presented at World Bank workshop, "Quantifying the Trade Effects of Technical Barriers and Standards: Is it Possible?" World Bank, Geneva.

Murina, M. and Nicita, A. (2017) Trading with conditions: the effect of sanitary and phytosanitary measures on the agricultural exports from low-income countries. *The World Economy*, doi 10.1111/twec.12368.

Roberts, D., Josling and Orden, D. (1999) *A framework for analysing technical barriers in agricultural markets*. Technical Bulletin No. 1876, U.S. Department of Agriculture, Economic Research Service, Washington D.C.

Sansford, C. (2002) *Quantitative versus qualitative: pest risk analysis in the UK and Europe including the European and Mediterranean Plant Protection (EPPO) System*. NAPPO International Symposium on Pest Risk Analysis Puerto Vallarta, Mexico, 2002.

Soliman T., Mourits M.C.M., Oude Lansink, A.G.J.M. and van der Werf, W. (2010) Economic impact assessment in pest risk analysis. *Crop Protection* 29, 517–524

Soliman, T., Mourits, M.C.M., van der Werf, W., Hengeveld, G.M., Robinet, C. *et al.* (2012) Framework for modelling economic impacts of invasive species, applied to pine wood nematode in Europe. *PLoS ONE* 7(9): e45505, doi: 10.1371/journal.pone.0045505.

U.S. Interagency Committee on Water Resources (1950) *Proposed practices for economic analysis of river basin projects (the "Green Book")*. Washington, District of Columbia, USA.

Wilson, N.L.W. and Antón, J. (2006) Combining risk assessment and economics in managing a sanitary–phytosanitary risk, *American Journal of Agricultural Economics* 88(1), 194–202.

WTO (World Trade Organization) (1995) *Legal texts: results of the Uruguay round of multilateral trade negotiations*. Cambridge University Press, Cambridge.

30

Knowledge Management

_____

Any fool can know. The point is to understand.

Albert Einstein

30.1. Introduction

Risk analysis is decision-making under uncertainty. Risk assessment strives to gather the best available scientific evidence to objectively describe existing risks so they can be managed efficaciously. Knowledge reduces uncertainty. Scientific evidence is the foundation of much knowledge. Knowledge is, therefore, a commodity that is crucial to successful phytosanitary risk management. National plant protection organizations (NPPOs) are knowledge-based organizations, they learn, remember, and act based on the best available information, knowledge, and practical experience. Knowledge work calls for more collaboration than other forms of work do. Knowledge is a commodity with some characteristics that make it different from other valuable commodities. Dakir (2011) says these characteristics include the following:

- using knowledge does not consume it
- transferring knowledge does not result in losing it
- knowledge is abundant, but the ability to use it is scarce
- much of an organization's valuable knowledge walks out the door at the end of the day.

NPPOs' reliance on knowledge has created an opportunity for an intentional and systematic approach to cultivating and sharing the world's phytosanitary knowledge base. One which can be bolstered by the valid and valuable lessons and best practices learned around the globe. To be successful in an increasingly global environment, NPPOs need to learn from their past errors as well as the successes and errors of others.

"Knowledge management is the process of capturing, distributing, and effectively using knowledge" (Davenport, 1994). Knowledge management is a deliberate and systematic approach to ensure full utilization of an organization's knowledge base, coupled with the potential of individual skills, competencies, thoughts, innovations, and ideas to create a more efficient and effective organization (Dakir, 2011).

This chapter explores the potential for knowledge management to support and extend the effectiveness of phytosanitary risk management globally, as a strategy for reducing uncertainty and for making better evidence-based decisions. It proceeds by first considering the nature of knowledge as a precursor to an extended description of knowledge management. The chapter then turns to the potential benefits of a knowledge management system for NPPOs individually and as a loose confederation of stakeholders.

30.2. What is Knowledge?

Fig. 30.1 presents a data-wisdom pyramid adapted from Ackoff (1989). Data consist of facts, figures,

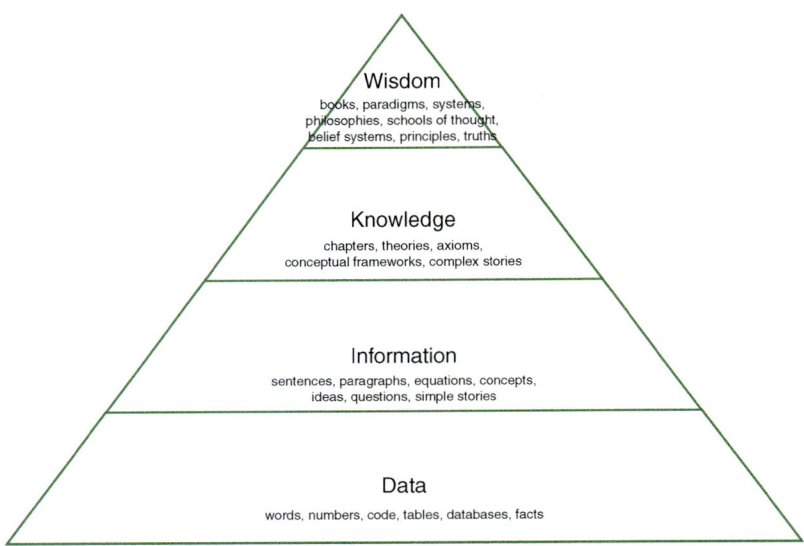

Fig. 30.1. A data, information, knowledge, wisdom pyramid based on Russell Ackoff's article (1989).

and symbols without any context. The number 125 is a piece of data, as is the name Tu Nguyen. Without anything else to define them, these two items of data are meaningless. Information is data that have been organized or processed in a useful way. It provides answers to "who," "what," "where," and "when" questions. So, pieces of information are "125 consignments of star fruit" and "Tu Nguyen is a CEO." Applied information provides knowledge which answers "how" questions. Knowledge gives us context and leads to understanding, i.e., an appreciation of the "why." "Tu Nguyen is the CEO of a Malaysian company that seeks to ship 125 consignments of star fruit annually." Knowledge gives us the power to take action. Finally, there is wisdom, an evaluated understanding. Wisdom helps us take the right action.

Three different types of knowledge are often articulated, they are explicit, implicit, and tacit knowledge. Koenig (2012) offers these definitions:

- *explicit:* knowledge that is set out in tangible form
- *implicit:* knowledge that is not set out in tangible form but could be made explicit
- *tacit:* knowledge that one would have extreme difficulty operationally setting out in tangible form.

Tacit knowledge is difficult to articulate and difficult to put into words, text, or drawings. It is knowledge that is in people's heads. Explicit knowledge represents content that has been captured in some tangible form such as words, audio recordings, or images. Implicit knowledge occupies an area between these two types. Implicit knowledge is in people's heads but it could readily be reduced to a tangible form.

The International Standards for Phytosanitary Measures (ISPMs) of the International Plant Protection Convention (IPPC) are examples of explicit knowledge. Implicit knowledge might include the manner in which the risk assessment and risk management tasks are divided by an NPPO that has not codified this division of labor. An example of tacit knowledge might be how a seasoned inspector knows where and what to look for when inspecting a consignment. The knowledge held by employees is one of the greatest assets of any NPPO. Dakir (2011) summarizes some of the major properties of explicit and tacit knowledge in Table 30.1.

30.3. Knowledge Management

So, what does knowledge management entail? It makes the organization's data and information available to the members of the organization through portals and with the use of content

Table 30.1. Properties of tacit and explicit knowledge from Dakir (2011).

Properties of tacit knowledge	Properties of explicit knowledge
Ability to adapt, to deal with new and exceptional situations	Ability to disseminate, to reproduce, to access and reapply throughout the organization
Expertise, know-how, know-why, and care-why	Ability to teach, to train
Ability to collaborate, to share a vision, to transmit a culture	Ability to organize, to systematize, to translate a vision into a mission statement, into operational guidelines
Coaching and mentoring to transfer experiential knowledge on a one-to-one, face-to-face basis	Transfer knowledge via products, services, and documented processes

management systems. Ruggles and Holtshouse (1999) identified the following key attributes of knowledge management:

- generating new knowledge
- accessing valuable knowledge from outside sources
- using accessible knowledge in decision-making
- embedding knowledge in processes, products, and/or services
- representing knowledge in documents, databases, and software
- facilitating knowledge growth through culture and incentives
- transferring existing knowledge into other parts of the organization
- measuring the value of knowledge assets and/or impact of knowledge management.

Koenig (2012) identifies three stages to knowledge management development. Stage 1 of knowledge management relies heavily on information technology. This is a "if we only knew what we know" stage. This includes the ability to access archived information, to carefully track and share information about ongoing cases, as well as developing databases based on risk monitoring activities, for example sampling results and interception data. This is a simple matter of identifying and gathering the intellectual capital of the NPPO.

Stage 2 arose in recognition of the fact that simply deploying new technology was not enough to effectively enable information and knowledge-sharing. Knowledge management implementation involves changes in the corporate culture, sometimes significant changes. Knowledge is power in many entities. Changes in corporate culture need to facilitate and encourage

information and knowledge-sharing can be major and profound. Two major themes from the business literature have emerged in this culture change stage of knowledge management. Senge's work on the learning organization (Senge, 1990) was first, followed by Nonaka and Takeuchi's (1995) work on how to discover and cultivate "tacit" knowledge. Both of these themes encompass the human factors encountered in implementing and using knowledge management as well as knowledge-sharing and communication. Communities of practice are a hallmark of this stage of development.

Koenig's third stage of knowledge management developed in particular from the awareness of the importance of the retrievability of content. This made the arrangement, description, and structure of content focal points. Knowledge is no good if people try to use it but can't find it. For the first time, taxonomies and content management emerged full blown as major topics.

Thus, learning from history, in order to benefit from knowledge management, NPPOs need to develop their IT capabilities, nurture the human and corporate culture that will be affected, and then turn their attention to taxonomy and content. These will be significant challenges for any NPPO, but individually successful NPPOs will be the best argument for extending knowledge management across NPPOs. NPPOs rely largely on scientific knowledge to identify and manage plant pests. They share a fundamental set of goals and are already loosely united by their allegiances to the Agreement on the Application of Sanitary and Phytosanitary Measures (SPS) and the IPPC. Even the unique value judgments considered in risk management decisions would be aided by a common knowledge base.

Ten basic rules

1. To be effective, knowledge management must address three essential components—people, process, and technology.
2. Knowledge management systems must engage every department in the organization.
3. An organization should first conduct a needs assessment to determine the goals, scope, and requirements of any knowledge management activities it is considering.
4. Before creating knowledge management systems, the organization should identify existing formal and informal knowledge sources.
5. The design of a knowledge management system should be based on desired results and on the organization's culture and expertise.
6. Effective knowledge management systems enable the organization to share its knowledge and discover inconsistencies in that knowledge.
7. Knowledge must be continually updated, revised, and built on to maintain a valid base.
8. Knowledge management systems must use the principles they encourage—sharing, cooperation, and continual growth.
9. No single knowledge management system will work for every organization.
10. Ultimately, knowledge management is not an exercise in managing. It is all about communicating existing knowledge and expertise so that others can continue to build on that knowledge.

(Atwood, 2009)

H.G. Wells (1938) provided a model for such collaboration. He called it "The World Brain" and said it would represent "a universal organization and clarification of knowledge and ideas." He anticipated the need for knowledge management, and some say the World Wide Web, when he spoke of "this wide gap between ... at present unassembled and unexploited best thought and knowledge in the world ... we live in a world of unused and misapplied knowledge and skill." It is exciting to anticipate how a collaboration of NPPOs could be one of the first global stewardships to make this World Brain a reality. Koenig (2012) describes three practical undertakings that are the essence of knowledge management. They are:

1. Lessons learned databases.
2. Expertise locations.
3. Communities of practice (CoPs).

Lessons learned databases attempt to capture and make accessible knowledge that has been operationally obtained but that typically would not have been captured in a fixed medium. This means capturing knowledge embedded in persons and making it explicit. The lessons learned concept broadens the notion of "best practices," which can be interpreted as meaning there is only one best practice in a situation. Furthermore, a best practice in North American culture might well not be a best practice in another

culture. "Lessons learned," what the military calls "after action reports," have become the most common hallmark phrase of early knowledge management development.

NPPOs rarely find or take the luxury of time to debrief a risk management action before the project team is disbanded and the team members are reassigned to new pressing priorities. Organizations, like NPPOs, that operate in a project team milieu need to pay very close attention to after action reports to harvest the lessons learned.

Koenig (2012) aptly describes the political and operational complexity of such a practice. Many of the questions about how such a system works are difficult to answer. Who decides what is or is not a worthwhile lesson learned? Is the system monitored or not? How long do items stay in the system? He finds most successful lessons learned systems have an active weeding or stratification process. This is necessary to keep the system new and fresh so usage and utility do not fall. By stratifying and cataloguing, items removed from the foreground can be archived and moved to the background but still made available.

An expertise location system is another important component of a knowledge management system. If knowledge resides in people, it is essential to enable employees to talk to those knowledgeable people to learn what the expert knows. Finding the right expert with the knowledge you need, when you need it, can be a problem. The basic purpose of an expertise locator system is to identify and locate those persons within an organization who have expertise in a particular area.

Three sources of data for an expertise locator system are identified by Koenig (2012). They are employee resumes, employee self-identification of areas of expertise, or by algorithmic analysis of electronic communications from and to the employee. Resumes are rather straightforward, providing access to education, experience, and publication records. Academia commonly uses the practice of annually updated resumes to maintain a current and up-to-date record. Employees can often self-identify as experts by filling out an online survey. Larger organizations may develop their own process for designating individuals as subject matter experts. These systems sometimes use an assessment checklist to qualify individuals. Email analysis typically monitors the subject matter of email and social networking electronic communications.

In order to prevent the burden of "consulting" from becoming too great for any given expert, commercially available software can be used to match queries for help with the experts who are available. This software has load-balancing schemes to avoid overloading any particular expert. These systems can also rank the degree of presumed expertise and will shift a request for help down the expertise ranking before the most expert personnel start to become overloaded. Requests can be flagged as a priority and the system can match higher priority requests with people of higher expertise rank.

The third element is CoPs. CoPs are defined as groups of individuals with shared interests that come together in person or virtually to tell stories, to share and discuss problems and opportunities, discuss best practices, and talk over lessons learned (Wenger, 1998; Wenger and Snyder, 1999). They require a domain, a community, and a practice. The social nature of learning within or across organizations is emphasized in these CoPs. It is challenging to replicate conversations around the water cooler in geographically distributed organizations or when people work online from home. The natural knowledge-sharing that occurs in social spaces can sometimes be replicated virtually in CoPs.

Hypothetical CoP interactions

Solving problems
"Can we work on this systems approach and brainstorm some ideas; I'm stuck."
Requests for information
"Where can I find the code to connect to the server?"
Seeking experience
"Has anyone dealt with this pest on this commodity before?"
Reusing assets
"We have a risk assessment we did just last year. I can send it to you and perhaps you can use some of it."
Coordination and strategy
"Can we coordinate our risk management measures to achieve better results?"
Building an argument
"How do people in other countries regulate this pest? This information will make it easier to convince our Ministry to make some changes."
Growing confidence
"Before we finalize these risk management measures, let me run it through the CoP first to see what they think."
Discussing developments
"What do you think of the new eradication treatment? Does it really help?"
Documenting projects
"Six countries have faced this problem, let's document their approaches."
Visits
"Can we come and observe your systems approach? We need to establish one in our country."
Mapping knowledge and identifying gaps
"Who already knows something about this pest and what are we missing? What other groups should we connect with?"

CoPs formed across NPPOs provide a rich variety of opportunities for risk assessors and risk managers alike. Organizing and maintaining a CoP is not easy. CoPs develop shared resources, experiences, stories, tools, ways of addressing recurring problems via websites, social media, and occasional face-to-face meetings. A good CoP is a shared practice. This takes time and sustained interaction.

30.4. Benefits of Knowledge Management

Knowledge management offers benefits to individuals, communities of practice, and organizations (Dakir, 2011; Garfield, 2014). Knowledge management helps the individual:

- do their job better and be more productive
- save time through better decision-making and problem-solving
- find information and relevant resources more easily
- build a sense of community bonds within the organization
- keep up to date
- makes scarce expertise widely available
- by providing challenges and opportunities to contribute.

Knowledge management helps CoPs:

- develop professional skills
- promote peer-to-peer mentoring
- avoid making the same mistake twice
- avoid redundant effort
- facilitate more effective networking and collaboration
- develop a professional code of ethics to which members can adhere
- provide methods, tools, templates, techniques, and examples
- develop a common language.

Knowledge management helps an organization:

- share information easily
- reuse ideas, documents and expertise
- take advantage of existing expertise and experience
- communicate important information widely and quickly
- drive strategy
- solve problems quickly
- diffuse best practices
- promote standard, repeatable processes and procedures
- improve knowledge embedded in products and services
- cross-fertilize ideas and increase opportunities for innovation
- share ideas to improve customer relationships
- better stay ahead of the competition
- build organizational memory
- accelerate delivery to customers
- show customers how knowledge is used for their benefit.

30.5. Summary and Look Forward

Here are five things to remember from this chapter.

1. Knowledge management enhances the evidence base of phytosanitary risk management.
2. Knowledge management is an effective aid to reducing uncertainty in phytosanitary risk management.
3. NPPOs possess explicit, implicit, and tacit knowledge, all of which can be harnessed by a knowledge management system.
4. Lessons learned databases, expertise locations, and CoPs are three practical knowledge management ideas for NPPOs.
5. Knowledge management benefits individuals, communities of practice, and organizations of all sizes.

The next chapter examines some of the unique phytosanitary risk management issues encountered when considering commodities for consumption.

30.6. References

Ackoff, R.L. (1989) From data to wisdom. *Journal of Applied Systems Analysis* 15, 3–9.
Atwood, C.G. (2009) *Knowledge Management Basics*. American Society for Training & Development, Alexandria, VA.
Dakir, K. (2011) *Knowledge Management in Theory and Practice*, 2nd edn. MIT Press, Cambridge, MA.

Davenport, T.H. (1994) Saving IT's soul: human centered information management. *Harvard Business Review, March–April*, 72(2), 119–130.

Garfield, S. (2014) 15 Knowledge management benefits. *LinkedIn*. Available at: https://www.linkedin.com/pulse/20140811204044-2500783-15-knowledge-management-benefits (accessed February 3, 2017).

Koenig, M.E.D. (2012) *What is KM?* Knowledge management explained. *KMWorld*. Available at: http://www.kmworld.com/Articles/Editorial/What-Is-.../What-is-KM-Knowledge-Management-Explained-82405.aspx (accessed February 3, 2017).

Nonaka, I. and Takeuchi, H. (1995) *The Knowledge Creating Company: How Japanese Companies Create the Dynamics of Innovation*. Oxford University Press, New York, New York.

Ruggles, R. and Holtshouse, D. (1999) *The Knowledge Advantage*. Capstone Publishers, Dover, New Hampshire, USA.

Senge, P.M. (1990) *The Fifth Discipline: The Art & Practice of the Learning Organization*. Doubleday Currency, New York, New York.

Wells, H.G. (1938) *World Brain*. Doubleday, Doran & Co. Garden City, New York, USA.

Wenger, E.C. (1998) *Communities of Practice: Learning, Meaning, and Identity*. Cambridge University Press, Cambridge, MA.

Wenger, E.C. and Snyder, W.M. (1999) Communities of practice: The organizational frontier. *Harvard Business Review*, 78(1), 139–145.

31

Commodities for Consumption

Trade has not ruined any nation yet.

Benjamin Franklin

31.1. Introduction

The International Plant Protection Convention (IPPC) as well as the laws and regulations of national plant protection organizations (NPPOs) refer to plants, plant products, and other regulated articles when describing the scope and objective of phytosanitary authority. This includes a wide range of items. Some of these items, such as empty shipping containers or the superstructure of vessels, may have no direct relationship to agricultural products but have demonstrated the potential to present a threat for pest introduction. Others, such as plants for planting and fresh fruit, are clearly agricultural products that we know can serve as a pathway for introducing harmful pests. This chapter examines the unique aspects of risk management for those items that are identified as commodities for consumption.

31.2. What is a Commodity for Consumption?

The primary pathways for pests of phytosanitary concern are commodities, passengers, and conveyances, as seen in Figure 31.1. Commodities are the agricultural goods or materials bought and sold as an article of commerce. They may be raw, unprocessed products such as grain or fresh fruits. They may also be products that have received various levels of treatment or processing such as frozen peas or lumber. The primary characteristics that identify a commodity for phytosanitary purposes is that it is an agricultural product with a consistent, identifiable nature that can be collected into a consignment and sold.

Commodity: A type of plant, plant product, or other article being moved for trade or other purpose.

(ISPM No 5, IPPC, 2019)

The IPPC definition (see box) is intentionally broad, however, to include items that may not be agricultural but are associated with agricultural pests. Brassware from areas where Khapra beetle occurs is an example. Brassware is obviously not an agricultural product but Khapra beetle, a serious pest of stored grain, has a high affinity for this product and especially the packaging associated with it, which is why brassware is subject to phytosanitary regulations in many countries.

Are fresh blueberries a commodity for consumption? Yes. Is blueberry jam? Juice? Also, yes. The difference from a phytosanitary standpoint is that jam and juice are not likely to be regulated products under phytosanitary authority because they are processed, whereas fresh blueberries

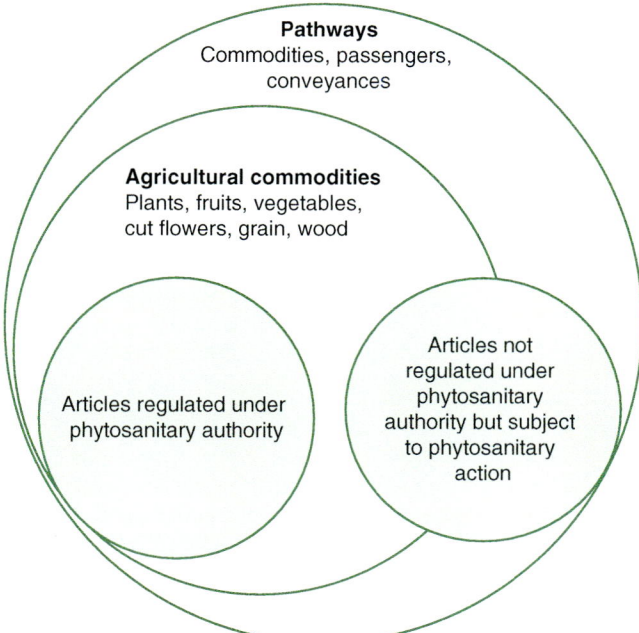

Fig. 31.1. Phytosanitary action status of agricultural commodities and other pathways.

would be. Thus, not all agricultural commodities are subject to phytosanitary regulation.

Fig. 31.1 summarizes this concept of a commodity for phytosanitary risk management. Some, but not all, commodities for trade are agricultural commodities. Some, but not all, of the agricultural commodities are regulated under phytosanitary authority. Commodities with specific entry requirements are usually regulated. All agricultural commodities may not be specifically addressed in regulations although all will be subject to regulatory action based on inspection.

31.3. How Does a Commodity Enter a Country?

For Customs purposes, consumption raises the question of how a commodity legally enters a country. Consumption, in this context, refers to a specific classification for an entry submitted to Customs for commercial, business, or personal purposes without any restrictions of time or use. Table 31.1 describes different types of consumption entries identified by U.S. Customs.

The type of entry has no bearing on the phytosanitary status of a consignment except to

the extent that it may affect Customs procedures that impact phytosanitary clearance processes. The significance of this for risk managers depends on understanding Customs procedures for the different entry types and the relationship of Customs procedures to phytosanitary clearance procedures. Notice for instance that informal entries have fewer Customs requirements except for those identified as high risk. While Customs deals with many types of risk, the question of *pest risk* is determined by phytosanitary risk managers and should be an important factor in Customs' determination.

The recently implemented World Health Organization (WTO) *Trade Facilitation Agreement* describes obligations for WTO Members on a level with the SPS (Agreement on the Application of Sanitary and Phytosanitary Measures) (WTO, 2017). The *Trade Facilitation Agreement* largely addresses Customs procedures in its aim to simplify clearance processes with concepts such as "the single window" and electronic entries. There is no mention of plant health or agriculture in the agreement, but the upshot of its implementation is that Customs is ultimately responsible for every border process. Under the agreement, all border agencies are bound to work through their

Table 31.1. Types of consumption entries (adapted from Customs and Border Protection (United States Food and Drug Administration, FDA, 2020).

Consumption entry type	Basic entry information
Mail	Importation via U.S. Postal Service (commercial senders) is subject to the same requirements and restrictions as importation by any other entry types.
Informal entry	Informal entries, as defined by CBP regulations, are usually valued at less than $2,500 (value subject to change), and usually do not require a bond. Some products are restricted from informal entry (e.g. high risk products), regardless of value.
Formal entry	Formal entries, as defined by CBP regulations, generally have an aggregate value of $2,500 or more and are required to be covered by a bond.

national Customs organization, including phytosanitary authorities. This means that requirements and processes applied at the border cannot be created or maintained by phytosanitary authorities except through collaboration with their national Customs organization. It is therefore essential for phytosanitary risk managers to build a strong and effective working relationship with Customs in order to effectively communicate plant health risk in the context of all other risks faced by border agencies, including bombs, drugs, contaminants, and the like, in order to effectively implement phytosanitary requirements at the border.

What is a single window?

The single window is a facility that allows parties involved in trade and transport to lodge standardized information and documents with a single entry point to fulfill all import, export, and transit-related regulatory requirements. If information is electronic then individual data elements should only be submitted once.

31.4. Commodity Identity

Every regulated commodity has an identity and corresponding requirements. From Customs' standpoint, this may involve tariffs, quotas, or any of a range of requirements put in place by other border agencies. From a phytosanitary standpoint, the identity of the commodity is crucial to not only the border clearance process, but the entire authorization process, beginning with the initial request for market access.

As an extreme example to demonstrate the point, consider the difference between a raw log and finished lumber. Both are wood, but the pest risk for each is vastly different. The import of raw logs would require strong risk management measures for the plethora of pests found associated with the trees in the forest, including those pests known to attack the tree as well as the many contaminant pests that are likely to be associated with an unprocessed forest product. Finished lumber, on the other hand, may be from the same forest and tree but we would expect far fewer pests and thus fewer and less stringent requirements because we know that processes have occurred which distinguish a raw log from finished lumber. It is illogical and a waste of resources to evaluate for the risk of raw logs when we know the commodity is finished lumber. Likewise, it is not legitimate for risk managers to subject lumber to the measures designed for raw logs.

The log–lumber comparison helps us to reflect on the importance of identity even if it is unlikely that risk managers struggle with such a clear situation. In practice, the questions around the identity of a commodity are much more subtle. For instance, the fresh bananas of commerce are always covered in cultivation. This is to preserve their clean appearance, to thwart many pests, and to reduce physical damage that commonly occurs to uncovered bananas. There is a clear difference in pest risk, so the question becomes whether the commodity is identified as simply "bananas," or does it need to be specifically identified as "bananas grown under covers in commercial plantations"?

Commercially produced mangos are always washed because they get stained with sap during harvest and the sap discolors the skin of the fruit. The same problem exists: is this simply "mango," or must they be identified as "washed mangos?" Tomatoes are even more complicated. Tomatoes may be imported with or without stems. This implies completely different pests with stems/leaves; unripe tomatoes that are not pink or red, or a specific variety of tomato, would also have different pests. Imagine an authorization process for blueberries based on the assumption that the commodity is fresh when the actual import is frozen. There are completely different risks, so proper commodity identification is crucial to the success of risk management.

The key to avoiding problems with the identity of the commodity is to be as specific as necessary to distinguish the commodity from similar commodities with a different pest risk. This identity follows the commodity through the market access request, the pest risk analysis, the risk management design, and finally the border requirements. In some cases, such as during the pest listing process in risk analysis, it may be necessary to reinforce the identity by ensuring that the pest list is limited to pests that can be associated with the commodity as identified. Otherwise, valuable resources are wasted on analysis and measures that are unnecessary and unjustified.

31.4.1. Commercial versus non-commercial

We have all seen dooryard fruits and vegetables and may have compared them to their "store bought" cousins; typically reflecting on how much better the fresh picked fruit from grandma's tree always tastes. But of course, grandma does not supply tons of fruit to foreign markets unless she has many more trees and a commercial production operation. One of the central tenets of commodity identity is distinguishing commercial and non-commercial sources. Ignoring this distinction leaves open the possibility that commodities from both dooryards and commercial operations may be imported under the same conditions; essentially assuming them to be equivalent risks. For some commodities this may indeed be the case. Ginger root, for instance, is imported from many countries without requirements other than being subject to inspection. Lots of ginger from grandma's garden is not a problem.

For many other commodities, however, it is extremely important to make the distinction which brings with it certain assumptions about cultivation, harvest, washing, culling, and packing that redefine the pest risk for the commodity. Bananas and mangos are good examples of commodities for which commercial production practices play an important part in pest risk management (see Chapter 24, this volume).

31.4.2. Perishable versus non-perishable

Agricultural commodities have a shelf-life. Commodities such as fresh fruits and vegetables, cut flowers, and plants require special handling and rapid transport to ensure their suitability for the market. Grains, nuts, and wood are examples of commodities that are far less perishable and require less care. The difference is important to risk managers for two reasons. First, fresher commodities generally have fresher pests. Second, the potential for ruining a consignment is much greater for highly perishable commodities. This means that the application of measures and the clearance process must be timely and appropriately gentle to avoid undue delay and damage. A shipment of wood may be held for a few days for inspection and treatment. A shipment of fresh flowers is likely to spoil under the same circumstances.

Perishable commodities, because they are sensitive, also tend to have minimal handling and processing. It is easy to understand why the washing and culling process for cut flowers must be much less rigorous than for mangos. As a result, some pests that might be removed in the mango packing process may be missed in the cut flower packing process. Rapid harvesting, packing, and shipping assure that pests escaping the process will quickly arrive at their new destination with their host in good condition. For commodities like cut flowers that are not consumed as food, the host may be discarded into the environment along with its pests, which are soon looking for new hosts. Risk managers need to take account of these important differences when deciding the requirements for perishable commodities.

Thus, in addition to the risk mitigation aspect, there is also a practical aspect of risk management that needs to account for the perishability and durability of the commodity to ensure that the selected measures result in effective protection while also preserving marketability of the commodity. Producers play a critical role in this process by supplying clean and high-quality products. There is also a role for the exporting NPPO in providing feedback to the producers, packers, and shippers on pests of concern and processes that increase the exportability of their products. Finally, it is incumbent on the NPPO of the importing country to provide timely, accurate, and complete information to the exporting country on problems encountered upon import. All this collaboration helps risk managers on both sides of the trade to better understand and address the pest issues that make trade safe and build confidence between trading partners.

31.4.3. Processing

Wheat grain is obviously very different than flour, and fresh blueberries are different than frozen blueberries, but what about raw versus peeled carrots? Both are fresh carrots, but the pest risk associated with peeled carrots is likely to be much less due to the peeling and additional handling required for peeling. How about chopped fresh mango, blueberries in a plastic clamshell, or mixed fresh fruit in prepared fruit salads?

Beyond the normal cultivation, harvesting, and packing processes associated with producing the commodity for commerce, there may also be processing required to transform the commodity into the desired product. Processing can range from dramatic change, as in making flour from grain, to subtle change such as careful selection and packing for premium blueberries in a plastic clamshell. ISPM 32, *Categorization of commodities according to their pest risk* (IPPC, 2016) is a helpful reference for risk managers to understand the relationship of processing to pest risk (See Chapter 24, this volume).

31.5. Packaging

Until a half-century ago, most agricultural commodities were shipped bulk. Today nearly all agricultural commodities except grain are containerized in one form or another, which also means that they are palletized and therefore must be in boxes, bags, or other packages that can be stacked. While greatly simplifying the shipping process, these innovations also raise different challenges for risk managers and inspectors. Commodities in packages can be more difficult to inspect because packages must be taken from the pallets and opened, which destroys the integrity of the pallet stack and can make some cargo unsellable. Randomized inspections may require complete devanning of the consignment from the truck or container to provide access to all the cargo. This requires valuable time and effort that increases shipping costs and delays movement. Finally, the conveyance itself represents an additional risk if it is found contaminated. These factors are part of the calculus for risk managers on both sides of the trade and require a common understanding of phytosanitary requirements and operational processes in place at the border.

Shippers generally understand and expect some loss of product and time from the border clearance process. The risk however can be unpredictable, ranging from zero to 100% depending on many factors but mainly the results of inspections. Commodities identified as high risk are likely to have more rigorous inspections. For this reason, it is important for exporters to establish and maintain a high level of vigilance regarding the phytosanitary status of commodities for a good phytosanitary track record. It also requires some insight regarding the inspection process. (See Chapter 20, this volume).

For some commodities, preclearance inspection can be a good option to expedite the clearance process. For instance, the Netherlands has a long history of cooperating with the United States on the inspection of plant bulbs prior to export. The cost of such offshore inspection is offset by exporters who benefit from expedited clearance at the border and the ability to ship high volumes of small, retail-ready packages with much lower likelihood of damage or delay.

31.6. Regulated Commodities

Different regulatory designs exist for authorizing the import of commodities. The two primary types are known as the blacklist and whitelist designs. A whitelist is a regulation that lists articles which are authorized and the conditions for their importation. A blacklist is a regulation that lists the articles which are prohibited or restricted and the universe of other articles is allowed entry, usually subject to a permit, a phytosanitary certificate, or both. In either case, commodities are always "subject to" inspection. This doesn't mean that the commodity is certain to be inspected, nor does it commit to a specific level of inspection.

Regulated commodities fall within the scope of a specific regulation. This does not mean that articles outside the scope of the regulations have no risk or cannot become regulated. For instance, the importation of fresh apples is likely to be covered under a regulation for fresh fruits, whereas the importation of used farm implements may fall under no specific regulation. Nevertheless, the tractor implements are likely to be of interest and perhaps raise greater concerns for risk managers because of their potential to be contaminated with soil. The result is that a consignment of tractor implements is likely to be scrutinized as much or more than a consignment of apples despite being "unregulated."

The blacklist approach is the most challenging design for risk managers. It requires extensive front-end analysis to determine which commodities to prohibit or restrict, and a mechanism for closely monitoring imports to identify other commodities that may need to be restricted. A blacklist is best used for a low-risk group of commodities with well-known pest concerns, e.g., grain.

The whitelist approach is most used by NPPOs and it is consistent with the design anticipated by Annex C of the SPS Agreement (Control Inspection, and Approval Procedures). The process (Fig. 31.2) begins with a commodity that is not authorized (see Chapter 25, this volume). A market access request is then made by the exporting country to the importing country. In a best-case scenario, this request will be accompanied by a complete description of the commodity, a comprehensive pest list, and proposals by the exporting country for risk mitigation. The importing country then prepares a pest risk assessment (PRA), evaluates risk management options, and decides the conditions for import. Once both trading partners agree on the parameters, the authorization is finalized according to the administrative processes of the government of the importing country.

It is important to note that risk managers on both sides of the trade have essential roles in this process. The more open, honest, and collaborative they are, the greater the likelihood of a successful program. Withholding relevant information, providing misleading or incorrect information, or simply not cooperating, jeopardizes the potential for success and increases the probability for pest introduction. A failed program reflects badly on the competency of both NPPOs

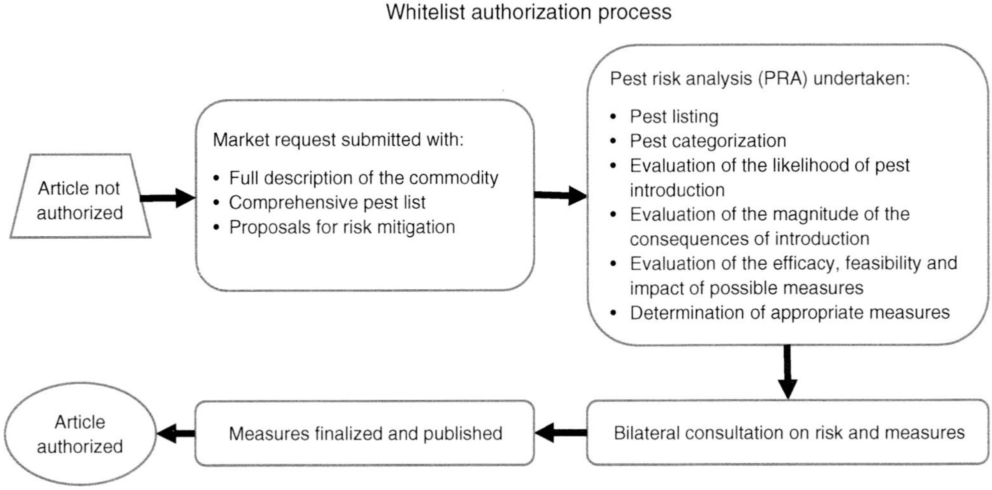

Whitelist authorization process

Fig. 31.2. Whitelist authorization process.

and degrades their confidence in each other for future programs.

Based on the discussion above, we can see that regulatory design has a significant effect on risk management from the standpoint of administrative arrangements. By the same token, there are documentation requirements that can have a similar effect on risk management from the operational standpoint. Included here are legal documents such as import permits and regulations, technical documents such as pest lists and pest risk analyses, and operational documents such as treatment and phytosanitary certificates. These provide important information for regulatory purposes corresponding to a particular function and taking a specific form—even formally identified as forms because their historical format has been an official paper document. Permits and phytosanitary certificates are arguably the best known and most used documents for the import and export of commodities. Both have a long history and strong institutional foundation as tools for risk management.

31.6.1. Permits

Permits take a wide variety of forms around the world, but two basic legal characteristics are common. First, permits demonstrate the importing country has legally extended an authorization to a private entity for the importation of regulated articles. The second common feature is that permits identify the conditions required by the importing country. Both are important for enforcement purposes, which also has some bearing on risk management. The ability to follow-up with penalties for violations of the regulations is a deterrent and important tool in the risk management toolbox. Beyond this, permits may also be used to control volume, offset costs through fees, identify commercial entities, and myriad other functions that may be associated with the legal control provided through linking regulatory authority to the importer and consignment(s).

The IPPC is silent on permits, but ISPM 20 (*Guidelines for a phytosanitary import regulatory system*) identifies permits as "import authorizations" and identifies two types: general and specific. According to the standard, General Import Authorizations are used when no specific requirements

are needed, or the requirements are generic and set out in regulations. General Import Authorizations do not require a permit or license, but consignments imported under general conditions are subject to inspection and verification upon import. Specific Import Authorizations include the typical import permit used to confer official consent for individual consignments or a series of consignments, including articles that are admissible and those which would otherwise be inadmissible. Despite the guidance provided by ISPM 20, there is very little harmonization among NPPOs on the design and use of permits.

The SPS Agreement identifies permits conceptually within Annex C: Control, Inspection, and Approval Procedures. The emphasis in this section of the Agreement is on transparency and a process that is fair and responsive to the exporting country; not a disguised barrier to trade. The aggregate of international guidance on permits is therefore focused on dissuading the use of authorization processes as trade barriers. Considerable space exists in risk management for the adaptation of import authorizations to digital systems for next-generation trade.

31.6.2. Phytosanitary certificates

Phytosanitary certificates are the best known and most ubiquitous document in the phytosanitary community after permits. The concept dates to the original draft of the IPPC in 1951 when it was agreed that exporting countries should accept responsibility for meeting the requirements of importing countries by officially "certifying" compliance using a standardized process and format. This single point of international harmonization later provided the basis for the negotiators of the SPS Agreement to adopt the IPPC as the phytosanitary standard-setting organization identified in the Agreement.

The *New Revised Text* of the IPPC includes updated model certificates as annexes and includes specific provisions for phytosanitary certificates in Article V (IPPC, 1997). In addition, the IPPC has produced ISPM 7 (*Phytosanitary certification system*) and ISPM 12 (*Phytosanitary certificates*) providing greater detail on certification form, format, and process. A key aspect of this guidance in both the Convention and ISPMs is acknowledgement of the possibility for electronic certification. Considerable effort has been devoted to developing such systems in recent years (see Chapter 19, this volume).

31.6.3. Big data for risk management

As mentioned above, the *Trade Facilitation Agreement* came into force for all WTO Members in 2017. The agreement places Customs in full control of the processes used for the exchange of information and documentation required for import, export, and transit of commodities in trade. The agreement institutionalizes the single window concept and digital systems as new obligations for all WTO Members.

The digital exchange of trade information and documentation in the single window system is not designed around forms and formats, but rather data. Seals, stamps, signatures and official letterheads are replaced by the secure designs within the electronic systems. Information is condensed to codes and "message sets" that are linked to databases with the relevant information and managed in real time. These systems have the advantage of enormously reducing entry and clearance time as well as errors and fraud associated with the manual processing of paperwork for the immense volume and diversity of goods moving in international commerce. Digital clearance systems also allow for real-time risk management by providing the capability for algorithms designed by risk managers to select inspection targets, specify inspection levels, check electronic permits and phytosanitary certificates, record pests and actions, and inform importers, exporters, and NPPOs of the status of a consignment—all in real time.

As NPPOs integrate into the single window system in the near term and fully digital information exchange in the longer term, we can envision new designs for permits and phytosanitary certificates that build on legitimate risk management objectives while also leveraging the digital environment for better information and analysis to inform and improve phytosanitary systems. To make this leap requires a fresh view of the roles of permits and phytosanitary certificates and a close partnership with Customs to facilitate integration. The historical notion of a particular

form or format must be abandoned in favor of key data elements that work in concert with databases and algorithms within "big data" systems to provide robust tracking, reporting, analysis, and real-time responses to the millions of actions, including phytosanitary actions, that must occur instantly in the next-generation trade environment. Risk managers are crucial to ensuring that the new systems not only capture transactional data that has been historically collected but also information that was previously unavailable to support risk management.

For example, most NPPOs already have access to data on trade volumes, number of shipments, the number of inspections, number of treatments, value, and other transactional data. However, simply combining this data with information on permit holders allows risk managers to see which importers cause problems with which commodities and pests, how often, in what ports, and how they compare to others. This can then provide the basis for targeting, prioritization, enforcement, and myriad other risk management activities which were previously not possible. The analysis and response can even be programmed to be automatic and run continuously behind the data as it is collected.

Customs organizations around the world are increasingly pressed by trade for timely implementation of the single window concept with digital systems for documentation. Astute risk managers will see an opportunity to reconsider the historical role, function, and form of key documentation requirements, whether they are needed, their future design, and strategies for evolution toward the next generation of risk management.

31.7. Summary and Look Forward

Here are five things to remember from this chapter.

1. The full and accurate identification of a commodity is essential for regulatory processes and risk management.
2. Processes, processing, and packaging associated with creating the commodity should be considered by risk managers.
3. Regulatory designs and the documentation requirements for regulated commodities play an important role in risk management.

4. Risk management is not one-sided. It works best when trading partners collaborate honestly and freely about the risks and solutions associated with commodities they trade.

5. Trade is rapidly transitioning from documents to data, creating enormous opportunities for risk managers to design more effective and efficient processes for collecting, evaluating, and reacting to pest risk.

The next chapter looks at the unique risk management challenges of genetically modified organisms, and especially their relationship to the fundamental concepts of pest risk in the SPS-IPPC framework.

31.8. References

IPPC (1997) *New revised text of the International Plant Protection Convention*. Secretariat of the International Plant Protection Convention (IPPC), Food and Agriculture Organization of the United Nations, Rome.

IPPC (2016) International Standards for Phytosanitary Measures, Publication No. 32, *Categorization of Commodities According to their Pest Risks*. Secretariat of the International Plant Protection Convention (IPPC), Food and Agriculture Organization of the United Nations, Rome.

IPPC (2019) International Standards for Phytosanitary Measures, Publication No. 5, *Glossary of Phytosanitary Terms*. Secretariat of the International Plant Protection Convention (IPPC), Food and Agriculture Organization of the United Nations, Rome.

United States Food and Drug Administration (2020) *Import Basics: Common Entry Types*. Available at: https://www.fda.gov/industry/import-basics/common-entry-types (accessed April 15, 2020).

WTO (2017) *Trade Facilitation Agreement*. World Trade Organization, Geneva.

32

Genetically Modified Organisms and Invasive Species

There doesn't seem to be any other way of creating the next green revolution without GMOs.

E.O. Wilson

32.1. Introduction

A risk management discussion on genetically modified organisms (GMOs) and invasive species in the same chapter may seem awkward on the surface, but these two topics share a common background in risk management through their relationship with the precautionary approach, also called the precautionary principle. This chapter examines the background and conceptual foundation that links these topics and aims to clarify their role in risk management from a phytosanitary standpoint.

32.2. Terminology

Before considering the specific issues, it is worthwhile considering some of the key terms, including GMOs, living modified organisms (LMOs), invasive species, and their relationship to a quarantine pest. *Terminology of the Convention on Biological Diversity in relation to the Glossary of phytosanitary terms* (Appendix 1 to ISPM 5, IPPC, 2016) is an excellent reference for this discussion.

32.2.1. Genetically modified organism and living modified organism terminology

In literal terms any organism that has had its genome modified by human activities, including conventional breeding, could be included under the terms GMO or LMO, but in practice the terms have become restricted to organisms that have acquired novel genetic material by the techniques of modern biotechnology. The *Cartagena Protocol on Biosafety* to the *Convention on Biological Diversity* (CBD, 2003) uses the term LMO with the following definition:

> "Living modified organism" means any living organism that possesses a novel combination of genetic material obtained through the use of modern biotechnology. Where:
>
> (a) "Living organism" means any biological entity capable of transferring or replicating genetic material, including sterile organisms, viruses and viroids.
> (b) "Modern biotechnology" means the application of:
> (i) in-vitro nucleic acid techniques, including recombinant DNA and direct injection of nucleic acid into cells or organelles
> (ii) fusion of cells beyond the taxonomic family, that overcome natural physiological reproductive or recombination barriers and that are not techniques used in traditional breeding and selection.
>
> (Article 3, *Use of Terms*)

The use of the term "living" in the Biosafety Protocol emphasizes the capacity to pass on or reproduce the genetic modification. This feature distinguishes the LMO from genetically engineered products such as a pharmaceutical or a food material produced using genetic engineering techniques or derived from an LMO (CBD, 2003).

The definition given in the *Protocol on Biosafety* is quite broad and equivalent to the common understanding of a GMO. However, it narrows the focus to exclude foods, feed, products for processing, and human pharmaceuticals from consideration.

32.2.2. Relationship between a quarantine pest, LMOs, and GMOs

The International Plant Protection Convention (IPPC) defines a quarantine pest as: "a pest of potential economic importance to the area endangered thereby and not yet present there, or present but not widely distributed and being officially controlled" (IPPC, 1999). The key points to note in this definition are the absence of a pest in an area and the potential for economic harm. Regarding economic harm, ISPM 5 (IPPC, 2016) includes "Supplement 2: Guidelines on the understanding of 'potential economic importance' and related terms including reference to environmental considerations." This guidance is dedicated specifically to addressing environmental and other risks which may not have a direct market value.

The scope of the IPPC includes all GMOs and is broader than the *Protocol on Biosafety* definition of an LMO. For example, consideration of any possible phytosanitary risks associated with the import of food, feed, or products for processing and pharmaceuticals are specifically excluded under the *Protocol*. The definition of regulated article in the IPPC would also allow

> Regulated article: Any plant, plant product, storage place, packaging, conveyance, container, soil, and any other organism, object, or material capable of harbouring or spreading pests, deemed to require phytosanitary measures, particularly where international transportation is involved.
>
> (IPPC, 1999)

the consideration of any potential phytosanitary risks associated with any GMOs including those not derived from plants.

In terms of assessment of the phytosanitary risks under the IPPC, the issue is not the technology that is used to create a new organism but the possible impact these changes might have on the pest status of the organism, i.e., its potential status as a pest and the risk associated with it. Therefore, LMOs and GMOs may present phytosanitary risks requiring risk management.

32.2.3. Invasive species terminology

The term "invasive species" refers to the ability of an organism to establish in a new area and to cause change or harm. Recognizing and controlling any species causing serious damage to the environment are important objectives. In the United States, the term officially debuted in policy with *Executive Order* (EO) *13112* (February 3, 1999).

"Invasive species" were defined in the EO as: "... alien species, whose introduction does or is likely to cause economic or environmental harm or harm to human health." Alien species means, "with respect to a particular ecosystem, any species, including its seeds, eggs, spores or other biological material capable of propagating that species, that is not native to that ecosystem." At first glance, the definition of invasive species seems like an appropriate foundation for such policies. However, gaps in logic become apparent when examining the definition more closely. For instance, the focus on alien species implies a sort of blacklist which would include beneficial non-native species and shifts attention away from the need to control all pests, regardless of origin that have substantial negative impacts on plant resources or the environment. There are also challenges defining an ecosystem and determining which species in an ecosystem are native depends on the definitions, area, and timeframe one uses. Likewise, the likelihood of environmental harm is subjective without an agreed upon scientific definition or methodology.

Beyond the United States, the issue becomes further complicated with a plethora of definitions and interpretations for invasive species across a range of international agencies and non-governmental organizations (NGOs). The

CBD is a central player with its own definition: "Invasive alien species" means an alien species whose introduction and/or spread threatens biological diversity.

The International Union for the Conservation of Nature (IUCN) definition tends to emphasize the impact invasive species have on native biological diversity, while the US definitions allow for a broader impact: "alien species" (non-native, non-indigenous, foreign, exotic) means a species, subspecies, or lower taxon occurring outside of its natural range (past or present) and dispersal potential (i.e. outside the range it occupies naturally or could not occupy without direct or indirect introduction or care by humans) and includes any part, gametes or propagule of such species that might survive and subsequently reproduce. "Alien invasive species" means an alien species which becomes established in natural or semi-natural ecosystems or habitat, is an agent of change, and threatens native biological diversity (IUCN, 2000).

The Global Invasive Species Program (GISP) uses a broader definition of invasive alien species, like the US definition: "Invasive alien species" are non-native organisms that cause, or have the potential to cause, harm to the environment, economies, or human health (GISP, 1999).

These definitions generally emphasize the impact on native biological diversity of an invasive species while the US definitions focus on "natural" ecosystems and by implication appear to exclude conventional agricultural production systems and therefore would exclude conventional agricultural pests.

3.2.4. Relationship between an invasive species and a quarantine pest

The definition of a quarantine pest and the standards developed under the IPPC allow potential environmental risks and other risks that may not have a direct economic value to be considered. For example, a weed species with the potential to cause damage to natural ecosystems is a legitimate issue for risk analysis and phytosanitary action under the IPPC. Many quarantine pests are invasive in agricultural systems and also have the potential to invade natural ecosystems. Invasive species, as defined above,

clearly fall within the scope of the IPPC and may require risk management. From a risk analysis standpoint, invasiveness is only one aspect of risk. The conceptual scope and relationship of a quarantine pest to risk and risk management fully accounts for invasiveness and much more.

32.3. Precautionary Approach

A key issue for risk managers is the role of precaution in the regulation of pest risks to plant health and the environment. A concept known as the precautionary approach, also as the "precautionary principle," emerged from the international framework for environmental protection to become a contentious issue in trade where the concept is not expressed in the same terms. The lack of clarity in this regard is often mistaken for a lack of precaution or of concern for the importance of precaution.

The precautionary approach was first institutionalized as a distinct concept in an international instrument when it was adopted as Principle 15 of the Rio Declaration (UN, 1992).

Principle 15 states that:

In order to protect the environment, the precautionary approach shall be widely applied by States according to their capability. Where there are threats of serious or irreversible damage, lack of full scientific certainty shall not be used as a reason for postponing cost-effective measures to prevent environmental degradation.

Higher profile attention was given to the precautionary approach in the *Cartagena Protocol* (CBD, 2003). Driven largely by issues surrounding the use of biotechnology, the Protocol reaffirms the precautionary approach within the context of risk assessment.

The precautionary approach, as institutionalized in multilateral environmental agreements, is widely held to be both valid and desirable. Likewise, it has not been technically, legally, or otherwise inconsistent with other relevant international objectives or obligations. Indeed, in the case of the IPPC, the concept existed and has been practiced by the phytosanitary community for many years before being labeled by the environmental community.

By virtue of its association with the principles of the *Rio Declaration* (UN, 1992), the precautionary approach has increasingly been referred to as a "principle." From the standpoint of a risk manager, it is difficult to envisage how the concept could be considered either an approach or a principle. It is more a process of accounting for uncertainty in decision-making based on a systematic evaluation of the evidence, a process that is integral to risk analysis.

Two key points must be made here. The first is that the precautionary approach as described in the *Declaration* and reaffirmed in the *Convention on Biological Diversity* and again in the *Cartagena Protocol* (CBD, 2003) is not necessarily incompatible with the IPPC or the World Trade Organization (WTO) Agreement on the Application of Sanitary and Phytosanitary Measures (SPS Agreement). The second important point is that the environmental agreements do not explicitly associate the application of the precautionary approach with the "failure" of risk analysis. Indeed, they are rather explicit about risk analysis as the basis for evaluating the available information. The central question is whether a determination regarding the adequacy of information is made before risk analysis is done, or whether the risk analysis is completed and becomes the basis for identifying the uncertainty.

Recent evolution of the precautionary approach has included significant rhetoric and diverse interpretations by governments as well as international and private organizations claiming to understand the application of the concept in practice and promoting regulatory systems that may not be consistent with existing mechanisms and international obligations. There is a growing impression that the precautionary approach is an alternative basis for regulatory decision-making to be used where there is judged to be insufficient information to undertake a risk-based approach. The implication is that the process of risk analysis would not be undertaken or completed if the lack of information led to some level of uncertainty that was deemed to be unacceptable.

This interpretation indicates a fundamental misunderstanding regarding the nature of risk analysis and its role in regulatory decision-making. It ignores the role of uncertainty in risk analysis. The adoption of such an interpretation is also inconsistent with the IPPC and other international instruments including the WTO *Agreement on the application of sanitary and phytosanitary measures*.

The term "precautionary measures" is not explicitly used or described in either the IPPC or the SPS Agreement. However, it may be argued that phytosanitary measures are by their nature precautionary depending on the influence of uncertainty in the judgment regarding acceptable risk. The concept of precaution based on uncertainty is therefore implicit in the application of proper risk analysis; however, the role of scientific principles and evidence in risk analysis has historically been given greater prominence than the role of uncertainty. This has resulted in a strong focus on "sufficient scientific evidence" without fully recognizing that uncertainty is inherent in all scientific evidence and proper risk analysis should account for uncertainty (see also Chapters 3 and 15, this volume).

If risk analysis is affected by different interpretations of the precautionary approach, then trade is also affected. The precautionary approach is overtly embedded in regulatory language underpinning phytosanitary guidance in the EU and can be found in some US legislation. More commonly, it is hidden in different regulatory processes without a by-name identification of the underlying concept. For example, a country may state that large uncertainties associated with a given commodity, e.g., nothing is known about the pests in country X, justify indefinite prohibitions and preclude the consideration of risk analysis and, thus, the usefulness of IPPC guidance.

In the WTO dispute cases of the US vs. EC on hormones in beef and Japan vs. New Zealand on Fireblight of apples, arguments were made that potential hazards existed in the absence of evidence and a precautionary approach justified strong measures. In both cases, WTO dispute settlement panels agreed that uncertainty was not sufficient condition for prohibition and an evidence-based approach, i.e., risk analysis, was necessary to justify measures (WTO, n.d.).

32.4. GMOs/LMOs

The concerns around GMOs in a phytosanitary context stem primarily from fears that plants genetically engineered through biotechnology can become harmful, either directly or through impacts on biodiversity. Although biotechnology is used for a range of other organisms, including insects, the primary focus from a risk management standpoint is the potential for genetically engineered plants to become weeds. Much of this can be attributed to spill-over from the controversies associated with GMO food products and potential food safety issues, but also a long history of introduced plants becoming harmful.

The risk analysis procedures of the IPPC recognize both phenotypic and genotypic characteristics as legitimate factors in risk assessment. Likewise, impacts on both cultivated plants and the environment are considered. The difficulty in risk analysis for GMOs/LMOs is collecting evidence that can be used to demonstrate that the traits which have been engineered into the plant make it a pest, or if it was already a pest, removes these pest traits.

Identification of the recipient and the donor organism is essential to begin, and a description of the genetic modification is needed. Credible laboratory and field observations are then required to establish the actual effects and predict the risk. The tendency, however, is to speculate on worst-case possibilities and then insist on extraordinary or unspecified information that proves otherwise. Another tactic, one that is widely practiced with phytosanitary irradiation (see Chapter 23, this volume), is creating an overly cumbersome and unjustified approval process. These leanings toward precautionary measures in the absence of evidence or legitimate rationale for different treatment, opens risk managers to challenges from industry and trading partners. It is an approach that has already proven unsupportable in the Dispute Settlement Body of the WTO (US-EC Dispute on Approval and Marketing of Biotech Products. WTO, 2008).

32.5. Invasive Species

Early practitioners of international plant protection and quarantine did not distinguish between different types of organisms causing negative effects. The term "pest" has long been considered sufficiently descriptive for injurious organisms that should be actively excluded or otherwise controlled, including arthropods, pathogens, molluscs, and weeds. The concept was deliberately designed to cover all pests of all plants and has been institutionalized in practice for nearly a century and in the IPPC for more than 60 years.

In recent decades, the concept of "nuisance species" associated with aquatic organisms morphed into "invasive species," a more emotive term to identify the same problems. In the 1990s, this terminology quickly spread across the environmental protection community and first began appearing in plant protection literature associated with weeds. Since then, the term "invasive species" has found its way into the mainstream lexicon of the phytosanitary community and firmly established itself as a political firebrand for environmentalists.

The environmental protection community originally embraced the term invasive species to contrast with the concept of pests which was assumed to ignore environmental concerns. This was due in large part to the long history of governments placing emphasis on agricultural protection over environmental protection rather than a flaw in the pest concept for plant protection. The perception of a serious threat was further amplified in the public and political psyche when coupled with the term "alien" in contrast to exotic or non-native.

Today the terms are freely mixed; sometimes considered synonymous and other times conflicting, depending on the context. In many cases, the actual meaning of the terms is lost, and usage is linked instead to the audience or desired impact. This blending of terms and meanings often results in confusion in a technical and regulatory context where the proper use of terms can have important scientific and legal implications.

Whether or not the environmental protection community recognizes it, invasive species that affect plant health and the environment, as defined by the terminology above, clearly fall under the scope of the IPPC and the longstanding concept of pests. Many quarantine pests are invasive in agricultural systems and also have the potential to invade natural ecosystems, but invasiveness is only one aspect of "pestiness." The IPPC concepts of pest and pest risk are more complete, robust, technically correct, and legally defendable for risk management purposes.

32.6. Summary and Look Forward

Here are five things to remember from this chapter.

1. The precautionary approach is a complicating factor with GMOs/LMOs and invasive species because of the tendency to speculate on risks that have no evidence to support their need for risk management.

2. Terminology is complicated by different definitions for overlapping concepts.

3. IPPC terminology and risk analysis procedures address invasive species and GMOs/LMOs in a phytosanitary context.

4. The key concern for risk managers is evidence of a pest risk.

5. Invasiveness is not equivalent to risk; risk is a more complete concept.

The next chapter considers the SPS agreement, the strength of measures and a reasonable relationship, and the risk management continuum as central concepts that break from the past and redefine the future of pest risk management.

32.7. References

CBD (2003) *Cartagena Protocol on Biosafety to the Convention on Biological Diversity*. Available at: http://bch.cbd.int/protocol/text/ (accessed January 17, 2020).

GISP (Global Invasive Species Programme) (1999) *Working definitions used by the Global Invasive Species Programme (GISP)* (UNEP/CBD/SBSTTA/6/INF/5 Annex II). Secretariat of the Convention on Biological Diversity, Montreal.

IPPC (1999) *International Plant Protection Convention, New Revised Text*. Secretariat of the International Plant Protection Convention (IPPC). Food and Agriculture Organization of the United Nations, Rome.

IPPC (2016) International Standards for Phytosanitary Measures, Publication No. 5: *Glossary of Phytosanitary Terms*. Secretariat of the International Plant Protection Convention (IPPC), Food and Agriculture Organization of the United Nations, Rome.

IUCN (2000) *Guidelines for the Prevention of Biodiversity Loss Caused by Alien Invasive Species*. SSC Invasive Species Specialist Group Approved by the 51st Meeting of the IUCN Council, Gland Switzerland, February 2000.

UN (United Nations) (1992) *Rio Declaration on Environment and Development,* Rio Conference on the Environment and Development. Available at: https://sustainabledevelopment.un.org/content/documents/Agenda21.pdf (accessed January 17, 2020).

WTO (2008) *DS 291: European Communities—Measures Affecting the Approval and Marketing of Biotech Products*. Available at: https://www.wto.org/english/tratop_e/dispu_e/cases_e/ds291_e.htm (accessed January 17, 2020).

WTO (n.d.) *Disputes by Agreement* (SPS Summary). Available at: https://www.wto.org/english/tratop_e/dispu_e/dispu_agreements_index_e.htm (accessed January 17, 2020).

33

A New Framework

The secret of change is to focus all of your energy,
not on fighting the old, but on building the new.

Socrates

33.1. Introduction

The fundamentals of pest risk management
have shifted greatly in the evolutionary transformation of early plant protection programs
into the phytosanitary systems we have today.
This chapter looks at central concepts that break
from the past and redefine the future of pest risk
management.

33.2. The SPS shake-up

A little more than a century ago, nations began
forming national regulatory programs for plant
quarantine based on inspection as the primary
means to exclude harmful plant pests. These
efforts became the seeds of the national and
international institutions for regulatory plant
protection that we know today. In the early decades, the old adage of an ounce of prevention being
worth more than a pound of cure (attributed to
Ben Franklin) fitted perfectly with the underlying philosophy. Given the potentially cata-
strophic consequences from the introduction of
harmful plant pests, a significant investment in
exclusion always seemed reasonable and indeed

necessary. The pest should be absent or killed to
eliminate the risk, and it was the responsibility
of the plant protection authorities to ensure that
regulatory programs were focused on these
objectives. This mindset resulted in a century of
expansion for national regulatory authorities
exercising port-of-entry controls based on inspection and treatment to prevent pest entry as
the primary risk management strategy.

International regulatory frameworks evolved
from the same frame of mind; placing protection in
the forefront but also promoting harmonization to
encourage cross-border collaboration. The early
part of the 20th century saw efforts for international phytosanitary collaboration sputter. The
need was recognized and accepted but the community was small and much larger political priorities
occupied the global stage. In the second half of
the century, after two world wars and a severe
economic depression, the world was ready for
new political and economic relationships.

The General Agreement on Tariffs and
Trade (GATT) was crafted from the Bretton
Woods Conference following World War II. The
seminal idea behind GATT was that liberalizing
trade would promote economic growth and reduce political tension. GATT marched through
eight rounds of negotiations between 1947 and
1995 as global efforts toward trade liberalization
stepped through stages that first emphasized free
trade (removing tariffs), then fair trade (removing
non-tariff and technical barriers), and finally

safe trade (measures to protect human, animal, or plant life or health).

A parallel trajectory began in plant health when the International Plant Protection Convention (IPPC) emerged in 1952 with the concept of phytosanitary certification. This marked the beginning of international collaboration and harmonization. By placing some of the burden for preventing pest spread on the exporting country, the IPPC made pest exclusion a responsibility shared by both importing and exporting countries, and certification became a globally shared exercise. These innovative concepts transformed pest risk management from a one-sided affair within the importing country into a collaborative business between trading partners.

National plant protection programs did not evolve much further over the following decades, but a collision course was set with Article XX:b provisions in GATT that allowed countries to put measures in place to protect "human, animal or plant life or health." This point was identified in the Uruguay Round of GATT negotiations as a priority for discipline in agricultural trade. In 1994, the carefully crafted Agreement on the Application of Sanitary and Phytosanitary Measures (the SPS Agreement) emerged as the result.

The SPS Agreement focused on international harmonization based on standards and specifically identified the IPPC as the organization responsible for providing this service to the phytosanitary community. The Uruguay Round negotiations also resulted in the establishment of the World Trade Organization (WTO) and a binding international dispute settlement mechanism. The potential to challenge trading partners with binding results had profound effects on the regulatory community and quickly brought a critical eye to traditional practices.

Historically, the balance between protection and trade had leaned strongly toward protection, i.e., "when in doubt, keep it out," but the SPS Agreement swung the pendulum toward trade liberalization with emphasis on a scientifically defendable rationale for restrictions. Regulatory decision-making that was traditionally rooted in distrust and worst-case assumptions found itself grappling with harmonization and struggling with the need for evidence and analyses to justify restrictions. Principles and procedures that followed from this fundamental shift in philosophy challenged many historical paradigms as evidenced by a series of SPS-based disputes in the WTO. Jurisprudence collected from these real-world challenges provided important insight into the implementation of SPS concepts in practice and the implications of a new framework for safe trade.

The upshot of this reset is a more sophisticated view of the relationship between exclusion and all other measures and conditions that affect the risk of pest introduction across the risk management continuum. The question is no longer only about the presence or absence of the pest in trade but rather its risk of introduction, whether mitigation measures are required and the justification for the strength of measures. Although the concept of exclusion still plays a key role in this picture, inspection at the border is only part of the strategy rather than the central strategy. The SPS Agreement tells us that measures designed for exclusion must be scientifically justified and the least restrictive for achieving the appropriate level of protection. Simply having protection as an objective does not provide carte blanche for all possible measures or any level of restrictions. The SPS Agreement requires a rational relationship between the measures which can be justified with scientific evidence and the risk. This point is repeatedly demonstrated in SPS jurisprudence (WTO, 2019).

The disciplines created by the SPS Agreement are designed to ensure that barriers to trade which have the objective of providing protection are not overly restrictive or politically motivated. It creates a regulatory focus on safe trade as a singular objective, recognizing that neither the extremes of exaggerated protection nor completely open trade are desirable. The role of risk management in this context shifts both philosophically and operationally from an emphasis on protection based on exclusion to a continuum approach that embraces exclusion as an element of safeguarding that is applied to the extent justified. From an operational standpoint this translates to a much stronger role for analysis and gathering data needed to understand where, when, how, and how strongly exclusion measures are applied.

33.3. Strength of Measures and Rational Relationship

Every government has the sovereign authority to determine whether phytosanitary measures are required and the action necessary to provide the appropriate level of protection. In cases where phytosanitary measures are deemed to be necessary, there are usually a few to several options that can be evaluated for their efficacy. A high level of flexibility is needed for emergency measures which are generally conservative, i.e., deliberately over-restrictive to compensate for the lack of information or time to fully evaluate the situation. Other measures must be based on international standards or pest risk analysis and are subject to challenge for their appropriateness.

Strength of measures refers to the restrictiveness of phytosanitary measures based on a range of risk management options from most to least restrictive. This concept evolved from the SPS Agreement, but it is not expressly identified as such in the Agreement. The terminology comes from the IPPC definition of pest risk analysis which is the basis for deciding whether phytosanitary measures are justified and "the appropriate strength of measures" (IPPC, 1999).

The metaphor of a sliding scale is useful for visualizing the concept of strength of measures but in any particular regulatory situation there may not be such an extensive range of risk management options available for a precise match to the risk level. This means that in some cases, a risk which is found to be low may require a measure designed for higher risk because there is no other feasible measure. Begging clarity is the criterion and process used for comparing and aligning measures with risk outside the odd instance where very similar situations may be compared for consistency.

The words "rational relationship" are not found in either the SPS Agreement or the IPPC but emerged from WTO jurisprudence as a common issue at the core of nearly all WTO disputes on SPS measures. The concept of rational relationship has two parts: first, that the measures have a demonstrable cause and effect relationship to the risk; second, that the strength of measures is consistent with the level of risk.

> Imagine requiring fruit to be treated for a pest found only on the plant's roots. Such a requirement would be challengeable because the measure is not designed for the pest of concern. Likewise, an overly restrictive measure such as prohibition would be challengeable if a less restrictive measure such as treatment is known to be effective. Both situations violate the concept of rational relationship.

The concept links to the primary principles of risk management—that phytosanitary measures are limited to what is necessary to protect plant health (necessity), consistent with the pest risk involved (managed risk), represent the least restrictive measures (minimal impact), on the basis of conclusions reached using an appropriate risk analysis (technical justification), and are promptly modified or removed as conditions change and new facts become available (modification) (IPPC, 2016). The basic concept is both simple and logical: measures are established and adjusted to be appropriate for the risk.

Understanding and institutionalizing the concepts of strength of measures and rational relationship are fundamental to the contemporary practice of pest risk management—not only for the proper implementation of SPS obligations and to avoid challenges, but perhaps more importantly to design rational, defendable phytosanitary measures that are linked to evidence of risk. The key to the proper implementation of these concepts is a focus on risk.

33.4. Risk Management Continuum

The idea of representing risk management as a continuum has evolved from the need to demonstrate a link between export and import processes under the aegis of safe trade and recognition that domestic programs provided critical support to both. This brings risk management to the point where it is not narrowly focused only on import and "exclusion by inspection," but rather includes the spectrum of plant protection activities directly or indirectly supporting safe trade. The risk analysis piece contributed by the SPS Agreement was crucial to this evolution because it shifted the focus of risk management from preventing pest entry to preventing pest establishment.

From a technical standpoint, these shifts represented a necessary update and incremental rise in sophistication around the practice of regulatory plant protection. From a practical and operational standpoint, this represented a sea change in the mind-set that guided inspection for nearly a century: the historical view that exclusion based on port-of-entry inspection and treatments is the bulwark of protection.

There is no question that inspection and treatment will always be important for pest risk management and preventing pest entry is a central strategy to prevent pest establishment. The challenges arise when we ask whether these strategies effectively and consistently achieve our objectives, and how well they align with the principles of safe trade. In the past, the effectiveness of risk management was measured by work accomplished: primarily the number of inspections and treatments performed and the number of pest interceptions. By analytically pursuing the question of risk management effectiveness, we realize the many shortcomings in focusing on inspection as the central strategy for safeguarding but continue to struggle with embracing the conceptual shift in the collective mind-set.

The risk management continuum reminds us that there are multiple possibilities across the supply chain for mitigating the risk of pest introduction, i.e., entry and establishment, beginning offshore and extending to surveillance for detection and emergency action after introduction. One piece of this is inspection at the port-of-entry, but also included within this continuum are offshore programs, regulatory policy designs including prohibitions, restrictions, controls, as well as trade negotiations or work plans, and numerous tools that enhance effectiveness such as X-ray, detector dogs, targeting algorithms, new detection technologies. Thinking broadly of risk management in a continuum and putting the range of options on the table, we find the flexibility to innovate with different regulatory designs that can take advantage of combined effects to create risk attrition which may reduce or eliminate our reliance on inspection or treatment and provide more effective risk management with fewer resources.

The key to making this work however, is not measuring the number of tools we have or the number of times we use them; it is understanding their effectiveness toward achieving our objective of safe trade. For this, we need to collect different data than in the past and place much more emphasis on analysis, especially statistical and economic analysis to provide robust support for prioritization linked to maximizing our ability to manage pest risk with the available resources.

The "inspection for exclusion" focus of the past was a blunt instrument; a one-size-fits-all approach with highly variable effectiveness. The risk management continuum in the SPS/IPPC framework offers more options, greater precision, and a contemporary model for the most effective implementation of plant protection programs.

33.5. Summary

Here are five things to remember from this chapter.

1. The historical concepts underpinning pest risk management have changed dramatically from the early days when regulatory plant protection was defined primarily by exclusion based on inspection and treatment.

2. The SPS Agreement, and jurisprudence associated with its implementation, have redefined risk management to include a strong analytical background linking phytosanitary measures to pest risk.

3. The emphasis on analysis to justify phytosanitary measures raises the need to shift risk management designs and information collection to focus on the risk of pest introduction.

4. The rational relationship of phytosanitary measures to risk and the justification for the strength of measures are central concepts for risk managers to adopt and practice.

5. The risk management continuum provides opportunities for more effective regulatory plant protection designs.

33.6. References

IPPC (1999) *International Plant Protection Convention, new revised text*. Secretariat of the International Plant Protection Convention (IPPC), Food and Agriculture Organization of the United Nations, Rome.

IPPC (2016) International Standards for Phytosanitary Measures, Publication No. 1: *Phytosanitary Principles for the Protection of Plants and the Application of Phytosanitary Measures in International Trade.* Secretariat of the International Plant Protection Convention (IPPC), Food and Agriculture Organization of the United Nations, Rome.

WTO Analytical Index (2019) Available at: http://www.wto.org/english/res_e/booksp_e/analytic_index_e/sps_e.htm (accessed September 20, 2019).

Index

CABI – who we are and what we do

This book is published by **CABI**, an international not-for-profit organisation that improves people's lives worldwide by providing information and applying scientific expertise to solve problems in agriculture and the environment.

CABI is also a global publisher producing key scientific publications, including world renowned databases, as well as compendia, books, ebooks and full text electronic resources. We publish content in a wide range of subject areas including: agriculture and crop science / animal and veterinary sciences / ecology and conservation / environmental science / horticulture and plant sciences / human health, food science and nutrition / international development / leisure and tourism.

The profits from CABI's publishing activities enable us to work with farming communities around the world, supporting them as they battle with poor soil, invasive species and pests and diseases, to improve their livelihoods and help provide food for an ever growing population.

CABI is an international intergovernmental organisation, and we gratefully acknowledge the core financial support from our member countries (and lead agencies) including:

Ministry of Agriculture
People's Republic of China

Australian Government
Australian Centre for
International Agricultural Research

Agriculture and
Agri-Food Canada

Ministry of Foreign Affairs of the
Netherlands

Schweizerische Eidgenossenschaft
Confédération suisse
Confederazione Svizzera
Confederaziun svizra
Swiss Agency for Development
and Cooperation SDC

Discover more

To read more about CABI's work, please visit: **www.cabi.org**

Browse our books at: **www.cabi.org/bookshop**,
or explore our online products at: **www.cabi.org/publishing-products**

Interested in writing for CABI? Find our author guidelines here:
www.cabi.org/publishing-products/information-for-authors/